# Studies in Logic
## Volume 106

# New Directions in Term Logic

Volume 99
Theories of Paradox in the Middle Ages
Stephen Read and Barbara Bartocci, eds

Volume 100
The Fallacy of Composition: Critical Reviews, Conceptual Analyses, and Case Studies
Maurice A. Finocchiaro

Volume 101
The Logic of Partitions. With Two Major Applications
David Ellerman

Volume 102
Bounded Reasoning Volume 1: Classical Propositional Logic
Marcello D'Agostino, Dov GAbbay, Costanza Larese, Sanjay Modgil

Volume 103
The Fertile Debate. Affective Exploration of a Controversy
Claire Polo

Volume 104
Argument, Sex and Logic
Dov Gabbay, Gadi Rozenberg and Lydia Rivlin

Volume 105
Logic as a Tool. A Guide to Formal Logical Reasoning
Valentin Goranko

Volume 106
New Directions in Term Logic
George Englebretsen, ed

Studies in Logic Series Editor
Dov Gabbay                                      dov.gabbay@kcl.ac.uk

# New Directions in Term Logic

Edited by
George Englebretsen

© Individual author and College Publications, 2024
All rights reserved.

ISBN 978-1-84890-462-0

College Publications
Scientific Director: Dov Gabbay
Managing Director: Jane Spurr

http://www.collegepublications.co.uk

---

All rights reserved. No part of this publication may be reproduced, stored in a retrieval system or transmitted in any form, or by any means, electronic, mechanical, photocopying, recording or otherwise without prior permission, in writing, from the publisher.

# CONTENTS

**Preface** .......................................... 1
  *Johan van Benthem*

**Introduction** .................................... 5
  *George Englebretsen*

**Notes on the Contributors** ...................... 33

**On Semantics for Modal Term Logic** .............. 37
  *José-Martín Castro-Manzano*

**Aristotelian Diagrams for TFL** .................. 59
  *José David Garcia Cruz*

**Relative Terms** ................................. 85
  *Peter Simons*

**Modality as a Property of Terms** ............... 121
  *Manuel Correia*

**Sommer's Syllogisms Unlimited** ................. 135
  *Wallace Murphree*

Disentangling Leibniz's Notion of 'Truth' . . . . . . . . . . . . . 149
   *Wolfgang Lenzen*

Propositional Term Logic and an Intensional Semantics . . . . 199
   *Roberts van Rooij and Elbert Booij*

A Term Logic of Justification for Epistemic Attitudes . . . . . 229
   *Fabien Schang*

On Terms and Types . . . . . . . . . . . . . . . . . . . . . . . . . . . 263
   *Luc Schneider*

The Range of Algebraic Decision in Term Logic . . . . . . . . . 305
   *Miles Rind*

The Role of the Copula in Reasoning . . . . . . . . . . . . . . . 343
   *Pei Wang*

How Linear are Englebretsen's Line Diagrams? . . . . . . . . . 357
   *Amirouche Moktefi*

Names and Definite Descriptions in Natural Logic . . . . . . . 385
   *Lawrence S. Moss*

Aristotle, Term Logic and QUARC . . . . . . . . . . . . . . . . . 427
   *Jonas Raab*

Singular Power: On the Logic of Singular Terms in Term
   Logic. References and Related Works . . . . . . . . . . . . 505
   *George Englebretsen*

# Preface

Johan van Benthem

The canonical narrative in the history of logic is eschatological, with Gottlob Frege as the pivotal figure leading us out of the land of stagnation and obscurity to the blazing light of modern mathematical-logical systems where propositions hold sway. The present book is a tribute to the highly original pioneering work of Fred Sommers who singlehandedly undercut this standard story. Sommers showed how traditional logic, understood in its own right as a logic of terms rather than propositions, and taken further in creative ways, has a future as much as a past. I still remember reading his "The Logic of Natural Language" in the early 1980s and being struck by this new voice, putting forward a radical new vision without rhetoric or animosity. Around the same time, I read George Englebretsen's exquisite little book "Three Logicians" which put all this in an appealing alternative historical line. I had the privilege of meeting Sommers in his later years during a visit of his to Amsterdam, and it also gives me special pleasure to write this brief preface at the request of George Englebretsen who has done so much to promote and extend the Sommers program over the years.

There are various ways of understanding the Term Logic program, its scope and its claims. The rich and diverse set of contributions to this book show many of these. Authors propose systems for representing valid inference that offer alternatives to first-order predicate logic using key ideas from traditional logic, they strive to stay closer to natural language rather than reforming it, and they aim for computationally simplicity where full-blown modern logical systems always have the Damocles sword of undecidability hanging over them. Authors also extend the syllogistic core with propositional reasoning, modality and knowledge, with other themes in linguistic semantics, and even with numerical reasoning. Finally, Term Logic also comes with two empirical angles in this book. Authors think of it in terms of cognitive psychology as being closer to observable human practice, and one contribution even suggests that it may fit better with the way modern AI systems understand the reasoning embedded in natural language.

With the themes in this book, the Term Logic program interfaces with various streams and topics that are usually considered bona fide parts of modern logic, such as the algebraic tradition, categorial grammar, Quine's predicate-functor logic, monotonicity inference, and the formal semantics of natural language. This rapprochement is no coincidence. Several historical pieces in this book fit with the large body of work by historians trained in modern logic who have left the eschatology behind and study afresh what is of value and sophistication in the classical, medieval and early modern history of logic. Perhaps this broader movement relativizes the most radical interpretation of Sommers' program as having to challenge a monolithic orthodoxy. In the same spirit of convergence, it seems clear that the authors in this book know their modern logic and use its techniques and insights when it suits their purposes.

On a personal note, my own interest in 'natural logic' since the 1980s has revolved around elucidating as much as possible of reasoning in natural language with as little as possible of logical formalism. A typical example of this line is the 'monotonicity calculus' which describes a substantial decidable subsystem of first-order logic that operates directly on linguistic syntax with low polynomial complexity. What natural logic requires is an open-minded attitude toward locating representation and inference modules in our natural reasoning, and in this, Term Logic is a highly congenial enterprise. Even so, it seems to me that there is a borderline where natural logic will need to transition into technical mathematical reasoning constructing proofs in more Fregean styles. In fact, there are many deep questions about this transition zone and the thresholds in computational complexity which lie in wait at various stages of extending the scope of the reasoning covered. I hope that this book will encourage mathematical and computational logicians to take a serious look at current Term Logic systems and get a deeper foundational view of what such systems can and (equally enlightening in the foundations of logic since the seminal work by Gödel, Tarski and Turing in the 1930s) what they cannot achieve.

## Preface

A grander hope of mine is this. The emergence of logic as a recognizable field of study is a remarkable human phenomenon with a rich history of several millennia, in different ages and in different cultures. Our discipline would benefit immensely from achieving one integrated perspective on its development, including its diversity and honoring the past in our Western tradition, but also, in the same spirit of respect, in the Chinese, Indian, and Islamic traditions of logical reflection. The mathematicians have long done this, and present students with a narrative that integrates Babylonia, Egypt, Greece, the Islamic World and China: why can't we? What this requires is, for starters, an open mind in reshaping standard narratives, and the spirit of this book is a valuable step in this direction.

# Introduction

GEORGE ENGLEBRETSEN
*Bishop's University*

Homo sapiens (that's us) have been around a very long time — about three-thousand centuries. And there is no reason to believe that during that time our sapiens ancestors were not just like us in being *rational animals*. We reason. We argue. We use our language — a lot. We put things we know, or think we know, together and come up with new things we know, or think we know. We put two and two together. As I said, we reason. We're pretty good at doing this, though some are better than others. We start doing this even before becoming adults. A few of us take a special interest in our reasoning. About twenty-five centuries ago, Aristotle seems to have been the first to do so. He made a systematic study and account describing the general principles of how reasoning works — how we use a small number of statements that we take to express some bits of information and, on the basis of these, we discover a new bit of information. He called what the process he used to account for this *analytics*, the analysis of statements into the elements, the *terms* that carry the *matter* of the statements, what they are about, and the form of those statements, how the material elements are put together to form the statements. We call what he was up to *logic* — indeed, *formal logic*. Because the elements of a statement that Aristotle focused on were terms, his system of formal logic is a *term logic*.

Not long (relatively speaking, of course) after Aristotle set the history of the systematic science of logic on its course, others got into the act. In the 3rd century bce, the logician Chrysippus developed an account of formal logic that differed from Aristotle's in certain ways. Most importantly, Chrysippus had the notion that the material elements of statements are entire clauses (expressions that could stand alone as statements on their own). These two rival views traded places as the dominant way of conceiving of formal logic for a time — even being conflated or confused with one another now and then. However, Aristotle's term logic, with many emendations, amendments, and, eventually, fairly substantial additions, came to dominate logic investigation until the final decade or so of the 19th century. Then something happened.

Well before the 19th century, it had become clear that standard logic initiated by Aristotle was vulnerable to challenges posed by at least three kinds of statements that could be found in use in logical arguments: those with singular terms, those with relational terms, and those that can be analyzed as complexes of simpler statements (the kind that had been scrutinized by Chrysippus). Medieval logicians felt the sting of such challenges, and then so did Leibniz in the 17th century. Leibniz made some halting progress with each such challenge. Still, by the late 19th century, logic was all set for what happened — the Fregean Revolution. Among the results of that revolution is modern predicate logic (MPL). Consequently, it has seemed to most logicians that the only way to maintain logical credentials was to embrace MPL or settle for the old traditional syllogistic logic, that old term logic. But, could there be a term logic, somehow strengthened and revitalized in significant ways, that could offer an alternative to MPL? Are the challenges faced by traditional term logic due to its very nature, or can a new term logic meet them in a clear, principled, and adequate way?

Ironically, a new version of term logic was formulated by one of MPL's most ardent champions, W.V.O. Quine (Quine [22, 23, 24, 25, 26, 27, 28, 29]). His Predicate Functor Logic (PFL) is a formal system that takes the material expressions of its language to be terms (though he still insisted on calling them 'predicates') and functors that apply to them, generating more complex terms (including sentences). As it turns out, PFL was never intended to be a formal logic built to challenge MPL. Instead, it was aimed at revealing just what is involved in the binding of individual variables by quantifiers in MPL. His method was to contrive a system that had all the expressive and inferential power of the standard logic, with the quantifier-variable mechanics replaced by new functors on terms (predicates) that could have the same results. It was a nice, even interesting and instructive exercise, but it was never meant to be a genuine term logic.

About the time Quine was building his term-functor logic, Fred Sommers was building his own (Sommers [31, 32, 33, 35, 36, 36, 38, 39, 40], Sommers and Englebretsen [41]). In his case, the result was a genuine term-functor logic (TFL). It is a system of formal logic that was intended as a strengthened, revitalized version of traditional logic, one that could meet the challenges of the old version. A closer look at the nature of (TFL), in its historical context, is in order.

..............................

A widely held belief among many ordinary people, as well as formal scientists such as mathematicians and logicians, is that one of the fundamental differences between natural language and formal languages is this: ambiguity is ubiquitous in the former but systematically prohibited from the latter, and, moreover, features such as ambiguity, vagueness, ellipsis, etc. are taken to exclude ordinary natural language as an appropriate vehicle for logical reckoning. This is a myth. Consider the meanings of each token of the numeral 5 in 555. Three numerical positions — three numerical meanings. Or consider the fact that the plus and minus signs are systematically ambiguous, sometimes standing for binary functors (addition, subtraction) and other times standing for unary functors. Thus in the expression '−7+5' the minus is unary (while the plus is binary) but in the expression '12−7' the minus is binary. It is obvious that ambiguity of expressions is not confined to natural languages. Moreover, ambiguity, at least for some elements of a formal language (e.g., arithmetic), is a source of expressive power.

Traditional formal logic, from Aristotle to the end of the 19th century, was a *term logic*. Generally, a term logic divides the lexicon of its formal language into simple material/categorematic expressions, i.e., terms, and formal/syncategorematic expressions, i.e., *functors*. The former have meanings (on their own, as was often said); the latter do not. *Simple* terms are homogeneous in the sense that they are not divided into distinct types (e.g., general, singular, mass, count, concrete, abstract, etc.) by the term logician. Functors are of two types: unary and binary. These are applied to expressions one or two at a time to form new, more complex terms. Every statement is a *complex* term using a binary functor.

Modern mathematical logic sees things quite differently. Logicians, following Frege, begin not with terms but entire sentences (or sentential clauses). The newer logic takes its lexicon to consist of material expressions and functors (functions). Most importantly, it divides the material lexicon into exclusive types: names and predicates. Formal expressions, logical constants, are of various types and apply always and only to entire sentences. The distinction between names and predicates is purely semantic, determined by the number of objects denoted. Names are singular expressions, used to denote single objects. Predicates are general, denoting more than one object. Mathematical logic further divides sentences into two types: atomic and molecular. An atomic sentence consists of a single predicate and an appropriate number of names; no formal expressions are involved (cf. [36]). A molecular sentence consists of one or more sub-sentences combined by the use of one or more sentential functions.

Consider a simple term logic whose only functors are a unary minus and a binary plus. The syntax is simple. If $X$ is a term, then so is $-X$; if $X$ and $Y$ are terms then so is $X+Y$. (For convenience we can ignore the use/mention distinction, letting expressions of the formal language name themselves.) Since these functors are specified recursively, we can (using parentheses in the standard way for grouping) generate any number of more complex terms (e.g., $-(-A), (A+B), -((-A)+B), (A+B)+(A+(-C))$, etc.). The plus functor is symmetric and associative (but not reflexive or transitive). This means that some simple derivations are possible (e.g., $B+A$ follows from $A+B$; $(A+B)+C$ follows from $A+(B+C)$. All of this is pretty simple — and not very impressive. More is required.

Modern mathematical logicians may have felt a certain disdain for natural language (not to mention the ordinary person's ability to use it for logical reckoning), but traditional logicians did not. Among their many aims, some even contrary to others (what else could be expected of an enterprise that was carried on intermittently, in many places, by many kinds of people, over a period of twenty-four centuries?), was a goal to build a formal model of natural language that would bring into relief those linguistic features that were essentially involved in ordinary reasoning. Our little formal language of unary minuses and binary pluses could hardly reflect much of natural language — not even just that much that is involved in logical reckoning.

# INTRODUCTION

Let's define two more functors. A unary plus is defined as follows: $+X = \mathrm{df} - (-X)$. A binary minus is defined as: $X - Y = \mathrm{df} - ((-X) + Y)$. By introducing a binary minus our formal language is rendered much more expressive and useful. The binary minus is both reflexive and transitive (but neither symmetric nor associative). So we can, for example, derive $Z - Y$ from $X - Y$ and $Z - X$; we can take $X - X$ to be a tautology. The formal features of our binary functors, then, allow us to make a number of types of derivations. Moreover, these particular formal features are among those applying to various natural language expressions and essential to ordinary reasoning. Thus, conjunctive expressions such as 'and' are, when used ordinarily, both symmetric and associative ('Sam and Nathan' = 'Nathan and Sam', 'It's cold and raining' = 'It's raining and cold', 'It's cold, windy and raining' = 'It's cold and windy and it's raining'). Expressions such as 'some....is/are...' are also symmetric and associative ('Some singers are actors' = 'Some actors are singers', 'Some performers who are actors and singers are dancers' = 'Some performers who are actors are singers and dancers'). Aristotle expressed these using (Greek versions of) 'belongs to some'; the medieval scholastic logicians expressed it using '$i$' (thus: '$XiY$', 'Some $Y$ is $X$), calling such statements '$I$ categoricals'. And just as our binary plus, via its formal features, reflects various natural language expressions sharing those features, the same holds for the binary minus. Ordinary expressions such as 'if....then' and 'every....is....' are both reflexive and transitive ('If it's raining then it's raining', 'Every fool is a fool' are tautologous; 'If it's raining then it's cold' and 'If it's cold then I'll stay home' jointly entail 'If it's raining then I'll stay home', and 'All logicians are philosophers' and 'Every philosopher is wise' jointly entail 'All logicians are wise'). Aristotle expressed these latter kinds of statements using 'belongs to every'; the scholastics used '$a$' ('$YaX$', 'Every $X$ is $Y$'), calling them '$A$ categoricals'. It is obvious that the i and a functors are versions of our binary plus and binary minus.

But this is still not enough. Consider the inference from $X + Y$ and $Z - X$ to $Z + Y$. This is the form of such simple valid inferences as 'Some $Y$ is $X$, every $X$ is $Z$; so, some $Y$ is $Z$ (e.g., 'Some logician is a philosopher, and every philosopher is wise; so, some logician is wise'). While the formative expressions in play here maintain their usual formal features, those features alone cannot account for the validity of such an inference. What is required is a rule of inference that, inspired again by Aristotle, the scholastics called the *dictum de omni*. In order to fully understand this rule we need first to follow the scholastic logicians for just a few more steps.

Consider a statement such as 'Every man is rational'. This was initially paraphrased as 'Rational belongs to every man', and parsed as 'Rational / belongs to every / man'. 'Rational' and 'man' are the (categorical) terms (literally, the *termini* of the statement), the material elements, and 'belongs to every' is the logical copula, the formative element, the glue that binds the two terms into a new, more complex, syntactical unit, a statement. (Our binary plus and binary minus are *logical copulae*.) Next the scholastics *split the copula*, parsing the statement anew as: 'Rational belongs to / every man'. Thus far, the original paraphrase and its new parsing were seen as admittedly unnatural. To render them closer to natural language, they next reordered the elements: 'Every man / rational belongs to', then reordered the second element to yield: 'Every man / belongs to rational'. Following Abelard, the scholastics then replaced the quite unnatural expression 'belongs to' with a grammatically appropriate version of 'to be' (which is what they officially termed the *logical copula*), to arrive finally at: 'Every man is rational', a perfectly natural statement consisting of two parts which they called the *subject* ('every man') and the *predicate* ('is rational'). The subject was further divided into the *quantifier* ('every') and the *subject-term* ('man') and the predicate was divided into the *qualifier*, still often called, misleadingly, the copula (some version of 'is') (see [11]) and the *predicate-term* ('rational'). What we call today *traditional logic* is this logic of quantified subjects and qualified predicates; it is the *subject-predicate* logic that Frege rejected in favor of the *function-argument* logic that is now in place.

The scholastic logicians were hardly content with applying their creativity and insight just to questions of logical syntax. They were even more devoted to issues of semantics. Vast numbers of (sometimes byzantine) semantic schemes were devised. These logicians were, at the very least, masters of distinctions. The core notion in such theories was that a term used in statements has meaning in the sense that it can *supposit*, stand for, something else. Theories of supposition tended to distinguish among a number of different ways a term could stand for something. Some of these distinctions continued to concern many logicians up to the 20th century and beyond: the distinction between the *connotation* of a term and its *denotation* (or its *sense* and *reference*, *Sinn* and *Bedeutung*) is one; another is the distinction between those terms in a statement that are *distributed* and those that are *undistributed*. It is this last that is of interest to us now. Roughly, a term used in a statement is said to be distributed in that statement just in case it stands for its entire denotation; otherwise it is undistributed. Terms that are universally quantified are distributed and terms that are particularly quantified are undistributed. That takes care of subject-terms, but what of predicate-terms? Here the idea is that a predicate-term is distributed in a statement whenever that statement entails a statement in which that term is universally quantified. For example, in 'Every logician is a philosopher' the term 'logician' is clearly distributed; 'philosopher' is undistributed because it is not universally quantified here, nor is it universally quantified in any statement entailed by the original statement. By contrast, in 'No logician is a fool' both 'logician' and 'fool' are distributed since the first is already universally quantified ('No logician is a fool' = 'Every logician is not a fool') and the second is distributed in 'No fool is a logician', which can be immediately derived from the original. As we said, this notion of distribution is rough. We will smooth it a bit later on.

The notion of distribution plays a central role in the application of the *dictum de omni* rule. The dictum has had a long and checkered history; it has been championed and challenged, and this in spite of the fact that it has been difficult to find agreement on even how to state it. For now we will say that the rule allows one to derive from a pair of statements, at least one of which is universally quantified, a new statement which is just like the other statement except that the predicate-term of the universal statement has replaced the subject-term of the universal statement where it occurs undistributed in the other statement. Let's return to the inference we set aside earlier: 'Some logician is a philosopher, and every philosopher is wise; so, some logician is wise'. This valid inference conforms to the *dictum*. At least one of the premises is universal. Its predicate-term ('wise') is substituted for its subject-term ('philosopher') in the other premise, where that subject-term ('philosopher') occurs undistributed, yielding the conclusion. Traditional logicians would say that the major term is substituted for the middle term in the minor premise to yield the conclusion. In fact, some 19th century logicians would describe this by saying that the tokens of the middle term have 'cancelled out each other', have been 'eliminated'.

We can use some of these traditional insights not only to augment and strengthen our own little formal language of pluses and minuses, but, in turn, to use this new formal language to cast clarifying light on those traditional notions of distribution and the *dictum*.

Let's begin our renovations by following the scholastic lead of *splitting* our logical copula, our binary functors. The expression $X + Y$ (the form of such expressions as 'Sam and Nathan' and 'Some $Y$ is $X$'), will be written with a split version of the binary plus. We will also render our formulations more natural by following their example of reordering. The split version of our binary plus will be written: $+\ldots+\ldots$, so that the split version of $X + Y$ is $+Y + X$. Note two things here: the two pluses of the split copula are not themselves copulae; they are simply fragments of the original (unsplit) copula. We can think of the first of the fragments as a 'quantifier' and the second as a 'qualifier'. Also note that the two fragments of the split copula are systematically ambiguous and that the ambiguity is benign, their different roles determined by position (just as in our positional numeric system). The split version of our binary minus will be written: $-\ldots+\ldots$, so that the split version of $X - Y$ is $-Y + X$. Here again we can think of the first fragment resulting from this split as a 'quantifier' and the second as a 'qualifier'. The quantifier resulting from splitting the binary plus is *particular*; that resulting from splitting the binary minus is *universal*. In all of this, we keep in mind Aristotle's observation that quantity and quality fundamentally characterize entire statements — not their terms themselves.

Given our two unary functors, we can see that any term is either *positive* (in the range of an even number or no unary minuses) or *negative* (in the range of an odd number of unary minuses). Since every statement is itself a (complex) term, every statement is either positive or negative. So, in any simple statement using a split copula there will be 5 formative elements: a sign indicating whether the statement is positive or nega-

# INTRODUCTION

tive, a quantifier, a sign indicating whether the subject-term is positive or negative, a sign indicating the qualifier, and a sign indicating whether the predicate-term is positive or negative. Let us say that the sign of a term being positive or negative indicates the term's *charge*. Every term, simple or complex (including entire statements), is charged. Further note that every quantifier is either particular or universal. The general form of any statement can be indicated as follows: $\pm[\pm(\pm S)+(\pm P)]$, which might be read: 'It is/isn't the case that some/every (non)$S$ is (non)$P$'. Two conventions can be adopted, both common in ordinary discourse. Unary pluses, signs of positive charge, can generally be safely suppressed. Also, the qualifier and the charge of the predicate-term can be amalgamated. This last means that when the predicate-term is positive we need only indicate the qualifier (now called a sign of *affirmation*); when the predicate-term is negative we need only use that sign as the qualifier (now called a sign of *denial*). It is important to keep in mind that only unary pluses can be suppressed; no minuses (unary or binary) can be suppressed, and no quantifiers (particular or universal) can be suppressed. This *streamlining*, suppressing certain unary pluses, in a formula reflects a corresponding practice in our natural language.

It is easy now to formulate the standard categorical statements: $A: -S+P, E: -S-P, I: +S+P, O: +S-P$. We could, as an example, formulate our inference about wise logicians as: $+L+P, -P+W; \therefore +L+W$. The contradictory of any statement is its negation. In other words, two statements are contradictory just in case they are exactly alike except for their charge. The negation of the $I$ categorical ($+S+P$) $is -[+S+P]$, which, once the external minus is distributed inside the brackets yields $-S-P$, an $E$ categorical. In like manner, O and A can be shown to be contradictory. We will make use of this practice of distributing external signs of charge.

Let us say that a statement has positive/particular *valence* whenever either its overall charge is positive and its quantifier is particular or its overall charge is negative and its quantity is universal; a statement has negative/universal valence whenever either its overall charge is positive and its quantity is universal or its overall charge is negative and its quantity is particular. Valence, in other words, can easily be determined by looking at the first two signs of its general form (even if the positive charge sign happens to be suppressed). If these two signs are the same (both plus or both minus) the statement is positive/particular in valence; if they are different the statement is negative/universal in valence.

Often pairs of statements can be formulated so that the two formulas are *algebraically equal*. For example, $-S + P$ and $-[+S - P]$ are algebraically equal. So are $+S + P$ and $-(-S) + P$.

The notions of valence and algebraic equivalence permit us to formulate a principle of *logical equivalence*: Two statements are logically equivalent just in case they have the same valence and are algebraically equal. Thus, in the two cases above, the first pair are logically equivalent but the second pair are not (remember that the external signs of positive charge are suppressed here).

Singular terms, terms denoting just one unique individual object, seem to pose a challenge to the term logician. Recall that for a term logic the lexicon of material expressions is homogeneous; any term, singular, general, mass noun, count noun, concrete, abstract, etc., can occur in a statement in any terminal position (i.e., as a subject-term or as a predicate-term). A consequence of this is that any term might appear sometimes quantified, other times qualified. All this is, of course, anathema for most modern mathematical logicians. Frege had made it strikingly clear that names (arguments, singular terms), including proper names and singular pronouns, are complete (saturated), while predicates (function expressions), are incomplete (unsaturated). Names refer to objects; predicates refer to concepts. Since an entire statement is complete, not containing any gaps to be filled by names, it is itself a name (of its truth-value). The contrast between names and predicates (and its corresponding contrast between objects and concepts) is absolute. In particular, no name (qua name) could ever be used as a predicate.

Consider the statement 'Aaron is tall'. For the mathematical logician this is a simple atomic statement analyzable into a (one-gapped) predicate, '....is tall' and a name fit to fill that gap, 'Aaron'. By contrast, the term logician eschews any atomic/molecular distinction. Every statement is a complex term, constructed from a pair of terms (themselves either simple or complex) bound together by a (usually split) logical copula. This means that even singular statements, like 'Aaron is tall', must have a logical form that reveals this tri-part structure. If 'is' is taken in this statement as a qualifier, then the question arises: where is the quantifier? Some traditional logicians took singular sentences to have an implicit particular quantifier, but most held that the implicit quantifier was universal (since then a singular subject would be distributed, referring to its entire denotation). Leibniz held that such sente

nces can be viewed as having either quantifier arbitrarily (Leibniz [18], but see also: Sommers [31, 32, 35, 39], Englebretsen [3, 5, 6, 8, 9, 10]. Since it doesn't matter, for logical purposes, which quantifier is understood (i.e., quantified singular terms can be said to have "wild" quantity), the quantifier is suppressed in natural language ('Some Aaron' = 'Every Aaron' = 'Aaron'). This turns out to be a good idea. Suppose I assert 'Aaron is tall' and 'Aaron is a boy', from which I want to derive 'Some boy is tall'. I can do this of course, but how does the term logician account for it? The mathematical logician gives an account making use of a rule of statement logic, Conjunctive Addition, and a rule for introducing an "existential" quantifier, Existential Generalization. The term logician, following Leibniz, simply lets the two occurrences of 'Aaron' have different quantifiers, and then applies the *dictum de omni*.

Mathematical logicians have serious reservations about the quantification of singular subjects, but these are nothing compared to their reservations about the qualification of singulars [14]. Since the term logician's lexicon is homogeneous (any term can appear in any logical role), term logic must admit singular terms to predicate-term positions. Frege's rock-hard distinction between names and predicates is happily ignored by a term logic.

Consider the statement 'Twain is Clemens'. It looks as if it consists of a pair of singular terms flanking a qualifier. But logicians from Frege and Russell onward have warned that when it comes to natural language appearances are often (usually) deceiving. Such logicians analyze our statement into a pair of singular terms flanking a very special predicate expression. While the 'is' plays virtually no role most of the time in modern systems of formal logic, it becomes suddenly all-important in cases such as this one. It is taken to be a sign of *identity*, a relation between an object and itself. Unlike other predicate expressions, this one is given its own notation ($=$) and special limitations are put on the interpretation of statements in which it is involved, and special rules are employed for inferences involving such statements. Suffice it to say, the term logician, armed with the understanding that, just as in ordinary language, singular terms can appear in any logical role (as arbitrarily quantified subject-terms or as qualified predicate-terms), has no need of any special 'identity theory'. A so-called identity statement such as 'Twain is Clemens' has the logical form 'Some/every $T$ is $C$'. That's it. "But surely," the modern logician will say, "the logical form of such a statement must reveal the fact the identity is an equivalence relation, and our use of '=' reminds one of this." If all that one means by an equivalence relation is that statements like 'Twain is Clemens' are reflexive, symmetric, and transitive, then nothing more is required of the term logician [32, 37, 4, 5, 41]. Consider these logical facts: Every $A$ is $A$. Some $A$ is $B$ if and only if some $B$ is $A$. If every $A$ is $B$ and every $B$ is $C$, then every $A$ is $C$. Now let $A, B$, and $C$ be any singular terms. What more is required?

Traditional term logic tried to fit all statements into one of the four classic categorical forms. Three kinds of statements seemed ill-suited for such a fit: singular statements, relational statements, and compound statements (e.g., conjunctions, conditionals, disjunctions). We've seen that singular statements (including so-called identities) pose no serious challenge. Things seem not nearly so easy when it comes to relationals. Categoricals are supposed to be comprised of a pair of terms (possibly themselves complex) joined by a logical copula. If the copula is split, then the statement is construed as a subject and a predicate. But now consider a statement such as 'Some general is losing every battle'. It seems to have too many terms; and two of them are quantified; and what about that relational term 'losing'? Freed from any analysis in terms of subjects and predicates, the mathematical logician, can analyze such a statement using statement functions such as conjunction and conditionalization, individual variables (pronouns), and quantifiers applied to entire statements and binding those pronouns. Still, such cycles and epicycles are not required in a logic of terms. Remember: every statement is a complex term; any term used in a statement may be a complex term; every complex term is a copulated pair of terms. Our sample relational sentence consists of a subject ('some general') and a predicate ('is losing every battle'). That predicate is itself complex, consisting in this case of a pair of terms ('losing' and 'battle') joined by an unsplit logical copula ('every'). In natural language expressions such as 'some' and 'every' (e.g., '$a(n)$', 'any', 'every', etc.) are used both as quantifiers and as (abbreviated) unsplit copulae. Still that's not enough.

Let's look at how modern logicians analyze a simple relational such as 'Romeo loves Juliet'. The word '....loves....' is taken as a predicate with two gaps (a 'two-place predicate') and the two names, 'Romeo' and 'Juliet' fill the two gaps. Symbolized: 'Lrj'. Notice that the order of the two names is important. Reversing the order yields not an equivalent statement but rather the converse 'Juliet loves Romeo' ('Ljr'). In this formal language the arguments (names or individual variables) play at least two logical roles: their number indicated the 'adicity' (number of gaps) of the relational predicate, and their order indicates the 'direction' of the relation (in this case, who is loving and who is being loved). Adicity, needless to say, is of no concern in a term logic. But it is certainly important for any formal language hoping to capture the logical features of natural language to be able to reflect relational direction. And this is easily done. We introduce into our formal language numerical subscripts in the following way: a common numerical subscript is attached to any two terms in a statement that are copulated in that statement or in any statement that it entails. Whenever two terms are the only terms in a statement the numerical subscripts can be suppressed. For example, the statement 'Every man is rational' could be formulated as $-M_1 + R_1$ (it doesn't matter what numerals are used as long as they are the same), and more simply as $-M + R$. Numerical subscripts do their heavy lifting when it comes to relationals. 'Romeo loves Juliet' would be formulated as $\pm R_1 + (L_{12} \pm J_2)$, while 'Juliet loves Romeo could be formulated as $\pm J_2 + (L_{21} \pm R_1)$ or even $\pm J_1 + (L_{12} \pm R_2)$. 'Some general is losing every battle' would be formulated as $+G_1 + (L_{12} - B_2)$. 'A man gave a rose to every woman who hated him' would be formulated as $+M_1 + ((G_{12} + R_2)_{13} - (W_3 + H_{31}))$. In such statements, where even the relational term is complex, we could simplify the formulation by amalgamating the subscripts on the relationals to give us $+M_1 + ((G_{123} + R_2) - (W_3 + H_{31}))$. Where originally 'gave a rose to' was taken as a two-place relation between the man and the women who hate him, we now treat 'gave' as a three-place relation between the man, the rose, and the women. It should be noted here that we have allowed '$H$' to stand for 'hate him', thus ignoring the independent logical role played by the pronoun 'him'. As it happens, this term logic takes such anaphoric pronouns in a manner that reflects their occurrences in natural language statements (and avoiding making them, in the guise of bound variables, ubiquitous in today's not standard formal language). But the treatment of pronouns is a story for later.

We've seen that pairs of terms that are copulated in a statement entailed by an original statement are (perhaps tacitly) subscripted in the original statement by a common numeral. From our rose-giving statement we can derive such statements as 'A man gave a rose', 'A man gave a rose to every women who hated him', etc., but we cannot derive, e.g., 'A man is a rose', since in its formulation, $+M_1 + R_2$, the terms share no common subscript.

Modern mathematical logicians take the logic of compound statements (the 'propositional calculus') to be primary, basic, with the predicate calculus resting on it. Traditional logicians tried hard to incorporate the logic of compounds into term logic, in effect, reversing today's standard order. For example, Leibniz thought that if he could construe compound statements such as conjunctions and conditionals as categoricals it would make his attempt to build a term logic much easier. Traditional logicians already recognized strong similarities between conjunctive and particular statements and between conditional and universal statements. Even Frege allowed that any statement, $p$, could be understood as implicitly ascribing a predicate 'is true', reading a conditional of the form 'If $p$ then $q$' as 'No case of $p$ being true is a case of $q$ being false' (more accurately: 'No case of $p$ standing for the True is a case of $q$ standing for the False'). At any rate, we have already seen that a term logic need not distinguish between different types of statements. Every statement is a complex term, a pair of terms (complex or not) bound together by a split or unsplit copula. Moreover, those copulae apply to any term-pair, simple or complex, sentential or non-sentential. A formula such as $+X + Y$ can be the form of a particular statement (e.g., 'Some singers are actors') or a conjunctive statement ('Its raining and it's cold'). All that matters, from the logical point of view, is that the formulation of a natural language statement reflect such formal features as reflexivity, transitivity, symmetry, etc. — and that is just what our pluses and minuses are meant to do. We saw that in the case of relationals both split and unsplit copulae are commonly used in natural language and are likewise available in our formal language. This availability is especially obvious in the case of natural language expressions used to form compound statements. We have, for example, both a split and unsplit versions of natural language connectives for conjunctions (e.g., 'both....and' as well as 'and'), disjunctions ('either....or' as well as 'or'), and conditionals ('if....then' as well as 'only if').

So the logic of compound statements can be incorporated into a general logic of terms. But in doing this one needs to take care. There are two important disanalogies between statements and other complex terms. On the surface, these disanalogies appear to reflect poorly on any prospects for fully incorporating statement logic into term logic. A proper understanding of the nature of sentential terms will show, however, that the logic of statements is a special branch of term logic [39].

> *The first disanalogy*: While a particular statement does not logically entail its corresponding universal, a conjunction does entail its corresponding conditional. '(Both) $A$ and $B$' entails 'If $A$ then $B$' but 'Some $A$ is $B$' does not entail 'Every $A$ is $B$'.
>
> *The second disanalogy*: While a particular affirmative is logically compatible with its corresponding particular negative, a conjunction is not compatible with the conjunction of its first conjunct and the negation of its second conjunct. 'Some $A$ is $B$' and 'Some $A$ is not $B$' are compatible, but '(Both) $A$ and $B$' and '(Both) $A$ and not $B$' are not compatible.

It seems that these disanalogies rest on the difference between sentential terms and non-sentential terms. If we formulate the statements in the first disanalogy we get $+A+B$ entailing $-A+B$ when the terms are read as sentential but not when they are read as non-sentential. A similar distinction seems to hold for the second disanalogy. $+A+B$ and $+A-B$ are compatible when the terms are read as non-sentential but not when they are read as sentential. But, just for a moment, consider what happens if our term $A$ is singular. In that case the first disanalogy evaporates. 'Some $A$ is $B$' *does* entail 'Every $A$ is $B$' when $A$ is singular since singular subject-terms, as Leibniz saw, can be given arbitrary quantity. The same holds in case of the second disanalogy. Generally, $+A+B$ and $+A-B$ can both be true – unless $A$ is singular, in which case they are logically incompatible. But how does this help? It suggests that the logic of singulars and the logic of compound statements are in the same boat when it comes to these disanalogies. The disanalogies disappear when singulars are introduced. If sentential terms could be construed as singular terms, then the disanalogies disappear altogether and the logic of compound statements, the logic of sentential terms, becomes just a special branch of term logic. But can sentential terms be construed as singulars?

Every statement is a sentence used to make a truth-claim. The claim is that the proposition being expressed by the sentence is true. True of what? Every statement is made relative to some specifiable (not always explicitly specified) *domain of discourse*. To make a statement is to claim something about the domain relative to which it is made – that's the truth-claim. In our ordinary use of language our default domain of discourse is simply what we (and we hope our audience) take to be the actual world (or some salient part of it). To say that some singers are actors is to claim that there are singing actors in the world. To say that a filly won the derby is to claim that the world (at least the equine part of it) has as one of its constituents a derby winning filly. Generally speaking, to make a statement of the form $+A+B$ is to claim that the world has at least one thing that is $A$ and $B$ as a constituent; to make a statement of the form $-A+B$ is to claim that the world has no thing in it that is $A$ but not $B$. A more revealing way of expressing these claims is this: $+A+B$ claims that something characterized by $A$ is characterized by $B$; $-A+B$ claims that everything characterized by $A$ is characterized by $B$. But, when $A$ and $B$ are sentential terms, what is being characterized? In ordinary discourse, what is being characterized is *the world*, the domain relative to which we are making our statement. We could read our formulas as 'Some $A$ world is $B$' and 'Every $A$ world is $B$'. But, as there is just one world, the actual world, these subject-terms are singular and have, therefore, arbitrary quantity. The logic of compounds statements, like the logic of singular terms, is just a special branch of a general term logic. Indeed, we can think of the logic of compound statements as a part of the logic of singular terms, which itself is part of the logic of terms.

Let's return, finally, to distribution and the *dictum*. As we saw, the traditional notion of distribution was less than perfectly clear. Most often, idea was that universally quantified subject-terms are distributed; particularly quantified subject-terms are not distributed. An unquantified term is distributed in a statement just in case that statement entails a statement in which that term *is* universally quantified, otherwise it is undistributed. Our formal language of pluses and minuses can make things clearer and easier. Recall the general form of any statement (with the copula split): $\pm[\pm(\pm S)+(\pm P)]$. As we saw, any term used in a statement is either in the range of an even number (or no) minuses or in the range of an odd number of minuses. Whether a term is distributed or undistributed in a statement is simply a matter of determining whether the number of minuses it is in the range of is odd or not. If it is odd, then the term is distributed; if it is even or zero, then the term is undistributed. "Briefly, a term in a sentence is distributed if it occurs only negatively, and undistributed if it occurs only positively" [16, p. 603]. For example, formulating the statement 'It is not the case that every logician is a philosopher', $-(-L+P)$, shows that 'logician', being in the range of an even number of minuses is undistributed, and 'philosopher', being in the range of an odd number of minuses is distributed. In our rose-giving example from above, our formulation reveals that 'man', 'gave', and 'rose' are undistributed and 'women' and 'hates' are distributed. Since singular terms can be quantified arbitrarily, when they are quantified in a statement they can likewise be taken to be distributed or undistributed arbitrarily (when they occur unquantified, of course, their distribution value is determined just as with any other term) (For more on term distribution see: [42, 19, 43, 34, 17, 15, 30, 7, 44, 41, 21, 16, 1, 20]).

Thus far, we have given the bare outlines of a formal language that has been built with the twin aims of simplicity and naturalness. But any formal language must be more than just relatively simple and natural. It must have sufficient *inference power*. The Fregean revolution in logic was so successful because, even though the new logic was far more complex than the old and (often aggressively non-natural), it was powerful. It could account for a very wide range of inferences in a systematic way. Nonetheless, it is possible to build a term logic that preserves much of the simplicity and naturalness of the old traditional logic, but which can match the power of the newer mathematical logic. There is not enough space here to lay out that entire system, but, by way of illustration, we can look at the central rule of inference it employs – the *dictum de omni*.

We saw earlier that our formal language preserves enough of the formal features of certain natural language expressions to allow a number of kinds of *immediate* inference. Our incorporation of singulars, relationals, and compound statements, allows us to extend this power of immediate inference even further. The *dictum* is a rule that governs *mediate* inference, the inference of a statement from a pair of statements already available (e.g., as premises, axioms, hidden assumptions, previously derived statements, etc.). It can be stated simply now, given our language of pluses and minuses (and thus distributed and undistributed terms).

> *Dictum de omni*: Given any universal statement, the predicate-term (whether positively or negatively charged) can be substituted for the subject-term of that statement in any other statement in which that subject-term occurs undistributed.

The first figure classic syllogisms satisfy this rule. And as Aristotle showed, all other valid classic syllogisms are reducible (derivable from) these, supplemented by the use of elementary rules of immediate inference. This is one reason that traditional logicians, including Leibniz, saw the *dictum* as *the* rule of mediate inference. But that's not enough. The rule applies in cases of inferences involving singulars, relationals, and compound statements. Consider an inference involving singulars that is usually taken to be outside the capacity of any term logic: 'Twain is Clemens, Twain was a Missourian; so Clemens was a Missourian'. We can formulate this as: $\pm T + C, \pm T + M \therefore \pm C + M$. In each statement the quantity (thus distribution) of the subject-term is arbitrary.

The term, $M$, affirmed of the *distributed* term $T$ in the second premise, is substituted for that term taken as *undistributed* in the first premise, yielding, via the *dictum*, $+M+C$, which is then immediately converted to the conclusion by virtue of the symmetry of the plus copula.

Logicians in the $19^{\text{th}}$ century (and of course earlier) worried about traditional logic's inability to account for an inference such as: 'Every horse is an animal; so, every head of a horse is a head of an animal'. Let's begin by formalizing (using $K$ for 'head of'): $-H+A \therefore -(K+H)+(K+A)$. In this case there is a hidden tautologous premise, something that 'goes without saying': 'Every head of a horse is a head of a horse' ($-(K+H)+(K+H)$). The *dictum* allows us to substitute $A$ for $H$ (explicit premise) for any undistributed occurrence of $H$ in another statement (viz., the second occurrence of $H$ in the hidden premise), yielding the conclusion. Consider next the inference: 'Carlee loves a boy; every boy admires some hockey player; every hockey player is poetic; whoever admires someone poetic is sensitive; therefore, Carlee loves someone who is sensitive'. We begin by formulating this as (nothing is lost here by suppressing subscripts): $\pm C+(L+B), -B+(A+H), -H+P, -(A+P)+S \therefore \pm C+(L+S)$. Applying the *dictum* to the second and third premises yields $-B(+A+P)$; applying it to this and the fourth premise yields $-B+S$; and applying it, finally, to this and the first premise gives us the conclusion.

Standard valid argument forms used as rules in most versions of today's mathematical logic can be shown to satisfy our *dictum*. Modus Ponens and Existential Generalization are just two examples. The first is formulated in our term logic as: $-P+Q$ ('Every $P$ world is $Q$'); $-(W+P)+(W+Q); \pm W+P$ ('The world is $P$'); so, $\pm W+Q$. We can even simplify this, making it more familiar, by suppressing the occurrences of 'the world', giving us: $-P+Q, +P \therefore +Q$. Here, since $Q$ applies to $P$ universally in the first premise, it can be substituted for $P$ wherever $P$ is undistributed, e.g., the second premise, to give us the conclusion. Existential Generalization purports to derive 'Something is $F$' directly from '$A$ is $F$' (where $A$ is a name). Our term logic takes such an inference to be mediate, with a second, hidden, innocuous premise ('$A$ is a Thing'): $\pm A+F, \pm A+T \therefore +T+F$. Again the *dictum* does the work here.

Earlier we saw that two statements are logically equivalent just in case they are both algebraically equal and share the same valence (i.e., both are positive/particular or both negative/universal). We can use these notions of algebraic equivalence and shared valence to specify the necessary and sufficient conditions for *argument validity*.

> *Validity*: An argument is valid if and only if the number of premises (explicit or otherwise) with positive/particular valence equals the number of conclusions with positive/particular valence (i.e., either one or zero) and the sum of the premises algebraically equals the conclusion.

The application of Validity amounts to a decision procedure for arguments. Once the argument is symbolized, it amounts to counting minus signs and using simple algebra. It is simple and fast.

It was hinted above that the term logic introduced here is a match in terms of inference power with the standard mathematical logic. In fact, it's possible to show that there are certain kinds of inferences that the standard logic, unlike this term logic, is powerless to adequately account for. We will briefly look at just two telling examples. What linguists call Passive Transformation, as in the inference of 'Juliet is loved by Romeo' from 'Romeo loves Juliet', is simple, natural, and quite common. Yet the best that modern logic can do to analyze such an inference is to formulate it in a decidedly less than illuminating fashion: $Lrj \therefore Lrj$. Our term logic cast more light: $\pm R_1 + (L_{12} \pm J_2) \therefore \pm J_2 + (L_{12} \pm R_1)$. Associative Shift, justifies the inference of 'Someone whom Socrates taught taught Aristotle' from 'Socrates taught a teacher of Aristotle'. Again, the standard formulation sheds no light at all: $\exists x(Tsx \& Txa) \therefore \exists x(Tsx \& Txa)$. Formulation by the term logic reveals the 'shift' involved: $\pm S_1 + (T_{12} + (T_{23} \pm A_3)_2) \therefore +(\pm S_1 + T_{12})_2 + (T_{23} \pm A_3)$.

This has been merely a sketch of the kind of term logic that can be built as a simpler, more natural, and as powerful (perhaps more so) alternative to the standard logic now in place. It seems to have a far better claim to being considered the logic of natural language – natural term logic.

Many logicians in the early days of the Fregean revolution took as settled fact that ordinary people were condemned to carry out their everyday reasoning tasks in the medium of natural language. Their idea was that proper reasoning, especially the kind associated with mathematics, can only be done in the medium of an artificial, logically constructed language, free of ambiguity, vagueness, etc., and supplied with explicit, carefully formulated rules of inference. The result would be a formal system that represents the way one *ought* to reason (rather than the way one actually reasons). Nevertheless, ordinary people do succeed in carrying out a variety of tasks involving logical reckoning. In fact, even children are quite adept at making simple inferences and recognizing inconsistencies. Ordinary people, including children, tend to reckon logically with speed and a high degree of accuracy in much of their daily life. And they do this without access to any artificial language designed for that purpose. They do it with what they have at hand – their natural, ordinary, everyday language. If the underlying logical features of natural language that are involved in reasoning are fairly simple and close to surface features of that language, then it is unlikely that they are best represented in the language of functions, quantified, variables, etc. It is far more likely that such features are similar to those proposed by a logic of terms along the line of the one sketched above.

............

Here is an attempt at a brief summary of Sommers' term logic:

*Sommers' term logic, also known as the theory of types or the ramified theory of types, is a system of formal logic developed by Willard Van Orman Quine and Nelson Goodman, building on the work of Giusepe Peano and Bertrand Russell. The theory is based on the idea that there are different "levels" or "types" of entities, each with its own set of properties and relationships, and that the logic used to describe these entities should reflect this hierarchy of types.*

*In Sommers' term logic, terms are the basic units of meaning, and the theory is concerned with the relationships between these terms and the properties they possess. The basic idea behind the theory is that there are different types of terms, such as individuals, properties, and relations, and that the relationships between terms are determined by the types of terms involved.*

> *One important aspect of Sommers' term logic is the distinction between "simple" terms, which represent basic entities, and "complex" terms, which represent relationships between entities. For example, the term "John" is a simple term, representing an individual, while the term "is taller than" is a complex term, representing a relationship between individuals.*
>
> *Another important aspect of Sommers' term logic is the idea of "type assignment", which assigns a type to each term in a formula based on its role in the formula. This allows for a more precise and consistent treatment of the relationships between terms, and ensures that the logic used to describe these relationships is appropriate to the types of terms involved.*
>
> *Sommers' term logic is used in a variety of formal systems, including set theory, modal logic, and type theory, and is an important tool in the development of formal systems for natural language semantics and computational linguistics. Theory of types has also been the subject of much philosophical debate, particularly in the context of the foundations mathematics and the relationship between language and reality.*
>
> *In conclusion, Sommers' term logic is a powerful and elegant system of formal logic that provides a precise and consistent way of describing the relationships between terms in formulas. The theory of types is an important tool in the development of formal systems for natural language semantics, computational linguistics, and the foundations of mathematics, and continues to be an active area of research and philosophical discussion.*

For anyone familiar with Sommers' term logic this is an outlandish, but humorous, characterization of that. It is wrong in so many ways, and downright confusing in others. So, it can simply be ignored. After all, it was AI-generated. Fortunately, all the essays collected in this volume represent the serious work of experts on contemporary developments in term logic ... and all of these experts are homosapiens.

............

INTRODUCTION

I shared nearly 50 years of friendship and scholarly cooperation with Fred Sommers. His devotion to getting beyond the "official" philosophical and logical theses was unrelenting. Yet, he was a fair critic. More importantly he was able to build more than a few new and valuable innovations in his accounts of linguistic-ontological structure, truth by correspondence, and logic. His enthusiasm (and his humor) helped me to become attracted to his work and then to work along with him. However, during the year or so before Fred died in 2014, he often confided in me that he had little reason to hope that his ideas, especially his work on logic, would be remembered after he and I were gone. He worried that few others had been eager to examine or promote his version of term logic (TFL). As it turns out, Fred was mistaken. There were already other philosophers and logicians who were willing to examine the idea of an expanded and strengthened logic of terms. A number of younger scholars from the fields of logic, philosophy, mathematics, linguistics, and computational theory have seen in such a term logic features that can be expanded, exploited, or explained in new ways. Much of this work has been done in recent years, fueled by dissatisfaction with the various aspects of today's standard first-order predicate logic. So, the prospects for the future of term logic are bright.

I had become familiar with many of these scholars' works, and had shared ideas with some of them. Back in the late 1990s I looked for a way to formulate a system of logic diagrams that would serve as a graphic analogue for TFL in the same way that Venn's diagrams had been meant to do the same for Boole's algebraic logic. Sommers said the he had tried this but failed to find a clear way forward and gave up. But, he encouraged me to continue my own attempts. My first results were presented in [12]. Fifteen years later, I came across the work of a relatively young Mexican philosopher-logician, José Martin Castro-Manzano. He had already published a number of interesting pieces, including some on logic diagrams. Thus began our steady stream of correspondence. Eventually, we (along with J.R. Pacheco-Montes) published a book on the history and philosophy of diagrams and the linear diagram system appropriate for TFL [13]. Perhaps more importantly, Castro-Manzano was doing more work dealing directly with TFL. This work was innovative and showed how TFL can be augmented and extended in various ways, especially, and more recently in [2]. It was that essay that initially inspired me to do something more to show that work on TFL and term logics in general is still the focus of new, wide-ranging, path-breaking ideas and studies. Thus I began to think about bringing examples of the best of these ideas together as a collection of original essays that would showcase the ways in which, in spite of skeptics, term logic is a lively and growing field of research – a century after the Fregean revolution. I shared my idea with a number of others, including Amirouche Moktefi, who has offered helpful advice and suggestions along with enthusiastic encouragement. An immediate response came from Castro-Manzano, who sketched out the possibility for conference presentations and publication of such a collection. Unfortunately, soon after this, he was faced with a long-term health issue, which kept him from continuing with the project until quite recently.

There is no doubt in my mind that Fred Sommers would have been relieved and gratified to learn of this kind of new work on term logic, that the worries that stressed him in his final year about the future of his own work on logic were unnecessary. The results here should be taken as a tribute to him and to what he began.

The authors of this collection of essays offer their deep appreciation and sincere thanks to Jane Spurr. Her efficient and tireless work of editing has made this book better than it otherwise would have been.

# References

[1] Alvarez, E. and Correia, M., 2012, "Syllogistic with Indefinite Terms," History and Philosophy of Logic, 4: 297-306.

[2] Castro-Manzano, J.-M., 2022, "On Mixing Term Logics," *Logics for New-Generation AI, Vol. 1*, Lino, B., Markovich, R., Yang, Y.N. (eds.), London: College Publications, pp. 6-23.

[3] Englebretsen, G., 1980, "Singular Terms and the Syllogistic," The New Scholasticism, 54: 68-74.

[4] Englebretsen, G.,1981, "A Note on Identity, Reference and Logical Form," Crítica, 8: 75-81,

[5] Englebretsen, G., 1982, "Do We Need Relative Identity?" Notre Dame Journal of Formal Logic, 23: 91-93.

[6] Englebretsen, G., 1983, "Reference, Anaphora, and Singular Quantity," Dialogos: 41: 67-72.
[7] Englebretsen, G., 1985, "Defending Distribution," Dialogos, 45: 157-159.
[8] Englebretsen, G., 1986a, "Singular/General," Notre Dame Journal of Formal Logic, 27: 104-107.
[9] Englebretsen, G.,1986b, "Czeżowski on Wild Quantity," Notre Dame Journal of Formal Logic, 27: 62-65.
[10] Englebretsen, G., 1988, "A Note on Leibniz's Wild Quantity Thesis," Studia Leibnitiana, 20: 87-89.
[11] Englebretsen, G., 1990, "A Note on Copulae and Qualifiers," Linguistic Analysis, 20: 82-86.
[12] Englebretsen, G.,1992, "Linear Diagrams for Syllogisms (with Relationals)," Notre Dame Journal of Formal Logic, 33: 37-69.
[13] Englebretsen, G., with Castro-Manzano, J.M. and Pacheco-Montes, J.R., 2020, Figuring It Out: Logic Diagrams, Berlin and Boston: De Gruyter.
[14] Frederick, D., 2013, "Singular Terms, Predicates and the Spurious 'Is' of Identity," *Dialectica*, 67: 325-343.
[15] Friedman, W.H., 1978, "Uncertainties over Distribution Dispelled," Notre Dame Journal of Formal Logic, 19: 653-662.
[16] Hodges, W., 2009, "Traditional Logic, Modern Logic and Natural Language" Journal of Philosophical Logic, 38: 589-606.
[17] Katz, B.D. and Martinich, A.P., 1976, "The Distribution of Terms," Notre Dame Journal of Formal Logic, 17: 279-283.
[18] Leibniz, G., 1966, "A Paper on 'Some Logical Difficulties'," Leibniz: Logical Papers, G.H.R. Parkinson (ed.), Oxford: Oxford University Press, pp. 115-121.
[19] Makinson, D., 1969, "Remarks on the Concept of Distribution in Traditional Logic," Noûs, 3: 103-108.
[20] Martin, J. N., 2001, "Proclus and Neoplatonic Syllogistic," Journal of Philosophical Logic, 30: 187-240.
[21] Parsons, T., 2006, "The Doctrine of Distribution," History and Philosophy of Logic, 27: 59-74.
[22] Quine, W.v.O., 1959, "Eliminating Variables Without Applying Functions to Functions," Journal of Symbolic Logic, 24: 324-325.
[23] Quine, W.v.O., 1960, "Variables Explained Away," Proceedings of the American Philosophical Association, 104:343-347.
[24] Quine, W.v.O., 1971, "Predicate Functor Logic," Proceedings of the Second Scandanavian Logic Symposium, J. Fenstand (ed.), Amsterdam:

North-Holland, pp. 309-315.
[25] Quine, W.v.O.,1972, "Algebraic Quantifiers and Predicate Functors," Logic and Art: Essays in Honor of Nelson Goodman, R. Rudner and I. Sheffler (eds.), Indianapolis: Bobbs-Merrill, pp. 214-238. Reprinted in The Ways of Paradox and Other Essays, revised and enlarged edition, Cambridge, MA: Harvard University Press, pp. 283-307.
[26] Quine, W.v.O., 1976a, "The Variable," The Ways of Paradox and Other Essays, revised and enlarged edition, Cambridge, MA: Harvard University Press, pp. 272-282.
[27] Quine, W.v.O., 1976b, "Algebraic Logic and Predicate Functors," The Ways of Paradox and Other Essays, revised and enlarged edition, Cambridge, MA: Harvard University Press, pp. 283-307.
[28] Quine, W.v.O.,1980, "The Variable and its Place in Reference," Philosophical Subjects: Essays Presented to P.F. Strawson, Z. van Straaten (ed.), Oxford: Clarendon Press.
[29] Quine, W.v.O.,1981, "Predicate Functors Revisited," Journal of Symbolic Logic, 46: 649-652.
[30] Rearden, M., 1984, "The Distribution of Terms," Modern Schoolman, 61: 187-195.
[31] Sommers, F., 1967, "On a Fregean Dogma," Problems in the Philosophy of Mathematics, I. Lakatos (ed.), Amsterdam: North-Holland.
[32] Sommers, F., 1969, "Do We Need Identity?" Journal of Philosophy, 66: 499-504.
[33] Sommers, F., 1970, "The Calculus of Terms," Mind, 79: 1-39. Reprinted in The New Syllogistic, G. Englebretsen, (ed.), 1987, pp. 11- 56.
[34] Sommers, F., 1975, "Distribution Matters," Mind, 84: 27-46.
[35] Sommers, F., 1976, "Leibniz's Program for the Development of Logic" Essays in Memory of Imre Lakatos, R.S. Cohen, P.K. Feyerabend, and M.W. Wartofsky (eds.), Dordrecht: D. Reidel.
[36] Sommers, F., 1981, "Are There Atomic Propositions?" Midwest Studies in Philosophy, 6: 59-68.
[37] Sommers, F., 1982, The Logic of Natural Language, Oxford: Clarendon Press.
[38] Sommers, F., 1990, "Predication in the Logic of Terms," Notre Dame Journal of Formal Logic, 31: 106-126.
[39] Sommers, F., 1993, "The World, the Facts, and Primary Logic," Notre Dame Journal of Formal Logic, 34: 169-182.
[40] Sommers, F., 2000, "Term Functor Grammars," Variable-Free Semantics, M. Böttner and W. Thümmel (eds.), Osnabrück: Secolo Verlag, pp. 68-89.

[41] Sommers, F. and Englebretsen, G., 2000, An Invitation to Formal Reasoning, Aldershot: Ashgate.
[42] Toms, E., 1965, "Mr. Geach on Distribution," Mind, 74: 428-431.
[43] Williamson, C., 1971, "Traditional Logic as a Logic of Distribution Values," Logique et Analyse, 14: 729-746.
[44] Wilson, F., 1987, "The Distribution of Terms: A Defense of the Traditional Doctrine," Notre Dame Journal of Formal Logic, 28: 349-454.

# Notes on the Contributors

**Elbert Booij**  Independent researcher and ILLC guest, University of Amsterdam, Amsterdam,The Netherlands.

**J.-Martin Castro-Manzano**  Professor of Philosophy, Universidad Popular Autónoma del Estado de Puebla, Puebla, Mexico. His research interests are in formal systems and philosophical engineering.

**José David Garcia Cruz**  Doctoral student, Pontifica Universidad Cathólica de Chile, Santiago, Chile. He obtained his undergraduate degree in Philosophy at Benemérita Universidad Autónoma de Puebla and his master's degree in Philosophy at Pontificia Universidad Cathólica de Chile. His current research is focused on ancient logic, medieval logic, dynamic-epistemic logic, and logical geometry time problem-solving, dynamic resource allocation, etc.

**Manuel Correia Machuca**  Professor of Philosophy, Pontifica Universidad Cathólica de Chile, Santiago, Chile. My research interests are in Ancient philosophy, Logic, and History of logic with a special reference to the Aristotelian tradition. I also have some proficiency in Leibniz's philosophy. As for particular subjects, I have studies and taught determinism and freedom, the induction problem, and fallacies. I have published a number of journal articles and book chapters.

**George Englebretsen**  Professor of Philosophy, emeritus, Bishop's University, Sherbrooke, Québec, Canada. His primary areas of research have been logic, history and philosophy of logic, philosophy of language, and formal ontology. He is the author or co-author of a large number of books, articles, reviews, etc.

CONTRIBUTORS

**Wolfgang Lenzen** Professor emeritus, University of Osnabrück, Osnabrück, Germany. He has been working in philosophy of science, epistemic logic, epistemology, philosophy of mind, applied ethics, and history of logic. His main book publications comprise Recent Work in Epistemic Logic(1978), Glauben, Wissen und Wahrscheinlichkeit (1980), Das sysem der Leibnizschen Logik (1990), Calculus Universalis (2004), and Abaelards Logik (2021). His current research interests are focused on medieval logic.

**Amirouche Moktefi** Senior Lecturer in Philosophy, Nurkse Department of Innovation and Governance, Tallinn University, Tallinn, Estonia. His research interests are in the history of science, with a focus on modern logic and visual reasoning. He has extensively published on these subjects and has co-edited several volumes, including Visual Reasoning with Diagrams (Springer, 2013) and The Mathematical World of Charles L. Dodgson (Lewis Carroll) (Oxford University Press, 2019).

**Larry Moss** Professor of Mathematics and Director of the Program in Pure and Applied Logic, Indiana University, Bloomington, Indiana, USA. Main research interests: areas of overlap between logic and computer science, and logic and linguistics, including studies of grammar formalisms, non-well-founded sets, the mathematics of language, foundational work on recursion, the semantics of programming, and the logic of natural language.

**Wallace Murphree** Professor of Philosophy, emeritus, Mississippi State University, USA. Author of a number of articles in philosophy of religion, and several articles and a book on an expanded syllogistic with numerical quantifiers. Created ReasonLines (at www.reasonlines.com), a schematical alternative to Venn diagrams for assessing syllogistic validity.

**Jonas Raab** Government of Ireland Postdoctoral Fellow, Trinity College Dublin, Dublin, Ireland. His areas of specialization are (meta-)metaphysics, philosophy of science, and history of philosophy (esp. Aristotle, and analytic philosophy). He received his PhD in 2021with a thesis on paraphrase strategies, ontological commitment, and explication. His interests span most of theoretical philosophy, focusing especially on methodological questions. Jonas has published on the history of logic, meta-ontology, and methodology.

**Miles Rind** Independent scholar, Seattle, Washington, USA. Miles Rind has published articles on Immanuel Kant's theory of aesthetic judgment, 18th-century British aesthetic theory, and philosophical logic. He has written, though not published, a textbook of formal logic based on the system of algebraic term logic developed by Fred Sommers.

**Luc Schneider** Associate Lecturer at the Institute of Philosophy, Saarland University, Saarbrücken, Germany, and collections curator for philosophy and religious sciences, National Library of Luxembourg. Research interests: Logic (formal and informal, as well as term logic and legal argumentation) and Metaphysics (mereology, theory of persistence, substance, properties, etc.).

**Fabien Schang** Independent scholar and secondary school professor of Philosophy, Lycée Alfred Mézières, Longwy, France. Main areas of research: formal philosophy, philosophy of logic, philosophical logics, theory of opposition, formal semantics, epistemology. Post-doctoral research on formal theory of (dis)agreement, a formal theory of logical relations, and conceptual engineering for philosophical concepts in moral theory, epistemology, and politics. Recent books include États de choses (Dialogue sur la politique) and forthcoming La logique du clivage gauche-droite: Essai de métapolitique.

**Peter Simons** Professor of Philosophy, emeritus, Trinity College Dublin, Dublin, Ireland, Metaphysician, Philosopher of Logic, Historian of Central European and Polish Logic and Philosophy, author or co-author of several books, of 300 articles on a wide range of topics.

CONTRIBUTORS

**Robert van Rooij** Professor of Logic and Cognition, Institute of Logic, Language and Computation (ILLC), University of Amsterdam, Amsterdam, The Netherlands.

**Pei Wang** Associate Professor of Instruction, Department of Computer and Information Sciences, College of Science and Technology, Temple University, Philadelphia, Pennsylvania, USA. Research interests: foundational principles of intelligence and cognition, categorical reasoning and learning, multiple-criteria decision-making, real-time problem-solving, dynamic resource allocation, etc.

# On Semantics for Modal Term Logic

José-Martín Castro-Manzano

## 1 Introduction

Mondialism is a novel metaphysical theory that highlights the importance of the world (Englebretsen, [9, 10]). According to it, the world, its constituents and their properties are real, but to be real is not exactly the same as being existent. For the mondialist, existence is not a property of individuals, but of worlds. Hence, to say that something exists is to ascribe a constitutive property (i.e. the presence of that thing) to a relevant domain of discourse.

Thus, the mondialist might say, for example, that while William of Occam is real, William of Baskerville is not; however—a mondialist may proceed—both exist in so far as they are present in some domain of discourse: the former, in what we call the real world; the latter, in the fictitious world created by Eco; relative to Eco's *The Name of the Rose*, William of Baskerville exists in the sense that he is present in it, but he is not real.

This short description, or rather depiction of mondialism may remind us, for instance, of free logic and its exploits (in particular, for the claim that existence is a special predicate satisfied for some non-empty domains), or noneism and its strange semantics (in particular, for the thesis that there are things that do not exist (Priest, [13])). But there are some important differences. For a start, *vis-à-vis* free logic, mondialism is related to a logic of terms, rather than first/second order logic (and hence existence is not a special predicate of individuals); and compared to noneism, for mondialism existence is a property of worlds, not of individual items (and hence existence is not a predicate of individuals *simpliciter*).

Mondialism, hence, is a distinct metaphysical theory, but what matters the most for our current purposes is its underlying semantics are related to a logic of terms. Given this context, in this contribution we attempt to use said mondial semantics in connection with modal term logic (Englebretsen, [8]) in order to define a structure that we call "Englebretsen model." An Englebretsen model, we will see, is a semantical device that provides meaning to modal term logic within the metaphysical framework of mondialism.

In order to achieve our goal we proceed in the following way: first, we briefly expound a couple of deductive bases (namely, Term Functor Logic, and Modal Term Functor Logic); then we introduce our main contribution (namely, some way to interpret Modal Term Functor Logic); and finally we conclude with some broad remarks.

## 2 Preliminaries

### 2.1 Term Functor Logic

Assertoric syllogistic—the logic at the core of traditional term logic—is a term logic that makes good use of categorical statements in order to capture a basic notion of assertion. A categorical statement is a statement composed by two terms, a quantity, and a quality. Typically, we say a categorical statement is a statement of the form:

$$\langle Quantity \rangle \, \langle S \rangle \, \langle Quality \rangle \, \langle P \rangle$$

where $Quantity = \{\text{All, Some}\}$, $Quality = \{\text{is (are), is not (are not)}\}$, and $S$ and $P$ are term-schemes. From the standpoint of Sommers & Englebretsen's (assertoric) Term Functor Logic ($TFL^\alpha$, from now on) (Sommers, [15]; Englebretsen, [6]; Englebretsen & Sayward, [7]), we say:

**Definition 1** (Categorical statement in $TFL^\alpha$). A categorical statement in $TFL^\alpha$ is a statement of the form:

$$\pm S \pm P$$

where $\pm$ are functors, and $S$ and $P$ are term-schemes.

We can then obtain the typical, four categorical statements: the universal affirmative, the universal negative, the particular affirmative, and the particular negative. The following are examples of categorical statements in said order:[1]

1. All logicians are smart $:= -L + S$

2. No logician is smart (i.e. all logicians are not smart) $:= -L - S$

3. Some logicians are smart $:= +L + S$

4. Some logicians are not smart $:= +L - S$

Given this language, TFL$^\alpha$ offers a sense of validity as follows (Englebretsen, [6, p. 167]):

**Definition 2** (Valid syllogism (in TFL$^\alpha$))). A syllogism is valid (in TFL$^\alpha$) iff:

1. The algebraic sum of the premises is equal to the conclusion, and

2. the number of particular conclusions (*viz.*, zero or one) is equal to the number of particular premises.[2]

And so, with this logic, we can model assertoric inferences like the one shown in Table 1.

Now, given the previous exposition, one could think the notion of validity for this logic only covers monadic or syllogism-like inferences,

---

[1]In this context, terms are those elements into which a statement can be divided, that is, into that which is predicated of something (i.e. the predicate) and that of which something is predicated (i.e. the subject), as Aristotle suggested (*Pr. An.* A1, 24b16–17); whereas functors are logical expressions. As Englebretsen [6, 9] explains, a term might be formed by the use of a single word or a complex of words. In English, for example, *smart*, and *logician* are terms, as well, as *taught Plato*, or *in the agora* are terms. Terms are what the medieval scholastic philosophers called *categoremata*; whereas functors are *syncategoremata*, that is, words that are not terms but are used to turn terms into more complex terms. In English, for example, *and*, *or*, *only if*, *if ... then*, *all*, *some*, *not*, *is*, and *is not* are functors. This is similar to our current, typical distinction between logical variables and logical constants.

[2]We must mention that this approach is not only capable of representing syllogistic inference, since it can also represent relational, singular, and compound statements with ease and clarity (Englebretsen, [6]), but for our current purposes, this exposition will suffice.

| Statement | TFL$^\alpha$ |
|---|---|
| 1. All persons are interesting. | $-P + I$ |
| 2. All logicians are persons. | $-L + P$ |
| ⊢ All logicians are interesting. | $-L + I$ |

Table 1: A valid assertoric inference

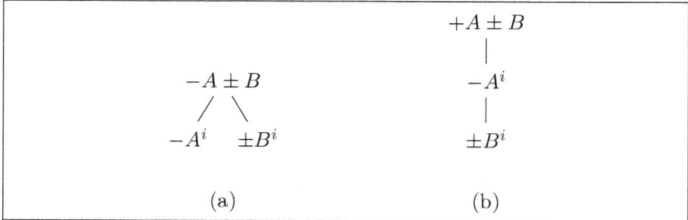

Figure 1: Expansion rules for TFL$^\alpha$

but as we hinted in a previous note, that would be a hasty conclusion. We can extend said notions of validity either by enlarging the rules of inference (Englebretsen, [6]) or by implementing tableaux proof methods (Castro-Manzano [3, 5]): we will follow the second path.

Thus we say, as usual, that a tableau is an acyclic connected graph determined by nodes and vertices. The node at the top is called the root. The nodes at the bottom are called tips. Any path from the root down a series of vertices is a branch. To test an inference for validity we construct a tableau which begins with a single branch at whose nodes occur the premises and the rejection of the conclusion: this is the initial list. We then apply the expansion rules that allow us to extend the initial list: consider Figure 1.

These rules behave as follows: after applying a rule we introduce some index $i \in \{1, 2, 3, ...\}$. For statements whose initial term has a minus, "-" (i.e. universal statements), the index may be any natural (Figure 1a); for statements whose initial term has a plus, "+" (i.e. particular statements), the index has to be a new natural if they do not already have an index (Figure 1b). Also, following TFL tenets, we assume the next rules of rejection: $-(\pm T) = \mp T, -(\pm T \pm T) =$

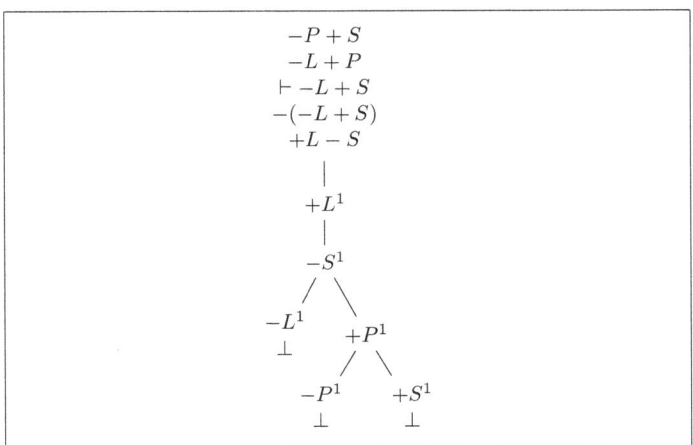

Figure 2: A valid assertoric inference

$\mp T \mp T, and -(--T--T) = +(-T)+(-T)$.

For this logic we say a tableau is complete if and only if every rule that can be applied has been applied; a branch is closed if and only if there are terms of the form $\pm T^i$ and $\mp T^i$ on two of its nodes; otherwise it is open. A closed branch is indicated by writing a $\bot$ at the end of it; an open branch is indicated by writing $\infty$. A tableau is closed if and only if every branch is closed; otherwise it is open. So, again as usual, a term $\pm T$ is a logical consequence of the set of terms $\Gamma$ (i.e. $\Gamma \vdash \pm T$) if and only if there is a closed complete tableau whose initial list includes the terms of $\Gamma$ and the rejection of $\pm T$ (i.e. $\Gamma \cup \{\mp T\} \vdash \bot$). With these definitions it is easy to prove that:

**Theorem 1** (Completeness for TFL$^\alpha$). An inference is valid in TFL$^\alpha$ iff there is a closed complete tableau for said inference (Castro-Manzano [3]).

As an example, consider the Diagram 2 for the inference shown in Table 1.

## 2.2 Modal Term Functor Logic

Englebretsen's Modal Term Functor Logic (TFL$^\mu$) —a formal version of modal syllogistic (Bocheński, [2]; McCall, [12]; Becker, [1]; Rini, [14]; Malink, [11])— tries to capture modality by extending TFL$^\alpha$ with $\Box$ and $\Diamond$. So, given a term $T$, TFL$^\mu$ allows the next combinations: $+\Box + T$ (i.e. $\Box + T$), $+\Box - T$ (i.e. $\Box - T$), $-\Box + T$ (i.e. $-\Box T$), $-\Box - T$, and, as usual, the operator $\Box$ is defined as $-\Diamond -$. Thus, we can say a *de dicto* modal statement is a statement of the form:

$$\langle\text{Modality}\rangle\ (\langle\text{Quantity}\rangle\langle S\rangle\langle\text{Quality}\rangle\langle P\rangle);$$

and a *de re* modal statement is a statement of the form:

$$\langle\text{Modality}\rangle\langle\text{Quantity}\rangle\langle S\rangle\langle\text{Quality}\rangle\langle P\rangle$$

where $Modality = \{\Box, \Diamond\}$, $Quantity = \{All, Some\}$, $Quality = \{\text{is (are)}, \text{is not (are not)}\}$, and $S$ and $P$ are term-schemes. Thus, formally:

**Definition 3** (Modal statement in TFL$^\mu$). *A modal statement in TFL$^\mu$ is a statement of one of the following forms:*

$$\mu(\pm S \pm P) | \pm S \pm P | \pm S \pm \mu P$$

*where $\pm$ are functors, $\mu$ is a modality, and $S$ and $P$ are term-schemes.*

Given this language, we have the next notion of validity:

**Definition 4** (Valid syllogism (in TFL$^\mu$). *A syllogism is valid (in TFL$^\mu$) iff:*

1. The algebraic sum of the premises is equal to the conclusion,

2. the number of particular conclusions (*viz.*, zero or one) is equal to the number of particular premises,

3. the conclusion is not stronger than any premise,[3] and

4. the number of *de dicto-*$\Diamond$ premises is not greater than the number of *de dicto-*$\Diamond$ conclusions.

| | | | |
|---|---|---|---|
| 1. $-M + \Box P$ | 1. $-M + \Box P$ | 1. $-M + \Box P$ | 1. $-M + \Box P$ |
| 2. $-S + \Box M$ | 2. $-S + \Box M$ | 2. $-S + \Box M$ | 2. $-S + M$ |
| $\vdash -S + P$ | $\vdash -S + P$ | $\vdash -S + \Diamond P$ | $\vdash -S + \Box P$ |
| 1. $-M + \Box P$ | 1. $-M + \Box P$ | 1. $-M + \Diamond P$ | 1. $-M + P$ |
| 2. $-S + M$ | 2. $-S + M$ | 2. $-S + \Box M$ | 2. $-S + \Box M$ |
| $\vdash -S + P$ | $\vdash -S + \Diamond P$ | $\vdash -S + \Diamond P$ | $\vdash -S + P$ |
| 1. $-M + P$ | 1. $-M + P$ | 1. $-M + P$ | 1. $-M + \Diamond P$ |
| 2. $-S + \Box M$ | 2. $-S + M$ | 2. $-S + M$ | 2. $-S + \Diamond M$ |
| $\vdash -S + \Diamond P$ | $\vdash -S + P$ | $\vdash -S + \Diamond P$ | $\vdash -S + \Diamond P$ |

Table 2: Some valid *de re* syllogisms

| | | |
|---|---|---|
| 1.$\Box(-M + P)$ | 1.$\Box(-M + P)$ | 1.$\Box(-M + P)$ |
| $M + \Box P$ | 1. $-M + \Box P$ | 1. $-M + \Box P$ |
| 2.$\Box(-S + M)$ | 2.$\Box(-S + M)$ | 2.$\Box(-S + M)$ |
| $\vdash \Box(-S + P)$ | $\vdash -S + P$ | $\vdash \Diamond(= S + P)$ |
| 1.$\Box(-M + P)$ | 1.$\Box(-M + P)$ | 1. $-M + P$ |
| 2. $-S + M$ | 2. $-S + M$ | 2.$\Box(-S + M)$ |
| $\vdash -S + P$ | $\vdash \Diamond(-S + P)$ | $\vdash -S + P$ |
| 1. $-M + P$ | 1. $-M + P$ | 1. $-M + P$ |
| 2.$\Box(-S + M)$ | 2. $-S + M$ | 2. $-S + M$ |
| $\vdash \Diamond(-S + P)$ | $\vdash -S + P$ | $\vdash \Diamond(-S + P)$ |

Table 3: Some valid *de dicto* syllogisms

These rules allow us to derive, for example, the valid *de re* and *de dicto* syllogisms shown, respectively, in Tables 2 and 3.

As a particular example, consider the inference shown in Table 4.

Now, following our previous exposition pattern, consider Diagram 3, which shows the tableaux rules for this logic (Castro-Manzano [4]). After applying a rule we introduce a superindex $i \in \{1, 2, 3, ...\}$ and we

---

[3]According to Englebretsen [8], there is a transitivity or "strength" of modal operators in such a way that: $\Box T$ implies $T\Box$, $T\Box$ implies $T$, $T$ implies $T\Diamond$, and $T\Diamond$ implies $\Diamond T$.

| Statement | TFL$^\alpha$ |
|---|---|
| 1. All persons are necessarily interesting. | $-P + \Box I$ |
| 2. All logicians are persons. | $-L + P$ |
| $\vdash$ All logicians are necessarily interesting. | $-L + \Box I$ |

Table 4: A valid modal inference

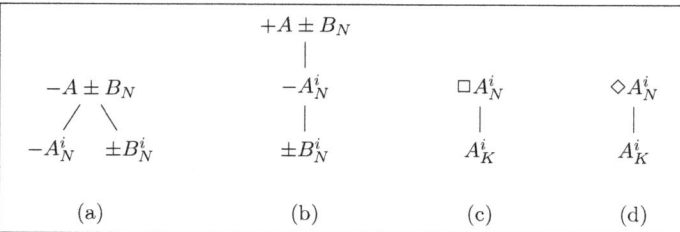

Figure 3: Expansion rules for TFL$^\mu$

let the subindex fixed as is. For statements whose initial term has a minus, the superindex may be any number; for statements whose initial term has a plus, the superindex needs to be a new number if they do not already have an index. Now to the next two rules: after applying a rule we introduce a subindex $K \in \{1, 2, 3, ...\}$ and we let the superindex fixed as is. For statements whose initial operator is $\Box$, the subindex may be any number; for statements whose initial term is $\Diamond$, the subindex has to be a new number if they do not already have an index.

For this logic we say a tableau is complete if and only if every rule that can be applied has been applied, a branch is closed if and only if there are terms of the form $\pm T_N^i$ and $\mp T_N^i$ on two of its nodes; otherwise it is open. A closed branch is indicated by writing a $\bot$ at the end of it; an open branch is indicated by writing $\infty$. A tableau is closed if and only if every branch is closed; otherwise it is open. As with the previous logics, we say a term$\pm T$ is a logical consequence of the set of terms $\Gamma$ if and only if there is a closed complete tableau whose initial list includes the terms of $\Gamma$ and the rejection of $\pm T$. So, following our exposition

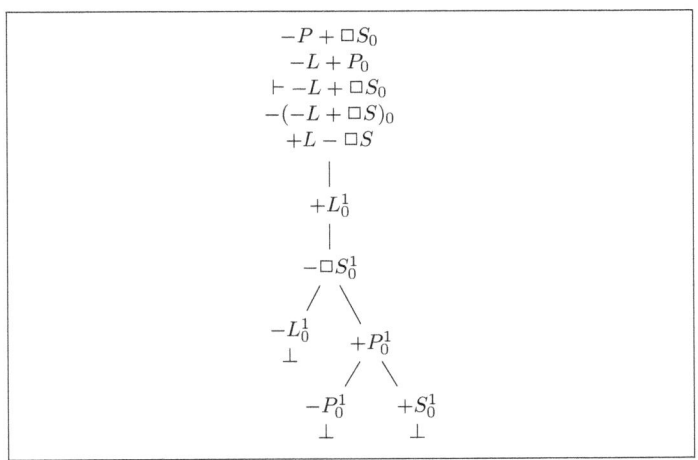

Figure 4: A valid modal inference

pattern, we can also prove that:

**Theorem 2** (Completeness for TFL$^\mu$). An inference is valid in TFL$^\mu$ iff there is a closed complete tableau for said inference (Castro-Manzano [4]).

Finally, consider Figure 4 for the inference shown in Table 4; and Figure 5 and 6 for further examples.

---

[4]In McCall's terminology, the letter $L$ stands for a categorical statement with an operator of necessity, $M$ denotes a categorical statement that includes a possibility operator, $Q$ denotes a categorical statement with a contingency operator, and $X$ denotes a categorical statement *simpliciter* (McCall, [12]).

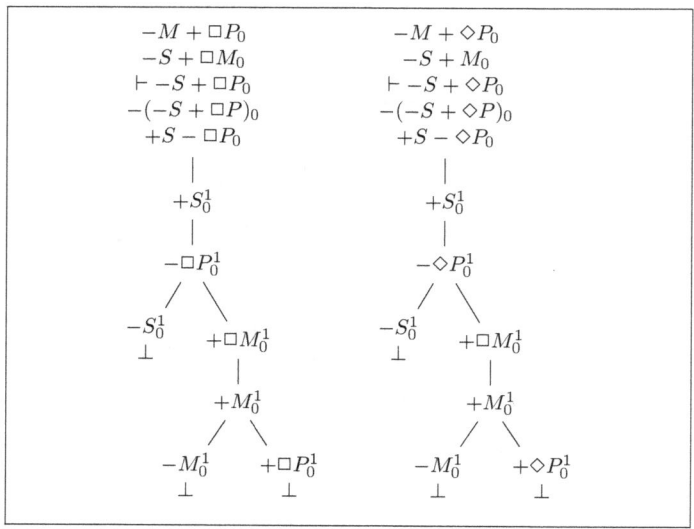

Figure 5: (A) Examples of valid *de re* syllogisms

## 3 Mondial and modal semantics

### 3.1 A primer on mondial semantics

As we said at the beginning, Sommers and Englebretsen's mondialism is a novel, distinct metaphysical theory that highlights the role of the world in the sense that, for the mondialist, existence is not a property of individuals, but of worlds. In mondial semantics, hence, statements are always made relative to some domain of discourse, and domains are understood as coherent totalities specified by their mutually compatible constituents.

Thus, domains can be defined as *consistent* sets of terms. When we say, for example, "William of Occam was excommunicated" our statement is made relative to some subset of the real world domain; but when

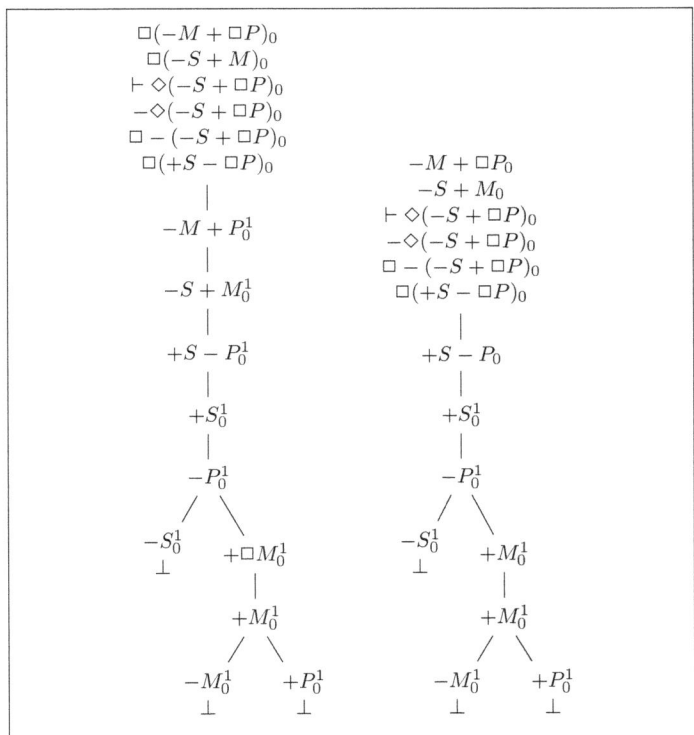

Figure 5: (B) Examples of valid *de dicto* syllogisms

we say, for instance, "William of Baskerville was excommunicated," our domain is not relative to the actual world, but rather to the world created by Eco. Consequently, in mondial semantics we can say both "Adso is a fool" and "Adso is not a fool" without contradiction as long as those statements are not both made relative to the same domain.

Thus, since on mondialism to be is to be a constituent of a domain of

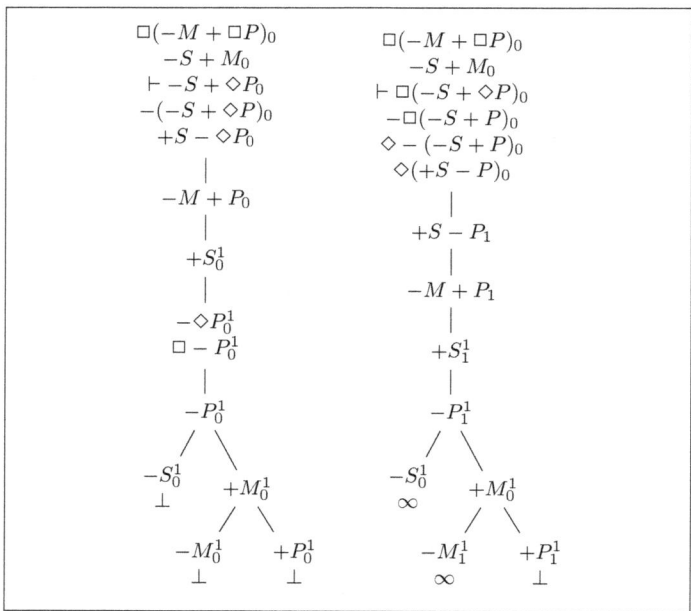

Figure 6: (A) Examples of a valid syllogism combining *de re* and *de dicto* modalities and an invalid *de dicto* syllogism.

discourse, then any categorical statement makes an implicit claim that some kind of items are (not) present in some given domain. To say, for example, with respect to the actual world, that "some latinx is edgy" is implicitly to claim that there is at least one edgy latinx in the actual world. To say, with respect to Eco's world, that "some monks died because of Jorge de Burgos' misdemeanors" is implicitly to claim that Jorge de Burgos, and some other monks, are present in that imaginary world.

According to Englebretsen [8], such implicit claims specify truth con-

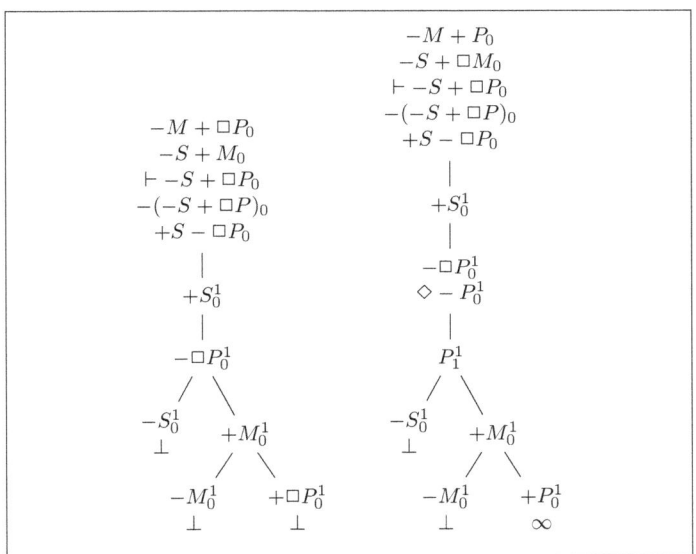

Figure 6: (B) Examples of a *Barbara* LXL *de re* syllogism, and a *Barbara* XLL *de re* syllogism.[4]

ditions. Thus, for example, to use "Some $S$ is $P$" relative to domain $d$ to make a statement is to characterize $d$ as having the property of being $SP$-ish, i.e. as having as one of its constituents an $S$ which is also $P$; and to use "All $S$ is $P$" relative to domain $d$ to make a statement is to characterize $d$ as having the property of not being $SP$-ish, i.e. as not having as one of its constituents an $S$ which is not $P$. The implicit truth claim of any categorical statement is that the domain which it denotes has the (constitutive) property it signifies (Table 5).

This characterization suggests that categorical statements are always made relative to a determinate domain, but there are some statements that are made with respect to no determinate domain. These statements are made relative to the domain of domains. To this group belong the

| Statement | Truth claim |
|---|---|
| $-S + P$ | $d$ is not $SP$ |
| $-S - P$ | $d$ is not $SP$ |
| $+S + P$ | $d$ is $SP$ |
| $+S - P$ | $d$ is $SP$ |

Table 5: Categorical statements in mondial semantics

| Statement | Truth claim |
|---|---|
| $\Box(-S + P)$ | $\cap d$ is not $SP$ |
| $\Box(-S - P)$ | $\cap d$ is not $SP$ |
| $\Box(+S + P)$ | $\cap d$ is $SP$ |
| $\Box(+S - P)$ | $\cap d$ is $SP$ |
| $\Diamond(-S + P)$ | $\cup d$ is not $SP$ |
| $\Diamond(-S - P)$ | $\cup d$ is not $SP$ |
| $\Diamond(+S + P)$ | $\cup d$ is $SP$ |
| $\Diamond(+S - P)$ | $\cup d$ is $SP$ |

Table 6: *De dicto* mondial semantics (adapted from (Englebretsen, [8]))

statements of *de dicto* modality. For example, the statement "possibly some $S$ is $P$" is a statement made with respect not just to the actual world domain, but rather with respect to any domain. It claims that $SP$-things exist in some undetermined domain: and hence it is modeled by joints (i.e. $\cup$). And similarly, a statement like "necessarily some $S$ is $P$" claims that $SP$-things are in every domain: and hence it is modeled by meets (i.e. $\cap$) (Table 6).

Statements of *de re* modality are always made relative to a determinable domain. We can say that a statement of the form "some $S$ is possibly $P$," made relative to the actual world, claims that there are $S$'s which are possibly $P$ in the actual world. If $d$ is a domain, let $d^*$ be a domain accessible from $d$. Now let $\cup d^*$ be the union of domains accessible from $d$, and let $\cap d^*$ be the intersection of domains accessible from $d$. To make a *de re* problematic statement relative to a domain $d$ is to claim implicitly that $\cup d^*$ has a specified property. To make a *de re* apodeictic statement relative to $d$ is to claim implicitly that $\cap d^*$ has a

| Statement | Truth claim |
|---|---|
| $-S + \Box P$ | $\cap d^*$ is not $SP$ |
| $-S - \Box P$ | $\cap d^*$ is not $SP$ |
| $+S + \Box P$ | $\cap d^*$ is $SP$ |
| $+S - \Box P$ | $\cap d^*$ is $SP$ |
| $-S + \Diamond P$ | $\cup d^*$ is not $SP$ |
| $-S - \Diamond P$ | $\cup d^*$ is not $SP$ |
| $+S + \Diamond P$ | $\cup d^*$ is $SP$ |
| $+S - \Diamond P$ | $\cup d^*$ is $SP$ |

Table 7: *De re* mondial semantics (adapted from (Englebretsen, [8]))

specified property (Table 7).

## 3.2 Coming to terms with trees

So far so good, but we think we can go a little bit further. Recall that, usually, we make use of Kripke models in order to provide interpretations for modal logics. Now, since, on the one hand, TFL$^\mu$ avoids some important syntactic features of first order logic (thus taking distance from classical and Kripkean interpretation techniques) but, on the other hand, it cannot escape the basic demands of formal semantics (for TFL$^\mu$ is not an uninterpreted system), and since mondialism can be understood in terms of sets of worlds, we can adapt some usual strategies of interpretation as to build an analogous concept of interpretation for TFL$^\mu$.

So, in order to accommodate the previous philosophical, mondial semantics, let us suggest a 4-tuple $\mathfrak{E} = \langle D, N, T, v \rangle$ where $D = \{d_0, d_1, d_2, ...\}$ is a set of domains of discourse (i.e. subindexes within the tableaux), $N$ is a set of natural numbers (i.e. superindexes within the tableaux), $T = \{\pm A, \pm B, \pm C, ...\}$ is a set of terms-and-functors, and $v : T \to N \times D$ is a function that assigns members in $T$ a pair defined by an natural and a domain. Call it an Englebretsen model. Now, let us exemplify this notion of interpretation or model by observing some tableaux for invalid inferences (Diagram 7), and let us focus on the open branches, since those branches, as usual, induce said interpretations.

From left to right, the first tableau corresponds to an invalid inference

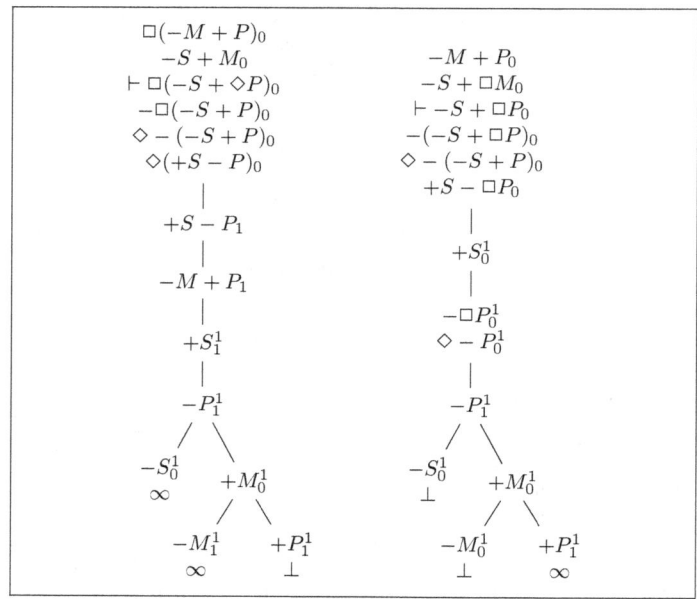

Figure 7: A couple of invalid modal syllogisms

with a pair of open branches. The first six lines but the conclusion define the initial list. The next line results from applying the rule for a *de dicto*-$\Diamond$ statement to the sixth line, hence moving from domain $d_0$ to domain $d_1$; the next line results from applying the rule for a *de dicto*-$\Box$ statement moving from $d_0$ to domain $d_1$; the next couple of lines results from applying the rule for a particular statement to the seventh line. Then the first split comes from applying the rule for a universal statement to the second line; and the second split is the result of the same process applied to the eight line. As a result of this process we obtain an open tableau, since $+S_1^1$ and $-S_0^1$, as well as $+M_0^1$ and $-M_1^1$ will not contradict each other. Similarly, the second tableau will not do.

**Example 1.** Now, in each case, by following the open branches of the tableau we can see the rejection of the conclusion leads to the acceptance of some consistent, coherent domains. In the first case we would have $\mathfrak{E} = \langle D, N, T, v \rangle$ as follows:

- $d = \{d_0, d_1\}$
- $n = \{1\}$
- $T = \{-S, +S, -M, +M, -P, +P\}$
    - 1st open branch: $v(-S) = (1, d_0), v(+S) = (1, d_1), v(-P) = (1, d_1)$
    - 2nd open branch: $v(-M) - (1, d_0), v(+S_=(1, d_1), v(-P) = (1, d_1), v(+M) = (1, d_0)$

Or in tabular form:

| $\mathfrak{E}$ | $d_0$ | $d_1$ | $N$ |
|---|---|---|---|
| 1st | $-S$ | $+S$ $P$ | 1 |
| 2nd | $-S$ $+M$ | $+S$ $-M$ | 1 |

This means that we have two interpretations that invalidate said inference. The first one stipulates that we require two domains, and one constitutive item such that $d_0$ is not $S$-ish (because it has one item that is not S as a constitutive property) but $d_1$ is $S$-ish but not $P$-ish. The second one indicates that we require two domains, and one item such that $d_0$ is not $S$-ish but is $M$-ish, and $d-1$ is $S$-ish but not $M$-ish.

At this point, however, maybe some informal, or rather visual/tabular versions of these interpretations are in order. Consider, then, for the first inference, the following table:

| $\mathfrak{E}$ | $d_0$ | $d_1$ |
|---|---|---|
|  | $-M+P$ $-S+M$ | $-M+P$ |

This makes sense within mondialist semantics. Since premise 1 states $\Box(-M+P)$, then it claims the intersection of domains includes $-M+P$, while premise 2 states that just domain $d_0$ is $-S + M$. Now, let us add the first counterexample (in order to distinguish the addition of the counterexample, we use italics):

| $\mathfrak{E}$ | | $d_0$ | $d_1$ | $N$ |
|---|---|---|---|---|
| | | $-M+P$ | $-M+P$ | |
| | $1^{st}$ | $-S+M$ | $+S$ | 1 |
| | | $-S$ | $-P$ | |

But notice that after adding the counterexample, we obtain what follows:

| $\mathfrak{E}$ | | $d_0$ | $d_1$ | $N$ |
|---|---|---|---|---|
| | $1^{st}$ | $-S+P$ | $-M$ | 1 |
| | | | $+S$ | |

Which shows that, at most, from that set of premises we can only obtain $-S + P$, but not the intended conclusion, which was $\Box(-S+P)$, for that would require having $-S + P$ in both domains.

**Example 2.** In the second example we would have $\mathfrak{E} = \langle D, N, T, v \rangle$ as follows:

- $D = \{d_0, d_1\}$
- $N = \{1\}$
- $T = \{+S, +M, -P. +P\}$
    - $1^{st}$ open branch: $v(+S) = (1, d_0), v(+M) = (1, d_0), v(-P) = (1, d_1), v(+P) = (1, d_o)$

| $\mathfrak{E}$ | | $d_0$ | $d_1$ | $N$ |
|---|---|---|---|---|
| | | $-S$ | | |
| | $1^{st}$ | $+M$ | $-P$ | 1 |
| | | $+P$ | | |

In this case we have just one interpretation that invalidates the corresponding inference. It stipulates that we require two domains, and one item such that $d_0$ is not $S$-ish, but is $M$-ish and $P$-ish; and $d_1$ is not $P$-ish. Now, following our previous exposition pattern, consider the following table representing the premises:

| $\mathfrak{E}$ | $d_0$ | $d_1$ |
|---|---|---|
| | $-M+P$ | |
| | $-S+M$ | $+M$ |
| | $+M$ | |

Since premise 1 states $-M+P$, then it claims $d_0$ is $-M+P$, while premise 2 states that $-S+\square M$ and so the intersection of domains from $d_0$ are $+M$. Now, let us add the counterexample :

| $\mathfrak{E}$ | $d_0$ | $d_1$ | $N$ |
|---|---|---|---|
| | $-M+P$ | | |
| | $-S+M$ | | |
| $1^{\text{st}}$ | $+M$ | $+M$ | 1 |
| | $-S$ | $+M$ | |
| | $+M$ | | |
| | $+P$ | | |

And after adding the counterexample, we obtain what follows:

| $\mathfrak{E}$ | $d_0$ | $d_1$ | $N$ |
|---|---|---|---|
| | $-S+P$ | | |
| $1^{\text{st}}$ | $+M$ | $+M$ | 1 |
| | $-S$ | | |
| | $+P$ | | |

Which shows that, at most, from that set of premises we can only obtain $-S+P$, but not the intended conclusion, which was $-S+\square P$, for that would imply having $+P$ in both domains.

## 4 Final remarks

In this contribution we attempted to develop some modal semantics by following Englebretsen's modialism and by using some typical tableaux techniques. In broad terms, we focused on the role domains of discourse play in modal logic; but having reached our goal, a very natural, annoying objection may appear: *nihil sub sole novum*. Indeed, probably the most annoying objection is the objection of novelty. What is the point of introducing a new sort of structure if we already have Kripke models? After all, Kripke models have been used extensively, and there is no need for some other structure. *Answer*: it is certainly true that Kripke models are the norm when dealing with modal logics, but we fail to see how this is a knock-out objection, let alone a knock-down. Having Kripke models is no restriction for implementing alternative solutions to alternative formal systems and, in this case, we have developed a structure that relates Englebretsen's modal term logic—-an alternative system—- with his mondialism, which is something that has not been done before, as far as we are aware. Of course, there is much we need to say about the potential of these structures, or mondialism for that matter, but that is material for another contribution.

# References

[1] A. Becker. *Die Aristotelische Theorie der Möglichkeitsschlüsse.* Wissenschaftliche Buchgesellschaft, 1968.

[2] J.M. Bocheński. *Formale Logik.* Orbis academicus; Karl Alber, 1962.

[3] Castro-Manzano, J.-Martín (2018), "A Tableaux Method for Term Logic", *Proceedings of the Latinamerican Workshop on New Methods of Reasoning 2018* 2264, pp. 1-14.

[4] Castro-Manzano, J.-Martín (2019), "Un método de árboles para la lógica de términos modal", *Open Insight* 12(23), pp. 165-180.

[5] Castro-Manzano, J.-Martín (2020), "Distribution Tableaux, Distribution Models", *Axioms* 9(41). [https://doi.org/10.3390/axioms9020041]

[6] G. Englebretsen. *Something to Reckon with: The Logic of Terms.* Canadian electronic library: Books collection. University of Ottawa Press, 1996.

[7] G. Englebretsen and C. Sayward. *Philosophical Logic: An Introduction to Advanced Topics.* Bloomsbury Academic, 2011.

[8] G. Englebretsen. Preliminary notes on a new modal syllogistic. *Notre Dame J. Formal Logic,* 29(3):381–395, 06 1988.

[9] G. Englebretsen. *Robust Reality: An Essay in Formal Ontology.* Philosophische Analyse / Philosophical Analysis. De Gruyter, 2013.

[10] G. Englebretsen. *Bare Facts and Naked Truths: A New Correspondence Theory of Truth.* Taylor & Francis, 2017.

[11] M. Malink. *Aristotle's Modal Syllogistic.* Harvard University Press, 2013.

[12] S. McCall. *Aristotle's Modal Syllogisms*. Studies in logic and the foundations of mathematics. North-Holland Publishing Company, 1963.

[13] G. Priest. *Towards Non-being: The Logic and Metaphysics of Intentionality*. Oxford University Press, 2016.

[14] A.A. Rini. Is there a modal syllogistic? *Notre Dame J. Formal Logic*, 39(4):554–572, 10 1998.

[15] F.T. Sommers. *The Logic of Natural Language*. Clarendon Library of Logic and Philosophy. Clarendon Press; Oxford: New York: Oxford University Press, 1982.

# Aristotelian Diagrams for TFL

José David García Cruz

jdgarcia2@uc.cl

## 1 Introduction

This paper presents a study of Aristotelian diagrams in Term Functor Logic (TFL). The main question to be solved is: What kind of diagrams exist in these logics, are they isomorphic to the diagrams of first order (FOL), modal (K) or classical propositional logic (CPL), or could new diagrams be defined in TFL?

The relation between TFL and Aristotelian diagrams has not been explored in depth in recent studies on TFL, possibly with a single exception in [9][1]. In that sense, it is relevant to investigate Aristotelian diagrams in TFL to show a number of cases that complement recent research in Logical Geometry. In particular, the topics that are developed and present some innovation are, certain versions of already known diagrams, such as $JSB$ hexagons, $SC$ hexagons, and Béziau octagon. An interesting contribution of our approach, inherited from Logical Geometry, is the semantic analysis of TFL logic expressions. Given the syntactic character of these logics, and their proximity to natural language, the issue of proposing a semantics for this logic has been minimally analyzed, with the exception of [6].

The paper is divided as follows. Section 2 contains the main definitions required to elaborate our proposal. We begin with a brief review of the origin and rebirth of TFL. Subsequently, we introduce the tableaux system for TFL. Finally, we present the main concepts of Logical Geometry. This section concentrates on presenting the definition of Aristotelian

---

[1] And that paper just uses Aristotelian diagrams, and concretely an octagon, to explore relations between logics, and not to study relations between fragments of formulas.

relations and bitstring semantics formulated by Lorenz Demey and Hans Smessaert[2].

Consequently, section 3 integrates the previous definitions in a series of Aristotelian diagrams with interesting characteristics. First, the Classical square is analyzed, whose main characteristic is that the subalternation arrows are inverted. This, as the reader may suppose, has as a consequence that the rest of the relations are modified, resulting in a diagram different from the classical one, where the universals are subcontrary and the particulars are contrary. This behavior is not new, it is already found in logics such as *Public Announcement Logic*, as well reported by [17]. In addition, the most common hexagons in Logical Geometry are presented, the $JSB$ and $SC$ hexagons. In both cases, the same behavior to that of the square is found. Thirdly, we present the Béziau's octagon. In all three cases, bitstring semantics is presented as denotational support for our analysis. Finally, section 4 closes the paper with a review of two possible consequences of our analysis. On the one hand, a proposal of Kripke semantics for **TFL**, and on the other hand, a reconstruction of the syllogistic with **CPL**.

## 2 Logical Technicalities

This section is divided into three parts. We will begin with a brief review of the origin and rebirth of *Term Functor Logics*, mentioning a few words about the current development that José Martín Castro-Manzano has elaborated. The second part contains our formulation of a base system, the logic we have called *assertoric-hypothetic Term Functor Logic* (**TFL**$_{av}$). In this part we reconstruct a Tableaux system and offer a formulation that allows us to integrate **TFL**-systems into Logical Geometry in order to study their Aristotelian diagrams. Finally, on the basis of our reconstruction, the **TFL** versions of Aristotelian opposition relations and bitstring semantics are presented.

---

[2]Cf. [23] and [33].

## 2.1 The Renaissance of Term Functor Logics

As we well know, Term Functor Logic originates with the work of logicians Fred Sommers and George Englebretsen[3]. The rationale for these systems lies, firstly, in the need to propose systems closer to natural language, which preserve the syntactic characteristics of the usual way of formulating propositions; that is, under the subject-predicate structure. Secondly, that they are consistent with the traditional formulation of the properties of the proposition, where we can highlight quantity and quality as determining signs of the type of proposition[4]. Finally, we can highlight one of the main aspects of these systems, the representation of linguistic phenomena that are difficult to represent in more popular systems, as FOL, CPL, K, etc. One of the most important advantages, with respect to the calculation and representation of deductions, is the rescue of the algebraic properties of signs. Deductions depend on a sum, and the deduction results in an equality between the conclusion and the set of premises.

Since its origin with Sommers and Englebretsen, logicians have remained suspicious about the representational and evidential capabilities of these systems. This omission is not accidental, if we take into account the developments that first-order logic allows, in the words of José Martín Castro-Manzano, this is due to the fact that first-order logic practically serves for everything[5]. This omission has not prevented the development of these systems, much less their flourishing. Several decades after their origin, with the work of José Martín Castro-Manzano, these logics are once again the focus of attention, not only from a scholarly point of view, but also taking these systems into new terrains. Castro-Manzano's case is special, and therefore, we will highlight the most relevant points of his project.

The main novelty proposed by Castro-Manzano, with respect to the traditional versions of TFL, is the Tableaux approach. This approach has allowed the formulation of a great variety of extensions of these systems,

---

[3] Cf. [34], [35], [25], [36], [24], and [25].

[4] A more complete list of properties of the propositions is: a) quantity, b) quality, c) mode, d) time, e) matter, f) indefinite subject, g) indefinite predicate. Cf. [13, p. 60 - 66].

[5] In a conference in the city of Puebla, José Martín Castro-Manzano expressed in very similar words this idea, from which we extract this paraphrase.

which not only emulate systems of classical logic, but also impose a novel path in the formulation of new logics. It is enough to explore the long list of papers in which these systems are presented. One of the main, and most original, is *Programming with Term Functor Logic* [11]. This work presents a language written in C, based on the triadic syntax of TFL, with capabilities similar to ProLog, but following the expressive structure of TFL. Besides presenting a technical benefit, this shows the expansive power of taking seriously the philosophical and scientific approach of these systems.

Other extensions consider modal aspects [8], numerical aspects [7], or relevance aspects [10], and all these extensions are synthesized in one of the most solid works by our author, *Mixing Colors, Mixing Logics* [9]. This work, not only presents an innovation with respect to possible contributions in Term Logic, but, in general terms, speaking of philosophical logic, it is a work that shows algebraic, analogical, and philosophical results simultaneously. By exploring the analogy between analysis and synthesis of color and analysis and synthesis of logics, Castro-Manzano proposes an analysis of the systems he has defined throughout this period of intense research. The renaissance of the TFL is manifested in this work, as a starting point to explore different research alternatives.

Our proposal starts from this point, we will try to carry out a basic study of the Aristotelian diagrams found in TFL. Since in Logical Geometry a large number of Aristotelian concepts have been systematized, our plan will be to integrate the logics we are now concerned with into this equally new research program, thus proposing a synthesis between contemporary work in Logical Geometry and TFL.

## 2.2 Tableaux System for TFL

In this subsection our version of assertoric-hypothetical term functor logic ($TFL_{\alpha v}$) is presented. We follow closely Tomazs Jarmuzek's book *Tableau Methods for Propositional Logic and Term Logic* [28]. All definitions presented here are adaptations of Jarmuzek's approach following José Martín Castro-Manzano's reconstruction of TFL[6]. The exposition start with $TFL_{\alpha v}$ language, then we continue with tableau concepts.

---

[6]Cf. [12], [5], [9], and [7].

Finally, the concept of tableau consequence will be presented.

**Definition 2.1.** We fix a set of *term schemes* $\mathfrak{T} = \{S, P, Q, ...\}$, and a set of *term functors* $\mathfrak{F} = \{-, +\}$. The language $\mathcal{L}_{av}(\mathfrak{T}, \mathfrak{F})$, which will usually be abbreviated as $\mathcal{L}_{av}$, is defined by the following BNF:

1. $\Theta ::= \pm T \mid \pm T \pm T \mid \pm \Theta \mid \pm \Theta \pm \Theta$ \hfill (where $T \in \mathfrak{T}$).

This language is basically the same as for assertoric term functor logic[7] with hypothetical (or propositional) expressions. The logic defined can be compared with Manuel Correia's reconstruction in [14], and with ancient and medieval expositions of Aristotelian logic, especially that of Boethius [4], Ammonius Hermia [1], and William of Sherwood [29]. In such expositions logic is presented in a language that integrates Aristotle's theory of the categorical proposition and Stoic's logical theory. Ammonius, for example, defines in his commentary [2, pp. 72, ll. 10 - pp. 75, ll. 15], an interesting debate on 'conjunctions' is presented taking some elements from Stoic propositional analysis. This system, therefore, corresponds to a formal core to start our analysis, but it also contains historical support. We need to continue with the remaining Tableaux definitions.

**Definition 2.2.** (Contradiction function) Function $\circ : \mathcal{L}_{av} \longrightarrow \mathcal{L}_{av}$ is defined by the following condition:
1. $\circ(\pm T) = -(\pm T)$.

**Definition 2.3.** (T-Inconsistency) The collection $X \subseteq \mathcal{L}_{av}$ will be called tableau inconsistent if the following condition is met:
1. $\exists \pm T \in \mathcal{L}_{av}$ s.t. $\pm T \in X$ and $\circ(\pm T) \in X$.
$X$ is t-consistent iff it is not t-inconsistent.

**Definition 2.4.** (Tableau Rule) Let $P(\mathcal{L}_{av})$ be the power set of language expressions. Let $P(\mathcal{L}_{av})^n$ be the Cartesian product $P(\mathcal{L}_{av}) \times ... \times P(\mathcal{L}_{av})$. A rule will be any subset $R \subseteq P(\mathcal{L}_{av})^n$ s.t. if $\langle X_1, ...X_i \rangle \in R$ then:
1. $X_1$ is consistent,
2. $X_1 \subset X_i$ (for each $1 \leq n$).

---
[7]Cf. [9, p. 70].

**Definition 2.5.** (T-Rules) The collection $\mathbb{R}_{av}$ of assertoric-hypothetical tableau rules, is defined by the following rules:

$$\frac{X \cup \{-S \pm P\}}{X \cup \{-S \pm P, +S\}, X \cup \{+S \pm P, \pm P\}} : a, e \qquad (2.1)$$

$$\frac{X \cup \{+S \pm P\}}{X \cup \{+S \pm P, +S, \pm P\}} : i, o \qquad (2.2)$$

**Definition 2.6.** (Branch) Let $K = \mathbb{N}$ and let $X \subseteq \mathcal{L}_{av}$. A Branch (branch beginning with $X$) will be any sequence $\phi : K \longrightarrow P(\mathcal{L}_{av})$ satisfying the following conditions:
1. $\phi(1) = X$,
2. $\forall i \in K$: if $i + 1 \in K$ then $\exists R \in \mathbb{R}_{av}$ and $\exists \langle Y_1, Y_2 \rangle \in R$, such that $\phi(i) = Y_1$ and $\phi(i+1) = Y_2$.

**Definition 2.7.** (Sub-branch) Let $\phi$ and $\psi$ two branches, if $\phi \subseteq \psi$ then, $\phi$ is a sub-branch of $\psi$.

**Definition 2.8.** (Maximal branch) Let $\phi : K \longrightarrow P(\mathcal{L}_{av})$ be a branch. We shall state that $\phi$ is a maximal branch iff:
1. $K = \mathbb{N}$,
2. $\neg \exists \psi$ such that $\phi \subseteq \psi$.

**Definition 2.9.** (Closed/open branch) A branch $\phi$ will be called closed if $\phi(i)$ is T-inconsistent (for some $i \in K$). A branch is open iff is not closed[8].

**Definition 2.10.** (Tableau) Let $X \subseteq \mathcal{L}_{av}$, $x \in \mathcal{L}_{av}$, and $\Phi = \{\phi_i, \phi_j, ...\}$ be a set of branches. The ordered triple $\mathbb{T} = \langle X, x, \Phi \rangle$ will be called a tableau for $\langle X, x \rangle$ iff $\Phi$ is a one-element subset of the set of branches beginning with set $X \cup \{\circ(x)\}$.

**Definition 2.11.** (Complete tableau) Let $\mathbb{T} = \langle X, x, \Phi \rangle$ a tableau for $\langle X, x \rangle$. We shall state that $\mathbb{T} = \langle X, x, \Phi \rangle$ is complete iff all branches contained in $\Phi$ are maximal. A tableau is incomplete is it is not complete.

**Definition 2.12.** (Closed/open tableau) Let $\mathbb{T} = \langle X, x, \Phi \rangle$ be a tableau for $\langle X, x \rangle$. $\mathbb{T}$ is closed iff all branches contained in $\Phi$ are closed. A tableau is open if it is not closed.

---

[8]Some corollaries follow from this fact, in specific 3.29 and 3.30 from [28].

**Definition 2.13.** (Assertoric-Hypotetical Tableau) An *Assertoric-Hypothetical Tableau* is a complete and closed tableau $\mathbb{T} = \langle X, x, \Phi \rangle$ constructed just with assertoric and hypothetical rules $\mathbb{R}_{\alpha v}$. The collection $\mathcal{T}_{\alpha v}$ is the collection of assertotic-hypothetical tableaus.

**Definition 2.14.** (Tableau validity) Consider a $x \in \mathcal{L}_{\alpha v}$. We will say that formula $x$ is T-valid, and we will write $\mathbb{T} \vdash x$, iff there is a complete and closed tableau $\mathbb{T} \in \mathcal{T}$, for $\langle \emptyset, x \rangle$.

**Definition 2.15.** (Tableau Consequence) Let $X \subseteq \mathcal{L}_{\alpha v}$ and $x \in \mathcal{L}_{\alpha v}$. Then:

1. $X \vdash_{\alpha v} x$ iff for all asertoric-hypothetical tableaus $\mathbb{T} \in \mathcal{T}_{\alpha v}$: if $\mathbb{T} \vdash X$ then $\mathbb{T} \vdash x$

If it is not the case that $X \vdash_{\alpha v} x$ we write $X \nvdash_{\alpha v} x$.

**Definition 2.16.** (Tableau Equivalence) Let $x, y \in \mathcal{L}_{\alpha v}$. Then:

1. we will say that $x$ and $y$ are equivalent, and we will write $x \equiv_{\alpha v} y$, iff $x \vdash_{\alpha v} y$ and $y \vdash_{\alpha v} x$.

An interesting property can be defined following Castro-Manzano's suggestion in [5, p. 40], in which the author presents a way to relate Boolean operations and Functors. The following definition syntheses this property.

**Definition 2.17.** (Boolean-Functor operation) Let CPL a Classical Propositional Logic defined as usual, with Boolean operations $\{\wedge, \vee, \rightarrow, \neg\}$, and truth-functional semantics '$\models$' (usually truth-tables). Let TFL$_{\alpha v}$ our term-functor logic defined as above. A Boolean-Functor mapping is a translation mapping $\tau : \mathcal{L}_{\mathsf{CPL}} \to \mathcal{L}_{\alpha v}$, such that:

$$\forall A, B \in \mathcal{L}_{\mathsf{CPL}}, \forall \theta, \kappa \in \mathcal{L}_{\alpha v}:$$
$$\tau(\neg A) = -\theta$$
$$\tau(A \wedge B) = +\theta + \kappa$$
$$\tau(A \vee B) = --\theta--\kappa$$
$$\tau(A \to B) = -\theta + \kappa$$

We will call the combination of functors in **TFL**-formulas (i.e. "$+...+...$, $--...--...$, etc.") Boolean-Functor operations.

**Definition 2.18.** (Logical System) The assertoric-hypothetic term functor logic is the tuple $\mathsf{TLF}_{\alpha v} = \langle \mathcal{L}_{\alpha v}, \vdash_{\alpha v} \rangle$.

To close this logical presentation we mention some interesting properties of this logic related to the square of opposition, by means of a theorem that summarizes some interesting results[9].

**Theorem 2.19.** *The following hold:*

1. $\vdash_{\alpha v} -(+S+P) + (-S+P)$
2. $\vdash_{\alpha v} -(+S-P) + (-S-P)$
3. $\vdash_{\alpha v} -(+(+S+P) + (+S+P))$
4. $\nvdash_{\alpha v} --(+S+P) --(+S+P)$
5. $\nvdash_{\alpha v} -(+(-S+P) + (-S+P))$
6. $\vdash_{\alpha v} --(-S+P) --(-S+P)$
7. $--(-S) --P, +S \vdash_{\alpha v} +P$
8. $-S+P, +S \vdash_{\alpha v} +P$
9. $+S+P \vdash_{\alpha v} -S+P$
10. $-S+P \nvdash_{\alpha v} +S+P$

*Proof.* All items follows by definitions 2.14 and 2.15. □

## 2.3 Aristotelian Relations and Bitstring semantics

In this subsection we will define the main concepts of Logical Geometry, Aristotelian relations and bitstring semantics. We will follow the standard definition proposed by Lorenz Demey and Hans Smessaert in many works, but specifically in [23], [19], [20], [21], and [33]. Our version adapts to the definition of logical system that we have proposed,

---
[9]These examples differ from the usual ones to a certain extent, since in general in works such as [12], [5], [6], [7], and [9], syllogistic reasoning and not Aristotelian relations are explored. For this reason it would be superfluous to repeat what is mentioned in these works.

considering that Aristotelian relations are defined in relation to a logical system possessing Boolean connectives. This last condition, as stated in the definition 2.17, is fulfilled without problem.

**Definition 2.20.** Let S be a TFL system with *Boolean-Functorial* operations and a tableau consequence relation $\vdash_S$. The *Aristotelian relations for* S are defined as follows: two formulas $\theta, \kappa \in \mathcal{L}_S$ are said to be

| | | | | |
|---|---|---|---|---|
| S-contradictory | iff | $\vdash_S -(+\theta+\kappa)$ | and | $\vdash_S --\theta--\kappa$, |
| S-contrary | iff | $\vdash_S -(+\theta+\kappa)$ | and | $\nvdash_S --\theta--\kappa$, |
| S-subcontrary | iff | $\nvdash_S -(+\theta+\kappa)$ | and | $\vdash_S --\theta--\kappa$, |
| in S-subalternation | iff | $\vdash_S -\theta+\kappa$ | and | $\nvdash_S -\kappa+\theta$. |

Furthermore, $\theta$ and $\kappa$ are said to be

| | | | | | |
|---|---|---|---|---|---|
| S-unconnected | iff | $\nvdash_S -(+\theta+\kappa)$ | and | $\nvdash_S --\theta--\kappa$ | and |
| | | $\nvdash_S -\theta+\kappa$ | and | $\nvdash_S -\kappa+\theta$. | |

With these concepts defined we can characterize the semantics of bitstrings adapted to the logic $\text{TFL}_{av}$. As with Aristotelian relations, we will first present the standard version of this concept and then adapt these ideas to $\text{TFL}_{av}$. In this respect, the formulation of classical propositional logic in TFL as presented in [5] will be of great use.

**Definition 2.21.** Consider[10] a logical system S as in Definition 2.20 and a finite fragment $\mathcal{F} \subseteq \mathcal{L}_S$. The *partition induced by* $\mathcal{F}$ *in* S, denoted $\Pi_S(\mathcal{F})$, is defined:

$$\Pi_S(\mathcal{F}) := \{\sum_{\varphi \in \mathcal{F}} \pm\varphi \mid \sum_{\varphi \in \mathcal{F}} \pm\varphi \text{ is t-consistent (relative to S)}\},$$

The elements of $\Pi_S(\mathcal{F})$ are called *anchor formulas*. The *Boolean closure of* $\mathcal{F}$ *in* S, denoted $\mathbb{B}_S(\mathcal{F})$, is defined to be the smallest set $C \subseteq \mathcal{L}_S$ such that (i) $\mathcal{F} \subseteq C$ and (ii) $C$ is closed under the Boolean-Functorial operations[11] (up to logical equivalence), i.e., for all $\varphi, \psi \in C$, there exist $\alpha, \beta \in C$ such that $\alpha \equiv_S +\varphi+\psi$ and $\beta \equiv_S -\varphi$.

---

[10]This and the following bitstring semantics definitions are adaptations of definitions from [22]. In this case we just make a TFL version of Definiton 4 in that work.

[11]Cf. Definition 2.17.

The reason to call $\Pi_S(\mathcal{F})$ a 'partition' is that, the anchor formulas are (i) jointly exhaustive, that is, for distinct $\alpha, \beta \in \Pi_S(\mathcal{F})$, $\vdash_S --\alpha--\beta$, and (ii) mutually exclusive, that is, $\vdash_S -(+\alpha+\beta)$ for distinct $\alpha, \beta \in \Pi_S(\mathcal{F})$[12].

It can be demonstrated that each formula within (the Boolean closure of) $\mathcal{F}$ is equivalent to a disjunction of anchor formulas. That is to say, for every $\varphi \in \mathbb{B}_S(\mathcal{F})$[13]:

$$\varphi \equiv_S \sum_{i \in \mathbb{N}} \{--\alpha_i \in \Pi_S(\mathcal{F}) | \vdash_S -\alpha + \varphi\}$$

We end this section defining Bitstring Semantics following, again, Lorenz Demey's characterization.

**Definition 2.22.** (Bitstring Semantics) The bitstring semantics $\beta_S^\mathcal{F}$: $\mathbb{B}_S(\mathcal{F}) \to \{0,1\}^{|\Pi_S(\mathcal{F})|}$ maps every formula $\varphi$ in $\mathbb{B}_S(\mathcal{F})$ onto its bitstring representation $\beta_S^\mathcal{F}(\varphi)$, which 'keeps track' of which formulas of $|\Pi_S(\mathcal{F})|$ enter into this disjunction[14].

# 3 Aristotelian Diagrams for TFL$_{\alpha v}$

## 3.1 Aristotelian Squares for TFL$_{\alpha v}$

We will begin this section by analyzing the classical Aristotelian square and then we will present some extensions well known in the specialized literature. Figure 1 shows the classical square, composed of the usual categorical propositions: universal affirmative $A$, universal negative $E$, particular affirmative $I$, and particular negative $O$.

This figure shows the usual configuration of the square, which comes, as we know, from the interpretation of certain passages of Aristotle's work, and which took this form until the works of Apuleius, Ammonius, and Boethius[15]. The picture shows the universals at the top and the particulars at the bottom, this feature is essential in the classical square, this is the big difference with the square of TFL$_{\alpha v}$, shown in Figure 2.

---

[12] Cf. [22, pp. 1708].
[13] Cf. [22, pp. 1708].
[14] Cf. [18, p. 126] and [22, pp. 1707 - 1708].
[15] Cf. [37], [1], and [4].

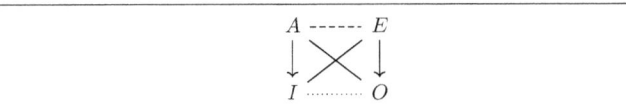

Figure 1: Abstract Classical Square

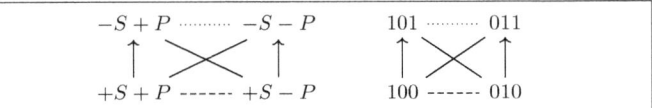

Figure 2: Classical Inverted Square in $\mathsf{TFL}_{\alpha v}$

In this diagram the relations are inverted, the subalterns point upwards, that is, from the $I(O)$ to the $A(E)$, and the contraries and subcontraries are interchanged, the particulars are contraries and the universals subcontraries. For obvious reasons the contradictories remain unchanged. This phenomenon is not new, and the most popular case is found in [17], but also in the more abstract case of arbitrary sequences of bitstrings as in [23] and [22]. Two versions of this inverted square are shown in Figure 3 and 4. The first case corresponds to the square of PAL and the second to a fragment of CPL with conjunction and conditional. In all three cases the propositions that we would consider universal are implied by the particulars, in the case of $\mathsf{TFL}_{\alpha v}$, it is interesting that the syllogistic is reconstructed with those conditions. We will study two possible variations that we can elaborate on this inverted square, initially we present the analysis of bitstrings.

In concrete, Fig. 2 shows the inverted square for $(\mathcal{F}_1, \mathsf{TFL}_{\alpha v})$. The fragment of formulas is $\mathcal{F}_1 = \{+S+P, +S-P, -S+P, -S-P\}$. The partition that is induced by this square is as follows:

$$\Pi_{\mathsf{TFL}_{\alpha v}}(\mathcal{F}_1) = \{ \begin{array}{rcl} \alpha_1 & := & +S+P \\ \alpha_2 & := & +S-P \\ \alpha_3 & := & +(-S+P)+(-S-P) \end{array} \}.$$

It is an easy exercise to prove that following items are satisfied:

1. $\vdash_{\alpha v} -(+(+S+P)+(+S-P))$

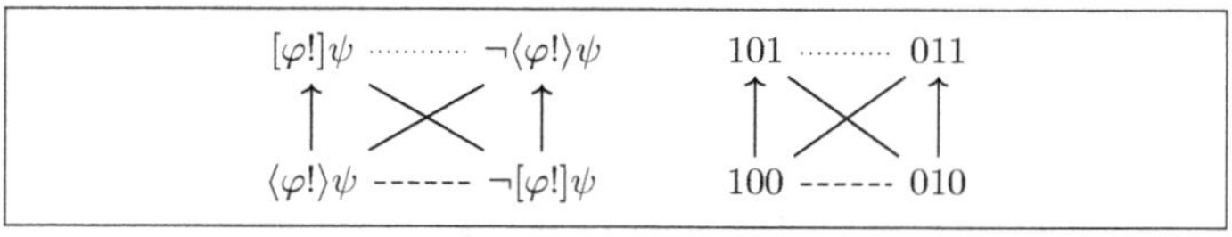

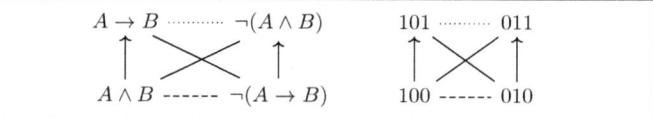

Figure 3: Classical Inverted Square in PAL

$$\begin{array}{ccc} A \to B & \cdots\cdots & \neg(A \wedge B) \\ \uparrow & \times & \uparrow \\ A \wedge B & \cdots\cdots & \neg(A \to B) \end{array} \qquad \begin{array}{ccc} 101 & \cdots\cdots & 011 \\ \uparrow & \times & \uparrow \\ 100 & \cdots\cdots & 010 \end{array}$$

Figure 4: Classical Inverted Square in CPL

2. $\vdash_{\alpha v} --(-S+P)--(-S-P)$

3. $\vdash_{\alpha v} -(+S+P)+(-S+P)$

4. $\vdash_{\alpha v} -(+S-P)+(-S-P)$

5. $\vdash_{\alpha v} -(+(+S+P)+(-S-P))$

6. $\vdash_{\alpha v} --(+S+P)--(-S-P)$

7. $\vdash_{\alpha v} -(+(+S-P)+(-S+P))$

8. $\vdash_{\alpha v} --(+S-P)--(-S+P)$

Items 1 and 2 represent contrary and subcontrary relations. Items 3 and 4 represent subalternation, and items 5 - 8 represent contradiction. The Boolean closure of this diagram is, as in classical version, $JSB$ hexagon. Next subsection shows the bitstring analysis.

## 3.2 Hexagons in TFL$_{\alpha v}$

In Fig. 5 is shown $JSB$ Hexagon[16] for $(\mathcal{F}_1, \mathsf{TFL}_{\alpha v})$. The fragment and partition are the same, the hexagon is the Boolean closure of the square. This means that every bit position is covered by all bitstrings, in

---

[16]Due to Jacobi, Sesmat and Blanche. Cf. [26], [27], [32], and [3].

other words, any other bit configurations is impossible. The remaining combinations are 111 and 000, top and bottom elements, respectively, but both converge in the middle of the diagram. This last property is due to the lack of relevance of top and bottom elements in Aristotelian relations, because all vertex must have contingent formulas.

To get the complete list of validities in $JSB$ hexagon, just add the previous square validities and the following:

1. $\vdash_{av} -(+(+S+P)+(+(-S+P)+(-S-P)))$
2. $\vdash_{av} -(+(+S-P)+(+(-S+P)+(-S-P)))$
3. $\vdash_{av} -(+(--(+S+P)--(+S-P))+(+(-S+P)+(-S-P)))$
4. $\vdash_{av} --(--(+S+P)--(+S-P))--(+(-S+P)+(-S-P))$
5. $\vdash_{av} -(+(-S+P)+(-S-P))+(-S+P)$
6. $\vdash_{av} -(+(-S+P)+(-S-P))+(-S-P)$
7. $\vdash_{av} --(--(+S+P)--(+S-P))--(-S+P)$
8. $\vdash_{av} --(--(+S+P)--(+S-P))--(-S-P)$
9. $\vdash_{av} -(+S+P)+(--(+S+P)--(+S-P))$
10. $\vdash_{av} -(+S-P)+(--(+S+P)--(+S-P))$

Items 1 and 2 fill contrary relations, while items 7 and 8 fill the triangle of subcontrary relations. Contradictory relations are completed by items 3 and 4. Finally, subalternations arrow are represented by items 5 and 6, and 9 and 10. A second way to extend inverted square is the $SC$ hexagon[17], to end this subsection bitstring analysis is presented.

Fig. 6 shows the $SC$ hexagon for $(\mathcal{F}_2, \mathsf{TFL}_{av})$. The fragment of formulas must be different because we need an additional contradictory axis to complete the diagram, concretely, the fragments is $\mathcal{F}_2 = \{+S+P, -S+P, +S-P, -S-P, +P, -P\}$. Additional elements are predicates of formulas in inverted square. The partition that is induced by this hexagon is as follows:

---

[17]Due to William of Sherwood and re-discovered by Tadeuz Czezowski. Cf. [30] and [15].

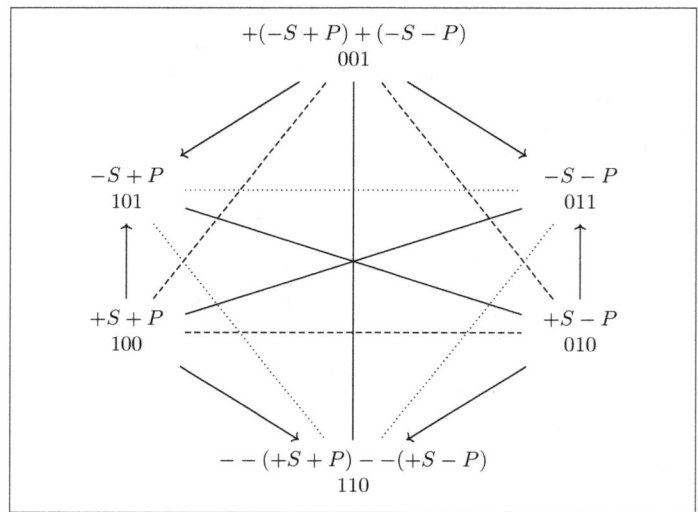

Figure 5: $JSB$ Hexagon in $\mathsf{TFL}_{\alpha v}$

$$\Pi_{\mathsf{TFL}_{\alpha v}}(\mathcal{F}_2) = \{ \begin{aligned} \alpha'_1 &:= +S + P \\ \alpha'_2 &:= +S - P \\ \alpha'_3 &:= +(+P) + (-S - P) \\ \alpha'_4 &:= +(-P) + (-S + P) \end{aligned} \}.$$

To obtain the complete list of validities we need to add the following to square's list:

1. $\vdash_{\alpha v} -(+(+P) + (-P))$

2. $\vdash_{\alpha v} --(+P) --(-P)$

3. $\vdash_{\alpha v} -(+(+P) + (+S - P))$

4. $\vdash_{\alpha v} --(+P) --(-S - P)$

5. $\vdash_{\alpha v} -(+P) + (-S + P)$

6. $\vdash_{\alpha v} -(+S+P)+(+P)$

7. $\vdash_{\alpha v} -(+(-P)+(+S+P))$

8. $\vdash_{\alpha v} --(-P)--(-S+P)$

9. $\vdash_{\alpha v} -(-P)+(-S-P)$

10. $\vdash_{\alpha v} -(+S-P)+(-P)$

Items 1 and 2 represent contradiction relation between new formulas, items 3 and 7 complete contary relations, and items 4 and 8 complete subcontrary relations. Items 5, 6, 9 and 10 complete subalternations. It will be noted that bitstring lengt increase in this case due to the inclusion of new formulas. If we look at anchor formulas in $\Pi_{\mathsf{TFL}_{\alpha v}}(\mathcal{F}_1)$ and $\Pi_{\mathsf{TFL}_{\alpha v}}(\mathcal{F}_2)$ just $\alpha_1$ and $\alpha_2$ remain the same in both partitions. Anchor formula $\alpha_3$ splits into two formulas in partition $\Pi_{\mathsf{TFL}_{\alpha v}}(\mathcal{F}_2)$, that is $\alpha_3$ splits into $\alpha_3'$ and $\alpha_4'$. Also note that formula $+(-S+P)+(-S-P)$ is equivalent to $-S$ and formula $--(+S+P)--(+S-P)$ is equivalent to $+S$. This equivalences led us to postulate a compound diagram joining the classical inverted square and the degenerated square of single terms showed in Figure 7. Next subsection analyses this Octagon.

### 3.3 Octagon in $\mathsf{TFL}_{\alpha v}$

In Fig. 8 is shown the Béziau's Octagon for $(\mathcal{F}_3, \mathsf{TFL}_{\alpha v})$. The fragment of formulas is $\mathcal{F}_3 = \{+S+P, -S+P, +S-P, -S-P, +P, -P, +S, -P\}$. The partition that is induced by this octagon is as follows:

$$\Pi_{\mathsf{TFL}_{\alpha v}}(\mathcal{F}_3) = \{ \begin{array}{lcl} \alpha_1'' & := & +S+P \\ \alpha_2'' & := & +S-P \\ \alpha_3'' & := & +(-S)+(+P) \\ \alpha_4'' & := & +(-S)+(-P) \end{array} \}.$$

There are no more validities that the previous, the remaining are unconnected relations that hold between single-term formulas represented by the following list:

1. $\nvdash_{\alpha v} -(+(+P)+(+S))$

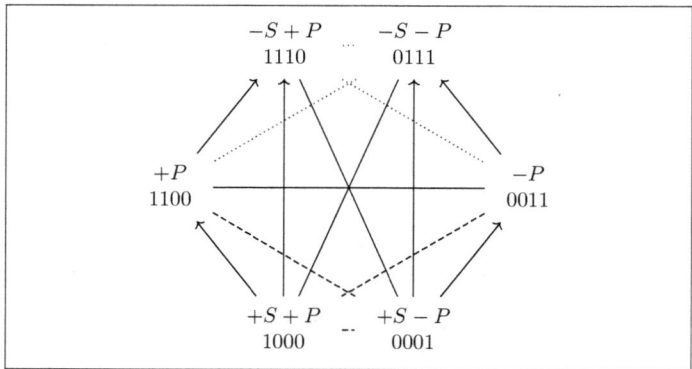

Figure 6: $SC$ Hexagon in $\mathsf{TFL}_{\alpha v}$

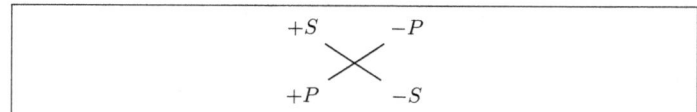

Figure 7: Degenerated square for single terms

2. $\not\vdash_{\alpha v} -(+(+P)+(-S))$

3. $\not\vdash_{\alpha v} -(+(-P)+(+S))$

4. $\not\vdash_{\alpha v} -(+(-P)+(-S))$

5. $\not\vdash_{\alpha v} --(+P)--(+S)$

6. $\not\vdash_{\alpha v} --(+P)--(-S)$

7. $\not\vdash_{\alpha v} --(-P)--(+S)$

8. $\not\vdash_{\alpha v} --(-P)--(-S)$

9. $\not\vdash_{\alpha v} -(+P)+(+S)$

10. $\not\vdash_{\alpha v} -(+P)+(-S)$

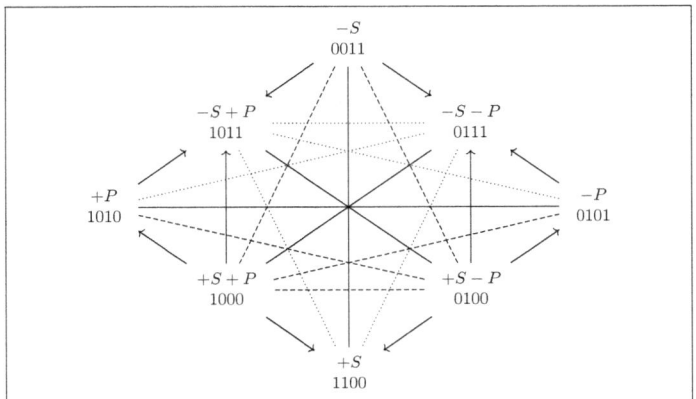

Figure 8: CPL Octagon in TFL$_{\alpha v}$

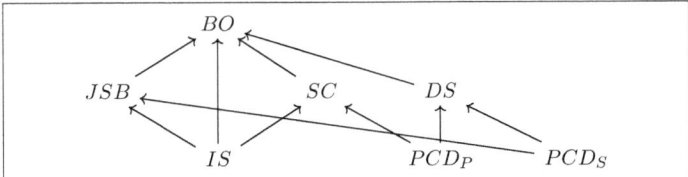

Figure 9: Morphisms between Aristotelian Diagrams in TFL$_{\alpha v}$

11. $\nvdash_{\alpha v} -(-P)+(+S)$

12. $\nvdash_{\alpha v} -(-S)+(-P)$

Note that this octagon is present in the previous fragment of formulas $\mathcal{F}_2$. And anchor formulas $\alpha_3''$ and $\alpha_3'$, and $\alpha_4''$ and $\alpha_4'$, are respectively equivalent, due to the equivalence between the previous mentioned formulas.

The abstract relations between these diagrams is shown in Figure 9. The main relation is the 'sub-diagram' relation, which elsewhere has been called infomorphism[18]. Béziau's octagon is at the top, because the rest of the diagrams are included in it. The square is at the bottom, because it is included in hexagons and octagon. As the image shows, the square belongs to both hexagons, and in turn the hexagons are included in Béziau's octagon. One thing to note is the interaction between these diagrams and the degenerate square in Figure 7. This square is not included in any of the diagrams other than the octagon, the union of this and the inverted square form the octagon, but the degenerate square is not related to the hexagons, although its parts are. Degenerated square is made by two $PCD$[19], one for each term, that is $+S, -S$ and $+P, -P$. These $PCD$ do belong to the hexagons, so they interact with the degenerate square as shown in the figure. Also the picture shows how each hexagon is defined as the union of a $PCD$ and the inverse square. These interactions can be described by the following equations[20].

- $JSB \cup SC = BO$
- $JSB \cap SC = IS$
- $\overline{JSB \cap SC} = DS$
- $PCD_S \cup PCD_P = DS$
- $IS \cup PCD_P = SC$
- $IS \cup PCD_S = JSB$
- $IS \cup DS = BO$
- $IS \cap DS = \emptyset$
- $JSB \cap SC = IS$

Finally, it is easy to see some similarities with propositional truth-tables and bitstrings in Béziau's octagon. Partition emulates distribution of truth-values for two arbitrary formulas as it is shown in Figure

---

[18]Cf. [22], and [16].
[19]Pair of Contradictories.
[20]With the following code: $JSB$:= JSB hexagon, $SC$:= SC hexagon, $BO$:= Beziau's Octagon, $IS$:= Inverse Square, $DS$:= Degenerated Square, $PCD_S$:= PDC for axis $+S, -S$, $PCD_P$:= PCD for axis $+P, -P$.

| | | | |
|---|---|---|---|
| 1 | 1 | $+S$ | $+S$ |
| 1 | 0 | $+S$ | $-P$ |
| 0 | 1 | $-S$ | $+P$ |
| 0 | 0 | $-S$ | $-P$ |

Figure 10: Truth-values and partition $\Pi_{\mathsf{TFL}_{av}}(\mathcal{F}_3)$

10. Bitstrings in Beziau's Octagon correspond to truth conditions for the respective connectives of **CPL**, and due to this correspondence we decided to call this Beziau's octagon **CPL** octagon. The way to compute bitstrings is as follows. If the formula has a positive subject all terms must be present in the anchor formula, and if the subject is negative just one terms is needed. For example, take formula $+S + P$. The first bitposition is 1 because both terms appear in anchor formula $\alpha_1''$, in the remaining anchor formulas that is not the case, therefore the remaining bitpositions are 0. This correspondence is kind of evident but we need to explore some results to try to make a deeper interaction between **TFL** and **CPL**.

## 4 Some Basic Results

In this section we present two results: a) a sketch of semantics for **TFL** and b) a reinterpretation of the syllogistic using the inverse translation function.

### 4.1 Semantics for $\mathsf{TFL}_{av}$

The semantics of $\mathsf{TFL}_{av}$ will be defined by means of Kripke structures. We will start by defining the basic models of the assertoric version, subsequently, we will outline an extension of this semantics to other versions of $\mathsf{TFL}_{av}$, namely, numerical.

**Definition 4.1.** (Models) A Kripke-Functor model is a tuple of the form $\mathbb{M} = \langle \mathcal{L}_{\mathsf{TFL}_{av}}, \mathcal{W}, R, v \rangle$, where:

1. $\mathcal{L}_{\mathsf{TFL}_{av}}$ is the language of term functor logic,

2. $\mathcal{W} = \{w, v, ...\}$ is a non-empty set of indexes,

3. $R \subseteq \mathcal{W}^n$ is an acessibility relation,

4. $v : \mathcal{L}_{\mathsf{TFL}_{\alpha v}} \to \mathcal{W}$ is a mapping assigning formulas to indexes.

**Definition 4.2.** (Semantics) Let $\mathbb{M} = \langle \mathcal{L}_{\mathsf{TFL}_{\alpha v}}, \mathcal{W}, R, v \rangle$ be a Kripke-Functor model, semantics is defined by the following conditions:

1. $\mathbb{M}, v \models -\varphi$ iff $\exists w \in \mathcal{W}$ & $Rvw : \mathbb{M}, w \not\models \varphi$,

2. $\mathbb{M}, v \models +\varphi \pm \psi$ iff $\exists w, u \in \mathcal{W}$ & $Rvwu : \mathbb{M}, w \models +\varphi$ and $\mathcal{M}, u \models \pm\psi$,

3. $\mathbb{M}, v \models -\varphi \pm \psi$ iff $\exists w, u \in \mathcal{W}$ & $Rvwu : \mathbb{M}, w \models -\varphi$ or $\mathcal{M}, u \models \pm\psi$.

The most basic type of model classes is the $\mathcal{M}_{cpl}$ which is characterized by the condition $|\mathcal{W}| = 1$. This implies that acessibility relations collapse into loops that only relate a single index. That is, $\forall v, w, u \in \mathcal{W} : v = w = u$. Which implies that the conditions of the definition 4.2 coincide with the usual definitions of the classical semantics of CPL.

We are in position to define a model-theoretic consequence.

**Definition 4.3.** Consider $\Gamma \subseteq \mathcal{L}_{\alpha v}$ and $\varphi \in \mathcal{L}_{\alpha v}$. Then:
$\Gamma \models_{\alpha v} \varphi$ iff for all classical models $\mathbb{M} \in \mathcal{M}_{cpl}$ and all valuation $v$ on $\mathbb{M}$: if $\mathbb{M}, v \models \Gamma$ then $\mathbb{M}, v \models \varphi$

This characterization can be used to impose conditions on the models and thus interpret some of the systems defined in the known literature. As an example we will take one presented in [7, p. 22].

*Example.* $(-_6 P + L), (-_{20} S + P) \models_\nu (-_{26} S + L)$.

*Proof.* We will prove by *reductio*. Let $\mathbb{N}$ be a numerical model[21]. Then:

1. If $\mathbb{N}, 0 \models -_6 P +_n L$ and $\mathbb{N}, 0 \models -_{20} S +_n P$, then $\mathbb{N}, 0 \models -_{26} S +_n L$.

---

[21] A numerical model $\mathbb{N}$ is a Kripke-Functorial model $\mathbb{N} = \langle \mathcal{W}, R, r, v \rangle$, where every element is the same as a basic Kripke-Functorial model, but $R \subseteq \mathcal{W}^3$ and $r \subseteq \mathcal{W}^2$. Both relations are defined as follows. Let $v, w, u \in \mathcal{W}$, then

1. $R(v, w, u)$ iff $\pm v = ((\pm w) + (\pm u))$

2. $r(v, w)$ iff $v \geq w$

2. Suppose that: $\mathbb{N}, 0 \not\models -_{26}S +_n L$.

3. By condition 1 in Def. 4.2, it follows that $\mathbb{N}, 0 \models -(-_{26}S +_n L)$.

4. Using functor product we get $\mathbb{N}, 0 \models +_{26}S -_n L$.

5. Using condition 2 in Def. 4.2 we get $R(0, 26, n)$ to obtain,

6. $\mathbb{N}, 26 \models +S$ and $\mathbb{N}, n \models -L$.

7. By precedence accessibility $r(26, 20)$ we obtain $\mathbb{N}, 20 \models +S$.

8. By composition of accessibility relations in steps 5 and 7 we update the accessibility to $R(0, 20, n+6)$, which lead us to work with next formula to obtain;

9. first: $\mathbb{N}, 20 \models -S$, which contradicts step 7, and

10. second: $\mathbb{N}, 6 \models +P$, which follows from composition of accessibility relations and update in step 8.

11. By condition 2 in Def. 4.2 we are able to make the access $R(0, 6, n)$ to get;

12. first: $\mathbb{N}, 6 \models -P$, which contradicts step 10, and

13. second: $\mathbb{N}, n \models +L$, which contradicts step 6.

14. By inconsistent information in steps 13, 12, and 9, it follows that step 2 is impossible, which proves the consequence.

$\square$

This example shows an interesting way to extend the semantics defined here. This basis leaves open several alternatives to explore models of already known term logics, namely, semantics for relational, modal, or relevant term functor logic; but even new systems, like many-valued, paraconsistent, linear, temporal, or dynamic term functor logics. But in addition, we can expand the spectrum of TFL systems using Kripke semantics, proposing a dialogue with non-classical logics.

## 4.2 Syllogistc with propositions

In this final section an interpretation of syllogistic in classical propositional logic is presented. It will be shown that the traditional modes are valid, and furthermore, that the logic TFL$_{\alpha v}$ is comparable to Manuel Correia's systematization in [14].

As noted above, our proposed semantics for TFL$_{\alpha v}$ suggests a link between logical connectives and combinations of functors. This has been expressed by the Definition 2.17. Moreover, Bitstrings semantics reinforces this fact, the partitioning of formulas for the Béziau's octagon fits in a synchronous way with truth tables. Taking advantage of these links we can go back to classical logic and try to formulate syllogistics with logical connectives. This idea has also been explored following other approaches in [31, p. 93] and [28, p. 68].

**Definition 4.4.** According to mapping in Definition 2.17, $\tau : \mathcal{L}_{\mathsf{CPL}} \to \mathcal{L}_{\alpha v}$ traditional categorical propositions are defined as follows:
$\forall A, B \in \mathcal{L}_{\mathsf{CPL}}$, assuming the natural invers mapping $\tau^-$, we have:

1. $\tau^-(-\theta + \kappa) = A \to B$,
2. $\tau^-(-\theta - \kappa) = A \to \neg B$,
3. $\tau^-(+\theta + \kappa) = A \wedge B$,
4. $\tau^-(+\theta - \kappa) = A \wedge \neg B$.

**Theorem 4.5.** *(Valid syllogistic moods) Let* CPL *Classical Propositional Logic. All traditional syllogistic moods are valid.*

*Proof.* All valid moods follows using translation between CPL and TFL$_{\alpha v}$ tableaux. For example, Figure 11 shows a closed and complete tableau for Barbara mood. All remaining moods can be proved case by case. □

# 5 Conclusion

We have presented the classical square of oppositions in the logic TFL$_{\alpha v}$. This logic has been defined from two approaches, by tableaux system, and in a more original way, with Kripke semantics. We have started

```
                    1. A → B
                    2. C → A
                    3. ¬(C → B)
                    4. C
                    5. ¬B
       ↓                              ↓
       6. ¬A                          B
       ↓      ↓                       ×
       7. ¬C  A
       ×      ×
```

Figure 11: CPL-Tableau for Barbara mood.

from a brief recapitulation of the work of José Martín Castro-Manzano, who takes up the project founded by Fred Sommers and George Englebretsen, and extends it by taking the study of Term Functor Logics to the formalization of several interesting linguistic variations. The original part of our work has been to try to integrate this type of logic into the Logical Geometry project.

As a first contribution to this project, we have presented some extensions of the classical square, which acquires a peculiar shape, present in some other systems (i.e. PAL, KF). As we have seen, the hexagons presented satisfy the usual semantic conditions, and correspond to the diagrams found in CPL. This link between TFL and CPL constitutes an interesting source of further research, as we saw with the example in the previous section. The next steps, as will become evident, will be to explore, on the one hand, the different varieties of TFL systems in Kripke semantics, to propose different classes of models suitable for each system, and thus to integrate model theory into the study of TFL. On the other hand, it is evident that Logical Geometry has much to contribute to this type of logics, and with this integration we hope to find more cases of Aristotelian diagrams with interesting properties.

# References

[1] Ammonius and Busse. Ammonius in aristotelis categorias commentarius. 1895.

[2] H. Ammonius, A. Busse, and K. P. A. der Wissenschaften zu Berlin. Ammonius in aristotelis de interpretatione commentarius.
[3] R. Blanché. *Structures Intellectuelles*. Vrin, 1966.
[4] A. M. S. Boethius. *Anicii Manlii Severini Boetii Commentarii in librum Aristotelis Peri hermēneias*, volume 2. Teubner, 1880.
[5] J. M. Castro-Manzano. Tableaux for murphree's numerical term logic. In *LANMR*, pages 1–12, 2019.
[6] J.-M. Castro-Manzano. Distribution tableaux, distribution models. *Axioms*, 9(2):41, 2020.
[7] J. M. Castro-Manzano. Murphree's numerical term logic tableaux. *Electronic Notes in Theoretical Computer Science*, 354:17–28, 2020.
[8] J. M. Castro-Manzano. Un método de árboles para la lógica de términos modal. volume 11, pages 160–180, 2020.
[9] J.-M. Castro-Manzano. Mixing colors, mixing logics. In *International Conference on Theory and Application of Diagrams*, pages 70–77. Springer, 2022.
[10] J.-M. Castro-Manzano. Toward relevance term logic. *Computación y Sistemas*, 26, 2022.
[11] J.-M. Castro-Manzano, L. I. Lozano-Cobos, and P. O. Reyes-Cardenas. Programming with term logic. 2018.
[12] J.-M. Castro-Manzano and P.-O. Reyes-Cárdenas. Term functor logic tableaux. *South American Journal of Logic*, 4(1):1–22, 2018.
[13] M. Correia. *La lógica de Aristóteles: Lecciones sobre el origen del pensamiento lógico en la antigüedad*. Ediciones Universidad Católica de Chile, 2003.
[14] M. Correia. La lógica aristotélica y sus perspectivas. *Pensamiento. Revista de Investigación e Información Filosófica*, 73:5–19, 2017.
[15] T. Czeżowski. On certain peculiarities of singular propositions. *Mind*, 64:392–395, 1955.
[16] A. De Klerck, L. Vignero, and L. Demey. Morphisms between aristotelian diagrams. *Logica Universalis*, pages 1–35, 2023.
[17] L. Demey. Structures of oppositions in public announcement logic. *Around and beyond the square of opposition*, pages 313–339, 2012.
[18] L. Demey. Boolean considerations on john buridan's octagons of opposition. *History and Philosophy of Logic*, 40(2):116–134, 2019.
[19] L. Demey. A hexagon of opposition for the theism/atheism debate. *Philosophia*, 47(2):387–394, 2019.
[20] L. Demey. Logic-sensitivity of aristotelian diagrams in non-normal modal

logics. *Axioms*, 10(3):128, 2021.

[21] L. Demey and S. Frijters. Logic-sensitivity and bitstring semantics in the square of opposition. *Journal of Philosophical Logic*, pages 1–19, 2023.

[22] L. Demey and S. Frijters. Logic-sensitivity and bitstring semantics in the square of opposition. *Journal of Philosophical Logic*, 52:1703 – 1721, 2023.

[23] L. Demey and H. Smessaert. Combinatorial bitstring semantics for arbitrary logical fragments. *Journal of Philosophical Logic*, 47:325–363, 2018.

[24] G. Englebretsen. *The New Syllogistic*. American University Studies. P. Lang, 1987.

[25] G. Englebretsen. *Something To Reckon With: The Logic of Terms*. Philosophica. University of Ottawa Press, 1996.

[26] P. Jacoby. A triangle of opposites for types of propositions in Aristotelian logic. *New Scholasticism*, 24:32–56, 1950.

[27] P. Jacoby. Contrariety and the triangle of opposites in valid inferences. *New Scholasticism*, 34:141–169, 1960.

[28] T. Jarmuzek. Tableau methods for propositional logic and term logic. 2020.

[29] N. Kretzmann. *William of Sherwood's introduction to logic*. U of Minnesota Press, 1966.

[30] N. Kretzmann. *William of Sherwood's Introduction to Logic*. Minnesota Archive Editions, 1966.

[31] V. Rooij. Uva-dare (digital academic repository) the propositional and relational syllogistic. 2012.

[32] A. Sesmat. *Logique II. Les Raisonnements. La syllogistique*. Hermann, 1951.

[33] H. Smessaert and L. Demey. The unreasonable effectiveness of bitstrings in logical geometry. *The square of opposition: a cornerstone of thought*, pages 197–214, 2017.

[34] F. Sommers. The calculus of terms. *Mind*, pages 1–39, 1970.

[35] F. Sommers. *The Logic of Natural Language*. Clarendon library of logic and philosophy. Clarendon Press, 1982.

[36] F. Sommers and G. Englebretsen. *An invitation to formal reasoning: the logic of terms*. Routledge, 2017.

[37] P. Thomas et al. *Apulei platonici Madaurensis De philosophia libri*, volume 3. in aedibus BG Teubneri, 1908.

# Relative Terms

Peter Simons
*Trinity College Dublin*

All the world's a stage,
And all the men and women merely players;
They have their exits and their entrances;
And one man in his time plays many parts,
His acts being seven ages.

Shakespeare, *As You Like It*, II,7.

*Dla Ewy*

## 1 Terms, Absolute and Relative

In traditional logic, terms are general nominal expressions which may denote one or several items, or none. Expressions like *person, cat, tree, star, table, river* denote enduring individuals; expressions like *kick, collision, explosion, collapse, speech* denote events and processes; expressions like *water, snow, butter, wood, potassium, lava* denote consignments of stuff; and expressions like *family, team, committee, platoon, orchestra* denote collectives. All these different kinds of terms tell us *what* something is that is of the relevant kind. I follow Peirce[1] in calling such terms *absolute*.

There are in addition many terms, not just in English but in other languages, which apply to and describe an item in the light of its relationship to something else. Examples include kinship terms such as *parent, sister, uncle, grandmother, cousin*; terms describing an office such as *teacher, captain, vice-president, bishop*; terms describing something standing in some less circumscribed type of relationship such as *donor, lover, enemy, listener, victim, violinist, traitor, subordinate, ingredient,*

---
[1] Peirce [16, p. 365].

*component*; terms describing determinable quantities such as *weight, temperature, distance, angle, viscosity*; and terms describing mathematical functions such as *sine, square root, scalar product.*[2] Terms of this general kind I call, again following Peirce, *relative*.[3] What Peirce calls relative terms or relatives are those which are, logically speaking, *functors*: they are not themselves absolute terms until one or more terms are inserted in the relevant argument slots.[4]

Relative terms may be subdivided according to the number of items related in the relation. *Binary* terms include *brother, lover, bishop, satellite, distance, angle*. *Ternary* terms include *donor, betrayer*; quaternary and higher numbered terms are rare: Peirce instances *winner-over (of —- from — to —-)*.[5] In theory there is no upper limit, but in practice most relative terms are binary or ternary.

## 2 Relations and Relative Terms

One of the principal gains of modern over traditional logic is the ease with which it handles relations. The early logic of relations, due to De Morgan,[6] Peirce and Schröder, made significant use of relative terms. It may seem old-fashioned in comparison with post-Frege–Peano–Russell predicate logic, but this older formulation has significant advantages. In the standard contemporary approach to logic via varieties of predicate calculus, predicates of one, two, three or more places are represented by predicate letters which combine with singular terms of the appropriate number to form atomic sentences. These are then used to apply to

---

[2]Predicate logic was invented by mathematicians for mathematics. But if one listens to mathematicians speaking their formulas out loud, they make abundant use of relative terms, for example, "the scalar product of vectors $x$ and $y$ is the product of the absolute values of $x$ and $y$ and the cosine of the angle between them", which contains no fewer than five relative terms.

[3]Peirce, loc. cit. "The second class embraces terms whose logical form involves the conception of relation, and which require the addition of another term to complete the denotation." Peirce distinguishes *simple* relative terms (two places) from *conjugate* relative terms (three or more places). We do not follow this terminology.

[4]Or they are derelativized: see Section 4 below.

[5]Peirce [16, p 366].

[6]For a comprehensive account of De Morgan on relations, see Merrill [14]. It is notable that when giving examples, De Morgan uses relative terms all the time.

such vernacular sentences as *Juliet loves Romeo, Judas betrayed Jesus to the authorities*; *Dublin is twice as far from Donegal as Limerick is from Killarney*. These gains are indisputable, but they come at a cost. In relational predications of predicate logic found in formulas such as *Rab, Sabc* etc., the fact that the predicate is schematic gives rise to two metaphysical issues about relations. The first is the *order problem*. Given that in the predications *Juliet loves Romeo* and *Romeo loves Juliet* we have the same three constituents, namely *Juliet*, *Romeo* and *loves*, and yet we know that these two sentences are logically independent of one another, so they can be opposite in truth value, what is it about the two predications that makes the difference? Is it the position or order of the two terms, or is this a superficial matter? If it is the order, then this leads to the second problem, the *How many relations?* problem. In *Romeo loves Juliet* the term *Romeo* occurs first, whereas in *Juliet is loved by Romeo* the term occurs last. These two sentences are (at least) logically equivalent, if not synonymous: they indeed appear to "say the same thing". A natural reaction is to say that *loves* and *is loved by* are distinct relations, because *Juliet loves Romeo* and *Juliet is loved by Romeo* are not equivalent. How then are *Romeo loves Juliet* and *Juliet is loved by Romeo* equivalent? The standard answer is that the two relations are each other's *converses*, which is a close logical connection. On the other hand, it seems as though there are not, so to speak, two facts, one of Romeo's loving Juliet and another of Juliet's being loved by Romeo. The problem ramifies as the number of places of a relation increases. For three places, there are six relations; in general, for $n$ places, there are $n!$ relations; for example, for six places, there are 720 relations, all corresponding to the one fact or situation. Yet how can this be just one fact, if the sentences do not have the same constituents? Is there a relation which is somehow neutral between these orders?[7] But if so, how does it then solve the order problem?[8]

The way out of this *impasse* is to note that actual relations, as distinct from schematic predicates, are not simply so-and-so-many-ary, but that their arguments play different parts or *roles*.[9] In the *loves* relation

---

[7] See Fine [9].

[8] For an overview of the issues, see MacBride [11].

[9] In a previous paper [20] I called these 'case roles', but I now think this invites confusion, so I am here calling them simply 'roles'.

these roles are the lover and the beloved, in the *gives* relation they are donor, gift and beneficiary, in the *converts* relation (meaning converting someone from one religion to another) they are converter, convertee, relinquished, adopted. In natural languages we do not always have terms for these different roles, but that is a contingent matter. For instance, there is no simple lexical term in English for a person who has a brother. Sometimes there will be two or more roles of like type: in the *is as old as* relation there are two equally-aged, in German *Gleichaltrige*.

In real relations then, we only need to consider the several roles, and in some cases, how many arguments there are to each role type. These are independent of order. So in the case where $a$ gives $b$ to $c$, the roles and those filling them are donor : $a$, gift : $b$, beneficiary : $c$. The order in which these are listed or mentioned is immaterial: all that matters is which individuals, which *actors*, fill them. The actors *qua* filling their roles can be called *players*. And the expressions for all these roles are nothing other than relative terms. As Shakespeare tells us, each actor plays many parts or roles, and not just successively, as in Jacques's speech, but simultaneously. The schoolboy is both a scholar (at some school) and a son (to two parents). One actor may fill the same kind of role more than once. King Henry VIII notoriously played the husband part six times successively, while the Prophet Muhammad had eleven (undisputed) wives, monogamously for most of his life, but in his final years as many as nine simultaneously. To complete the picture, we may consider non-relational predicates: here the term describes the role, and the actor is simply the bearer or player of that part. Jacques's man is at one time young, small, and weak; later bearded; later full-bellied; still later lean, and so on, but a human and male throughout.

## 3 Notation

To represent relational propositions by means of relatives, instead of writing such formulas as

*Rab, aRb, Sabc* etc., we write (for binary relatives) formulas like

$$(G : a \; H : b)$$

where the upper-case letters give the correlated relatives and the lowercase letters their arguments or actors. The whole enclosed in parentheses amount to the proposition that $a$ is $G$ to $b$'s $H$; for example

(Lover : Juliet Beloved : Romeo)
(Monarch : Naruhito Monarchy : Japan)

For ternary and higher arity relations the number of relatives is duly increased:
$$(L : a\ M : b\ N : c)$$
for example

(Donor : Emma Gift : \$1,000 Beneficiary : ASPCA)

The groups $G\ H$ etc. are a *unity* corresponding to a single relation: they are not independently variable. The *parent* role correlates to the *child* role; the *larger* to the *smaller*; the *north of* to the *south of*, the *donor*, *gift* and *beneficiary* roles likewise form a package. To be evenhanded towards the various roles, we should talk about the *parent–child* relation, the *larger–smaller* relation, and so on. The order of relatives and their players in our notation is arbitrary and has no semantic or logical significance. In natural languages, when one role is agentive, it tends to be placed first (in SVO and SOV languages) or second (in VSO languages), but this is not a logical matter, and the usual order can be overruled by use of the passive, which in our notation is not significant.

Because of this unity, there is no suggestion that employing relative terms rather than relational predicates constitutes a *reduction* of relational to non-relational predicates.

Relational predications, however formulated, are indispensable to thought. The prevalence of relatives in the formulation of many relational truths does however recall the controversy about whether Leibniz believed relational truths to be *eliminable* in favour of non-relational ones, or merely *grounded* in terms of these. Leibniz does say that the two predications *David is a father* and *Solomon is a son* do not entail that Solomon is the son of David. Rather, we need a further connection: that David is a father *insofar as* (*quatenus*) Solomon is a son, or that David is a father *and because of that* (*eo ipso*) Solomon is a son. The non-extensional connectives bring the two into relation, though Leibniz regards relations as not real but merely ideal.[10]

When considering the logic of statements about groups of relatives we are dealing with relations, but in a different way from usual. All the logical results regarding converses of relations, and their analogues for more than two places, drop away. It is of course possible to formulate the usual logical properties of relations, such as symmetry, reflexivity, transitivity, linearity etc. using the notation introduced, but this is basically a rewriting exercise. It becomes more interesting when we consider such operations as relative products and ancestrals, which we come to in following sections.

One distinction which can be made, that the standard analysis does not, concerns symmetry. In predicate logic, a binary relation is symmetrical when it holds in both directions whenever it holds in either:

For all $x$ and $y$ : if $xRy$ then $yRx$.

In most cases, when a relation is symmetrical, this is because the same role type is filled twice, for example:

For all $x$ and $y$ : if (Sibling : $x$ Sibling : $y$) then (Sibling : $y$ Sibling : $x$)

and in this case the result is trivial, because order of roles does not signify. But in some cases, we have symmetry, or what we might call *reciprocity*, despite the roles being different. Let us define point $a$ on the Earth's surface as being *due west* of point $b$ if it is possible to reach point $a$ from point $b$ by travelling due westwards. The converse of *due west* is of course *due east*. In our notation there are two distinct roles

(Occident : $a$ Orient : $b$)

Yet, because the Earth is a convex solid approximating a sphere, going westwards and going eastwards reach all the same points on a given latitude, so it is true that

---

[10]The literature on Leibniz on relations is extensive and I do not intend to be drawn into it. For a representative entry, see Burdick [5].

For all $x$ and $y$: if (Occident : $x$ Orient : $y$) then
(Occident : $y$ Orient : $x$).[11]

The relationship is symmetrical (and in this case reflexive) despite there being two distinct roles.[12]

## 4 Derelativization

It is possible to use relative terms in an absolute way, as in *Jane is a grandmother, Justin is an archbishop, Emma is a donor*. It is easy to see how these come about: Jane is a grandmother *of* some grandchild; Justin is archbishop *of* some archdiocese; Emma is donor *of* something *to* some person, institution, or cause. In predicate logic, such derelativization is handled by the existential (particular) quantifier: *for some x, Jane is grandmother of x*, and so on. But in natural language, the same function is performed not by variable-binding operators but by *quantifier phrases* such as *someone, something, everyone, nothing, always, never, everywhere,* etc. Such phrases may consist of more than word, as *some grandchild, some archdiocese, any charity* and so on. Leaving numerical determiners aside, the quantifier words like *every, some, no, most, few* form what linguists call a *closed class*, while the nouns they consort with form an *open class*. In derelativization, it is quantifier phrases that close or bind up places in a relation, and they do so with minimal scope, so can occur within the context of a proposition rather than being operated from outside as in predicate logic. We then arrive at logically derived propositions:

(Grandmother : Jane) = (Grandmother: Jane Grandchild: some grandchild)
(Archbishop : Justin) = (Archbishop : Justin Archdiocese : some archdiocese)

---

[11] Giving a concert in California, the great Victor Borge said, "I just came out from the East Coast, and I was amazed that ... the further west you go in this country, the closer you get to the ... Far East." [*Looks puzzled.*]

[12] Normally, if body $a$ orbits body $b$, then $b$ does not orbit $a$. The roles Orbiter and Orbited are distinct. But in the case of binary star systems, each star orbits the other, so for this limited domain, the relation is symmetrical.

(Donor : Emma) = (Donor : Emma Gift : some gift Beneficiary : some beneficiary)

Such derived propositions follow from more specific propositions, such as

(Grandmother : Jane Grandchild : Roddy)
(Archbishop : Justin Archdiocese: Canterbury)
(Donor : Emma Gift : $1,000 Beneficiary : ASPCA).

Relations with three or more places may be *partially* derelativized, giving rise to such examples as

Emma is a donor to the ASPCA
The ASPCA is the beneficiary of a gift of $1,000
Emma donated $1,000.

Another source of derelativization is the transfer of direct objects to qualify nouns, as in *rock climber, sheep farmer, blood donor*:[13]

(Rock climber : Grace) = (Climber : Grace Climbed : some rocks)
(Sheep farmer : George) = (Farms : George Farmed : some sheep)
(Blood donor : Tony) = (Donor: Tony Gift: some of Tony's blood Beneficiary: some person)

## 5 Internal Booleans

The propositional connectives of conjunction, disjunction, negation etc., and the usual quantifiers *for all* and *for some* have been treated so far as operating on propositions. But Boolean constants operate on other parts of speech than just sentences. A brother is a male sibling, male *and* sibling; a sibling is a brother *or* sister; the US and UK flags are red

---

[13] These last analyses are not quite exact, because in the attributive noun phrase there is usually a connotation of frequency, habit or occupation. I climbed rocks once, but cannot thereby count as a rock climber. A rock climber will generally climb rocks often: I do not intend to repeat the experience.

*and* white *and* blue; a vegetarian is a *non*-meat-eater; the Earth rotates *and* orbits the Sun.

Conjunction and other Booleans have long been applied to terms, indeed from Boole himself through to Leśniewski, and they figure in set theory as union, intersection and complement. It is therefore reasonable to form complex terms such as *tall Dutch female violinist* by conjoining terms directly: using '∩' for term conjunction we have

(tall Dutch female violinist : $a$) ↔ (Tall ∩ Dutch ∩ Female ∩ Violinist : $a$) ↔ (Tall : $a$) ∧ (Dutch : $a$) ∧ (Female : $a$) ∧ (Violinist : $a$) ↔ (Tall : $a$) ∧ (Native : $a$ Country : Netherlands) ∧ (Female : $a$) ∧ (Player : $a$ Instrument : violin).

Obviously this can be turned back into a more familiar sentential conjunction by noting that $(G \cap H : a) \leftrightarrow ((G : a) \wedge (H : a))$, enabling us to shuttle back and forth between term and propositional conjunctions. Several roles are jointly filled by one actor, which is important in the next section.

# 6 Relative Products

The relative product was first highlighted by De Morgan and was developed by Peirce and Schröder. Relative products occur frequently in kinship terms: an uncle is a brother of a parent; a grandparent is a parent of a parent; a cousin is a child of a sibling of a parent; a stepmother is a wife of the father of $x$ but is not the mother of $x$. Relative products can be formed on the fly using genitives, or 'of' in English, e.g. lover of a benefactor of, cleaner of a house of the mother of a friend of the husband of a daughter of.[14] Relative products are dealt with in predicate logic by quantifying:

$a$ is a brother of a parent of $b$

being rendered as

For some $x$, $a$ is a brother of $x$ and $x$ is a parent of $b$,

---

[14]The example is not fictitious: it's how my daughter mediated my cleaner.

in symbolic definition[15]

$$a(R;S)b \leftrightarrow \exists x(aRx \wedge xSb)$$

But in natural language, what binding is required is done within the phrase: no one would imagine that the 'cleaner' example above is understood by speakers through there being five existential quantifiers which then get dropped from the front of the whole clause. The relative product was one of the first operations recognised in the algebra of relations, predating predicate logic, it has a special case in functional composition, and it is easily expressed in natural language. By contrast, its dual, the relative sum, defined as[16]

$$a(R+S)b \leftrightarrow \forall x(aRx \vee xSb)$$

is rarely met with and has no or virtually no natural uses.

How is the relative product to be understood if we dispense with prefixed quantifiers? Consider again the example *brother of a parent*. The 'brother of' relation has two roles: the brother, and the second role, taken by the brother's sibling, which lacks an English term but which for the nonce we'll call *the brothered*.[17] The 'parent of' relation has two roles and they do have their own terms, namely 'parent' and 'child'. In the context of a particular case, the roles are not abstract types, but tokens filled by their actors. So, to take a concrete example, King William IV of the United Kingdom and Hanover was a brother of Prince Edward, Duke of Kent, who was a parent of Queen Victoria. We have here two cases:

---

[15]The semicolon notation is due to Peirce and Schröder and is generally preferred to the vertical stroke '|' used by Russell and Whitehead. Occasionally the functional composition symbol '∘' is used.

[16]Schröder [18, p. 29]. A relative sum is the negation of a relative product. For any $a$ and $b$, if $a$ is *not* an uncle of $b$ then $a$ is the relative sum of non-brother and non-parent of $b$: anything of which $a$ is a brother is not a parent of $b$. For instance, any pencil and anything else stand in this relation. It is hard to imagine a sensible use for this idea, and none seems to have emerged.

[17]The word 'brothered' does exist in English but its meanings do not include this. Idiomatically, if we want to make the brothered the subject of a sentence, we can say *Anne has Charles as brother* as an alternative to *Charles is brother of Anne*. Derelativizing, *Anne has a brother* is the nearest we get in idiomatic English to *Anne is brothered*. The 'has' locution is a sort of general converse to 'is' as copula: *Quebec is part of Canada* corresponds to *Canada has Quebec as part*.

(Brother : William Brothered : Kent)

and

(Parent : Kent Child : Victoria)

and four roles. In this case, the roles of Brothered and Parent are filled by the same actor, Kent, something represented in predicate logic by two occurrences of the same bound variable. The roles are in our representation glued together by hosting the same actor: we thus have a dual role of Brothered–Parent and this dual role is passed through and over when we consider the outer roles of Uncle and Uncled. The dual possession of roles is not in principle different from the dual possession of the roles of Male and Sibling in the case of Brother, or Female and Parent in the case of Mother. In the cleaner example, there are five such dual roles chained together in sequence. For obvious reasons, there are few lexical entries for the relation or relatives involved, and even in the kinship case, although unitary lexicalization as 'uncle' exists because the kind of compound relationship is common; as we have seen, not all roles have their own term, and it is notorious that differentiation and lexicalization of kinship terms vary considerably across languages. In the English cumulative nursery rhyme *The House that Jack Built* there are twelve relations chained together. Between Jack and the farmer's "Horse, Hound and Horn", there are eleven intermediate actors, each filling two roles, double-filling twenty-two out of the twenty-four involved. The eleven intermediate actors are named, which is part of the fun, but in total there are no fewer than 66 relations in play.

In the standard algebra of relations, if we represent the converse of relation $R$ by $R^{\smile}$, it needs to be specially shown that the converse of a product is the product of the converses in the opposite order:

$$(R\,;\,S)^{\smile} = S^{\smile}\,;\,R^{\smile}$$

but when roles untie relations from linearity, this can go without saying. All we need is a notation for the dual occupation of roles. We simply conjoin the two roles together using term conjunction $\cap$:

$$(G : a\,H : b) \wedge (L : b\,M : c) \leftrightarrow (G : a\,H \cap L : b\,M : c)$$

which gives us a new *three-place* relation, and this can then be partially derelativized:

($G : a\ H \cap L :$ something $M : c$)

to return us to two places, but with a new relation. In our example

(Brother : William Brothered ∩ Parent : Kent Child : Victoria)

which derelativizes to the Uncle–Uncled relation

(Brother : William Brothered ∩ Parent : someone Child : Victoria)

Since in our notation, order plays no part, this eliminates the converse issue. The associativity of relative product likewise drops out of the symbolism: take the example

(Child : Victoria Parent ∩ Sibling : Victoria of Saxe-Coburg-Saalfeld Sibling ∩ Parent : Ernest of Saxe-Coburg-Gotha Child: Albert)

When we quantify and eliminate the dual role players

(Child : Victoria Parent ∩ Sibling : someone Sibling ∩ Parent : someone Child: Albert)

we can mix the order of the roles any way. We lose the information as to which child is child of which sibling. But this is what we would expect, since all the new binary relation tells us is that Victoria and Albert are (first) cousins, and we know siblings cannot be identical.[18] Likewise if $a$ loves $b$ and $d$ loves $c$ and $b$ and $c$ are siblings, we do not know after quantification which lover loves which sibling – again, to be expected: the binary relation between $a$ and $d$ is symmetrical. If we turn it around so that $c$ loves $d$ then the roles of $a$ and $d$ are different in the chained complex, but that does not entail that $a \neq b$.

---

[18]Provided, of course, that the siblings did not incestuously have the children in common for it to be sure that Albert and Victoria are cousins and not siblings or even identical! Be cautious about the aristocracy.

Where our notation differs from the standard one is that because of the ways in which roles may be shared, any two relations have not one but *four* relative products: given relations $(G : - H : -)$ and $(K : - L : -)$[19] there are four ways of something sharing two roles:

$(G : a\ H : b)$ and $(K : b\ L : c)$
$(G : a\ H : b)$ and $(K : c\ L : b)$
$(G : a\ H : b)$ and $(K : a\ L : c)$
$(G : a\ H : b)$ and $(K : c\ L : a)$

so

$(G : a\ H \cap K : b\ L : c)$
$(G : a\ H \cap L : b\ K : c)$
$(G \cap K : a\ H : b\ L : c)$
$(G \cap L: a\ H : b\ K : c)$

which gives four products

$(G : a\ H \cap K : \text{something } L : c)$
$(G : a\ H \cap L : \text{something } K : c)$
$(H : b\ G \cap K : \text{something } L : c)$
$(H : b\ G \cap L : \text{something } K : c)$

These correspond to the $2 \times 2$ converses in the standard way of working. For example, if we have equivalences between order-directed and role-directed representations given by

$a\ R\ b \leftrightarrow (G : a\ H : b)$
$c\ S\ d \leftrightarrow (K : c\ L : b)$

then $a\ (R\ ;\ S)\ b \leftrightarrow \exists\ x (a\ R\ x \land x\ S\ b)$

$\leftrightarrow (G : a\ H \cap K : \text{something } L : b)$

The other three combinations indicate the products involving converses of $R$ and/or $S$.

---

[19] A dash in an actor slot represents that the place is there but is not filled.

## 7 Exploiting an Alternative Notation

For *binary* relations, a way of making the unity of the roles more graphic is to link then literally with a tie: thus, instead of

$$(G : a\, H : b)$$

we write
$$a : G - H : b$$

As before, order plays no part. The proposition above can equally well be displayed as
$$b : H - G : a$$

It is the difference between roles rather than linear order of mention that gives the relation direction. Where there is no difference in roles, as in

$$a : \text{sibling} - \text{sibling} : b$$

the symmetry becomes especially apparent.

The link notation lends itself effectively to displaying certain theorems. For example, a relative product may be elegantly notated as

$$a : G - H : \text{something} : K - L : b$$

and associativity of relative products is shown rather than said by simply chaining products together:

$$a : G - H : \text{something} : K - L : \text{something} : M - N : b.$$

One of De Morgan's more interesting if abstruse theorems about relative products is what he calls 'Theorem K'. Suppose every $P$ of a $Q$ of anything is an $R$ of that thing; it then follows that every $P^{\smile}$ of a non-$R$ of anything is a non-$Q$ of that thing, and also that every non-$R$ of a $Q^{\smile}$ of anything is a non-$P$ of that thing.[20] This particularly mind-twisting result is almost transparent in the link notation. Noting that

---

[20]De Morgan [10, p. 224]. See also Prior [17, pp. 153–4]. The '$K$' recalls the syllogisms 'Baroko' and 'Bokardo', so spelled, which Aristotle reduced to mood Barbara by an indirect proof using – as here – partial contraposition.

by the nature of the premise, all lower-case variables may be considered universally quantified, from

$$(a : G - H : b : K - L : c) \to a : M - N : c$$

by contraposing the consequent with either of the conjoined propositions of the antecedent we get

$$\sim (a : M - N : c) \wedge (a : G - H : b)) \to \sim (b : K - L : c)$$

and

$$((\sim (a : M - N : c) \wedge b : K - L : c)) \to \sim (a : G - H : b)$$

and by replacing propositional by role-negation and rearranging, these become[21]

$$(b : H - G : a : \bar{M} - \bar{N} : c) \to b : \bar{K} - \bar{L} : c$$

and

$$(a : \bar{M} - \bar{M} : c : L - K : b) \to a : \bar{G} - \bar{H} : b$$

respectively.

Relative products do not have to be binary of course. If Jim is Emma's brother, then Jim is the brother of a donor of $1,000 to the ASPCA, combining a binary with a ternary relation. However, the linear link notation only works well for binary relations. For relations of higher arity, a better graphic representation is a circle representing the relational link with various "stubs" for the roles, each with a "docking point", for their players.

## 8   Reflexivization

Another way in which relational predications lose a place is by reflexivization. Such absolute terms as *narcissist*, *suicide*, *patricide*, *patriot* apply to a single individual in virtue of a relation they bear to themselves, a relation of a kind which can be borne to others. In predicate logic this is covered by repetition of terms: $a$ loves $a$, $a$ killed $a$, $a$ killed

---

[21] Term negation is indicated here by overlining. Later we use a prefixed functor 'N'.

the father of $a$, $a$ loves the country of $a$. In natural language, repetition is avoided in good part by the use of anaphoric pronouns, which is too large a theme to explore here, but also by use of reflexives: *Donald loves himself, John killed himself, Oedipus killed his own father, Benjamin loves his own country*. The simplest way in which reflexivization can occur with relative terms is when only one definite nominal occurs, and the second occurrence is replaced by a reflexivizer *self*. This gives

(Narcissist : $a$) $\leftrightarrow$ (Lover : $a$ Beloved : self)
(Suicide : $a$) $\leftrightarrow$ (Killer : $a$ Victim : self)

The *patricide* case is a little more complex:

(Patricide : $a$) $\leftrightarrow$ (Killer : $a$ Victim $\cap$ Male $\cap$ Parent : someone Child : self)

Likewise

(Patriot : $a$) $\leftrightarrow$ (Lover : $a$ Beloved $\cap$ Country : some country Native : self)

There are more complex issues in this general area such as anaphoric reference, cross-reference, discontinuous elements, definite descriptions, plurality, and expressions like 'only' and 'else', which require treatment, but we are ducking them. The reflexive operator '-self' is logically closely related to the combinator W, which doubles occurrences of a variable:[22]

W$fx > fxx$

and there is evidence that, unlike predicate logic, natural languages employ vernacular versions of combinators:[23] passivization, for example, switches order like the combinator **C**. Interestingly, with the relative term representation of relational predications, **C** is not needed, at least for basic cases.

---

[22] Curry and Feys[8, p. 152].
[23] Steedman [21, 22].

# 9  Ancestrals

When several instances of the same relation are chained together, as in parent, grandparent, great-grandparent, and so on, again all we have is the number of intermediate people, leaving just the actors at the ends of the chain. That leads to the idea of ancestrals. Though De Morgan had the idea of the ancestral, he did not define it. Modern predicate logic came of age with Frege's definition of the ancestral of a relation. His (second-order) definition of the ancestral is ingenious but has two drawbacks. It quantifies over *all* properties, which is a potentially dangerous thing to do (think 'heterological'). And secondly, it goes well beyond what the semantics of any given case requires. Consider by way of contrast the following *inductive* definitions:

> Any parent of John is an ancestor of John
> Any parent of an ancestor of John is an ancestor of John
> Nothing else is an ancestor of John

> Any child of John is a descendant of John
> Any child of a descendant of John is a descendant of John
> Nothing else is a descendant of John

> The mother of John is a matrilineal ancestor of John
> The mother of any matrilineal ancestor of John is a matrilineal ancestor of John
> Nothing else is a matrilineal ancestor of John

None of these even hint at quantification over all properties: the main issue is one of matching the initial relation with its ancestral, something that is occasionally lexicalized, as here, but not always or even often. Inductive or recursive definitions have often been frowned upon as not "true" definitions, because they do not have the definiens and definiendum on opposite sides of a biconditional, but that is to take a blinkered and excessively restricted view of definitions. In applications, especially in programming, inductive definitions are not only acceptable; they are often the only way in which ancestrals can be practically handled.

The standard definition of the ancestral is applied either to $R$ or to the converse of $R$. But we don't have converses, so we must specify "direction of travel". There are two ways to define ancestrals from a binary relation, and they correspond to the two directions in which we may progress. Take as an example

1. If (Parent : $a$ Child : $b$) then (Ancestor : $a$ Descendant : $b$)
2. If (Parent : $a$ Child : $b$) and (Ancestor : $c$ Descendant : $a$) then (Ancestor : $c$ Descendant : $b$)
3. Nothing else is an ancestor of $b$
4. If (Parent : $a$ Child : $b$) and (Ancestor : $b$ Descendant : $c$) then (Ancestor : $a$ Descendant : $c$)
5. Nothing else is a descendant of $a$.

These linked inductions capture *both* directions without difficulty. In general

1. If $(G : a\ H : b)$ then $(G^* : a\ H^* : b)$
2. If $(G : a\ H : b)$ and $(G^* : c\ H^* : a)$ then $(G^* : c\ H^* : b)$
3. Nothing else is a $G^*$ of $b$
4. If $(G : a\ H : b)$ and $(G^* : b\ H^* : c)$ then $(G^* : a\ H^* : c)$
5. Nothing else is an $H^*$ of $a$.

The new relation $(G^* : -\ H^* : -)$ then cuts both ways as far as ancestrals are concerned. But that something is in the ancestral of one direction does not entail that it is in the ancestral of the other. John has a line of matrilineal ancestors but, *qua* male, no matrilinear descendants.

## 10  Ancestrals of Higher Arity Relations

Ancestrals of ternary and higher arity relations are not often considered, mainly because not all roles in such relations iterate naturally. But they can arise. Consider again matrilineal ancestry. Jane and her parents form a parental triad:

(Mother : Anne Father : Tom Child : Jane)

Merging mother and daughter roles repeatedly in a chain gives a new triad consisting of matrilineal descendant Jane, some matrilineal ancestor of Jane, and this ancestor's male co-parent of a daughter who is also Jane's matrilineal ancestor. We can do the same with patrilineal triads by merging the father and child roles repeatedly. For obvious reasons the co-parent roles do not merge interestingly. The corresponding descendant cases give us two more sensible ancestrals.

A theoretically more interesting case is this. Let $a$, $b$ and $c$ be three points on a circle with $a$ and $b$ separated by angle $\alpha$, and $b$ and $c$ by angle $\beta$. That gives a ternary relation between $a$, $b$ and $c$. Let $\theta$ be an irrational angle, that is, one which is not a rational fraction of a complete turn of $2\pi$ radians. Let $T_\theta$ be the transformation which takes points on the circle to points an angle $\theta$ away in one direction, say clockwise. Applied to $a$, $b$ and $c$ its values are three points $T_\theta(a), T_\theta(b)$ and $T_\theta(c)$ which stand in the same relation, since a rotation through $\theta$ is a circle automorphism preserving the angles $\alpha$ and $\beta$. Now repeat $T_\theta$ any finite number of times. That gives us again three points $a'$, $b'$ and $c'$ standing in the same relation. Because $\theta$ is irrational, these three points never coincide with $a\ b\ c$ or any of the other preceding triplets.[24] To obtain an ancestral, consider the relation between $a$ and $b'$ and $c'$. That is a new relation, and $a$ stands in the first ancestral of the original relation to any two such points $b'$ and $c'$. It stands in the reverse ancestral to points $b''$ and $c''$ obtained by rotations of $T_\theta^{-1}$, i.e., in the opposite direction. Likewise, we could take as fixed the points $b$ or $c$, giving a total of six ancestrals.[25] Increasing the number of original points allows for relations of arbitrarily high arity. For an $n$-ary relation, the number of ancestrals definable by this method is $n(n-1)$, each with its own suite of $n$ relative terms. For the three-point case, and letting $\gamma = 2\pi - (\alpha + \beta)$, we can call the three role terms $\gamma\angle\alpha$ (which is the role of $a$), $\alpha\angle\beta$ (which is the role of $b$), and $\beta\angle\gamma$ (which is the role of $c$), making the relation as applied to these three points $(\gamma\angle\alpha : a\ \alpha\angle\beta : b\ \beta\angle\gamma : c)$. The theoretical significance of this is that the numbers of relations, relative terms and their many ancestrals are all uncountably infinite, since $\alpha$ and $\beta$ may

---

[24]This is so even if $\alpha$ or $\beta$ is an integral multiple of $\theta$. One or other of the points is always "new".

[25]We leave inductive definitions of these ancestrals as an exercise.

have uncountably many values, and this repeats *mutatis mutandis* for higher arities.

For ternary and higher arities it is possible to imagine "mixed" ancestrals, for instance alternating the roles of father and mother in the genealogical case, but such variations are likely to be of combinatorial rather than genuinely theoretical significance.

## 11  Plurality

Terms in term logic differ from those in standard predicate logic in that they can be plural, that is, denote more than one thing. The same goes for relative terms. Relations among pluralities are commonplace in natural language, though a relative latecomer to predicate logic.[26] The Capulets hated the Montagues; the Cubs beat the Indians in the 2016 World Series; the Brontë sisters wrote seven novels; the Austrians outnumber the Swiss; the parents of my ancestors are my ancestors, and so on:

(Haters: the Capulets Hated : the Montagues)
(Outnumber : the Austrians Outnumbered : the Swiss)

The parents–child relation and the parents–children relation are two for which we do have vocabulary.

(Child : Lily Parents: Anthony and Beth)

is an example, as is

(Children : Charles, Anne, Andrew and Edward Parents : Elizabeth and Philip).

The "upwards" ancestral of the first relation holds between Lily and her grandparents, Lily and her great-grandparents, and so on, i.e., each generation of her ancestors. The other (descendant, "downwards") ancestral of the second case gives us Elizabeth's and Philip's grandchildren, great-grandchildren, and so on, i.e., each generation of their descendants.

Several interesting mathematical propositions receive simple formulations in our representation, such as the following due to Cantor and Polignac respectively:

---

[26] See for example Oliver and Smiley [15], Carrara et al., eds., [7].

(Outnumber : the sets of natural numbers Outnumbered : the natural numbers)
(Equinumerous : the natural numbers Equinumerous : the pairs of twin primes).

When a relation is symmetric, it is often expressed conjunctively: *Jim and Liz are siblings*, *Donald and Joe are compatriots*. That gives an alternative way to express such relationships, using plurals:

(Siblings: Jim and Liz)
(Natives : Donald and Joe Country : some country).

When there are more than two individuals in the relationship, plurals can be used to finesse the fact that the number can vary:

(Siblings : Maria, Elizabeth, Charlotte, Branwell, Emily, Anne)
(Co-signatories : the 56 signers Signed : the US Declaration of Independence).

## 12  Spatial and Temporal Qualifications, Tense, Aspect, Modals

Not all languages have tenses as obligatory modifiers of verbs, and those that do vary in the tenses they possess. Temporal modification is optional in many cases and languages. Similar remarks go for aspect (such as perfective), and modal auxiliaries such as *could*. Tense, even where obligatory (as in Indo-European languages) is not an additional role in a predication, but a predicate modifier. In *Stanley met Livingstone* the past tense is not a third role alongside the two "meeters",[27] but modifies the predicate to indicate that the meeting took place in the past relative to the time of utterance.

By contrast, in a sentence like *Stanley met Livingstone on 10 November 1871 at Ujiji*, the prepositional phrases indicating place and time of the meeting *do* in effect add two roles to the relation, because the place and time mentioned are further "actors", better, arguments, in the situation:

---
[27] The word does exist with this meaning in English.

(Meeter 1 : Stanley  Meeter 2 : Livingstone  Place : Ujiji
Time : 10 November 1871).

Unlike the two "meeter" roles, which are essential to lexicalization using the verb *meet*, the place and time roles are optional to such predications, though of course ontologically any meeting must take place somewhere at some time. I have simply added the two roles to a "flat" predication, though there may be some justification for structuring the roles differently. Prepositional phrases in English cover a wide range of situations, not just those of spatial and temporal locations. Sometimes they are essential to the predication, as in *He strode across the stage*; sometimes they join clauses, as in *She raised the baton before giving the downbeat*. The variety of kinds of preposition (or in some languages, postpositions and/or cases) and their phrases means there may not be a single way to treat them all. It is perhaps enough for present purposes to note that prepositional phrases of place and time are a very frequent concomitant to relational predications. Among those worthy of treatment is the philosophers' 'at $t$' qualifier, much derided by tense logicians but meaningful and easy to integrate into a role theory of relatives.

## 13 Copulas and Complexes

To turn a list of roles and their players into a predication we need a copula: Charles *is* a parent of William, Emma *is* a donor to the ASPCA, London *is* a city, Rupert *has* a sister. In our principal notation this binding function is performed by placing the roles and their players inside a pair of parentheses, and to repeat the point, it is the same in whatever order the roles and their players are listed. The parentheses, because they work in the same way for lists of different length, are in effect a multigrade copula, or a scheme for various analogous copulas.

In these predications, the copula is playing two distinct roles. One is that of binding the various roles and their players together – hence the very term 'copula'. But the other role is that of marking assertion. These roles can come apart. In sentences such as

I wonder whether Jane is a grandparent
It is unlikely that Terry is a good golfer
Helga is hoping to be a musician

the embedded clauses

> Jane is a grandparent
> Terry is a good golfer
> Helga [will] be a musician

are not entailed by the complete sentence, despite employing a binding copula (finite, or, in the last case, infinite).

This raises the question whether we can, or indeed should, separate these two functions for logical purposes. Here is a suggestion as to how this might go. To make the difference clear, I shall replace the parentheses of the copula used to date by square brackets: thus $(G : a\,H : b)$ becomes $[G : a\,H : b]$, and so on.[28] That no longer counts as an assertion. The simplest way to make one is to add a special assertion marker, and for this we co-opt the assertion sign used by Frege, Whitehead and Russell. Thus $\vdash [G : a\,H : b]$ has the same meaning as $(G : a\,H : b)$. We may consider '$\vdash$' as being the only verb in its sentence. If we are to render a formula like $\vdash [G : a\,H : b]$ into a natural language sentence, it would have to be something like

> *a in the role of G in relation to b in the role of H* is a fact

where 'is a fact' is our sole verb[29] and the rest corresponds to what we might call a *complex*. The idea of a complex is taken from early Russell.[30] In the Introduction to *Principia Mathematica* he writes:[31]

> We will give the name of "a *complex*" to any such object as "*a* in the relation *R* to *b*" or "*a* having the quality *q*" or "*a* and *b* and *c* standing in the relation *S*." Broadly speaking, a complex is anything which occurs in the universe and is not simple. We will call a judgment *elementary* when it merely

---

[28] In the link notation for binary relations, we can simply leave the link as a copula but divest it of its assertoric force.

[29] This idea, and the locution "is a fact" (*ist eine Thatsache*) are found in Frege's *Begriffsschrift*, § 3.

[30] It occurs earlier in Meinong, which may be where Russell got the idea: cf. Meinong [12, pp. 387–94]. It also reverberates through Wittgenstein's dealings with Russell: cf. Simons [19]. It is usually assumed that such a theory of complexes is unworkable, but I suggest that judgment is hasty.

[31] Whitehead and Russell [23, p. 44].

asserts such things as "$a$ has the relation $R$ to $b$," "$a$ has the quality $q$" or "$a$ and $b$ and $c$ stand in the relation $S$." Then an *elementary* judgment is true when there is a corresponding complex, and false when there is no corresponding complex.

The analyses we are presenting, with terms rather than verbs carrying the principal semantic weight, is a variant of Russell's idea, the main difference being precisely this redistribution of semantic material away from predicates to terms.

In the case Russell's elementary judgments, it is also possible to consider a dual sentence-forming operator of *rejection* or denial. Then

$$\dashv [G : a\, H : b]$$

says of the complex that it is not a fact, or, as Meinong might have put it, it is an unfact (*eine Untatsache*).[32] This duality of acceptance and rejection has its origin in Brentano,[33] though the extent to which it can be extended beyond atomic (elementary) judgments is a moot point, which will not be taken up here. In such a sentence as

$\dashv$ Lover : Donald Beloved : Hillary]
Donald does not love Hillary

the prefixed rejection operator works as a negator. That does not make it a "negative copula" however: the copula (binder) is the same as in the positive case, being that symbolized by the square brackets in both cases. Since the copula binds without producing a sentence it is a

non-verb, non-sentential copula: only in conjunction with the acceptance or rejection operator do we have something doing the work of a sentential copula.

## 14 Existentials

Brentano (correctly) saw it as a virtue of his acceptance/rejection theory of judgment that it could effortlessly cope with kinds of judgment that

---

[32]Meinong [13, p. 95], where he talks of an objective being unfactual (*untatsächlich*).
[33]Brentano [2, 213 ff.], [3, 210 ff].

the traditional subject–predicate theory could not: these included so-called subjectless judgments like "It is raining" or "It is hot", as well as positive and negative existential judgments such as "There are horses" and "There are no unicorns". Unfortunately, Brentano, this time adhering to tradition, did not make a distinction between general terms like *horse, rain, unicorn*, and definite terms, whether singular, like *King Charles III*; plural, like *Whitehead and Russell*; or descriptively definite like *the discoverer of continental drift* and *the apples in this barrel*. Our insistence on this distinction comes out in the $G : a$ notation, where '$G$' stands in for a general term and '$a$' for a definite term.[34] We cannot congruously have $\vdash [G]$ or $\vdash [a]$ or their negatives. However, all is not lost. Assertions and denials of existence, whether general or definite, can be expressed in our notation with a small deviation from Brentano, and exploiting the previously used narrow scope quantifier phrases. To say *There is at least one horse* we have

$\vdash$[Horse : something]

while to deny that there are unicorns we have

$\dashv$[Unicorn : something].

To accept the existence of Julius Caesar we have

$\vdash$[Some kind : Julius Caesar]

or perhaps, allowing for the idea that all individual terms have an implicit kind as part of their meaning:

$\vdash$[Human being : Julius Caesar].

For those ontologists of fiction who agree that there is no human being called 'Sherlock Holmes' but insist that there is a fictitious character so called, we can offer

$\dashv$[Human being : Sherlock Holmes] $\wedge$ $\vdash$ [Fictitious character : Sherlock Holmes]

as a way for them to have their ontological cake and eat it.

---

[34] In a categorial grammar, definite terms would be assigned to the *name* category N, while general terms are assigned to the *common noun* category C. Relative terms, though semantically incomplete, are syntactically common nouns: they occur "naked" as such when derelativized.

Brentano also (correctly) claimed to be able to represent the standard four forms of categorical syllogistic. Here is how this can be done, using term conjunction (∩) and term negation (N):

⊢[Man ∩ Philosopher : something]    Some man is a philosopher.
⊢[Man ∩ N (Philosopher) : something]    Some man is not a philosopher
⊣[Man ∩ Donkey : something]    No man is a donkey
⊣[Man ∩ N (Mortal) : something]    All men are mortal.

Note that whereas Brentano located existence (and its denial) solely within the acceptance and rejection operators, in our alternative formulation the local quantifier phrase also contributes to the meaning, so we are affirming or denying something's being something, which is after all what existential sentences are about.

[34] In a categorial grammar, definite terms would be assigned to the *name* category N, while general terms are assigned to the *common noun* category C. Relative terms, though semantically incomplete, are syntactically common nouns: they occur "naked" as such when derelativized.

## 15 Grammar

General terms fall into the grammatical category of *common nouns*. Definite terms by contrast fall into the category of *names* or (better) *nominals*. A relative term requires completion by one or more nominals (or quantifiers) in order to yield a compound absolute term. As an indication that the formalism here presented is perfectly amenable to treatment by categorial syntax, I will give an example. Let $\alpha\beta\gamma$ be syntactic categories. The functor category taking arguments of categories $\beta$ and $\gamma$ and returning a result of category $\alpha$ will be denoted $\alpha\langle\beta\gamma\rangle$. We regiment so that functors occur before their arguments. The simple English sentence 'Romeo loves Juliet' has a straightforward analysis (we are not here worrying about tense):

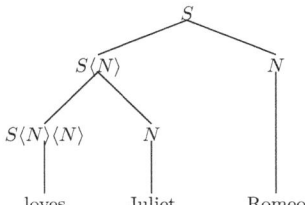

while one of our alternative renderings of the same thought comes out as

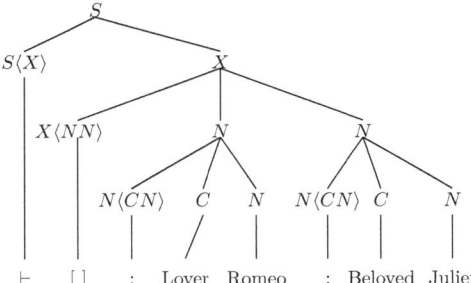

The latter is indeed more complicated, though the former has an extra level of structure if we include the (for English) obligatory tense. Also, avoiding the order problem, the English analysis combines the verb with the direct object first, as happens in around 90% of the world's languages.[35] The second analysis however separates out the copulative and assertive functions, and order is irrelevant. The new category $X$ is for expressions for complexes. They cannot be simply named, or else '⊢ [a]' would be well-formed, which it is not. The colon symbol, which looked initially just like a bit of punctuation, turns out to have a signification: like the generic 'as' in English, it combines role and actor, and the resulting higher $N$ is the *player*, the actor *as* filling the role, while the

---
[35]Baker [1, 93ff.].

brackets are the copula, gathering the players together in a complex, of category $X$, where as usual the order of players is not significant. If we were to render this analysis in words, it might go something like

> Romeo *as* Lover *in relation to* Juliet *as* Beloved *is accepted.*

The English sentence, to which is of course the standard predicate-logical analysis is close, is obviously the more streamlined of the two, which is not surprising, given that in the term-logical rendering the verbs have been reduced to just two, and the semantic "meat" is redistributed to the terms. This discrepancy in complexity is lessened when more complex sentences involving prepositional phrases are considered.

If instead of the square brackets and the acceptance and rejection operators we analyse the sentence with round brackets, then these turn out to be simply copulas, general-purpose verbs, of categories $S\langle N \rangle$, $S\langle NN \rangle, S\langle NNN \rangle$, etc. The category $X$ of expressions for complexes is not then needed and the analysis has only three levels, the same as the English sentence.

## 16  Equivalence in Extension of the Two Formulations

Let $R \subseteq D^n$ be any $n$-ary relation in extension on the domain $D$ (where $R$ holds among individuals only).

Define $D_i$, the $i^{\text{th}}$ domain of $F$, as the set $\{x_i | \exists x_1 \ldots x_{i-1}, x_{i+1} \ldots x_n (Rx_1 \ldots x_{i-1}, x_i, x_{i+1} \ldots x_n)\}$[36]

Let role terms $G_i, 1 \leq i \leq n$ be such that for all $a_1 \ldots a_n$

$$(G_1 : a_1 \ldots G_i : a_i \ldots G_n : a_n) \leftrightarrow Ra_1 \ldots a_i \ldots a_n$$

where each $G_i$ has $D_i$ as its derelativized extension. For example, if $Rabc$ means '$a$ gives $b$ to $c$' then $G_1$ is donors, $G_2$ is gifts, and $G_3$ is beneficiaries.

Let $\pi$ be any permutation of the integers 1 to $n$. There is a relation $\pi(R)$, the $\pi$-induced converse of $R$ (one of $n!$ including $R$ itself) such that

$$\pi(R)a_{\pi(1)} \ldots a_{\pi(i)} \ldots a_{\pi)n)} \leftrightarrow Ra_1 \ldots a_i \ldots a_n$$

---
[36]Bocheński and Menne [4, 23.42 p. 110].

so
$$G_{\pi(1)} : a_{\pi(1)} \ldots G_{\pi(i)} : a_{\pi(i)} \ldots G_{\pi(n)} : a_{\pi(n)}) \leftrightarrow (G_1 : a_1 \ldots g_i : a_i \ldots G_n : a_n).$$

Since the role-based representation is permutation-invariant, the former is not just equivalent to the latter but is a mere notational variant of it. In the alternative complex formulation

$$[G_{\pi(1)} : a_{\pi(1)} \ldots G_{\pi(i)} : a_{\pi(i)} \ldots G_{\pi(n)} : a_{\pi(n)}] = [G_1 : a_1 \ldots G_i : a_i \ldots G_n : a_n]$$

For example, if $(G_1 : a_1 \; G_2 : a_2 \; G_3 : a_3) \leftrightarrow Ra_1a_2a_3$ and $\pi$ cycles 123 to 231 so

so $\quad \pi(R)a_2a_3a_1 \leftrightarrow Ra_1a_2a_3$
then $\quad (G_2 : a_2 \; G_3 : a_3 \; G_1 : a_1) \leftrightarrow (G_1 : a_1 \; G_2 : a_2 \; G_3 : a_3)$
and $\quad [G_2 : a_2 \; G_3 : a_3 \; G_1 : a_1] = [G_1 : a_1 \; G_2 : a_2 \; G_3 : a_3].$

If *Rabc* again means '*a* gives *b* to *c*' then $\pi(R)bca$ means '*b* is given to *c* by *a*' and [Donor : *a* Gift : *b* Beneficiary : *c*] is the Russellian complex "*a*-giving-*b*-to-*c*", which may be designated in any of six different ways.

## 17 Propositional Attitudes

The linguistic account of so-called propositional attitudes is that they are expressed by verbs which govern an embedded 'that' clause or similar sentential object, such as

> Othello believes that Desdemona loves Cassio
> Sven hopes that Sweden will win the World Cup
> Mary wants to be a doctor.

The naïve interpretation of these, as indicated by their name, is that they consist in a relation between a person and a proposition, which is an abstract semantic entity denoted by the 'that'- or similar object clause, and which is the meaning of the relevant clause or sentence. The attitude is a mental act or disposition of its bearer that somehow reaches beyond itself to this proposition. Scepticism about the existence of such propositions, and about the way in which mental goings-on might intend them, lay behind one concerted attempt to find an alternative account of the phenomena. This was Bertrand Russell's multiple relation theory of propositional attitudes. Russell held variants of this theory between 1910 and 1919. Here is an early formulation:[37]

> [A] judgment does not have a single object, namely the proposition, but has several interrelated objects. That is to say, the relation which constitutes judgment is not a relation of two terms, namely the judging mind and the proposition, but is a relation of several terms, namely the mind and what are called the constituents of the proposition.

Russell's motivations were several. Firstly, he wished to avoid Meinongian unfacts or objective falsehoods as worldly denizens on a par with facts or objective truths; secondly, he wished to avoid propositions as mysterious Fregean senses, a theory he had rejected in 'On Denoting'; thirdly, he wanted to ensure it is possible to think (judge, believe etc.) falsely as well as truly; and finally, he wanted to uphold a correspondence theory of truth:[38]

> [W]e may define *truth*, where such judgments are concerned, as consisting in the fact that there is a complex *corresponding* to the discursive thought which is the judgment. That is, when we judge "$a$ has the relation $R$ to $b$", our judgment is said to be *true* when there is a complex "$a$-in-the-relation-$R$-to-$b$," and it is said to be *false* when this is not the case.

A series of problems that arose during Russell's adherence to his theory eventually made it, in his own view, impossible simultaneously to achieve all these aims, and so he abandoned the theory in favour of behaviorism and the bizarre theory of neutral monism.[39]

I shall demonstrate that recasting the account of relations using relative terms in the way proposed in this paper enables a solution to be found to the problems confronting a theory of Russell's type, so that he need not have given up on the theory. This constitutes a retrospective vindication of the methodological and ontological insights which motivated him in the first place.

---

[37] Whitehead and Russell [23, p. 43], (first published 1910).
[38] Ibid.
[39] For a comprehensive account of the twists and turns of Russell's theories, see Candlish [6].

We need to make one adjustment before we start. For Russell, the constituents of a proposition, for example the proposition that Jack loves Jill, are the three actual items Jack, Jill, and the relation of loving. We shall pull back from this kind of commitment. What we seek is an *analysis* of sentences like 'Othello believes that Desdemona loves Cassio' which satisfies Russell's requirements, leaving ontological questions aside. 'Desdemona loves Cassio' is represented as

(Lover : Desdemona Beloved : Cassio)

The arguments in Othello's believing relation will still be Desdemona and Cassio, and the relation of loving. The last is represented as the scheme

(Lover : – Beloved : –)

But Desdemona and Cassio do not appear in Othello's believing as *actually* Lover and Beloved respectively, since this would rule out false thought. Rather they are *believed by Othello* (falsely, in the story) to be Lover and Beloved respectively. The roles in the believing relation are *lifted* from the embedded relation but modified by their new context: we thus obtain

(Believer : Othello Believed Relation : (Lover : – Beloved : –)
Believed-as-Lover : Desdemona Believed-as-Beloved : Cassio)

as our representation of 'Othello believes that Desdemona loves Cassio'. The believing relation contains two more arguments than the believed relation, one for the believer, another for the embedded relation, so that believing, as in Russell, is multigrade: believing that Socrates was a philosopher has three terms, whereas believing that London is further from Edinburgh than Dublin is from Cork has six terms. Similar remarks go for other attitudes such as judging, hoping, wanting, fearing etc.

Russell's insistence that actual entities be the relata of the relational predication means he cannot give a straightforward account of the truth

Le Verrier believed that Vulcan is closer to the Sun than Mercury

because there is no such object as Vulcan. On our ontologically light-touch approach we can apply the straightforward analysis, while noting that

> Vulcan is closer to the Sun than Mercury

is false, precisely because there is no such thing as Vulcan.

Such believings, judgings etc. will, as Russell requires, be prone to error; they will be true just when the arguments in the lifted terms actually fill the corresponding roles in the unlifted relation, and with other verbs, the embedded relation need not be one that is affirmed to exist. For example, if we replace 'Othello' by 'Iago' and 'Believe' etc. by 'Disbelieve', then clearly Iago is not committed to Desdemona loving Cassio. Neither Frege's mysterious propositions nor Meinong's incredible unfacts are to be found in the analysis. Finally, since no clause corresponding to the embedded sentence 'Desdemona loves Cassio' is to be found in the analysis, the vexed issue of so-called non-extensional contexts simply does not arise.

It is unclear to what extent this theory can be extended from elementary propositions such as these to logically more complex ones. No attempt to do so will be made here, and it is in any case unclear how far Russell intended such an extension. His own descriptive theory of proper names would however require at least some enlargement of the multiple relation theory to account for the truth of the Le Verrier sentence, 'Vulcan' being presumably equivalent in meaning to a definite description like 'the planet closer to the Sun than Mercury whose gravitational influence accounts for the anomalies in Mercury's orbit'. There is of course nothing to stop us from adding the standard variable-binding quantifiers to our account to facilitate greater expressive and logical capacity.

## 18 A Leaner Solution

The previous section followed Russell in treating the "naked" relation as one of the terms of the whole attitude sentence. But the relation is recoverable from the lifted terms, and the main verb acts as its own combinator, modifying the embedded relative terms. A leaner representation dispenses with the relation as an item:

> (Believer : Othello Believed as Lover : Desdemona Believed as Beloved : Cassio)

and this works for different arities:

(Surprised : John Surprising as Donor : Emma Surprising as Gift :
$1,000 Surprising as Recipient : ASPCA)
(Wanter : Mary Wanted as Doctor : self)

This account also has some traction in representing subjectless verbs taking sentential objects, as *It is unlikely that Desdemona loves Cassio*:

(Unlikely as Lover : Desdemona Unlikely as Beloved : Cassio)

and *It is more probable that Iago hates Othello than that Desdemona loves Cassio*

(More Probable as Hater : Iago More Probable as Hated : Othello
Less Probable as Lover : Desdemona Less Probable as Beloved : Cassio)

## 19 Conclusion

De Morgan developed his theory of relations to subsume syllogistic as a special case. Since the variables of syllogistic were terms, it was both theoretically and linguistically natural for
 him to use relative terms. When Peirce continued the study of relations, he too naturally employed relative terms in his examples and theorising, and this was carried further by Schröder. As we have seen, using relative terms instead of relational predicates allows different aspects of the theory of relational predication to be highlighted. The linearity of spoken and written language constrains relational talk to rely variously on word-order and/or cases and inflections, and/or prepositional (or postpositional) phrases, to accommodate the variety of actors and roles in relational situations and predications. Logical theory is not bound by this linearity, as Frege's two-dimensional notation and other experiments testify.

Highlighting relative terms as expressive of roles allows the constraint of linearity to be relaxed, albeit at the cost of brevity, while, if a non-linear representation is pursued, it comes at the cost of typographical simplicity, itself largely tied to linearity. The order-indifference of relations as distinct from relational predicates in predicate logic dispenses with converses, but complicates the theory of relative products and ancestrals. It can be harnessed to support a multiple-relation theory of propositional attitudes. All told, employing relative terms in the analysis of relational predications sheds new light on these, which may count as a recommendation.

# References

[1] Baker, M. C. 2002. *The Atoms of Language*. Oxford: Oxford University Press.

[2] Brentano, F. 1973. *Psychology from an Empirical Standpoint*. London: Routledge.

[3] Brentano, F. 1956. *Die Lehre vom richtigen Urteil*. Bern: Francke.

[4] Bocheński, J. M. and Menne, A. 1973. *Grundriß der Logistik*. 4th ed. Paderborn: Schöningh.

[5] Burdick, H. 1991. What Was Leibniz's Problem about Relations? *Synthese* **88**, 1–13.

[6] Candlish, S. 1996. The Unity of the Proposition and Russell's Theories of Judgement. In: R. Monk and A. Palmer, eds., *Bertrand Russell and the Origins of Analytical Philosophy*. Bristol: Thoemmes, 103–135.

[7] Carrara, M., Arapnis, A. and Moltmann, F., eds., 2016. *Unity and Plurality: Logic, Philosophy, and Linguistics*. Oxford: Oxford University Press.

[8] Curry, H. B. and Feys, R. 1974. *Combinatory Logic*. Vol. I. Amsterdam: North-Holland.

[9] Fine, K. 2000. Neutral Relations, *Philosophical Review*, **199**, 1–33.

[10] De Morgan, A. 1966. *On the Syllogism and other Writings*, ed. P. Heath. London: Routledge.

[11] MacBride, F. 2020. Relations. *The Stanford Encyclopedia of Philosophy* (Winter 2020 Edition), E. N. Zalta (ed.), https://plato.stanford.edu/archives/win2020/entries/relations/\T1\textgreater.

[12] Meinong, A. 1971. Über Gegenstände höherer Ordnung und deren Verhältnis zur inneren Wahrnehmung. In: *Abhandlungen zur Erkenntnistheorie und Gegenstandstheorie*, ed. R. Haller (*Alexius Meinong Gesamtausgabe* II). Graz: Akademische Druck- u. Verlagsanstalt, 377–480.

[13] Meinong, A. 1977. *Über Annahmen*, ed. R. Haller. (*Alexius Meinong Gesamtausgabe* IV). Graz: Akademische Druck- u. Verlagsanstalt.

[14] Merrill, D. 1990. *Augustus De Morgan and the Logic of Relations.* Dordrecht: Kluwer.
[15] Oliver, A. and Smiley, T. 2013. *Plural Logic.* Oxford: Oxford University Press.
[16] Peirce, C. S. 1984. Description of a Notation for the Logic of Relatives, resulting from an Amplification of the Conceptions of Boole's Calculus of Logic. In: *Writings of Charles S. Peirce. A Chronological Edition.* Volume 2, 1867–1871. Bloomington: Indiana University Press, 359–429. (First published 1870.)
[17] Prior, A. N. 1962. *Formal Logic.* 2nd ed. Oxford: Clarendon.
[18] Schröder, E. 1895. *Algebra und Logik der Relative, der Vorlesungen über die Algebra der Logik 3. Band.* Teubner.
[19] Simons, P. 1985. The Old Problem of Complex and Fact. *Teoria* **5**, 205–225.
[20] Simons, P. 2020. Aspects of the Grammar and Logic of Relative Terms. *Przegląd Filozoficzny –Nowa Seria* **28**, 73–89.
[21] Steedman, M. 1988. Combinators and Grammars. In: R. T. Oerhle, E. Bach & D. Wheeler, eds., *Categorial Grammars and Natural Language Structures.* Dordrecht: Reidel, 417–442.
[22] Steedman, M. 2018. The Lost Combinator. *Computational Linguistics* **44**, 613–629.
[23] Whitehead, A. N. and Russell, B. 1927. *Principia Mathematica.* Cambridge: Cambridge University Press. 2nd ed. (First ed., 1910).

# MODALITY AS A PROPERTY OF TERMS

### MANUEL CORREIA
*Pontificia Universidad Católica de Chile*

---

It has been suggested that one of the traditional rules of categorical syllogism is sufficient to transform the Aristotelian syllogistic into an ampler and more decidable theory than it was.[1] This key rule is about the predicate and says that in every negative premise the predicate is taken universally and in any affirmative premise the predicate is particularly taken. This rule together with obversion allows introducing consistently indefinite terms in syllogistic. Let us take an example:

1. $S$ is $P$
2. $S$ is not $P$
3. $S$ is not-$P$.

According to this rule, in (1) the predicate is distributed. In (2) it is undistributed. And in (3) the predicate, 'not-$P$' is distributed as it is the predicate of an affirmative proposition. But this makes $P$ undistributed in (3). And it must be so because (2) and (3) are equivalent propositions by obversion.

This key rule allows the introduction of indefinite or infinite terms in the syllogistic theory and validates syllogisms with indefinite terms with the rest of the traditional rules of syllogism. Let the following syllogism be:

4. Every non-$P$ is $T$
   Some non-$S$ is non-$P$
   Then, Some $T$ is not $S$.

---

[1] Cf. [2, pp. 297–306].

Traditional syllogistics cannot directly validate (4), but if this key rule is added to the three traditional rules,[2] it could. Indeed, in (4) the middle term non-$P$ is undistributed in the major premise –since it is under the effect of a universal quantifier– and distributed in the minor premise –since it is the predicate of an affirmative proposition. Moreover, it is not the case that both premises are particular propositions. And no term in the conclusion is more extensive than in the premises –since $T$ is particular in the conclusion and distributed in the major premise as the predicate of an affirmative proposition. And $S$ is undistributed in the conclusion and so it is in the minor premise, –since 'non-$S$' is particular because it is under the particular quantifier and so $S$ must be universal (or if "Some non-$S$ is non-$P$" is converted you will have "Some non-$P$ is non-$S$" and obverted, you will have "Some non-$P$ is not $S$", where $S$ is undistributed because it is the predicate of a negative proposition).

The use of these rules is not restricted to syllogistic theory though, for it can prove (in)validity to any conclusion either immediate or mediate (with any number of premises), so that indefinite terms can be introduced to the entire Aristotelian logic without restriction. Moreover, since these rules can also resolve (in)validity of hypothetical syllogisms, Aristotelian logic also extends its limits to the extent of refusing the last-century idea that Aristotelian logic is only a fragment of conditional logic.[3]

However, while assertoric or non-modal logic can extend its results, modal propositions have not been evaluated yet, and we do not know whether modal propositions and their conclusions agree with our results. Accordingly, I would like to fill this gap by showing how these rules can give a general explanation

---

[2] These traditional rules are reformulated in [2], thus: In every valid syllogism, $Ax\ P$: no conclusion comes from only particular premises. $Ax\ Q$: The middle term must be alternatively particular and universal. $Ax\ L$: The terms in the conclusion must have the same quantity as in the premises.

[3] The idea that Aristotelian logic is just a fragment of conditional logic was propagated by Łukasiewicz and Tarski within the general framework of acid criticism. For instance: Jan Łukasiewicz [6, p. 99]: "We also know that the logic of propositions is of much greater importance than the thin fragment of the logic of terms which takes shape in Aristotle's syllogistic. The logic of propositions is the basis of all logical and mathematical systems. We must thank the Stoics for laying the foundations of this admirable theory". Alfred Tarski [12, p. 19]: "The new logic surpasses the old in many respects, not only because of the solidity of its foundations and the perfection of the methods employed in its development, but mainly on account of the wealth of concepts and theorems that have been established. Fundamentally, the old traditional logic forms only a fragment of the new, a fragment moreover which, from the point of view of the requirements of other sciences, and of mathematics in particular, is entirely insignificant."

of Aristotle's modal syllogistic of *An Pr* I, 8-22. However, since this task would involve a much larger project, I will focus on discussing the most abstruse problem of mixed modal syllogisms, the so-called problem of the two Barbaras.[4] The aim is to show how these rules can also be useful to prove (in)validity within Aristotelian modal logic. Thus, the first problem is to explain how a rule controlling quantity could control modality. The second problem is to show how the rules are applied to modal syllogisms. The third problem is to show whether these rules find some textual support in Aristotle's logical writings.

The outline of this paper will follow these three questions. And it will be divided into the following six sections: I Quantity and quality. II The pure modal syllogistic. III The mixed modal syllogisms. IV The mixed modal syllogisms NA/N, AN/N: The *de re* view. V The mixed modal syllogisms NA/N, AN/N: The *de dicto* view. VI Final remarks.

## Quantity and modality

In Aristotelian logic, the categorical or simple proposition has some properties.[5] But these properties are the properties of terms. Quantity affects the subject term of a proposition since the quantifier affects the subject term. Thus,

---

[4]I do not know who was the first to call this problem so, but in secondary literature has become common. For example, see [4, pp. 381–395].

[5]The ancient commentaries on Aristotle's *De Interpretatione*, especially those by Boethius and Ammonius Hermiae, from VI AD, make the most complete treatment of it. The latter accepts nine properties for categorical propositions: two or three terms, modal or non-modal, definite or indefinite subject, definite or indefinite predicate, quality, quantity, time, and matter. He remarks that these properties are behind all the oppositions Aristotle defines in *De Int* (*pasin hai en Peri Hermeneias paradedomenai antiphaseis: in Int.* p. 219, 20-21, Busse). They also count the number of propositions from these properties. Let us adopt the following convention: (i) [UPIS]: division by quantity; (ii) [Tense]: division by verbal tense; (iii) [Matter]: division by matters of proposition; (iv) [Mod.]: by modality; (v) [IS]: by indefinite subject; (vi) [IP]: by indefinite predicate. Thus:

(a) 1 ('a man walks') ×4 [UPIS] ×3 [Tense] ×3 [Matter] ×2 [IS] = 72.

(b) 1 ('a man is just') ×2 [IS] ×2 [IP] = 144.

(c) 4 (basic species) ×4 [UPIS] ×3 [Tense] ×3 [Matter] ×3 [Mod.] = 432.

(d) 8 (basic species) ×4 [UPIS] ×3 [Tense] ×3 [Matter] ×3 [Mod.] = 864.

For (a) cf. Ammonius *in Int.* p. 90, 21 - p. 91, 3; (b) p. 160, 17-32; (c) and (d) p. 218, 30 - p. 219, 24.

the reason why we say that a proposition is universal or particular is that the subject term (which must be a universal concept) is universally or particularly taken.[6] Tense affects the verb. Matter affects the relation between the subject and the predicate. Indefinite propositions are indefinite because one of the terms either the subject or the predicate is indefinite. Quality seems to affect the proposition, but it depends on the negative or indefinite terms since it changes according to obversion. For instance, "Some *S* is *P*" is affirmative, but "Some *S* is not not-*P*", which is its obverse, is negative, but they are equivalent. Modality is, however, a problem, because it seems to affect the proposition (*de dicto* interpretation) and it sometimes seems to affect the predicate (*de re* interpretation). It would be rare that modality was an exception, and we could examine whether this is so or not. Perhaps, modality could also be reduced to a property of the terms.

If we assume it, some interesting consequences will follow, because accepting that the syllogism is a valid assertoric syllogism, it will be modally valid if and only if the terms in the conclusion have the same modality as in the premises. Thus, in general, to examine the validity of modal syllogisms, we now simply add the assessment of whether the terms in the conclusion have the same modality as in the premises. As simple as that, but if we are to accept less modal force in the conclusion terms, then we can also add the axioms that Aristotle mentions in *De Interpretatione* 13, namely:

5. $\Box p \to p$

6. $p \to \Diamond p$

7. $\Box p \to \Diamond p$

These three modal axioms can give us some room to accept that when the terms in the conclusion have less modal force than the premises, the syllogism can also be accepted as valid. In other words, the terms in the conclusion cannot have more modal force than in the premises.

---

[6] Aristotle in *De Int* 7, 17a38-17b1, defines what are universal and what are particular things (*pragmata*). But then, he distinguishes to state universally of a universal that something holds or does not from to state not universally of a universal that something holds or does not (*De Int* 7, 17b2-6). There are things to distinct here. (i) Things or facts can be universal or singular in the sense of *Categories* 5, 2a11-18. And (ii) things that are universal can be predicated (or taken) universally or particularly, which is a way to assume quantification for universal subjects. But (iii) only universal subjects can be quantified, for singular subjects do not accept quantification (namely, expressions like "Every/no/Some Socrates" are not allowed).

## The pure modal syllogistic

Aristotle's treatment of modal syllogism in *An Pr* I, 8-22 assumes that modal syllogisms are pure or mixed. Intuitively, this distinction relates to whether the premises and the conclusion are all of one type or not. Pure modal syllogisms are those in which all have only one kind of modality, either assertoric, contingent, possible, or necessary.[7]

If we define **M** as one of the modalities, we can represent a pure modal syllogism as follows:

8. **M**: *Every S is P*
   **M**: *Some T is S*
   **M**: *Some T is P*

One of the reasons why pure modal syllogisms are easy to deal with is the fact that they are valid or invalid just as the correspondent non-modal syllogisms are. For instance, Darii is a valid assertoric or non-modal syllogistic mood of the first syllogistic figure: if every premise is modified by the same modality, the conclusion will continue being modally valid.[8] Pure modal syllogisms are treated in *An Pr* I, 8, 14, 17, and 20. As in non-modal syllogisms, pure imperfect modal syllogistic moods of the second and third figures will be valid if they can be reduced to the first figure, that is, the premises or the conclusion can be converted either simply or by accident. The rules of conversion are the same as those already defined for assertoric or non-modal propositions.

Now, pure modal syllogisms can also be ruled by modal balance, that is, by the correspondence between the modalities that affect the predicate in the conclusion and the major term in the major premise. This term is the same in the first syllogistic figure whether the syllogism is pure or mixed. The identification of this term is crucial to reducing valid moods from the third and second figures to the first and perfect figure. On the one hand, there is a manifest modal balance in pure modal syllogisms of the first figure and in every imperfect mood when reduced to the first figure (Barbara, Celarent, Darii, and Ferio). Let us take an example:

---

[7] In theory, modalities should amount to contingent, possible, necessary, and impossible, but Aristotle in *An Pr* I, 8-22 only refers to Assertoric, Necessary and Contingent.

[8] The only difference he adds is irrelevant to our case, for it relates to the impossibility of using demonstration *per absurdum* in moods Baroco (second figure) and Bocardo (third figure), and the convenience of using a proof by *ecthésis*. cf. *An Pr* 30 a 10-15.

|   | *Celarent* | *Celantes* |
|---|---|---|
| 9 | **N** No *S* is *P* | **N** No *S* is *P* |
|   | **N** Every *H* is *S* | **N** Every *H* is *S* |
|   | **N** Then, No *H* is *P* | **N** Then, No *P* is *H* |

In this Celantes of the first indirect figure, *P* is modally taken in the major premise, but not in the conclusion, because here *H* is modally taken. In contrast, in Celarent, *P* is modally taken in the major premise and so it is in the conclusion, establishing a perfect balance.

Celantes is not modally valid until we know that the Celarent from which it comes is modally valid. This is the aim of the theory of the reduction of imperfect moods, either assertoric or modal, to the perfect ones of the first figure: if they are not reducible to a modal mood of the first direct figure, they cannot be modally valid, which shows that the function of reduction is of the highest importance to modal syllogistic.

## The mixed modal syllogisms

Aristotle's *An Pr* I, 9-13; 15-16; 18-19, combines modality by using assertoric and necessary modalities (**A**, **N**), assertoric and possible modalities (**A**, **P**), and possible and necessary modalities (**P**, **N**).[9] In the following, I will focus on the first figure especially in the case of the so-called two Barbaras problem, which combines **A** and **N**.

Although Aristotle does not explain it, it is useful to know that each modal pair yields six ternary sets, and only some of them are modally conclusive. Aristotle only discusses two cases but, as I will show, he is aware of the rest. The six ternary sets are **AN/A, NA/A, NA/N, AN/N, AA/N, NN/A**. From here, Aristotle discusses only the third and fourth which is, I think, correct because the other modal moods can be easily rejected based on his general law of *An Pr* I, 12, which says that a valid conclusion is assertoric (**A**) only if both premises are assertoric. And a conclusion can be necessary (**N**), though not always, if one of the premises is necessary. According to this law, only the third (NA/N) and the fourth (AN/N) are the critical cases.

---

[9] A complete list of pure and mixed modal syllogisms can be seen in Smith [11, pp. 229–235] and Ross [10].

## The mixed modal syllogisms NA/N, AN/N: The *de re* view

The *de re* modal reading can be seen in the following formula: "*S* is *modally P*", where *S* and *P* are the terms of the proposition. As such the terms can have many other logical properties.[10] The variable *modally* stands for at least three general modalities.[11] A *de re* reading in which **M** is a modality, '*S*' a subject, '*P*' a predicate, 'is' a sign of the quality of the *dictum* (the modalized part of the modal proposition) should be read as follows:

**M** *S* is *P*, where **M** affects directly to *P*.

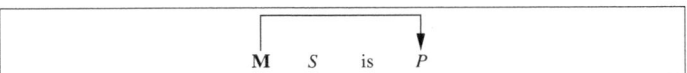

In modern modal logic, the question of whether the modality must be put inside or outside the quantifier has been identified as to whether modality should be read in a *de re* or *de dicto* sense. Aristotle should resolve this issue by maintaining that modality is inside the quantifier if he intends to be consistent with his opinion that the Barbara NA/N is valid. Aristotle's NA/N Barbara is as follows:

10. Every *S* is necessarily *P*.
    Every *T* is *S*
    Then, every *T* is necessarily *P*.

Aristotle gives this **NA/N** problematic syllogism as valid in his *An Pr* I, 9. The assertoric structure is valid as it is a Barbara, namely, the first perfect syllogism he validates. The three rules we have used to validate syllogisms, in general, can validate it: *T* is universal in the minor premise, and so it is in the conclusion. *P* is distributed in both the conclusion and the major premise.

---

[10] *S*, the subject, is a name, universal or singular, definite or indefinite, universally taken or particularly taken. And *P* is universal (or wider than the subject term), definite or indefinite, undistributed or distributed, and now modally taken or not modally taken.

[11] Aristotle in *De Int* 12 and 13 [1] recognizes four general modalities: contingent, possible, impossible, and necessary, but he identifies contingency and possibility. In *An Pr* he distinguishes contingency from possibility, which has produced other kinds of confusion in his exposition.

$S$ is universal in the major premise and distributed in the minor premise. To validate its modal load, we examine its modality: we note that $P$ in the conclusion is modalized just as in the major premise, namely, $P$ has been affected in both cases by 'necessarily'. Accordingly, the modal syllogism is valid and the emergence of the modality in the premises and the conclusion is perfectly balanced.

Let us examine the other mixed modal syllogism that Aristotle here gives as invalid. It is the following syllogism AN/N:

11. Every $S$ is $P$
    Every $T$ is necessarily $S$
    Then, every $T$ is necessarily $P$.

It is evident that $P$ in the conclusion is affected by the modality of necessity, but it is not so in the major premise, where $P$ is free of modality. All the quantitative parameters are according to the three rules: $S$, the middle term, is once universally taken and once particularly taken. $T$ is universal in the conclusion and so it is in the minor premise. And $P$ is particularly taken or distributed in both the conclusion and the first premise. On the other hand, the syllogism has not only particular premises.

Our results are similar to those of Aristotle. Now it is important to note that when *de re* modality is considered, the terms in question are only two: the predicate in the conclusion and the major term, which in the first figure is the predicate of the major premise.

Conversion is the instrument Aristotle uses to find modal balance in his theory. This explains why Aristotle rejects some modal mixed moods in the second and third syllogistic figures. For instance, he rejects the following Cesare of the second figure (*An Pr* I, 10, 30 b 7-19):

12. **N** No $S$ is $P$
    **A** Every $H$ is $P$
    Then, **N** No $H$ is $S$.

Even if the assertoric syllogism is valid, there is no modal balance in its terms. Indeed, the asymmetry is manifest, for in the conclusion **N** applies to $S$, but in the major premise **N** applies to $P$. Asymmetry entails that the axiom of modality (if it can be called so) has been broken, which entails that it is invalid.

But this **NA/N** Cesare could be valid if it can be reduced to the Celarent mood of the first figure. To make this reduction one needs to know how to convert modal propositions, which Aristotle teaches in *An Pr* I, 3. He accepts that a necessary, universal negative proposition (i.e., the canonical *E*) is simply converted, as the corresponding assertoric one, which allows him to convert the major into "**N** No *P* is *S* ", which forms the correspondent Celarent of the first figure, and which is valid, because of modal balance. Thus, Aristotle uses conversion to reestablish modal balance in the crucial terms, which are the predicate of the conclusion and the major term in the first figure. However, Baroco of the second figure does have a different destiny. The following modal syllogism is a **NA/N** in Baroco:

13. **N** *Every S* is *P*
    **A** *Some H* is not *P*
    Then, **N** *Some H* is not *S*

Since in this modal Baroco the major premise cannot be simply converted, there is no way to maintain the modal balance between *S* in the major premise and *S* in the conclusion. Symptomatically, Aristotle in *An Pr* I, 30 a 10-14 goes to use proof by *ekthesis* to validate Baroco and Bocardo.

## The mixed modal syllogisms NA/N, AN/N: The *de dicto* view

The *de dicto* interpretation, which is mainly present in contemporary modal logic, rejects both moods. Aristotle, however, only rejects the AN/N mood and accepts NA/N. If we try to prove these mixed modal syllogisms, we will confirm the contemporary result. Thus:

14. **N**: Every *S* is *P*
    **A**: Every *T* is *S*
    **N**: Every *T* is *P*

To be more effective in the *de dicto* logical proof, let us write this syllogism in a modal quantified language in its pure fashion first and let us try to prove in a reflexive frame:

15. $\Box\forall(Sx \to Px)$
    $\forall x(Tx \to Sx)$
    $\Box\forall x(Tx \to Sx)$

Dem.
1.                     $\neg\Box\forall x(Tx \to Sx)$
2.                     $\Box\forall x(Sx \to Px)$
3.                     $\forall x(Tx \to Sx)$

4.                     $Ta \to Sa$          3 ∀E
5.                     $Sa \to Pa$          2 Reflexivity

6.          $\neg Ta$                           $Sa$

7.    $\neg Sa$             $Pa$          $\neg Sa$            $Pa$
8.    $\neg\forall x(Tx \to Sx)$    $\neg\forall x(Tx \to Sx)$    ⊗    $\neg\forall x(Tx \to Sx)$      1 new world
9.    $\neg Tb \to Sb$          $\neg Tb \to Sb$               $\neg Tb \to Sb)$
10.   $Tb$                    $Tb$                        $Tb$
11.   $\neg Sb$               $\neg Sb$                   $\neg Sb$

12.   $\forall x(Sx \to Px)$      $\forall x(Sx \to Px)$        $\forall x(Sx \to Px)$
13.   $Sa \to Pa$            $Sa \to Pa$                 $Sa \to Pa$

14.   $\neg Sa$    $Pa$       $\neg Sa$    $Pa$           $\neg Sa$    $Pa$
15.   $\neg Sb$    $Pb$       $\neg Sb$    $Pb$           ⊗       $Pb$
       ...          ...          ...         ...         ...

    The result of this tree shows that six branches are open, which implies that the original argument is invalid. It should be noted that the invalidity of this mixed modal syllogism is due to the instantiation of the variables and not to a contradiction of the predicate letters and even less to a contradiction of the form of the argument. In other words, the branches are left open because we cannot assume that '$a$' instantiates $T$ and $S$, because if such an assumption could be made, the argument would be valid. On the other hand, in this proof, we have assumed a reflexive frame, but it must be noted that if we had assumed a S5 frame, the result would not change, and the syllogism is still invalid. Thus, it shows that Aristotle's opinion and contemporary modal logic are not in the same argumentative line.

    Aristotle's first disciples and commentators, Theophrastus and Eudemus, say that NA/N and AN/N do not respect the *peiorem* syllogistic principle saying that in every valid syllogism, the conclusion follows the weaker premise, the principle which makes valid only AN/A, where the assertoric (A) is the weakest and is inherited to the conclusion (the other premise is necessary (N) and it is strong). Alexander of Aphrodisias, the great commentator on Aristotle, agreed with Aristotle's associates. In modern times, Łukasiewicz [5, p. 191], S. McCall [7], and Englebretsen [4] have defended Aristotle though by different reasons.

What explanation can we give? Is it reasonable to believe that the father of modal logic was wrong about this relatively gross error? According to our discussion, the simplest answer is to state that Aristotle thinks that modality is a property of terms, and so he reads this modal syllogism in a *de re* modality. It is worth noting that in assuming this interpretation we are not alone, because our results agree with those given by G. Englebretsen in his [4, pp. 390–392].

## Final remarks

There are some textual remarks giving support to what has been discussed here. The first remark is that Aristotle works its modal syllogistic based on only the valid assertoric syllogisms he has already defined in previous chapters (I, 2-7). Thus, in our technical terms, modal balance can be proved only after the rules of quantitative symmetry has been confirmed. However, the fact that pure modal syllogistic determines its conclusions along with assertoric syllogistic does not make modal syllogistic superfluous. The counterexample is the mixed modal syllogistic. Moreover, Aristotle in *De Int* 12 (21b26-32) knows that truth and falsity can be modified by the modality that attains to the categorical proposition. Hence, he defends that to deny a modal proposition is to pose the negative particle before the modality, not before the verb or the predicate (*De Int* 12, 21 b 32-34).

There is also a need to call attention that Aristotle in *An Pr* I, 9, does not say that there is a problem in combining two different modalities in the premises of an assertoric syllogism. Rather, he says that sometimes happens that (*sumbainei de pote*) the syllogism that holds one necessary premise and one assertoric premise concludes necessarily. So he makes us aware of the order of combination that **N** needs to have in the set (**A, N**), since the order cannot be whatever (*plen oukh hopoteras etukhen*: *An Pr* I, 9, 16-17).

Second, Aristotle says that in the first figure, the major term in syllogism is the predicate of the conclusion. Hence, the major premise is the one that contains the predicate of the conclusion. This is the meaning of the obscure expression Aristotle adds immediately in the text ("but only in the premise containing the major term", *alla tes pros to meizon akron*). Accordingly, he assumes that the major term in the syllogism (*meizon akron*), which is both the predicate in the first figure and the predicate term in the conclusion, must be in modal balance, namely, either one having the same modality. We have extended his principle by saying that the predicate of the conclusion can have a minor modality according to the axioms he accepts in *De Interpretatione* 13 ($\Box p \to p; p \to \Diamond p$; and $\Box p \to \Diamond p$). Aristotle seems to remember it though when in *An Pr* I, 9, says that AN/N could be valid if the conclusion in N is limited to an assertoric conclusion because he is claiming here that a true necessary proposition entails a true assertoric proposition, but not vice versa.[12]

The main advantage of identifying the predicate of the conclusion to the major term is to avoid confusion in case the premises must be interchanged. What Aristotle does is define the major term as the term predicated in the conclusion. He does not say that the major term is instead the term in the major premise, but that the major term identifies with the predicate of the conclusion. Even more, this definition avoids the confusion that could arise because of Aristotle's way of writing syllogisms. He used to write an assertoric syllogism by putting the predicate first and the subject later. For instance, in *An Pr* I, 4, 37-39, he writes: [13]

16.
N $A$ is said of every $B$   (*To A kata pantos tou B*)
A $B$ is said of every $G$   (*To de B kata pantos tou G*)
N Then, $A$ is said of every $G$   (*Anagké kategoreisthai to A kata pantos tou G*).

Now, if we transform this way of expressing the syllogism into the Boethian

---

[12] According to Alexander of Aphrodisias commentary on Aristotle's *Prior Analytics* (*in An Pr* 124, 8-17), Theophrastus conceived this reduction of modality as a case of the rule later known as *peiorem*: in any valid syllogism, the conclusion must follow the weakest modality in the premises. Thus, Theophrastus' criticism of Aristotle entails that the validity of NA/N and AN/N could be restored if in these syllogisms the conclusion is in A. In Alexander's report, Theophrastus criticizes Aristotle's forgetfulness of the rule of *peiorem* that he used successfully in assertoric syllogism. [15].

[13] Among the ancient authors, Apuleius in his *Peri Hermeneias* (Thomas ed.) usd to write the syllogism in this mirror way. In modern times, Smith [11], and Mueller [8].

way,[14] the scheme will be the following and no difference will be made if the modal **N**, **A** are distributed as in the following Barbara:

17. **N** Every *B* is *A*
    **A** Every *G* is *B*
    **N** Then, every *G* is *A*.

As seen, in both schemes the predicate of the conclusion and the major term are **N**-modalized.

# References

[1] Ackrill, J.L. 1963. *Aristotle's Categories and De Interpretatione*. Translation with notes and glossary. Oxford: Clarendon Press.

[2] Alvarez, E. & Correia, M. 2012, "Syllogistic with Indefinite Terms", *History and Philosophy of Logic*, 33, 4, pp. 297-306.

[3] Busse, A. 1895. *Ammonii In Aristotelis De Interpretatione Commentarius*. (Ed.), Commentaria in Aristotelem Graeca, vol. iv, 4.6, Berlin, 1895.

[4] Englebretsen, G. 1988. "Preliminary notes on a new modal syllogistic". *Notre Dame Journal of Formal Logic* 29, 3, pp. 381-395.

[5] Łukasiewicz, J. 1951. *Aristotle's Syllogistic from the Standpoint of Modern Formal Logic*, 1st ed. Oxford: Clarendon Press.

[6] Łukasiewicz, J. 1975. *Estudios de Lógica y Filosofía*. Selección y traducción de A. Deaño, Biblioteca Revista de Occidente, Madrid: Grefol.

[7] McCall, S. 1963. *Aristotle's Modal Syllogistic*, Amsterdam: North-Holland.

[8] Mueller, I. & Gould, I. 1999. *Alexander of Aphrodisias on Aristotle's Prior Analytics 1. 14-22*. Ithaca/New York: Cornell University Press.

[9] Ross, W. D. (ed.), 1949. *Aristotle's Prior and Posterior Analytics*, Oxford: Clarendon Press.

[10] Ross, W. D. 1957. *Aristotle's Prior and Posterior Analytics*. A revised text with introduction and commentary. Oxford: Clarendon Press.

[11] Smith, R. 1989. *Aristotle's Prior Analytics*. Translation with notes. Indianapolis: Hackett.

[12] Tarski, A. 1946. *Introduction to Logic and the Methodology of Deductive Sciences*. Oxford: Oxford University Press.

---

[14] I mention especially to W.D. Ross [10] and in ancient times to Boethius' *De syllogismo categorico* p. 45,14-46,6 (Thomsen Thörnqvist ed.).

[13] Thomas, P. 1908. *Apulei Opera Quae Supersunt*, vol. 3. *Apulei Platonici Madaurensis de Philosophia Libri, Liber* ΠΕΡΙ ΕΡΜΕΝΕΙΑΣ. Thomas, P. (ed.), Leipzig: Teubner.

[14] Thomsen Thörnqvist, C. 2008. (ed.) *Anicii Manlii Seuerini Boethii De syllogismo categorico*. A critical edition with introduction, translation, notes, and indexes. Studia Graeca et Latina Gothoburgensia LXVIII, University of Gothenburg 2008: Acta Universitatis Gothoburgensis.

[15] Wallies (1883): *Alexandri in Aristotelis Analyticorum Priorum Librum I Commentarium*, M. Wallies (Ed.), in *Commentaria in Aristotelem Graeca*, vol. 2.1, Berlin 1883.

# Sommers's Syllogisms Unlimited

Wallace A. Murphree

## 1 Introduction

While once the categorical syllogism was considered central to logic it has since become relegated to a subordinate status and is now often considered a rather insignificant portion of the First Order Predicate Calculus which subsumes it. However, Fred Sommers, followed by George Englebretsen and others, have initiated a movement in which the centrality of the syllogism is revived, and the remarks that follow below are offered in support of this revival. (See Kelley [1, chapter 14] for a textbook presentation of Sommers's method.)

However those of us who wish for the revival do not wish for a restoration of the conditions that allowed its loss of status, and one great weakness I find with its traditional posture is that it was too narrow — too limited — in areas in which it might have led the way. I shall consider two areas below, namely, *Complementary Terms* and *Numerical Propositions*, and attempt to show how Sommers's approach is quite adequate in each. However, I shall begin by criticizing the traditional rules of the syllogism as a segue into these more serious issues.

## 2 Problematic Rules

Although traditional sets of rules of validity vary somewhat from text to text, the standard sets include requirements of distribution and of quality for the premises and also for the conclusions. In addition an optional rule of quantity allows the inference of particular conclusions from universal premises in five special cases.

Although these traditional rules serve perfectly to separate valid from invalid forms within the group of 256 syllogisms to which they are normally applied, their application nevertheless often seems haphazard and misleading. For example, when a syllogism is found to be in violation of a rule one would expect thereby to know what is really wrong with the syllogism and what needs to be done to fix it. And sometimes this is the case as, for example, when it is seen that in AIA-1,

MaP
SiM
SaP,

S is distributed in the conclusion but not in its premise (Illicit Process of the Minor). Then one knows to "undistributed" the S in the conclusion to change the argument into the valid AII-1:

MaP
SiM
SiP.

And surely this is the way rules should always work but, unfortunately, this is not the case. For example, in syllogism AOI-1,

MaP
SoM
SiP,

the rule of negation for the conclusion is violated; but when it is "corrected" to become AOO-1,

MaP
SoM
SoP,

it violates a distribution rule for the conclusion (Illicit Process of the Major). And such misleading signals are common in the application of the traditional rules. For example, premise sets AO_ 1, IO_1 & 2, OI_2 & 4, and OA_-4 all conform to the rules required for the premises, so if the completed syllogism is invalid it would seem the problem has to be that the wrong conclusion was adopted. However in these cases every possible conclusion stands in violation of some rule or another! For example, when considering what conclusion might follow from AO_-1, we find that...

AOA-1 violates the rule of negation and distribution for the minor,
AOE-1 violates the rule of distribution for the major and the minor,
AO I-1 violates the rule of negation, and
AOO-1 violates the rules distribution for the major.

And there are no other conclusions possible.

And, again, such is also the case with IO_1& 2, OI_2& 4, and OA_-4 as well. They conform to the rules for the premises, but simply yield no conclusion at all. In each case the syllogism is correctly judged invalid by one traditional rule or another, but the infraction(s) it is charged with is completely unrelated to the error it commits.

So although the traditional rules successfully separate the valid from the invalid 256 syllogisms to which they apply, they do so only in a heuristic way and contingent way. However, Sommers's method makes the same separation between the 256 members but does this quite systematically in a way that elucidates the logical relationship involved.

Following David Kelly's presentation, I shall refer to Sommer's technique as the Cancellation Method. As a prerequisite this method requires that either all the propositions of the argument be universal, or that one (and only one) premise and the conclusion be particular. Then the argument is valid if its conclusion is arithmetically equivalent to its premises when its middle terms are cancelled out.

## 3 Complementary Terms and the Syllogism

The 256 syllogisms to which the traditional arguments apply are selected by exhausting the possible ways the categorical propositions can be combined.

That is, first each premise is required to be one of the original A, E, I, or O forms or one of their converses. Accordingly the major premise then has to be

MaP, MeP, MiP, MoP, PaM, PeM, PiM, or PoM

while the minor must be

SaM, SeM, SiM, SoM, MaS, MeS, MiS, or MoS;

and then the conclusion must be one of the original A, E, I or O forms, but not one of their converses.

These amount to 256 (8x8x4) in all. And as was mentioned above, the traditional rules for the syllogism pick out the valid ones of these from the rest. Of these fifteen are valid as they stand while five more are valid when membership is presupposed for the appropriate term. Of course, this very same assessment of the 15 (or 20) from the 256 is made by the First Order Predicate Calculus, the Venn diagram, the ReasonLines schematics that I have developed (www.reasonlines.com) and by Sommers's Cancellation method.

Any one of these syllogisms comprises exactly three terms, namely the major (P), the middle (M), and the minor (S), and each of these occurs twice: PP, MM, and SS. Of course other terms can be added (UU, VV, etc.) to create an extended argument or sorites, but it then can be reduced to a series of syllogisms. Now for every term, $T$, there is a complementary term, non$T$, ($T'$), and tradition forbids the use of both $T$ and $T'$ in the same syllogism. However, there is no logical fallacy necessarily involved when both a term and its complement are used in the same argument; rather, then it is simply not a syllogism by definition. However, since the benefits of using both in the same argument can be substantial I suggest the syllogism be redefined to include that possibility. Perhaps rather than thinking of $T$ and $T'$ as separate terms in this context it would be better to think of "$T$ and $T'$" as composing a "term-set," and then by definition to allow a syllogism to be composed of one term from each of the three different term-sets. Then either specific member of the term-set "$T$ and $T'$" might be referred to as "distinct T" or "distinct T'."

This of course is not an attempt to "change logic," but rather to expand the reach of the syllogism to include more logic — which it should have included all along. With this stipulation, both occurrences of the major term can now be $P$ or both $P'$ or one $P$ and the other $P'$, and the same holds for the middle and minor terms. Then a syllogism would comprise 3, 4, 5, or 6 distinct terms from three different term-sets. Then when the traditional three-term requirement becomes a "three term-set" requirement (i.e., so a distinct term and its distinct complement together count as just one term-set) then the possible permutations of ingredient sentences for a syllogism are radically increased. That is, now there are eight A-form possibilities for the major premise, viz.,

$PaM, PaM', P'aM, P'aM', MaP, MaP', M'aP$, and $M'aP'$,

and, likewise, there are eight E-, I-, and O-forms, and this leaves 32 (8x4) possible major premises. Likewise, there are now 32 possible minor premises, which leave a total of 1,024 premise sets (32²). And finally, allowing also 32 possible conclusion sentences (that is, allowing converses as well as original conclusion sentences), there are then 32,768 (323) possible three-sentence, categorical argument versions in all! That is, allowing the complementary terms increases the number of syllogisms by 128 fold–from 256 to 32,768!

There are eight subgroups of this totality that are composed of only three distinct terms, and each subgroup contains 256 members. That is, there is the one subgroup that contains no complementary distinct terms at all, and seven others that contain one, two, or three complementary distinct terms as follows:

$S'MP$,
$SM'P$,
$SMP'$,
$S'M'P$,
$SM'P'$,
$S'MP'$, and
$S'M'P'$.

The traditional rules apply to each of these groups and, without existential presuppositions, pick out only 120 syllogisms in all as being valid. That is, they pick out the 15 that employ no complementary terms (the traditional set), and then 15 each from the seven other combinations of just three distinct terms. Moreover, for each set of 15, five additional forms are determined to be valid when appropriate existential presupposition is made, and this produces a total of 160 valid syllogisms in all.

However, the First Order Predicate Calculus, the Venn diagram (using all eight logical spaces), my ReasonLines schematics, and Sommers's Cancellation method pick out each of these as well. *And each of these other four methods also identify another large group of valid forms from those not judicable by the traditional rules.*

So, while the traditional rules are applicable to these eight sets of 256 (=960) they are completely inapplicable to the remaining 31,808, i.e., to those which comprise four, five, or six distinct terms. For example, the rules confirm the validity of traditional Barbara but identical arguments (i.e. made from equivalent sentences) cannot be so adjudicated. That is, the traditional rules confirm the validity of the argument presented in column 1 below (having 3 distinct terms) but not those in column 2, 3, and 4 (which have 4, 5, and 6 distinct terms respectively) although they are all equivalent. Indeed, again, they all can easily be shown to be so by the First Order Predicate Calculus, the

Venn Diagram, the ReasonLines schematics, and by Sommers's Cancellation method.

*Barbara* and equivalent argument forms

| Column 1 | Column 2 | Column 3 | Column 4 |
|---|---|---|---|
| 3 Terms | 4 Terms | 5 Terms | 6 Terms |
| $MaP$ | $MeP'$ | $MaP$ | $MaP$ |
| $SaM$ | $SaM$ | $SaM$ | $SeM'$ |
| $SaP$ | $SaP$ | $P'aS'$ | $P'aS'$ |

That is, only in column 1 above is the syllogism shown to be valid by the traditional rules while it is shown to be valid in each column by Sommers's method. Specifically, the premises and conclusions are all universals and the conclusion remains arithmetically equivalent to the premises after the middle terms cancel each other out in each case.

$$-M + P \quad -M - (-P) \quad -M + P \quad -M + P$$
$$-S + M \quad -S + M \quad -S + M \quad -S - (-M)$$
$$-S + P \quad -S + P \quad -(-P) + (-S) \quad -(-P) + (-S)$$

In addition to its reach to nonstandard syllogisms, Sommers's method has other benefits. For example, in dealing with sorites one can see at a glance whether the quantity requirement is satisfied and then one can simply cancel all the middle terms out and see if the extreme terms remaining are equivalent to the conclusion. Of course, one could derive a valid conclusion in a step-by-step fashion for a formal proof as below, but if the objective is simply to determine whether the argument is valid then just scratching out the middle terms might be enough.

1. $-V + U$ — Premise
2. $-(-V) - W$ — Premise
3. $+W - X$ — Premise
4. $-(-Y) - (-X)$ — Premise
5. $-Y - Z / + U - Z$ — Premise/Conclusion
6. $-W + U$ — 1 & 2 Cancel $-V$ & $--V$
7. $+U - X$ — 3 & 6 Cancel $W$ & $-W$
8. $-X - (-Y)$ — 4 &7 Cancel $-X$ & $--X$
9. $+U - Z$ — 5 & 8 Cancel $-Y$ & $--Y$

Furthermore, it was mentioned earlier that the traditional approach points to five additional syllogisms that are valid when existential assumptions are made for the appropriate terms, and that an optional consideration of quantity is included in its rule set to accommodate these. However, in Sommers's method, rather than adding an additional rule, the treatment involves adding an existential premise to make the presupposition explicit, and then testing the resulting sorites by the rules.

Sommers's clever way of adding the existential claim that, say "some $M$'s exists," is to enter $+M + M$ as a premise. So, for example, AAI-3 in not valid without the presupposition that some M's exists because the middle terms do not cancel out.

$$\begin{array}{r} -M + P \\ \underline{-M + S} \\ +S + P \end{array}$$

However, when $+M + M$ is added (as below) the two sets of $M$'s cancel leaving the premises and conclusion equal.

$$\begin{array}{r} +M + M \\ -M + P \\ \underline{-M + S} \\ +S + P. \end{array}$$

Sometimes the presupposed existence is carried into the conclusion, as in AAI-4. Here the required treatment involves adding the existential premise, $+P + P$, to make the presupposition that some $P$ exists explicit.

$$\begin{array}{r} +P + P \\ -P + M \\ \underline{-M + S} \\ +S + P \end{array}$$

Then the $M$'s cancel as does one set of $P$'s, but the uncanceled, presupposed $P$ then becomes the predicate of the conclusion.

## 4 Numerical Propositions and the Syllogism

I think that categorical propositions are inherently numerical. In fact, it has become customary to indicate particular quantifiers numerically, i.e., by "at least one," and there should be no difficulty in using "zero" as the quantifier for the universal propositions. That is the no quantifier for the E proposition would be the zero outright (i.e., No $S$ are $P$ = Zero $S$'s are $P$) and the all quantifier of the A proposition carries "with zero exception" in its very meaning. So the basic propositions might have been expressed as

A: $S$'s with zero exception are $P$,
E: Zero $S$ are $P$,
I: At least one $S$ is $P$, and
O: At least one $S$ is not $P$,

without any loss of meaning. Perhaps this did not occur because the early Greek and Roman number systems did not include the zero. But whatever the reason I wonder if the failure to recognize this has not stifled the development of what would have been a very comprehensive system of thought. In fact, I arranged the basic proposition forms into an expandable pattern by prefixing the modifiers of "at least," "at most," "all but $n$," and "none but $n$" to the Subject-copula-Predicate stem. (See Murphree [2].) These and their obverses are listed below. (Also see Szabolcsi[3] for an alternative treatment of the numerical approach.)

| Traditional Form | Expandable Forms |
|---|---|
| $SaP$   All $S$ are $P$ | At least (all but) zero $S$'s are $P$ |
| $SeP'$  No $S$ are non$P$ | At most (none but) zero $S$'s are non$P$ |
| $SeP$   No $S$ are $P$ | At most (none but) zero $S$'s are $P$ |
| $SaP'$  All $S$ are non$P$ | At least (all but) zero $S$'s are non$P$ |
| $SiP$   Some $S$ are $P$ | At least (none but) one $S$ is $P$ |
| $SoP'$  Some $S$ are not non$P$ | At most (all but) one $S$ is non$P$ |
| $SoP$   Some $S$ is not $P$ | At most (all but) one $S$'s are $P$ |
| $SiP'$  Some $S$ is non$P$ | At least (none but) one $S$ is non$P$ |

For what it's worth, the four traditional propositions do represent natural starting places because their claims are as great as possible. That is, the claims that more than all or that fewer than zero $S$'s are $P$ are necessarily false while, on the other hand, the claims that at least zero $S$'s are (not) $P$ are necessarily true. So both the traditional universals and the particulars stand at the head of their infinitely long lines of sequential contingent values. And the traditional logic that holds for these initial values holds just as truly for each of the logical relationships in the infinite sequences that continue from them, although they are typically not recognized as such.

In common parlance "at least," "at most," "all but" and "none but" are usually omitted when they are not needed for clarification; however, they are included systematically here even though they may seem redundant or even a bit confusing in some cases. Also it may be objected that the $A$ sentence should be "*Exactly* all but zero $S$'s are $P$" rather than "At least all but zero..."; however, since "exactly all" means "At least all" *and* "at most all" and since upon reflection it is clear that "At most all but zero $S$ are $P$" is a tautology this conjunct goes without saying; accordingly, it is the "At least all but zero..." conjunct that carries the meaning of the quantifier.

Although A and E forms with quantifiers greater than zero are not fully universal they will still be referred to as such in what follows for convenience, and the I and O forms with quantifiers greater than one will still be referred to as particulars. Then the affixed numerical values to each traditional propositions will indicate its extended orientation.

| Universals | | Particulars | |
|---|---|---|---|
| $0SaP$ | $0SeP$ | $1SiP$ | $1SoP$ |
| $1SaP$ | $1SeP$ | $2SiP$ | $2SoP$ |
| $2SaP$ | $2SeP$ | $3SiP$ | $3SoP$ |
| $3SaP$ | $3SeP$ | $4SiP$ | $4SoP$ |
| $4SaP$ | $4SeP$ | $5SiP$ | $5SoP$ |
| ... | ... | ... | ... |
| $xSaP$ | $xSeP$ | $xSiP$ | $xSoP$ |

Given this nomenclature, the traditional *Barbara* and *Darii* appear numerically as:

| *Barbara* | *Darii* |
|---|---|
| $0MaP$ | $0MaP$ |
| $0SaM$ | $1SiM$ |
| $0SaP$ | $1SiP$ |

while expanded values can be

| *Barbara* | *Expanded* | *Darii* | Expanded |
|---|---|---|---|
| $2MaP$ | $22MaP$ | $3MaP$ | $33MaP$ |
| $3SaM$ | $13SaM$ | $9SiM$ | $49SiM$ |
| $5SaP$ | $35SaP$ | $6SiP$ | $16SiP$ |

and full generality can be indicated by variables:

*Barbara General*   *Darii General*

$xMaP$   $xMaP$
$ySaM$   $ySiM$
$x+ySaP$  $y-xSiP.$

And the same expansion can be made with the same ease to every valid syllogism, so that each of them turns out to be but one of an infinite number of valid instances of that form. That is, not only do *Barbara* and *Darii* have an infinite number of instantiations, but so do *Cesare, Datisi, Bacardo, Dimaris*, and all the rest. And in addition, the same holds for each of the other categorical arguments that are not recognized at all by the traditional approach.

I find this is awesome. But perhaps of even greater interest is how this numerical expansion and Sommers's system fit so perfectly together.

| Universals | | Particulars | |
|---|---|---|---|
| $0SaP - 0S + P$ | $0SeP - 0S - P$ | $1SiP + 1S + P$ | $1SoP + 1S - P$ |
| $1SaP - 1S + P$ | $1SeP - 1S - P$ | $2SiP + 2S + P$ | $2SoP + 2S - P$ |
| $2SaP - 2S + P$ | $2SeP - 2S - P$ | $3SiP + 3S + P$ | $3SoP + 3S - P$ |
| $3SaP - 3S + P$ | $3SeP - 3S - P$ | $4SiP + 4S + P$ | $4SoP + 4S - P$ |
| $4SaP - 4S + P$ | $4SeP - 4S - P$ | $5SiP + 5S + P$ | $5SoP + 5S - P$ |
| ... | ... | ...... | ...... |
| $xSaP - xS + P$ | $xSeP - xS - P$ | $xSiP + xS + P$ | $xSoPxS - P$ |

Here the numerical quantification does not complicate the process of drawing conclusions from premises or testing the arguments for validity. Rather, they seem to be made for each other! This can be seen as the numerical renditions of Barbara and Darii above are cast into Summers's notation below:

*Barbara*   *Darii*

$-0M + P$   $-0M + P$
$-0S + M$   $+1S + M$
$-0S + P$   $+1S + P$

| *Barbara* | Expanded | *Darii* | Expanded |
|---|---|---|---|
| $-2M + P$ | $-22M + P$ | $-3M + P$ | $-33M + P$ |
| $\underline{-3S + M}$ | $\underline{-13S + M}$ | $\underline{+9S + M}$ | $\underline{+49S + M}$ |
| $-5S + P$ | $-35S + P$ | $+6S + P$ | $+16S + P$ |

$$\begin{array}{ll} \textit{Barbara General} & \textit{Darii General} \\ -xM + P & -xM + P \\ \underline{-yS + M} & \underline{+yS + M} \\ -(x+y)S + P & +(y-x)S + P \end{array}$$

Furthermore, the same holds for each of the other traditional syllogisms when it is expanded numerically, and for each other valid categorical arguments whether it is accepted by the traditional approach or not. In addition, this approach automatically works for sorites as well, as the problem below shows:

1. $-3B + A$      Premise
2. $+19B + C$      Premise
3. $-4C - D$      Premise
4. $-7(-D) + E / + 5A + E$      Premise/Conclusion
5. $+16A + C$      1 & 2 Cancel $-B$ & $+B$ (-3 +19)
6. $+12A + D$      3 & 5 Cancel $-4C$ & $+C$ (-4 +16)
7. $+5A + E$      4 & 6 Cancel $-D$ & $+D$ (-7 + 12)

Here the conclusion is derived in a step-by-step fashion as successive middle terms cancel leaving $+A$ and $+E$ and the arithmetic conclusion of five. Or, the cancelling middle terms can simply be scratched out and negative quantifiers can be summed and subtracted from the one positive number, viz: (19-(3+4+7)).

It was pointed out earlier that Sommers's system presupposes membership by introducing it as an additional premise and then treating the whole as a sorites. Again, that is one adds "there exists an $M$" for example, by entering "Some $M$ is an $M(+M + M)$," and then the traditional AAI-3 becomes the sorites:

$+M + M$
$-M + P$
$-M + S$
$+S + P$.

Here the two sets of M's cancel each other out and leave the subject and predicate terms. Of course the examples from the mere plus/minus approach work with the numerical quantifiers as well:

$$
\begin{aligned}
+M + M &= +1M + M \\
-M + P &= -0M + P \\
\underline{-M + S} &= \underline{-0M + S} \\
+S + P &= +1S + P
\end{aligned}
$$

Now with numerical propositions existential premises and conclusions can be much more flexible because one can presuppose the existence of any amount whatever. So, for example, the syllogism below corresponding to AAI-3 is invalid as it stands

$$
\begin{aligned}
-2M + P \\
\underline{-4M + S} \\
+3S + P
\end{aligned}
$$

because the middle terms do not cancel, and even if they did there would still be numerical value to deal with. But both issues can be solved by presupposing the existence of at least 9 $M$'s. Then the argument becomes valid as both sets of $M$'s cancel out and the universal quantifiers are subtracted from the presupposition.

$$
\begin{aligned}
-2M + P \\
-4M + S \\
\underline{+9M + M} \\
+3S + P
\end{aligned}
$$

That is, all but 2 of the 9 $M$'s (=7) are assigned to P, and all but 4 of the 9 $M$'s (=5) are assigned to S; so 7 plus 5 (=12) of the 9 $M$'s presupposed are assigned to S and/or P, which is to say the 3 are assigned to both. Of course the weaker conclusions of +2S+P and +1S+P are also entailed. The general formula for the strongest conclusion is:

$$
\begin{aligned}
-xM + P \\
-yM + S \\
\underline{+zM + M} \\
+z - (x + y)S + P
\end{aligned}
$$

But the amount presupposed must be greater than the numerical quantities of the combined universal premises for otherwise the conclusion would be necessarily true.

In addition to the possibility of presupposing the existence of greater memberships than ever before, the numerical quantification also admits the possibility of maximum presuppositions, and Sommers's notation makes this natural and easy. That is, whereas "there exist at least x M's" can be accommodated by entering "at least $x$ $M$'s are $M$," or "$+xM + M$," it accommodates "there exist at most $x$ $M$'s" by entering "at most $x$ $M$s are $M$," or "$-xM - M$."

Whereas the minimum presupposition is required for universals to yield a conclusion, a maximum presupposition is required for particular premises to yield a conclusion. For example, the form below corresponding to III-3 is the invalid

$$\begin{array}{r} +8M + P \\ +4M + S \\ \hline +2S + P \end{array}$$

since its middle terms don't cancel and its arithmetic doesn't add up. However, with the added premise that there exists at most 10 $M$'s

$$\begin{array}{r} +8M + P \\ +4M + S \\ -10M - M \\ \hline +2S + P \end{array}$$

the argument becomes valid. The general formula for this step is also

$$\begin{array}{r} -xM + P \\ -yM + S \\ +zM + M \\ \hline +z - (x + y)S + P. \end{array}$$

Here the maximum amount presupposed must not be less than the quantification of any particular premise for then the premises would be inconsistent. That is, for example, if a maximum presupposition of 10 $M$'s is made ($-10M - M$) then the premise that at least $11M$'s are $P$ ($+11M + P$) is inconsistent, and so is any premise with a greater quantification.

# 5 Conclusion

Earlier I charged the traditional syllogism with being too limited, and I have tried to show this to be the case. Specifically I have proposed there is no good reason to limit the count of syllogisms to 256 when there are thousands and thousands that are just as worthy of attention. Also I have tried to show that the numerical limitation of the syllogism to the interplay between zero and one is an arbitrary limitation and that the infinite body of the logic lies on the side not taken. The result has been that study of the syllogism has been so restricted that it was left no room to grow or develop. Instead it was thought to be (pretty nearly) complete early on and the best it could do was to hold its own.

But I certainly believe that this is not so, and I think Sommers's system sheds enough light on the situation to let it be clearly seen; also, I think it sheds enough light to show the direction further important development can take.

## References

[1] Kelley, D. The Art of Reasoning, expanded ed. W.W. New Norton, New York, 1990.
[2] Murphree, W. Numerically Exceptive Logic: A Reduction of the Classical Syllogism, Peter Lang, New York, 1991.
[3] Szabolsci, L. Numerical Term Logic, (Englebretsen, G. ed.) Lewiston, New York: The Edwin Mellen Press 2008.

# DISENTANGLING LEIBNIZ'S NOTIONS OF »TRUTH«

WOLFGANG LENZEN

## 1 Introduction

In a recently published essay M. Malink and A. Vasudevan ('M&V', for short) presented a comprehensive reconstruction of "The Logic of Leibniz's *Generales Inquisitiones*" ('GI', for short)[1]. The main aim of their paper was to show that Leibniz had "developed a symbolic calculus of terms that is capable of underwriting all valid modes of syllogistic and propositional reasoning" (p. 686). Somewhat more specifically, M&V's reconstruction yielded:

(1) A term logical system (called the "non-propositional fragment of Leibniz's calculus") which "is both sound and complete with respect to the class of Boolean algebras" (p. 721).

(2) A more comprehensive propositional logic (called "Leibniz's calculus") which "is both sound and complete with respect to the class of auto-Boolean algebras" (ibid.).

The "auto-Boolean" character of the system amounts to the assumption:

(3) Every true proposition $\alpha$ "coincides with 1", i.e. every true $\alpha$ is equivalent to a tautology, and every false proposition $\beta$ "coincides with 0", i.e. every false $\beta$ is equivalent to a self-contradiction (p. 710).

---

[1] Cf. Malink/Vasudevan [11]. All subsequent page-references without further bibliographical data refer to this paper.

As will be argued in this paper, result (1) *accords* with Leibniz's conception of logic, but (2) – and in particular (3) – *does not*. It is true, though, that Leibniz uses notions of truth and falsity *both* with respect to *terms and* with respect to *propositions*. On the one hand, he defines a term or *concept* $A$ as »false« whenever $A$ contains a contradiction (like $Y$ Not-$Y$); hence:

(4) Every »false« *term* $A$ is (provably) equivalent to a self-contradictory term.

Since Leibniz considers a concept $B$ as »true« if and only if ('iff', for short) $B$ is not »false«, a »true« $B$ need not be equivalent to a *tautological* term; rather:

(5) Term $B$ is »true« iff $B$ is *not equivalent* to a self-contradictory term.

Now every »normal« term $A$ (like 'man', 'animal', 'rational', etc.) is such that neither $A$ nor its negation, Not-$A$, contains a contradiction. Hence, in many cases:

(6) Both term $A$ and its opposite, Not-$A$, may be »true«.[2]

On the other hand, Leibniz's logic is clearly based upon a »classical«, two-valued understanding of the truth (and falsity) of *propositions*:

(7) Every *proposition* $\alpha$ is *either* true *or* false. Therefore, whenever $\alpha$ is true, its opposite, Not-$\alpha$ , must be false.[3]

In view of the discrepancy between (6) and (7), it seems necessary to disentangle Leibniz's notions of »truth« as applying to terms and

---

[2]M&V acknowledge this point in fn. 59: "... both *animal* and *non-animal* are true terms, i.e. neither of them contains a contradiction". Leibniz usually took it for granted that all basic terms are self-consistent. Cf., e.g., GP 7, 214: "In omnibus tamen tacite assumitur terminum ingredientem esse Ens". Parkinson translates this as follows: "In all of them, however, it is tacitly assumed that the ingredient term is an entity" (LLP, p. 119).

[3]M&V acknowledge this point when stressing that "... Leibniz subscribes to the law of bivalence in the GI: Every proposition is either true or false ..." (p. 710).

to propositions, respectively. Moreover, in order to arrive at a faithful reconstruction of Leibniz's logic, it is important to distinguish between two kinds of »principles«:

- The proper logical laws, i.e. propositions which are logically valid with respect to the semantics that Leibniz had in mind;
- Certain metalogical and metaphysical principles which guided Leibniz in developing his logical calculus.

In section 2, the historical background, on which Leibniz's logic is based, will be sketched. In section 3, the basic laws of Leibniz's »algebra of concepts« will be summarized, and in section 4, the underlying semantics will be elaborated. Section 5 is devoted to Leibniz's »Metalogic«. In the concluding section 6, Leibniz's »true« conception of the laws of propositional logic will be analysed.

## 2 The theory of the syllogism

The traditional theory of the syllogism may roughly be defined as the logic of the four categorical forms

- Universal affirmative proposition (UA) Every $S$ is $P$     $(SaP)$
- Universal negative proposition (UN):    No $S$ is $P$        $(SeP)$
- Particular affirmative proposition (PA): Some $S$ is $P$     $(SiP)$
- Particular negative proposition (PN):    Some $S$ isn't $P$   $(SoP)$.

According to the laws of *opposition*, the particular propositions are equivalent to the negations of corresponding universal propositions:

OPP 1    $SiP \leftrightarrow \neg(SeP)$
OPP 2    $SoP \leftrightarrow \neg(SaP)$.

As these formulas show, we use '$\neg$' and '$\leftrightarrow$' as symbols for propositional negation and equivalence, respectively. Furthermore, in what follows, '$\wedge$', '$\rightarrow$' and '$\rightarrow$' shall be used to symbolize conjunction, disjunction, and implication. At this moment we can leave open whether '$\rightarrow$' '$\leftrightarrow$' are to be understood in the sense of *material* or *strict* implication and equivalence, respectively.[4] Leibniz's understanding of these operators will be discussed in section 6.

The traditional laws of *subalternation* state that the universal propositions entail their particular counterparts:[5]

SUB 1   $SaP \to SiP$
SUB 2   $SeP \to SoP$.

Another group of »simple« inferences deals with the *conversion* of subject and predicate. According to traditional doctrine, the UN and the PA admit of a straightforward, »simple« conversion, while the UA can only be converted »accidentally« (i.e., by »weakening« the quality of the proposition from a universal to a particular one):

CONV 1   $SeP \to PeS$
CONV 2   $SiP \to PiS$
CONV 3   $SaP \to PiS$.[6]

The »Scholastic« theory of the syllogism differs from its »Aristotelian« counterpart mainly by the treatment of *negative* (or »infinite«) *terms*. Technically speaking, this amounts to the introduction of a new operator which, when applied to an arbitrary term $T$, yields the negative term Not-$T$. In what follows the negation of $T$ shall not be symbolized by the sign for propositional negation, '¬', but rather by means of a new symbol '∼' as '∼ $T$'. Like the negation of propositions, however, also the negation of terms satisfies the law of *double negation*:

DNEG   $\sim\sim T = T$.

Furthermore, the UA may be »converted by *contraposition*« according to law:

CONTRA   $SaP \to \sim Pa \sim S$.

---

[4]Whenever '→' is the *main operator* of a formula like OPP 1, it doesn't matter whether it is taken as a material or as a strict implication.

[5]According to the contemporary interpretation, $SaP$, i.e. $\forall x(Sx \supset Px)$, entails $SiP$, i.e., $\exists x(Sx \wedge Px)$, only under the assumption that $S$ is not empty: $\exists x(Sx)$. This issue will be discussed further below.

[6]CONV 3 is not really an extra principle of *conversion* but rather a corollary of SUB 1 (in conjunction with CONV 2).

Finally, by means of the laws of so-called *obversion*, the negative propositions $SeP$ and $SoP$ may be transformed into corresponding affirmative propositions (with a negated predicate):

OBV 1    $SeP \leftrightarrow Sa \sim P$
OBV 2    $SoP \leftrightarrow Si \sim P$.

Besides these »simple« laws there are the proper *syllogistic inferences* which lead from two premises $\alpha$, $\beta$ to a conclusion $\gamma$. The best-known examples of valid syllogisms are:

BARBARA    $MaP, SaM \Rightarrow SaP$
CELARENT    $MeP, SaM \Rightarrow SeP$
DARII    $MaP, SiM \Rightarrow SiP$
FERIO    $MeP, SiM \Rightarrow SoP$.

As Leibniz showed in "De Formis syllogismorum mathematice definiendis", all other valid moods can be derived from these »perfect« syllogisms by means of the principle of *regress*:

REGR    If a conclusion $\gamma$ *logically follows* from premises $\alpha$, $\beta$, then, if $\gamma$ is false, at least one of the premises must be false.[7]

In the course of »axiomatizing« the theory of the syllogism, Leibniz made use of the following »identities«:

ALL    $BaB$
SOME    $BiB$.

While ALL really is an »identical«, i.e. logically true proposition, SOME only holds when the term $B$ has a non-empty extension.[8]

---

[7] This principle goes back to Aristotle who, according to Kneale [4, p. 41] said that "if a number of propositions jointly entail a conclusion, then, if the conclusion is false, at least one of the premises must be false".

[8] SOME follows from ALL via SUB 1, but subalternation equally holds only if the subject term is non-empty. In addition to the affirmative identities ALL and SOME, Leibniz sometimes took corresponding "negative identities" like $(Be \sim B)$ or $(Bo \sim B)$ into account.

## 3 Expanding the theory of the syllogism into an algebra of concepts

The main achievement of Leibniz as a *logician* consists in having generalized the theory of the syllogism into a much more powerful logic of concepts. This generalization basically comprises three steps. First, Leibniz discovered a simple way to dispense with the traditional operators 'every', 'some' and 'no'. Instead of 'Every $A$ is $B$' he just says '$A$ is $B$' or '$A$ contains $B$'. The logical properties of the containment relation are easily determined. Already in 1676, Leibniz put forward the "absolute identical proposition $A$ is $A$" together with the "hypothetical identical proposition: If $A$ is $B$, and $B$ is $C$, then $A$ is $C$".[9] Hence the containment-relation is both *reflexive* and *transitive*. In what follows, we formalize this relation as

$AcB$     ('$A$ is $B$' or '$A$ contains $B$'). ][10]

The above-quoted laws thus take the form:

CONT 1    $AcA$
CONT 2    $AcB \land BcC \rightarrow AcC$.[11]

The second step of the generalization of the theory of the syllogism consists in the introduction of the operator of »addition« or *conjunction* of concepts which Leibniz usually denotes by mere *juxtaposition* in the form '$AB$'.[12] As he explained in an early draft of a logical calculus, it follows from the very meaning of conjunctive juxtaposition that $AB$ contains $A$ (and similarly that $AB$ contains $B$) because "$AB$ wants to express just this, namely that which is $A$ and which also is $B$"[13]:

---

[9] Cf. "De Elementis cogitandi" in Academy Edition (AE, for short) series VI, vol. 3, p. 506.

[10] M&V instead use the set-theoretical symbol '⊃'.

[11] These laws are the immediate counterparts of the syllogistic principles ALL and BARBARA.

[12] Only in the context of the so-called Plus-Minus-Calculus, Leibniz also used the symbol '+'.

[13] Cf. AE VI, 4, 148: "$AB$ est $A$ pendet a significatione huiusmodi compositionis literarum. Hoc ipsum enim vult $AB$, nempe id quod est $A$, itemque $B$". The editors of AE guess that the paper was written around 1678/79.

CONJ 1 $\quad ABcA$
CONJ 2 $\quad ABcB$.

Furthermore, as Leibniz pointed out in a paper of 1679, the conjunction operator is both *symmetric* and *idempotent*:

CONJ 3 $\quad AB = BA$
CONJ 4 $\quad AA = A$.[14]

Leibniz also recognized that in addition to principles CONJ 1, 2 which state that a "compound predicate can be divided into several", conversely "different predicates can be joined into one; thus if it is agreed that $A$ is $B$, and (for some other reason) that $A$ is $C$, then it can be said that $A$ is $BC$. For example, if man is an animal, and if man is rational, then man will be a rational animal".[15] Hence one gets as another law:

CONJ 5 $\quad AcB \wedge AcC \rightarrow AcBC$.[16]

In Leibniz's riper calculi (such as the GI) this law will usually be strengthened into an *equivalence*: "That $A$ contains $B$ and $A$ contains $C$ is the same as that $A$ contains $BC$" (GI, § 102; my emphasis):

CONJ 6 $\quad AcB \wedge AcC \leftrightarrow AcBC$.

In earlier drafts of a logical calculus, Leibniz preferred to use *containment* as a *primitive* operator and to define *coincidence* of concepts as mutual containment:

ID 1 $\quad A = B \leftrightarrow AcB \wedge BcA$.[17]

In later works (especially in the essays of August 1690) he preferred to take *coincidence* as *primitive* operator and to *define containment* by means of principle

CONT 3 $\quad AcB \leftrightarrow A = AB$.[18]

---

[14] Cf. GP 7, p. 222, or LLP, p. 40: "It must also be noted that it makes no difference whether you say $AB$ or $BA$, for it makes no difference whether you say 'rational animal' or 'animal rational'. The repetition of some letter in the same term is superfluous, and it is enough for it to be retained once".

[15] This quotation is from a paper of around 1679; cf. LLP, p. 40.

[16] Cf. also the variant in § 103 GI: "If $A$ is Not-$B$ and $A$ is Not-$C$, this is the same as '$A$ is Not-$B$ Not-$C$' ".

[17] Cf. § 30 GI: "That $A$ is $B$ and $B$ is $A$ is the same as that $A$ and $B$ coincide".

[18] Cf. § 83 GI: "Generally, '$A$ is $B$' is the same as '$A = AB$' ".

The third and last step of the generalization of the theory of the syllogism consists in elaborating the laws of *conceptual negation* far beyond the traditional principles DNEG and CONTRA. On the one hand, Leibniz put forward several variants of a *law of consistency* to the effect that a concept $A$ cannot *contain*, or even *coincide with*, its own negation. While M&V symbolize the negation of term $A$ as '$\overline{A}$', we continue to use instead the symbol '$\sim A$'; furthermore we use '$\mathfrak{c}\!\!\!/$' and '$\neq$' for symbolizing the negation of the relations '$c$' and '$=$'. Thus the laws of consistency take the following form:

$$\text{CONS 1}^* \quad A \mathrel{\mathfrak{c}\!\!\!/} \sim A^{19}$$
$$\text{CONS 2} \quad A \neq \sim A.^{20}$$

On the other hand, Leibniz introduced a *new operator* of *self-consistency* of concepts by defining that $A$ is *possible* iff $A$ does not contain a contradiction (like $Y$ Not-$Y$):

> $A$ Not-$A$ is a contradictory term. That which does not contain a contradictory term, i.e. $A$ Not-$A$, is possible. That which is not $Y$ Not-$Y$ is possible. (GI, § 2)

Besides the phrase '$A$ *is possible*' ("$A$ est possibile"), Leibniz uses many other locutions to express the self-consistency of $A$:

- '$A$ *is a being*' ("$A$ est ens");
- '$A$ *is a thing*' ("$A$ est res");
- '$A$ is *true*' ("$A$ est verum"), or simply
- '$A$ *is*' ("$A$ est").

While M&V symbolize $A$'s »truth« or self-consistency as '$\mathbf{T}(A)$', we prefer to take the formula

---

[19]Cf. § 43 GI: "It is false that $B$ contains Not-$B$; or $B$ does not contain Not-$B$". The '*' behind the »name« 'CONS 1' is meant to indicate that the principle is not entirely valid.

[20]Cf. § 11-13 GI: "A proposition false in itself is '$A$ coincides with Not-$A$'. From this it is inferred that it is false that Not-$A$ coincides with $A$. It is also inferred that it is true that $A$ does not coincide with Not-$A$". Similarly, § 45: "It is false that $B$ and Not-$B$ coincide".

$\mathbf{P}(A)$ ('$A$ is possible').

The trivial law that the conjunction of $A$ and Not-$A$ is impossible[21] thus takes the form $\neg \mathbf{P}(A \sim A)$.

There are several ways to render Leibniz's »definition« of the operator $\mathbf{P}$ more precise. His characterization of a possible term $A$ as "that which is not $Y$ Not-$Y$" makes use of an »indefinite concept« $Y$ which has to be interpreted as being bound by a corresponding *quantifier*. Somewhat more exactly, since $A$ is *impossible* iff *there exists* some $Y$ such that $AY \sim Y$, one has to set:

Poss 1   $\mathbf{P}(A) \leftrightarrow \neg \exists Y (A \mathfrak{c} Y \sim Y)$.

Instead of the »*indefinite* concept« $Y$, however, one may also choose an arbitrary *definite* concept, e.g. $B$:

Poss 2   $\mathbf{P}(A) \leftrightarrow A \mathfrak{c} B \sim B$.[22]

As a matter of fact, one might even put $B = A$ itself, thus obtaining the equivalence $\mathbf{P}(A) \leftrightarrow A \mathfrak{c} A \sim A$, which, in view of CONT 1, i.e. because of $A \mathfrak{c} A$, may be simplified as follows:

Poss 3   $\mathbf{P}(A) \leftrightarrow A \mathfrak{c} \sim A$.

Hence term $A$ is *possible* if and only if it *doesn't contain its own negation*. Although apparently Leibniz never realized that the condition $\mathbf{P}(A)$ is *equivalent* to the requirement $A \mathfrak{c} \sim A$, he at least noticed that *if $A$ is possible, then $A$ cannot contain its own negation*:

> It is false that $A$ contains Not-$A$; or, $A$ does not contain Not-$A$. [...] This is also evident in another way. $A$ contains $A$ (by 37); therefore, it does not contain Not-$A$, otherwise it would [contain $A$ Not-$A$ and thus] be impossible.[23]

---

[21] Cf. §§ 32-34 GI: "$B$ Not-$B$ is impossible; or, if $B$ Not-$B = C$, $C$ will be impossible. So if $A = $ Not-$B$, $AB$ will be impossible. That which contains $B$ Not-$B$ is the same as impossible; or $EB$Not-$B$ is the same as impossible".

[22] Cf. § 194 GI: "A false term is one which contains opposite terms, $A$ not-$A$. A true term is not false." Leibniz certainly was aware of the fact that $\mathbf{P}(A)$ can *alternatively* be defined by Poss 1 and by Poss 2. Cf. the related proof of the equivalence between CONT 3 and CONT 4 discussed in section 5 below.

[23] Cf. § 43 GI or LLP, p. 59. Leibniz's variable '$B$' has been replaced by '$A$'.

Hence, unlike Cons 2, Cons 1* has to be *restricted* to self-consistent terms:

Cons 1   $\mathbf{P}(A) \to A\mathfrak{c}{\sim}A$.

Similarly, the related principle $AcB \to A\mathfrak{c}{\sim}B$ only holds if $A$ is possible:

> $A$ is $B$, therefore $A$ is not Not-$B$. For let it be true that $A$ is Not-$B$, assuming that this is possible. Now, $A$ is $B$ (by hypothesis), therefore $A$ is $B$ Not-$B$, which is absurd.[24]

Thus, one obtains as another law of consistency:

Cons 3   $B(A) \to (AcB \to A\mathfrak{c}{\sim}B)$.

Let it be noted in passing that the propositional counterparts of Cons 1, 3, i.e. formulas $\Diamond \alpha \to \neg(\alpha \to \neg\alpha)$ and $\Diamond \alpha \to ((\alpha \to \beta) \to \neg(\alpha \to \neg\beta))$, are particularly interesting for so-called connexive logic which is usually characterized by adopting a non-classical axiom like "Aristotle's thesis", $\neg(\alpha \to \neg\alpha)$, or "Boethius' thesis", $(\alpha \to \beta) \to \neg(\alpha \to \neg\beta)$.[25]

As was mentioned earlier, within Leibniz's logic of terms the categorical forms can be represented by the following schema:

*Schema 1*

UA   $AcB$        UN   $Ac \sim B$
PA   $A\mathfrak{c}{\sim}B$   PN   $A\mathfrak{c}B$.

The crucial »law« of consistency, $AcB \to A\mathfrak{c}{\sim}B$, therefore expresses the same as the »law« of subalternation, Sub 1: If the UA, $AcB$, is true, then the UN, $Ac \sim B$ is false, and hence the PA, $A\mathfrak{c}{\sim}B$, is true. From the perspective of modern 1st order logic, the validity of this inference presupposes that the subject $A$ is *non-empty*. From the perspective of the Leibnitian calculus, however, the same inference »only« presupposes that the subject $A$ is *self-consistent*! This point will be further clarified in section 4.

To conclude our discussion of the operator $\mathbf{P}$, let it be noted that Leibniz never seriously considered the following principle:

---

[24]Cf. § 91 GI or LLP, p. 68.
[25]Cf. Lenzen [8, 9].

Poss 4   $(A \sim A)cB$.[26]

This law (which M&V, p. 708, refer to as the "law of explosion") is the *term-logical* counterpart of what in *propositional* logic is called "ex contradictorio quodlibet":

ECQ   $\alpha, \neg\alpha \Rightarrow \beta$.

The latter inference which leads from a contradictory pair of *propositions* $\langle\alpha, \neg\alpha\rangle$ to an arbitrary conclusion $\beta$ was well known in medieval logic, but Leibniz questioned its validity. In his excerpts from Caramuel's book *Leptotatos* he considered the "curious argument" by means of which the conclusion 'A circle has four angles' is derived from the premises 'Peter is running' and 'Peter is not running'[27]. Although the deduction is based on two impeccable formal principles[28], Leibniz thought it would be a fallacy ("Videtur esse sophisma"). Anyway, counterpart Poss 4, according to which the conjunction of the contradictory *terms* $A$ and $\sim A$ contains any term $B$, can easily be *derived* from another principle which Leibniz had set up himself, *viz.*:

Poss 5   $AcB \leftrightarrow \neg\mathbf{P}(A \sim B)$.[29]

Because of the "vital role that this principle plays in Leibniz's calculus", M&V dubbed it "Leibniz's principle" (p. 704). Given the validity of "Leibniz's principle", formula Poss 4 is easily proved, for it becomes equivalent to $\neg\mathbf{P}((A \sim A) \sim B)$. However, since $((A \sim A) \sim B)$ contains $(A \sim A)$, it immediately follows that $((A \sim A) \sim B)$ is *impossible*!

Leibniz sometimes felt tempted to interpret the »second order« concept 'is possible' as an »ordinary« concept, i.e. to understand '$A$ est Ens' as if the subject $A$ would *contain* the predicate 'Ens'. *If* this conception were correct, one might transform the formula $A$cEns (by means of principle CONT 3) into the equation $(A = A$ Ens$)$. Leibniz actually

---

[26] This principle entails that any two inconsistent terms coincide with each other. Clearly, if, for arbitrary $B$, $(A\sim A)cB$, then also $A \sim A)c \sim B$, hence $(A \sim A)c(B \sim B)$. The converse $(B \sim B)c(A \sim A)$ follows in the same way; thus altogether one obtains $(A \sim A) = (b \sim B)$.

[27] Cf. AE VI 4, p. 1334-1343.

[28] Caramuel uses the inference schemata: $\alpha \Rightarrow (\alpha \vee \beta)$ and $(\alpha \rightarrow \beta), \neg\alpha \Rightarrow \beta$.

[29] In the concluding § 200 GI, Leibniz formulated a variant of Poss 5 (with $\sim B$ replacing $B$) as follows: "If I say $AB$ *is not* [*possible*], it is the same as if I were to say $A$ *contains Not-B*".

considered this formula in "Difficultates quaedam logicae" as a representation of condition $\mathbf{P}(A)$.[30] However, this attempt is bound to fail! In general, for every $A$, either $A$ or $\sim A$ is self-consistent. If the self-consistency of $A$ and of $\sim A$ could correctly be expressed by '$A$cEns' and '($\sim$ self-contradictory concept $A \sim A$ were self-consistent! For, according to simple laws of conjunction, $(A \sim A)cA$ and $(A \sim A)c \sim A$; hence, by the transitivity of '$c$', one would obtain $(A \sim A)c$ Ens!

## 4 Extensional semantics for Leibniz's »intensional« algebra of concepts

As the foregoing example has illustrated, a rational reconstruction of Leibniz's logic must not take every law and principle, which Leibniz put forward in his writings, at face value. Some of his »laws« have to be rejected because they immediately lead to a contradiction (such as the attempt to interpret '$A$ est Ens' as a predication of the form '$A$cEns'). Others do not necessarily lead to outright inconsistencies but stand in conflict with other fundamental principles of Leibniz's logic. E.g., in the GI and in several other papers, Leibniz considered the »law« that '$A$ is not $B$' ("$A$ non est $B$") would be the same as '$A$ is Not-$B$' ("$A$ est non-$B$"):

NEG 1\*    $A \mathfrak{c} B \leftrightarrow Ac \sim B$.[31]

However, Leibniz himself recognized more than once that this »law« cannot be valid. On the one hand, the implication from right to left only

---

[30]Cf. GP 7, 213-4, where Leibniz first »translates« the condition '$AB$ is impossible' into the inequality $AB \neq AB$ Ens: "$AB$ est non Ens, etiam sic exprimi poterit: non aequivalent $AB$ et $AB$ Ens." Next he »translates« the condition '$AB$ is possible' (which may be viewed as a formalization of the PA 'Some $A$ is $B$') into the equation '$AB = AB$ Ens': "Et similiter: quoddam $A$ est $B$, id est $AB$ est Ens, etiam sic exprimi poterit $AB$ et $AB$ Ens aequivalent." A bit later, the tacit assumption that the subject term $A$ is self-consistent is »translated« as follows: "quia $A$ et $A$ Ens aequivalent ex hypothesi". Another inadequate conception of 'est Ens' or 'est res' may be found in § 155 GI where Leibniz formalizes the assumption $\mathbf{P}(A) \wedge \mathbf{P}(B)$ by the somewhat awkward equations "$A = R$ et $B = (R)$ (seu $A$ et $B$ sunt res)".

[31]Cf. § 82 GI: "It is also possible to say that '$B$ isn't $A$' is the same as '$B$ is not-$A$' ". The long story of Leibniz's cardinal mistake of mixing up '$A$ isn't $B$' and '$A$ is Not-$B$' has been analysed in Lenzen [6].

holds when the subject $A$ is *possible* (see CONS 3 above). On the other hand, the validity of the converse implication (from left to right) requires a stronger presupposition: the subject $A$ not only has to be *consistent*, it has to be »*maximally consistent*«, i.e. $A$ must be an *individual concept*. Already in the "Calculi Universalis Investigationes" of 1679, Leibniz had pointed out:

> [...] if two propositions are given with exactly the same *singular* [!] subject, where the predicate of the one is contradictory to the predicate of the other, then necessarily one proposition is true and the other is false. But I say: *exactly the same [singular] subject*, for example, 'This gold is a metal', 'This gold is a not-metal' (cf. AE VI, 4, 217-218).

The crucial issue here is that NEG 1* holds only for individual concepts like, e.g., 'Apostle Peter', but not for *general concepts* as, e.g., 'man'. The text-critical apparatus of AE reveals that Leibniz was somewhat diffident about this point. He began to illustrate the above rule by the correct example "if I say 'Apostle Peter was a Roman bishop', and 'Apostle Peter was not a Roman bishop'" and then he went on to generalize this law erroneously for arbitrary terms: "or if I say 'Every man is learned' 'Every man is not learned'." Finally, he noticed this error: "Here it becomes evident that I am mistaken, for this rule is not valid."

According to *Schema 1*, '$AcB$' and '$Ac\sim B$' represent the UA and UN, respectively. Therefore, the implication $Ac\sim B \to AcB$ amounts to the assumption that if the UN is false, i.e. if the PA is true, then the UA is true. As Leibniz noted in § 92 GI, this *converted* inference of subalternation clearly is invalid:

> The inference 'If $A$ is not Not-$B$, then $A$ is $B$' is invalid. That is, it is indeed false that every animal is a not-man, but it does not follow from this that every animal is a man.

In order to arrive at a truthful, consistent reconstruction of the Leibnitzian calculus, it seems indispensable first to elaborate the *semantics* of the algebra of concept which Leibniz himself had in mind. Let us start with the basic relation of conceptual containment, $AcB$! The choice of the expression 'contains' derives from the so-called »intensional« point of view which Leibniz explained, e.g., in the *New Essays concerning Human Understanding*:

> The common manner of statement concerns individuals, whereas Aristotle's [manner] refers rather to ideas or universals. For when I say *Every man is an animal* I mean that all the men are included among all the animals; but at the same time I mean that the idea of animal is included in the idea of man. 'Animal' comprises more individuals than 'man' does, but 'man' comprises more ideas or more attributes: one has more instances, the other more degrees of reality; one has the greater extension, the other the greater intension.[32]

With the help of the set-theoretical relations '$\subseteq$' and '$\supseteq$', the law of *reciprocity* of extension and »intension« can be symbolized as follows:

REZI 1   $\text{Ext}(A) \subseteq \text{Ext}(B) \leftrightarrow \text{Int}(A) \supseteq \text{Int}(B)$.[33]

This law entails that two concepts which have the same extension also have the same intension:

REZI 2   $\text{Ext}(A) = \text{Ext}(B) \rightarrow \text{Int}(A) = \text{Int}(B)$.

In contemporary philosophy of language, the intension of a linguistic expression is usually interpreted as its *meaning*. Given *this understanding*, principle REZI 2 becomes *invalid* because one can easily construct

---

[32]Cf. Book IV, ch. XVII, § 8 of *Nouveaux Essais de l'Entendement*, i.e. GP 5, p. 469.

[33]A referee of another paper of mine was puzzled by the use of the *set-theoretical* operator '$\subseteq$' for formalizing the law of reciprocity. He was wondering: "What properly means that $\text{Int}(A)$ *set-theoretically* includes $\text{Int}(B)$? How can this be possible?" Well, the answer is: If one follows the traditional approach as described, e.g., in Arnauld & Nicole's *Logique de Port Royal* and »defines« the intension (or »comprehension«) of a concept $A$ as the *set* of all »*attributes*« which are contained in $A$, then $\text{Int}(A) \supseteq \text{Int}(B)$ iff for every attribute $C$: If $C \in \text{Int}(B)$, then $C \in \text{Int}(A)$, i.e. whenever $B$ contains $C$ (in the sense of $BcC$), then $A$ also contains $C$. Cf. Arnauld/Nicole [1, p. 59]: "J'appelle *comprehension* de l'idée, les attributs qu'elle enferme en soi, & qu'on ne lui peut ôter sans la détruire, comme la comprehension de l'idée du triangle enferme extension, figure, trois lignes, trois angles, & l'égalité de ces trois angles à deux droits, &c.".

predicates $A$ and $B$ which have the same extension but not the same intension. To quote a famous example of Quine, it seems plausible to assume that (at least on our planet) all animals with a heart have a kidney, and vice versa.[34] Therefore the predicates 'animal with heart' and 'animal with kidney' have the same extension, while their meanings differ widely. But this discrepancy between the modern and the traditional conception of intension fails to justify the verdict of Couturat who thought that the »intensional« treatment was bound to remain confused and vague and at any rate inferior to the extensional treatment as invented by Boole.[35] As Leibniz never got tired of stressing, the logical relations between concepts can be conceived *alternatively* in an extensional and in an »intensional« way.[36]

The intended extensional interpretation of conceptual *conjunction*, $AB$, is uncontroversial. Something, $x$, has the conjunctive property $AB$ iff $x$ has both properties $A$ and $B$; hence the extension of $AB$ is just the *intersection* of the extensions of $A$ and of $B$:

SEM 1   $\text{Ext}(AB) = \text{Ext}(A) \cap \text{Ext}(B)$.

Similarly, according to Leibniz, an object $x$ has the »privative« or negative property $\sim A$ iff $x$ fails to have the positive property $A$. Hence the extension of $\sim A$ is the set-theoretical *complement* of the extension of $A$:

---

[34] Cf. [14, p. 21].

[35] Cf. [2, p. 387].

[36] Cf., e.g., "Elementa Calculi" of April 1679 translated in LLP, p. 20-21: "For example, the concept of gold and the concept of metal differ as part and whole; for in the concept of gold there is contained the concept of metal and something else – e.g. the concept of the heaviest among metals. Consequently, the concept of gold is greater than the concept of metal. (12) The Scholastics speak differently; for they consider, not concepts, but instances which are brought under universal concepts. So they say that metal is wider than gold, since it contains more species than gold, and if we wish to enumerate the individuals made of gold on the one hand and those made of gold on the other, the latter will be more than the former, which will therefore be contained in the latter as a part in the whole. [...] However, I have preferred to consider universal concepts, i.e. ideas, and their combinations, as they do not depend on the existence of individuals. So I say that gold is greater than metal, since more is required for the concept of gold than for that of metal [...]. Our language and that of the Scholastics, then, is not contradictory here, but it must be distinguished carefully".

SEM 2   $\text{Ext}(\sim A) = \overline{\text{Ext}(A)}$.

As a corollary it follows that the extension of the contradictory concept $A \sim A$ is the intersection of $\text{Ext}(A)$ and $\text{Ext}(\sim A) = \overline{\text{Ext}(A)}$; this intersection, however, is just the empty set:

SEM 3   $\text{Ext}(A \sim A) = \varnothing$.

According to REZI 1, the »intensional« statement '*A contains B*' is true iff the *extension* of $A$ is *included* in the extension of $B$:

SEM 4   $AcB = \text{t iff } \text{Ext}(A) \subseteq \text{Ext}(B)$.[37]

According to *Schema 1*, the UA and the UN can be represented by '$AcB$' and by '$Ac\sim B$', respectively, while the PA and the PN take the form '$A\mathfrak{c}\sim B$' and '$A\mathfrak{c}B$'. Altogether, then, rules SEM 1-4 entail the following truth-conditions:

*Schema 2*

| UA | $\text{Ext}(A) \subseteq \text{Ext}(B)$ | UN | $\text{Ext}(A) \cap \text{Ext}(B) = \varnothing$ |
|---|---|---|---|
|    | i.e. $\forall x(Ax \supset Bx)$ |    | i.e. $\neg \exists x(Ax \wedge Bx)$ |
| PA | $\text{Ext}(A) \cap \text{Ext}(B) \neq \varnothing$ | PN | $\text{Ext}(A) \not\subseteq \text{Ext}(B)$ |
|    | i.e. $\exists x(Ax \wedge Bx)$ |    | i.e. $\exists x(Ax \wedge \neg Bx)$. |

*Schema 2* shows that the extensional interpretation of the »intensional« logic of concepts, as Leibniz himself had it in mind, widely agrees with the contemporary interpretation of the categorical forms. The only difference concerns the interpretation of the underlying *universe of discourse*, **U**. While in modern 1st order logic, **U** is normally assumed to be a set of *existing* individuals, in Leibniz's opinion **U** should rather be viewed as the set of all *possible* individuals.[38] Somewhat more exactly, in §§ 144-148 GI Leibniz explained that all (categorical) propositions can be interpreted either »existentially« or »essentially«, i.e. either ranging over the narrower set of *actually existing* individuals or over the larger set of *possible individuals*. Furthermore, all categorical propositions can be transformed from the ordinary conception "tertii adjecti" (where the predicate $P$ takes the *third* position as in '$S$ is $P$') into a conception

---
[37] As a corollary it follows that concepts $A$, $B$ are "the same" iff their extensions coincide: $A=B = \text{t iff } \text{Ext}(A) = \text{Ext}(B)$.

[38] This view if defended above all in "Difficultates quaedam logicae" – cf. GP 7, 211-217 or LLP, p. 115-121.

"secundi adjecti" (where $P$ takes the *second* place and the »copula« will be put behind $P$):

> (145) From every proposition *tertii adjecti* a proposition *secundi adjecti* can be made, if the predicate is compounded with the subject into one term and this is said to be or to exist, i.e. is said to be a thing, either an arbitrary or an actually existing one.[39]
>
> (146) The particular affirmative proposition 'Some $A$ is $B$', transformed into a proposition *secundi adjecti*, will be '$AB$ is', i.e. $AB$ is a thing, namely either a possible or an actual existing one, depending on whether the proposition is essential or existential.[40] [...]
>
> (148) The particular negative proposition 'Some $A$ is not $B$' will be transformed into a proposition *secundi adjecti* as follows: '$A$ Not-$B$ is', i.e. $A$ which is not $B$ is a certain thing, - possible or actual, depending on whether the proposition is essential or existential.

Hence the »*essential*« reading of a PA or PN only requires that there is a *possible* individual $x$, while the »*existential*« reading requires that there is an *actually existing* individual $x$ which has the corresponding properties. In order to formalize this distinction, one might introduce besides the operator $\mathbf{P}(A)$, which expresses that $A$ is a possible thing ("res possibilis"), another operator, $\mathbf{R}(A)$, expressing that $A$ is an actually existing thing ("res actualiter existens"). Furthermore one might introduce a certain distinguished term, say $E$, to designate the *concept* of *existence* such that its extension is just the *subset* $\mathbf{E}$ of all actually existing individuals:

SEM 5   $\text{Ext}(E) = \mathbf{E}$.

---

[39] Cf. the original version (AE VI, 4, 780) vs. Parkinson's translation which simplifies Leibniz's "esse vel existere, hoc est [...] esse res sive utcunque, sive actu existens" somewhat misleadingly into: "to exist, i.e. [...] to be a thing, whether in any way whatsoever, or actually existing" (LLP, p. 81).

[40] Here, again, Parkinson translates Leibniz's condition "$AB$ est, hoc est, $AB$ est res nempe vel possibilis vel actualis" a bit misleadingly as: "$AB$ exists [!], i.e. $AB$ is a thing – either possible or actual".

It then turns out that the operator **R** can be defined as follows:

EXIST 1   $\mathbf{R}(A) \leftrightarrow \mathbf{P}(AE)$.

According to POSS 3, $A$ is possible iff $A$ doesn't contain its own negation. Thus it follows from SEM 2, 4 that $\mathbf{P}(A)$ is true iff the extension of $A$ is not included in its own complement, which means that $A$'s extension (within domain **U**) is non-empty:

SEM 6   $\mathbf{P}(A) = $ t iff $\text{Ext}(A) \neq \varnothing$.

In other words, concept $A$ is possible iff there is a *possible* individual which has property $A$! Similarly, $\mathbf{R}(A)$, i.e. $\mathbf{P}(AE)$, is true iff the extension of the conjunctive concept $AE$ is not empty. But $\text{Ext}(AE) = \text{Ext}(A) \cap \text{Ext}(E)$, and $\text{Ext}(E)$ itself is the set of all *actually existing* individuals, **E**. Hence $\text{R}(A)$ is true iff there exists an actually existing individual which has property $A$:

SEM 7   $\mathbf{R}(A) = $ t iff $\text{Ext}(A) \cap \mathbf{E} \neq \varnothing$.

Altogether, then, one obtains the following representation of the "essential" version of the categorical forms by means of propositions "secundi adjecti":

*Schema 3*

| | | | |
|---|---|---|---|
| PA$_{\text{ess}}$ | $\mathbf{P}(AB)$ | PN$_{\text{ess}}$ | $\mathbf{P}(A{\sim}B)$ |
| | i.e. $\exists x(Ax \wedge Bx)$ | | i.e. $\exists x(Ax \wedge \neg Bx)$ |
| UA$_{\text{ess}}$ | $\neg\mathbf{P}(A{\sim}B)$ | UN$_{\text{ess}}$ | $\neg\mathbf{P}(AB)$ |
| | i.e. $\neg\exists x(Ax \wedge \neg Bx)$ | | i.e. $\neg\exists x(Ax \wedge Bx)$.[41] |

The "existential" version can accordingly be formalized as follows:

*Schema 4*

| | | | |
|---|---|---|---|
| PA$_{\text{exist}}$ | $\mathbf{R}(AB)$ | PN$_{\text{exist}}$ | $\mathbf{R}(A{\sim}B)$ |
| | $\exists x(Ex \wedge Ax \wedge Bx)$ | | $\exists x(Ex \wedge Ax \wedge \neg Bx)$ |
| UA$_{\text{exist}}$ | $\neg\mathbf{R}(A{\sim}B)$ | UN$_{\text{exist}}$ | $\neg\mathbf{R}(AB)$ |
| | $\neg\exists x(Ex \wedge Ax \wedge \neg Bx)$ | | $\neg\exists x(Ex \wedge Ax \wedge Bx)$.[42] |

---

[41] This schema was formulated in § 151 GI as follows: "We have, therefore, propositions *tertii adjecti* reduced as follows to propositions *secundi adjecti*: 'Some $A$ is $B$ gives '$AB$ is a thing'- 'Some $A$ is not $B$' gives '$A$ Not-$B$ is a thing'. 'Every $A$ is $B$' gives '$A$ Not-$B$ is not a thing'. 'No $A$ is $B$' gives '$AB$ is not a thing'."

[42] The latter conditions can be transformed into $\forall x(Ex \wedge Ax \supset Bx)$ and $\forall x(Ex \wedge$

In the standard historiography of logic, it is often maintained that Leibniz's term logic would be *much weaker* than Boolean algebra.[43] In contrast, M&V argued that already the "non-propositional fragment of Leibniz's calculus [...] is both sound and complete with respect to the class of Boolean algebras"[44]. As regards *soundness*, it is easy to verify that the laws of conjunction, containment, negation, and possibility (as summarized in section 3) are *validated* by the extensional semantics reconstructed in this section. Furthermore, as regards *completeness*, it had been shown in Lenzen [5] that Leibniz's »intensional« algebra of concepts is equivalent (or isomorphic) to the ordinary extensional Boolean algebra of sets. This proof of completeness, however, presupposed the validity of certain laws or inferences of *propositional logic*, a point which has become a target of M&V's criticism and which shall be discussed in section 6 below.

## 5 Leibniz's philosophy of logic

Before dealing with Leibniz's propositional logic, we have to survey the underlying *metalogical* and *metaphysical* principles. As was mentioned in section 1, Leibniz always adhered to the »classical«, *two-valued* conception of logic. The *principle of bivalence* can be summarized by the statement that each proposition is *either* true *or* false.[45] Alternatively, one may split up this requirement into two sub-principles. The *law of*

---

$Ax \supset \neg Bx$); hence they correctly express that all *actually existing* $A$ *are* $B$ (or are not-$B$), respectively.

[43]Cf. in particular the verdict in [4, p. 337]: "When he began, he intended, no doubt, to produce something wider than traditional logic. [...] But although he worked on the subject in 1679, in 168[6] and in 1690, *he never succeeded in producing a calculus which covered even the whole theory of the syllogism*" (my emphasis).

[44]Cf. thesis (1) in section 1 above.

[45]The 'either ... or' is to be understood as an *exclusive* disjunction. Ishiguro [3, p. 49] pointed out that Leibniz probably would have admitted exceptions from this principle, especially with regard to propositions such as 'the greatest number is even' which don't have a (well defined) subject. As a matter of fact, Leibniz was inclined to think that propositions with a self-contradictory subject either don't have any truth-value at all or should generally be regarded as false. Thus in §§ 152-155 GI he considered to restrict even identities such as $A = A$ or $AB = BA$ to the case where $A$ (or $AB$) is possible!

*excluded middle* ("tertium non datur") says that *there is no third* truth-value besides True and False. Hence if $\alpha$ is not true, $\alpha$ is false, and if $\alpha$ is not false, $\alpha$ is true. Using '**T**' and '**F**' as metalinguistic predicates of truth and falsity, this principle can be formalized as follows:

T<sub>ND</sub>  For every proposition $\alpha$ : $\mathbf{T}(\alpha)$ or $\mathbf{F}(\alpha)$.

In addition to this law, the proper "law of contradiction", or better: *Law of excluded Contradiction*, says that no proposition is *both* true *and* false:

L<sub>OEC</sub>  For no proposition $\alpha$ : Both $\mathbf{T}(\alpha)$ and $\mathbf{F}(\alpha)$.

In a paper of around 1686, Leibniz defended these principles as follows:

> Above all I assume that every proposition (i.e., both an affirmation and a negation) is either true or false. And if an affirmation is true, then the negation is false; if the negation is true, then the affirmation is false. If it is negated (in a true way) that something is true, then it is false; and if it is negated that something is false, it is true. [...] Similarly, if it is true that something is false, or if it is false that something is true, then it is false; if it is true that something is true, and if it is false that something is false, then it is true. All these assertions use to be gathered under the one name *Principle of contradiction*.[46]

Hence for Leibniz the "principle of contradiction" comprises besides T<sub>ND</sub> and L<sub>OEC</sub> also the corollaries:

C<sub>OR</sub> 1  $\mathbf{T}(\mathbf{F}(\alpha))$ iff $\mathbf{F}(\mathbf{T}(\alpha))$ iff $\mathbf{F}(\alpha)$
C<sub>OR</sub> 2  $\mathbf{T}(\mathbf{T}(\alpha))$ iff $\mathbf{F}(\mathbf{F}(\alpha))$ iff $\mathbf{T}(\alpha)$.

---

[46]Cf. AE VI, 4, p. 804: "Ante omnia assumo Enuntiationem omnem (hoc est affirmationem aut negationem) aut veram aut falsam esse, et quidem si vera sit affirmatio, falsam esse negationem; si vera sit negatio, falsam esse affirmationem. Quod verum esse negetur, (vere scilicet) falsum esse; et quod falsum esse negetur, verum esse [...] Similiter, quod falsum esse verum sit, aut verum esse falsum sit, id falsum esse; quod verum esse verum sit, et quod falsum esse falsum sit, verum esse. Quae omnia sub uno nomine *Principii contradictionis* comprehendi solent."

These principles express that the truth-predicate is in a certain way *redundant*; and they also contain a metalinguistic version of the principle of *double negation*. Anyway, Leibniz's propositional operator 'non' satisfies the ordinary truth-table:

| $\alpha$ | $\neg\alpha$ | $\neg\neg\alpha$ |
|---|---|---|
| T | F | T |
| F | T | F |

Moreover, the principle of contradiction constitutes the basis for Leibniz's conception of *valid inferences* and for his acceptance of the inference scheme of so-called »regress«:

VALID   An inference from premises $P_1,..., P_n$ to the conclusion $Q$ is valid iff it can't be the case that all $P_i$ are true and yet $Q$ is false.[47]

REGRESS   If the conclusion $Q$ logically follows from premises $P_1, ..., P_n$, then, if $Q$ is false, at least one of the $P_i$ must be false.[48]

In the special case where the number of premises is just 1, VALID yields the following criterion:

ENTAIL   Proposition $\alpha$ logically implies (or entails) proposition $\beta$ iff it can't be the case that $\alpha$ is true and yet $\beta$ is false.

Similarly, for n=1, REGRESS yields the inference schema of *modus tollens*:

---

[47] In connection with his development of a semantics of »*characteristic numbers*«, Leibniz incidentally formulated the following criterion for the validity of an inference: "Si nosse volumus an aliqua [ratiocinatio] procedat vi formae, videmus an contradictorium conclusionis sit compatibile cum praemissis, id est an numeri [characteristici] reperiri possint satisfacientes simul praemissis et contradictoriae conclusionis; quodsi nulli reperiri possunt concludet argumentum vi formae" (AE VI, 4, 256).

[48] Cf., e.g., "De Formis Syllogismorum Mathematice Definiendis" of around 1684: "In Regressu utimur hoc principio quod conclusione existente falsa (hoc est contradictoria ejus existente vera), et una praemissarum existente vera, altera praemissarum necessario debeat esse falsa, seu contradictoria ejus debeat existere vera. Supponit ergo Regressus principium contradictionis" (AE VI, 4, 499).

TOLLENS  If $\alpha$ logically entails $\beta$ and if $\beta$ is false, then $\alpha$ must be false.

In her exposition of "Leibniz's Logic and Philosophy", Ishiguro [3, p. 48] attributed to Leibniz the view that the entire logic would be

> [...] based on the 'Principle of Identity' or [!] 'Principle of Contradiction' which says that '$A$ is $A$' or '$A$ is not not-$A$'. All proofs proceed by reducing propositions to identities [...] or contradictions.

Here, however, *three different* principles are conflated: that of *contradiction*; that of *identity*, and that of *provability*. Ishiguro tried to justify the identification of the former two by a quotation from Leibniz's discussion of the fundamental principles of Descartes' philosophy. Since Descartes had emphasized the epistemological importance of the "Cogito [ergo sum]" as a "*first truth*", Leibniz pointed out to the distinction between *truths of reason* and *truths of fact*. While the "Cogito" may be accepted as a "first truth" *of fact*, there are also "first truths" *of reason*:

> The first of the truths of reason is the principle of contradiction, or, what comes to the same thing, that of identity, as also Aristotle rightly observed.[49]

This incidental remark may hardly be taken as a proof that the principle of contradiction was really *the same* as the "principle of identity". For Leibniz, the principle of *contradiction* basically coincides with the principle of *bivalence*. In contrast, the principle of *identity* specifies that, because of their logical form, certain propositions are "first truths of reason":

> The first true propositions, however, are those which are commonly called *identicals*, such as '$A$ is $A$', 'Not $A$ is not $A$'; 'if the proposition $L$ is true, it follows that the proposition $L$ is true' [etc.].[50]

---

[49]Cf. GP 4, 357: "Veritatum rationis prima est principium contradictionis vel quod eodem redit identicorum, quemadmodum et Aristoteles recte animadvertit."

[50]Cf. AE VI, 4, 804: "Sunt autem verarum propositionum primae quae vulgo dicuntur *identicae*, ut $A$ est $A$, non $A$ est non $A$; si vera est propositio $L$, sequitur quod vera est propositio $L$".

Moreover, Ishiguro's formulation of the principle of identity "'$A$ is $A$' or '$A$ is not not-$A$'" remains in need of some clarifications. First, the »copula« 'is' may be understood either as expressing the *predication* '$A$ is $A$', i.e. '$A$ contains $A$', or as expressing a real *identity*, '$A = A$':

IDEN 1  (a) Each proposition of the form '$AcA$' is logically true;
(b) Each proposition of the form '$A = A$' is logically true.

Both »halves« of IDEN 1, of course, are valid. Second, although in the quoted passage Leibniz himself listed only »positive« identities like '$A$ is $A$' and 'Not-$A$ is Not-$A$', he certainly would have subscribed to the following principle which takes »negative identities« of the form '$A$ isn't Not-$A$' into account:

IDEN 2  (a) Each proposition of the form '$Ac\sim A$' is logically false;
(b) Each proposition of the type '$A = \sim A$' is logically false.

As was shown in section 3, however, principle IDEN 2(a) is not entirely valid but has to be restricted to *self-consistent* terms!

Furthermore, as Leibniz's example with "proposition $L$" shows, one may consider related principles where the variable does not stand for a *term*, but for a *proposition*. In this case, again, two interpretations have to be distinguished. In general, the »identity« or »coincidence« of two propositions means that $\alpha$ and $\beta$ are *logically equivalent*. In an analogous way the »containment« relation between propositions has to be understood in the sense that $\alpha$ entails $\beta : (\alpha \to \beta)$. The counterparts of IDEN 1, 2 thus have to be formulated as follows:

IDEN 3  (a) Each proposition of the form $(\alpha \to \alpha)$ is logically true.[51]
(b) Each proposition of the form $(\alpha \leftrightarrow \alpha)$ is logically true.

IDEN 4  (a) Each proposition of the form $(\alpha \to \neg\alpha)$ is logically false;
(b) Each proposition of the form $(\alpha \leftrightarrow \neg\alpha)$ is logically false.

Although Leibniz doubtlessly would have subscribed to these principles, IDEN 4(a), again, is valid only for self-consistent propositions. For, according to ECQ, $(\alpha \wedge \neg\alpha)$ entails *every* proposition $\beta$, hence it entails

---

[51] In Leibniz's example this corresponds to the tautological conditional: 'if $L$ (is true), then $L$ (is true)'.

in particular $\beta = \neg(\alpha \wedge \neg\alpha)$. Thus, *self-inconsistent* propositions entail their own negation.

Ishiguro saw an immediate connection between the principles of contradiction, the principle of identity, and the assumption that "all proofs proceed by reducing propositions to identities [...] or contradictions" (p.48). The latter "provability thesis" was put forward by Leibniz, e.g., in § 132 GI:

> Every true proposition can be proved; for since, as Aristotle says, the predicate is [contained] in the subject, or the concept of the predicate is involved in the concept of the subject when that concept is completely understood, then it must be possible for a truth to be shown by the analysis ["resolutio"] of terms into their values, i.e. those terms which they contain.[52]

If we pick up Leibniz's earlier distinction between "truths of reason" and "truths of fact", two »halves« of the provability thesis may be distinguished:

PROV 1    All truths of reason can be proved.

PROV 2    All truths of fact can be proved.[53]

PROV 1 appears to be unproblematic not only with respect to *logically*, but also with respect to *analytically* true propositions such "The part is smaller than the whole". The following quotation illustrates how Leibniz thought that such a proof proceeds. It starts with the *definition*:

> Something is smaller if and only if it is equal to a part of the other (larger). The proof runs as follows: The part is equal to a part of the whole (namely, equal to itself, by the axiom of identity). What is equal to a part of the whole is smaller than the whole (by the definition of 'smaller'), therefore the part is smaller than the whole. Quod erat demonstrandum. (AE VI, 4, 805).

---
[52]Cf. LLP, p. 77, and p. 62: "Every true proposition can be proved."

[53]Rescher [15, p. 23] paraphrases this principle as "The Principle of Sufficient Reason: Every True Proposition is Analytic".

More generally, a Leibnitzian proof of a proposition $\alpha$ consists of a finite set of definitions (of the terms occurring in $\alpha$ ) and a finite chain of logical moves by means of which $\alpha$ is reduced to an "identity", i.e. to an »axiomatic« logical truth. Given this understanding of a proof, however, it is very hard to believe that the other principle PROV 2 might also be correct. Consider, e.g., the proposition that today (i.e. on March $1^{st}$, 2018) the temperatures in my home town (i.e. in Osnabrueck, Germany) dropped below -10° C? How should it be possible to reduce this *contingent* proposition with the help of »definitions« to an »*identity*«? In order to understand Leibniz's (attempted) solution of this problem, it will be helpful to consider another general principle of his philosophy of logic which was already mentioned incidentally in § 132 GI. It is the famous "*praedicatum inest subjecto*"-principle saying, roughly:

PINS    In every true proposition the predicate is contained in the subject.

A major problem of this principle is that, from a grammatical point of view, not every proposition has a well-defined predicate, $P$, and a well-defined subject, $S$, at all. In particular, *relational propositions* like 'Cain is the son of Adam and Eve', or '$7 + 4 > 10$' do not seem to fall under the scheme '$ScP$'. Leibniz tried to solve this problem by assuming that, e.g., 'Adam is the father of Cain' can be *transformed* into a *logical complex* of propositions of type '$ScP$'. According to Rescher (1979: p. 56), this proposition is reducible:

[...] to the fact (1) that Adam has two properties

[1.1.] Being a father

[1.2] Being a father in virtue of (*propter*) Cain's being a son,

and (2) the fact that Cain has the two properties

[2.1] Being a son

[2.2] Being a son in virtue of Adam's being a father.

[...]

And Leibniz holds that this [analysis] is perfectly general. Whenever a relation obtains between two substances, i.e.

$aRb$, there will have to be purely descriptive (i.e. *nonrelational*) properties $P$ and $Q$ such that $aRb$ is logically equivalent with the conjunction $(Pa$ & $(Pa$ @ $Qb))$ & $(Qb$ & $(Qb$ @ $Pa))$, where @ stands for "is attributable to" (i.e. represents the *eo ipso* or *propter* connective.

However, the idea of *reducing* propositions with a dyadic (or more generally $n$-adic) predicate to a logical complex of propositions with monadic predicates faces serious problems. As a result of modern proof theory, the full calculus of $n$-adic predicate logic is known to be definitely »stronger« than monadic predicate logic. From this it follows that not every relational proposition of the form $R(a_1,...,a_n)$ can be reduced to a *truth-functional* complex of monadic propositions $P_1(a_1)$, ..., $P_n(a_n)$. Nevertheless, Leibniz's reduction procedure might be viable after all, since it is based on a peculiar logical operation which evidently is *not* a truth-functional but rather a *modal* connective, namely the *causal* relation that a proposition $\alpha$ is true *because* ("propter"), or *in virtue of the fact that* ("eo ipso"), another proposition $\beta$ is true. Unfortunately, Leibniz never worked out the logic of this operator in any detail. Therefore, this topic shall not be pursued further here.[54] Let us rather concentrate on those problems which remain if we restrict principle PINS to the »poorer« language of syllogistic logic:

PINS$_{\text{SYLL}}$  In every true proposition of the language of the theory of the syllogism, the predicate is contained in the subject.

In his correspondence with Arnauld, Leibniz defended this principle with great emphasis by maintaining that *always*:

> [...] in every true affirmative proposition, whether necessary or contingent, universal or particular, the notion of the predicate is in some way included in that of the subject. *Praedicatum inest subjecto;* otherwise I do not know what truth is.[55]

---

[54] An extensive discussion of Leibniz's theory of relations may be found in Mugnai [13].

[55] Cf. the French original in GP 2, 56; the English translation has been adopted from Look [10].

Every true UA, of course, is a paradigm of a proposition whose predicate is contained in the subject. Similarly, since 'No $S$ is $P$' is reducible to 'Every $S$ is Not-$P$', every true UN equally satisfies PINS. Furthermore, *singular propositions* may be taken to satisfy PINS because in Leibniz's logic the fact that an individual, $a$, has a property, $P$, can be represented by the proposition that $a$'s complete *individual-concept*, $A$, *contains* $P$. If the algebra of concept is extended by »indefinite concepts«, the idea of a (maximally) *complete* concept becomes *definable* in a straightforward way. As Leibniz pointed out, e.g., in § 72 GI, the »completeness« of the individual-concept $A$ amounts to the condition that whenever concept $B$ is *compatible* with $A$, $B$ is already contained in $B$: $\mathbf{P}(AB) \to AcB$.[56] Letting '$\mathbf{I}(A)$' abbreviate '$A$ is an individual-concept' and using '$\forall$' as a symbol for the universal quantifier (ranging over *concepts*), Leibniz's criterion can be formalized as follows:

IND 1    $\mathbf{I}(A) \leftrightarrow \forall Y(\mathbf{P}(AY) \to AcY)$.[57]

Although, then, besides *universal* propositions also *singular* propositions may be regarded as satisfying principle $\text{PINS}_{\text{SYLL}}$, *particular* propositions remain problematic. Consider the example 'Some animals are rational' which, of course, is *true*, because there exists an entire *species* of animals, namely *homo sapiens*, which (by »definition«) is rational. More generally, in "Elementa Calculi" (an essay of April 1679 devoted to the task of developing the semantics of »characteristic numbers«), Leibniz explained:

> (18) But in the particular affirmative proposition it is not necessary that [...] the concept of the subject should in itself contain the concept of the predicate; it is enough that [...] *the concept of some instance or species of the subject should contain the concept of the predicate*, even though it is not stated expressly what the species is.

---

[56] Cf. LLP, p. 65: "So if we have $BY$, and the indefinite term $Y$ is superfluous [...] then $B$ is an individual". A detailed discussion of Leibniz's conception of individual concepts may be found in Lenzen (2004).

[57] Somewhat more exactly, in order to warrant the self-consistency of $A$, one has to add the condition $\mathbf{P}(A)$! Alternatively, one might define that $A$ is an individual-concept iff for every concept $Y$: $A$ *either* contains $Y$ *or* $A$ contains $\sim Y$.

More specifically, Leibniz entertained the assumption:

PA 1*  'Some $A$ is $B$' is true if *and only if* there exists a "species" $Y$ such that $YA$ contains $B$.[58]

In §§ 19, 20 of this paper Leibniz tried to *prove* PA 1*, but in the end he had to acknowledge that his argument was based on a "paralogism":

> [...] for I see that a particular affirmative proposition holds even when neither term is a genus or species, such as [in the previous example] 'Some animal is rational' provided only that the terms are compatible.[59]

The *prima facie* obscure remark about the compatibility of the terms will be discussed soon. First, however, let it be noted that not only in April 1679 but also in the GI of 1686 Leibniz *went on* to represent the PA 'Some $A$ are $B$' by the formula '$YAcB$'. However, '$YA$' is then no longer interpreted as a *species* of $A$, but more neutrally as 'some $A$'. Thus in § 48 GI Leibniz explained: "'$AY$ contains $B$' is a particular affirmative proposition in respect of $A$", i.e. he adopted the following principle:

PA 2*  'Some $A$ is $B$' is true iff there exists an »indefinite concept« $Y$ such that $YAcB$.

In § 89 GI he suggested to formalize the PA instead with the help of *two* »indefinite concepts« as an *equation*: "Let us consider the particular affirmative 'Some animal is a man': $AY = BZ$".[60] Hence:

PA 3*  'Some $A$ is $B$' is true iff there exist $Y$ and $Z$ such that $YA = ZB$.

The equivalence between PA 2* and PA 3* rests on a law according to which *universal* affirmative propositions can be formalized as follows:

---

[58] Cf. LLP, p. 23: "the concept of the subject with some addition (i.e. the concept of a species of the subject) will contain the concept of the predicate". The 'if'-part of PA 1*, of course, is valid.

[59] Cf. LLP, p. 24.

[60] In § 159 GI this idea was formulated even more succinctly: "$YA = ZB$ is the particular affirmative"; cf. LLP, pp. 59, 68, and 82; for the sake of uniformity, some of Leibniz's term letters have been renamed.

CONT 4    $AcB \leftrightarrow \exists Y(A = YB)$.[61]

As a matter of fact, already in § 16 GI Leibniz had pointed out that the UA can *either* be formalized (according to CONT 3) as $A = AB$ *or* (according to CONT 4) with the help of an »indefinite concept« as $\exists Y(A = YB)$.[62] Therefore principles PA 2* and PA 3* are deductively equivalent.

However, both principles are easily seen to be *inadequate* because otherwise *every* PA would have to be rated as *true*! In view of CONJ 1, $BAcB$ is a tautology. Hence, by way of existential generalization, there *always* exists a $Y$ such that $YAcB$, namely $Y=B$! Similarly, according to CONJ 3, $BA = AB$; hence by twofold application of existential generalization it follows that there *always* exist $Y$, $Z$ such that $YA = ZB$. Now this problem has a surprisingly simple solution. As Leibniz himself came to realize, criterion PA 2* must be modified by adding the requirement that $Y$ is *compatible* with $A$:

textscPA 2    'Some $A$ is $B$' is true iff there exists a concept $Y$ such that $YA$ *is self-consistent* and $YAcB$.[63]

Similarly, PA 3* should be modified as follows:

PA 3    'Some $A$ is $B$' is true iff there exist $Y$ and $Z$ such that $YA$ *is self-consistent* and $YA = ZB$.[64]

These criteria can further be *simplified* by requiring that the terms $A$, $B$ themselves are *compatible*. For if $A$ is compatible with $B$, i.e. if $\mathbf{P}(AB)$,

---

[61] Accordingly, in § 178 GI principle SUB 1 was formalized as follows: "If $A = YC$, then $ZA = VC$"; cf. also the more explicit version in § 191: "If a universal affirmative is true, the particular affirmative is also true; i.e. if $A$ contains $B$, some $A$ contains $B$. For $A = XB$ [...] therefore $ZA = ZXY$ (from the nature of coincident terms). Let $ZX = V$ (arbitrarily), then $ZA = VB$."

[62] Cf. LLP, p. 56, fn. 1: "It is noteworthy that for '$A = BY$' it is also possible to say '$A = AB$'".

[63] Cf., e.g., § 190 GI: "*Particularis Affirmativa*: quoddam $A$ est $L$, idem est quod $A$ cum aliquo addito sumtum continere $L$ [...] posito scilicet $AL$ esse rem seu terminum verum qui non implicat opposita ut $X$ non-$X$".

[64] For reasons of symmetry, one might want to add that also $ZB$ is self-consistent, but this requirement is redundant since it follows from $\mathbf{P}(YA)$ in connection with $YA = ZB$.

then there exists a $Y$, namely $Y = B$, such that $\mathbf{P}(YA)$ and $YA\mathrm{c}B$. If, conversely, there exists a $Y$ such that $\mathbf{P}(YA)$ and $YA\mathrm{c}B$, then (in view of CONT 3) $YA = YAB$, which immediately entails that $\mathbf{P}(YAB)$; but $YAB$ contains $AB$, therefore it follows that $\mathbf{P}(AB)$. Altogether, then, Leibniz's incidental remark in § 20 of "Elementa Calculi" suggesting that a PA is true "provided only that the terms are compatible", turns out to be entirely *correct*:

PA 4   'Some $A$ is $B$' is true iff '$AB$' *is self-consistent.*

This criterion, by the way, only repeats what has already been explained in connection with *Schema 3*: 'Some $A$ are $B$' may be represented in Leibniz's logic by the formula $\mathbf{P}(AB)$!

Interestingly, one can go even a step further and establish a connection between the Leibnitzian criteria PA 2-4 and the contemporary truth-condition of a PA in the sense of $\exists x(Ax \wedge Bx))$! According to the semantics developed in section 4, $\mathbf{P}(AB)$ is true iff the extension of $AB$ is not empty, i.e. iff there exists an *individual y* (within the set of all possible individuals, $\mathbf{U}$) which has the conjunctive property $AB$. This condition can be »translated« into the language of Leibniz's logic as follows:

PA 5   'Some $A$ is $B$' is true iff there exists an *individual-concept*
       $Y$ which contains $AB$.

This criterion was anticipated in §§ 68-70 GI where Leibniz was dealing with the question of how one can be sure that the definition of a complex concept (as a conjunction of several other concepts) proceeds correctly, i.e. doesn't involve a contradiction:

> [...] for if I set $A = EFG$, then I must know not only that $E$, $F$, and $G$ are singly possible, but also that they are compatible with each other. But it is evident that this cannot be done except by experience [...]

> (69) So it is one of my first principles that terms which we discover to exist in one and the same subject do not involve a contradiction. Or, if $A$ is $B$ and $A$ is $C$, then $BC$ is possible, i.e. does not involve a contradiction.

(70) God judges about the possibilities of things solely from the experience of his mind, without any perception of anything else.

On the one hand, Leibniz is here defending the *epistemological* view that we humans can recognize the compatibility of two properties $B$, $C$, only by *perceiving* an (actually existing) individual, $a$, which has both properties.[65] God's infinitely powerful mind, in contrast, doesn't need any perception but realizes the compatibility of $B$ and $C$ by a purely intellectual reflection on $a$'s complete concept, $A$. On the other hand, Leibniz puts forward the *logical* principle that a conjunction $BC$ is possible iff there exists a "subject", i.e. a complete individual-concept $A$, which contains both $B$ and $C$:

Poss 6  $\mathbf{P}(BC) \leftrightarrow \exists Y(\mathbf{I}(Y) \wedge Y c BC)$.[66]

Since, according to *Schema 3*, formula $\mathbf{P}(BC)$ represents the PA 'Some $B$ are $C$', law Poss 6 gives us both a *confirmation* and an *explanation* of the validity of PA 5. The self-consistency of $BC$ is necessary and sufficient for the truth of 'Some $B$ are $C$', because $\mathbf{P}(BC)$ (semantically) guarantees the existence of a (possible) individual $y$ which has property $BC$, i.e., in other words, the existence of an individual concept $Y$ which contains $BC$! Hence Leibniz's criteria PA 2, 3, 4, and 5, are not only mutually equivalent, but they also conform with the contemporary interpretation according to which 'Some $B$ are $C$' is true iff there exists an individual $y$ which has both properties.

In the end, then, Leibniz happened to discover the correct truth-conditions for particular affirmative propositions. Unfortunately, however, criteria PA 2-5 can hardly be considered as a corroboration of principle PINS. The »predicate« $C$ of the PA 'Some $B$ are $C$' is neither contained in the »subject« $B$, nor can it properly be regarded to

---

[65] Since perceptions always take place in the actual world, the result not only confirms $\mathbf{P}(BC)$, but even $\mathbf{R}(BC)$.

[66] Since Leibniz doesn't explicitly speak of *singular* subjects here, it might seem more appropriate to specify his condition by (*) $\exists Y(YcBC)$; but for the sake of self-consistency, one has to add the requirement that $Y$ is possible: (**) $\exists Y(\mathbf{P}(Y) \wedge YcBC)$. This condition however, turns out to be logically equivalent to (***) $\exists Y(\mathbf{I}(Y) \wedge YcBC)$.

be contained in the "subject with something added", $YB$. Although the elliptic expression '$YBcC$' (in PA 2) suggests otherwise, from a logico-grammatical point of view '$YB$' signifies no proper *subject* at all. As the full formula '$\exists Y(\mathbf{P}(YB) \wedge YBcC)$' makes clear, what is at stake is rather the *existence* of a concept $Y$, which is *compatible* with the »subject« $B$ and which, in conjunction with $B$, contains »predicate« $C$.[67] As PA 5 reveals, the only real *containment* which is involved here is that of '$BC$' being contained in the (unspecified) individual-concept $Y$! Hence the famous "praedicatum-inest-subjecto"-principle turns out to be much less correct than Leibniz thought.

This verdict also applies to the provability thesis according to which "every true proposition can be proved". In §§ 133-134 GI, Leibniz made the following distinctions:

> (133) A true necessary proposition can be proved by reduction to identical propositions, or by reduction of its opposite to contradictory propositions; hence its opposite is called impossible.

> (134) A true contingent proposition cannot be reduced to identical propositions, but is proved by showing that if the analysis is continued further and further, it constantly approaches identical propositions, but never reaches them. Therefore, it is God alone, who grasps the entire infinite in his mind, who knows all contingent truths with certainty.

If one takes the word 'proof' in the strict sense of § 133 as a "reduction to *identical* propositions", then neither *singular* nor *particular* propositions can ever be *proved*. The fact that a certain woman, $a$, has a certain property, $B$, say of being taller than 2 meters, cannot be shown (at least not by us humans) by "reducing" it to an identity. We don't have an infinite mind which allows us to "analyze" $a$'s individual-concept $A$ and check whether it contains predicate $B$. We can at best find out by

---

[67]The French quotation marks around 'subject' and 'predicate' are meant to remind the reader of the difference between a UA and a PA. In '$SaP$' the *former* concept is the *subject* and the *latter* is the *predicate*. But in '$SiP$' this grammatical definition doesn't work because '$SiP$' is convertible to '$PiS$'.

*experience* whether $a$ is taller than 2 meters or not.[68]

The claim put forward in § 134 that one might prove a contingent proposition by analysing it and finding out that it "approaches an identity" must therefore at best be rated as a *metaphor*. Such a metaphor may have been helpful for the *metaphysician* Leibniz to understand how God was able to know the truth of everything. But from a methodological point of view, this metaphor should better be rejected because it blurs the important distinction between necessary and contingent propositions: Only the former but not the latter can really be *proved*, i.e. reduced to tautologies. In contradiction to M&V's thesis (3), it follows at any rate that *not every* true proposition is provably equivalent to a tautology and not every false proposition is provably equivalent to a self-contradiction. Hence, for Leibniz, propositional logic must never have the character of an "auto-Boolean algebra".

## 6 Leibniz's propositional logic

M&V's reconstruction of Leibniz's propositional logic heavily relies on a brilliant idea that was formulated in the GI as follows:

> (75) If, as I hope, I can conceive all propositions as terms, and all hypotheticals as categorical, and if I can treat all propositions universally, this promises a wonderful ease in my symbolism and analysis of concepts, and will be a discovery of the greatest importance. [...]

> (137) We have, then, discovered many secrets of great importance for the analysis of all our thoughts and for the discovery and proof of truths. We have discovered [...] how absolute

---

[68]The situation concerning *particular* propositions like 'Some women are taller than 2 meters' is even worse. Here we don't have the slightest idea *which* individual $a_i$ (from the possibly infinite set of all women $a_1$, $a_2$, ...) we should select before investigating whether it has the crucial property. Maybe God "who grasps the entire infinite in his mind" knows that it is exactly individual $a_i$ whose complete concept $A_i$ contains predicate $B$. But, as Leibniz admitted in § 69 GI, the only way for us humans to verify the compatibility of two terms (such as 'being a woman' and 'being taller than 2 meters') is to discover by *experience* a subject $a_i$ which actually has these properties.

and hypothetical truths have one and the same laws and are contained in the same general theorems, so that all syllogisms become categorical [...]

(189) Our principles, therefore, will be these [...] Sixth, whatever is said of a term which contains a term can also be said of a proposition from which another proposition follows.

The main message is that each law stating that a certain *concept A contains* another *concept B* can be transformed into a law stating that a certain *proposition α implies* another *proposition β* . This close *parallel* between the *containment* relation among concepts and the *implication* relation among propositions had already been pointed out in the "Notationes Generales" written between 1683 and 1685 where Leibniz explained that just as '$A$ is $B$' is true, "when the predicate is contained in the subject", so a *conditional* '*If $A$ is $B$, then $C$ is $D$*' is true, "when the consequent is contained in the antecedent".[69] In later writings this idea was compressed into formulations such as "a proposition is true whose predicate is contained in the subject *or more generally* whose consequent is contained in the antecedent"[70].

Such a slogan, however, must not be taken to mean that all *and only* those propositions would be true "whose predicate is contained in the subject or whose consequent is contained in the antecedent". As we saw in section 5, principle PINS holds only under certain restrictions. Not *every true* proposition can be reduced to the form $ScP$; this applies in particular to a PA like 'Some women are taller than 2 meters'. As a matter of fact, not even *every logically true* proposition can be reduced to such a form! E.g., principle CONS 2, $A \neq \sim A$, is equivalent to $\neg((Ac \sim A) \wedge (\sim AcA))$, i.e., $\neg(Ac \sim A) \vee \neg(\sim AcA)$. Hence this tautology only reduces to a disjunction of propositions which *negate* that the respective subject contains the respective predicate! Similarly, the propositional law $\neg(\alpha \leftrightarrow \neg\alpha)$ cannot be reduced to a proposition "whose consequent is contained in the antecedent".

In close connection with the ingenious discovery of the parallel between term-logical containment, $AcB$, and propositional implication,

---

[69]Cf. AE VI, 4, p. 551.
[70]Cf. C, p. 401.

$\alpha \to \beta$, Leibniz put forward the idea of introducing a new kind of terms which M&V refer to as "*propositional terms*". Thus in § 138 GI he explained:

> If the proposition $A$ is $B$ is considered as a term [...], there arises an abstract term, namely $A$'s being $B$. And if from the proposition $A$ is $B$ the proposition $C$ is $D$ follows, then from this there comes about a new proposition of the following kind: $A$'s being $B$ is (or contains) $C$'s being $D$; or, in other words, the $B$-ness of $A$ contains the $D$-ness of $C$ [...].

On the one hand, Leibniz here maintains that for any universal affirmative proposition $AcB$ one can form the abstract term 'That $A$ contains $B$'. In the subsequent §§ 139-143 he further inquired the possibility of forming abstract terms also for other propositions like 'Some man is learned', 'Some man is not learned' or 'No man is a stone', but he didn't find a satisfactory solution. Nevertheless M&V took it for granted that:

> [...] in Leibniz's view, the expressions for propositional terms can be derived from the expressions for the corresponding propositions in a systematic manner, so that the latter can be put in one-to-one correspondence with the former. We will use corner-quotes to indicate the application of this one-to-one mapping. Thus, for example, if $A = B$ is a proposition, the corresponding propositional term is $\ulcorner A = B \urcorner$. (p. 694)

In what follows, we adopt M&V's convention and formalize the propositional term associated with an arbitrary proposition $\alpha$ as $\ulcorner \alpha \urcorner$.

On the other hand, in the quoted § 138 Leibniz suggested that *two* such propositional terms, e.g. 'That $A$ contains $B$' and 'That $C$ contains $D$', may be put into relation to each other by means of the same containment relation that obtains between ordinary terms like $A$, $B$, $C$, etc. Hence in M&V's symbolism the example "$A$'s being $B$ is [...] $C$'s being $D$" takes the form:

(\*)  $\ulcorner AcB \urcorner c \ulcorner CcD \urcorner$ .

M&V further considered admitting also "'mixed' propositions relating propositional to non-propositional terms, such as the following" (pp. 694/5):

(**) $\ulcorner AcB \urcorner c\ C$.

However, both from a syntactic and from a semantic point of view such liberalization doesn't make much sense. Take, e.g., $A =$ 'man', $B =$ 'rational' and $C =$ 'animal'. How should one understand a »proposition« like

(***) 'Man's being rational' contains animal?

The grammatical »subject« of (***), i.e. the propositional term 'Man's being rational', is not an ordinary *concept* which might meaningfully be said to *contain* (or not to contain) another *concept* like 'animal'. This »subject«, if viewed as something like a *proposition*, might at best be said to *imply* (or entail) another *proposition*, but the »predicate« 'animal' is a *concept* and not a *proposition*. Or, to describe this difficulty in semantic terms: The assumption that a certain subject, $S$, *contains* the predicate 'animal' requires that the extension of $S$ be contained in the extension of 'animal', i.e. in the set of all animals. But the extension of the *propositional term* 'Man's being rational' can hardly be understood as being contained in the set of all animals (or in any other subset of the universe of discourse, **U**)! M&V attempted to justify the admission of propositions of type (**) by pointing out that:

> [...] Leibniz countenances propositions such as *man's being animal is a reason* and *man's being animal is a cause*. [...] If so, then Leibniz allows for mixed propositions such as $\ulcorner man\ c\ animal \urcorner c\ reason\ [\ldots]$.[71]

The state of affairs *man's being animal* may certainly be considered as a *reason* or *cause* for another state of affairs, e.g. for the *truth* of 'Socrates is an animal'. However, 'is a reason' must not be conceived of as a »first-order« concept on a par with 'man', 'animal', or 'rational'. It is rather a »second-order« relation that can only obtain between two propositional terms but not between a propositional term on the one hand and a »first-order« term on the other. Hence, whenever all three

---

[71]Cf. p. 695, fn. 28; M&V's symbol '⊃' for the containment relation between terms has been replaced by our 'c'.

terms $A$, $B$, $C$ are ordinary concepts, "mixed" proposition of type (**) will simply be meaningless.[72]

Furthermore Leibniz's »equation« of the containment-relation between concepts and the implication-relation between propositions – as presupposed in formula (*) – must not be taken at face value. After all, the ordinary (categorical) proposition '$AcB$' is true iff the *extension* of $A$, i.e. a certain set of possible *individuals*, is contained in the extension of $B$. But this semantic requirement is not directly applicable to the 'c'-relation among propositional terms. That a proposition $\alpha$ »contains« another proposition $\beta$ would rather have to be modelled by requiring that the *intension* of $\alpha$, i.e. the set of all *possible worlds* in which $\alpha$ is true, is contained in the intension of $\beta$.

Besides in §§ 138-143 GI, Leibniz apparently made no other attempt to develop the logic of propositional terms any further. Therefore, it is somewhat difficult to judge how far he would have subscribed to M&V's reconstruction. However, *conjunction* appears to be entirely unproblematic. Just as the juxtaposition of two concepts $A$, $B$ yields the conjunctive term $AB$, so the juxtaposition of two propositional terms $\ulcorner \alpha \urcorner$, $\ulcorner \beta \urcorner$ yields the conjunctive propositional term $\ulcorner \alpha \urcorner \ulcorner \beta \urcorner$ which is so to speak the abstract counterpart of the conjunctive proposition $\alpha \wedge \beta$. Hence, e.g., the law of transitivity of the containment-relation, CONT 2, can be »translated« as follows:

CONT $2_{\text{prop}}$     $\ulcorner AcB \urcorner \ulcorner BcC \urcorner c \ulcorner AcC \urcorner$.

Furthermore, it seems safe to assume that just as the coincidence of two concepts, $A=B$, is nothing but their mutual containment, so also the coincidence of two propositional terms may be interpreted as their mutual implication and hence be formalized by '='. The characteristic law of conceptual conjunction, CONJ 6, can then be »translated« into:

CONJ $6_{\text{prop}}$     $\ulcorner AcB \urcorner \ulcorner AcC \urcorner = \ulcorner AcBC \urcorner$.

However, the *negation* of propositional terms raises serious problems. M&V think that Leibniz's explanation in § 32a GI "If $B$ is a proposition,

---

[72] This critique, however, is not decisive because M&V admitted: "It would be equally open to us to develop Leibniz's calculus in a language that does not allow for the unrestricted mixing of propositional and non-propositional terms" (p. 695, fn. 29).

non-$B$ is the same as $B$ is false, or, $B$'s being false" would justify the following "principle of propositional privation":

PPP   The negation of a propositional term $\ulcorner\alpha\urcorner$ is the propositional term That $\ulcorner\alpha\urcorner$ is »false«.

Since the expression »false« here refers to a *term* and not to a *proposition*, it has to be taken in the sense of definition (4) (cf. section 1 above), i.e. '*that* $\ulcorner\alpha\urcorner$ *is* »*false*«' means more specifically that $\ulcorner\alpha\urcorner$ is *logically false* or *self-inconsistent*. While M&V formalize this condition as '$\mathbf{F}(\ulcorner\alpha\urcorner)$', in our symbolism it is expressed by the formula '$\neg\mathbf{P}\ulcorner\alpha\urcorner$' which better reveals that the »falsity« of a propositional term amounts to its being *impossible*!

If one sets in particular $\alpha_1 = (BcC)$ and $\alpha_2 = (B = C)$, the main instances of PPP can be formalized as follows:

$PPP_1$   $\sim \ulcorner BcC\urcorner = \ulcorner\neg\mathbf{P}(\ulcorner BcC\urcorner)\urcorner$

$PPP_2$   $\sim \ulcorner B = C\urcorner = \ulcorner\neg\mathbf{P}(\ulcorner B = C\urcorner)\urcorner$.[73]

Thus, the negation of the propositional term *that $B$ contains $C$* is the propositional term *that it is impossible that $B$ contains $C$*; and the negation of the propositional term *that $B$ coincides with $C$* is the propositional term *that it is impossible that $B$ coincides with $C$*.

»Re-translating« these principles into a language without the operator of propositional abstraction, but enriched with the modal operator '$\Diamond$' (for 'it is possible that'), one would obtain:

$PPP_3$   $\neg(BcC) \leftrightarrow \neg\Diamond(BcC)$

$PPP_4$   $\neg(B = C) \leftrightarrow \neg\Diamond(B = C)$.

Some instances of these »laws« appear quite justified. E.g., in the special case where $C$ is identified with $B$, the negation of the *tautological* proposition $BcB$ may well be taken as equivalent to the modal proposition that it is *impossible* that $BcB$. For, as Leibniz remarked in § 32 GI:

---

[73]Since M&V have only '=' as primitive operator (while 'c' is defined according to CONT 3), they do without PPP 1 and state only PPP 2 "as follows: $\overline{\ulcorner A = B\urcorner} = \ulcorner\mathbf{F}(\ulcorner A = B\urcorner)\urcorner$" (p. 700).

If $\beta$ is a proposition, Not-$\beta$ is the same as that $\beta$ is false [...]. Not-$\beta$, if $\beta$ is understood as a proposition *in necessary matters* [!], is either necessary or impossible.[74]

However, in *non*-necessary matters, principles PPP 3, 4 lead to rather unwelcome results. Let us consider again our earlier example and assume that some women are taller than 2 meters. Setting $B =$ 'being a woman' and $C =$ 'being *not* taller than 2 meters', the identity '$B = BC$' happens to be false. But is this identity therefore *necessarily* false? Wouldn't it be more appropriate to say that the concepts 'being a woman' and 'being a woman which is not taller than 2 meters', might *possibly* coincide?

As a matter of fact, the problems inherent in PPP 3, 4 are much more fatal. Take any standard example of a pair of propositions such that *both* the PA *and* the PN are true, e.g. 'Some men are learned', 'Some men are not learned'. If one sets $B =$ 'man' and $C =$ 'learned', the UA, $BcC$, and the UN, $Bc\sim C$, both become *false*. According to PPP 3, these propositions therefore are impossible, i.e. they entail their own negation:

(i) $(BcC) \to \neg(BcC)$

(ii) $(Bc \sim C) \to \neg(Bc \sim C)$.

Since, trivially, each proposition entails itself, one further obtains:

(iii) $(BcC) \to ((BcC) \land \neg(BcC))$

(iv) $(Bc \sim C) \to ((Bc \sim C) \land \neg(Bc \sim C))$.

However, according to principle ECQ, two self-contradictory propositions are provably equivalent:

(v) $((BcC) \land \neg(BcC)) \leftrightarrow ((Bc \sim C) \land \text{neg}(C \sim C))$.

Hence, together with (iii), (iv), it follows that the UA and the UN are provably equivalent:

(vi) $(BcC) \leftrightarrow (Bc \sim C)$!

---

[74] Cf. AE VI, 4, p. 753, fn. 18, my emphasis; Leibniz's letter '$B$' has been replaced by '$\beta$'.

Altogether, then, our initial assumption that 'Some men are learned' while also 'Some men are not learned' gives rise to the conclusion that, necessarily, 'Every man is learned' if and only if 'No man is learned'!

This odd result can be generalized: *Any* two false propositions $\alpha$, $\beta$ become *provably equivalent*! For if $\alpha$ and $\beta$ are false, they are *necessarily* false, i.e. $\alpha$ is equivalent to $(\alpha \wedge \neg\alpha)$ and $\beta$ is equivalent to $(\beta \wedge \neg\beta)$. But, on account of ECQ, a self-contradictory proposition entails any other proposition, hence in particular $(\alpha \wedge \neg\alpha)$ entails $(\beta \wedge \neg\beta)$, so that altogether $\alpha$ entails $\beta$ (and vice versa)! But then also two arbitrary *true* propositions $\gamma$, $\delta$ are provable equivalent, because if $\gamma$ and $\delta$ are true, then $\neg\gamma$ and $\neg\delta$ are false, hence it follows that $\neg\gamma$ and $\neg\delta$ are provably equivalent, so that also $\gamma$ and $\delta$ must be provably equivalent.

We thus have verified M&V's thesis that, within their reconstruction of Leibniz's logic, "every false propositional term coincides with 0 and every true propositional term coincides with 1" (p. 710) and consequently that "all true propositional terms coincide [with each other]" (p. 710, fn. 96)! In terms of modal logic, this "auto-Boolean" feature can be paraphrased by saying that every true proposition is *necessarily* true while every false proposition is *necessarily* false:

PPP$_5$    $\alpha \to \Box\alpha$

PPP$_6$    $\neg\alpha \to \Box\neg\alpha$.

This entails that there are *no contingent* propositions. Therefore M&V's calculus – technically brilliant as it may be – cannot be accepted as an adequate reconstruction of *Leibniz's* logic. M&V tried to justify PPP 6 by referring (in fn. 96, p. 710) to § 40 GI where Leibniz had said: "A true proposition is one which coincides with '$AB$ is $B$'". At first sight, this statement appears to maintain that all *and only those* propositions are true which "coincide" (or can be reduced to) identical propositions. However, in the continuation of § 40 it becomes clear that Leibniz only *reflected* whether the thesis of the *necessity* or provability of all true propositions really holds. He first expressed his *belief* ("puto") that the thesis "can also be applied to non-categorical propositions", but in § 41 he dug deeper into the consequences of this assumption:

> (41) Therefore, since a false proposition is one which is not true (by [§] 3), it follows (by [§] 40) that a false proposition

is the same as a proposition which does not coincide with '$AB$ is $B$', or a false proposition is the same as one which cannot be proved.

As our earlier discussion of principle PROV 2 has shown, the latter consequence in *untenable*! There are *many* propositions, especially those of type PA, which cannot be *proved*, i.e. reduced (in a finite number of steps) to identities, but which can nevertheless be recognized as *true*, *viz.* by empirical observation. Thus, towards the end of § 41 Leibniz himself came to admit: "Propositiones facti non semper probari possunt a nobis", i.e. *propositions of fact cannot* always *be proved* by us.

In Leibniz's philosophical works one finds overwhelming evidence that he would never have accepted a logic in which propositions can neither be contingently true, nor contingently false. Such a result is entirely incompatible with his various attempts to define the truth-conditions of contingent propositions. Thus in § 61 GI he characterized true contingent propositions by the condition that their analysis requires an infinite resolution.[75] Similarly, in § 134 GI he explained that a true contingent proposition "constantly approaches identical propositions without ever reaching them". Also in many other papers Leibniz stressed that not every true proposition is necessarily true:

> A truth is either necessary or contingent. A necessary truth can be recognized by a finite number of substitutions, i.e. by a commensurable coincidence; a contingent truth by an infinite [number of substitutions], i.e. by an incommensurable coincidence. 'Commensurable' corresponds to 'explicable'; 'incommensurable' to 'inexplicable'. A necessary truth is one whose truth is explicable; a contingent one whose truth is inexplicable.[76]

---

[75] Cf. LLP, p. 61: "True contingent propositions are those which need an analysis continued to infinity".

[76] Cf. C, p. 408: "Verum est vel necessarium vel contingens. Verum necessarium sciri potest per finitam seriem substitutionum seu per coincidentia commensurabilia, verum contingens per infinitam, seu per coincidentia incommensurabilia. Explicabile conferemus commensurabili, inexplicabilis incommensurabili. Verum necessarium est cujus veritas est explicabilis; contingens cujus veritas est [inexplicabilis]". The last word has been corrected by Couturat from Leibniz's 'explicabilis'.

Although, as was argued in section 5, thesis PROV 2 is very problematic and should better regarded only as a metaphor, the very fact that Leibniz was *trying to show* that "truths of fact" can be »proven« by means of an infinite analysis clearly shows that for him *contingently true propositions do at any rate exist*.[77]

M&V tried to mitigate the problems arising from PPP by maintaining that, for Leibniz, a proposition which entails a contradiction need not necessarily be *impossible*:

> In [Leibniz's] view, every false propositional term contains a contradiction; an impossible propositional term has the further property that one of the contradictions it contains can be disclosed through a finite as opposed to an infinite process of analysis. (p. 719)

To support this thesis, M&V referred to § 130[bis][78] which shall be quoted here, together with the preceding § 130, in full length:

> (130) [a] A true proposition is one which can be proved. [b] A false proposition is one which is not true. [c] An impossible proposition is one into which a contradictory term enters. [d] A possible proposition is one which is not impossible. [e] Is every universal negative proposition, therefore, impossible? [f] It seems that it is so because it is understood of concepts, and not of existing things; [g] thus if I say that no man is an animal, I do not understand this of existing men alone. [h] But from this it will follow that what is denied of

---

[77]This is further evidenced by § 66 GI where Leibniz was wondering: "Is everything true which cannot be proved false, or everything false which cannot be proved true? What, then, of the cases where neither of these hold? It must be said that true and false propositions can always be proved, at any rate by an analysis which is carried out to infinity. But then it is contingent, i.e. it is possible that it is true, or that it is false."

[78]As Couturat first noted in C, p. 387, fn 3, Leibniz erroneously numbered two paragraphs '130'. In LLP, pp. 76-77, Parkinson adopted this double numbering while in the first critical edition of the GI, i.e. in Schupp (1982), the §§ were differentiated as '130' and '130a'. M&V instead renumbered them as '130a' vs. '130b'. We prefer to follow the conventions of the Academy Edition (cf. AE VI, 4, p. 775) and speak of '130' vs. '130[bis]'.

some individual, such as Peter, is necessarily denied of him. [i] Therefore it must be denied that every universal negative is impossible; [j] and in reply to the objection, it can be said that the fact that $A$ contains Not-$B$ is proved either by a perfect proof or analysis or only by an imperfect analysis which must be continued *ad infinitum*. [k] Therefore it is certain, but not necessary, for it can never be reduced to an identity, or its opposite to a contradiction.

Sentence 130[a] contains a brief formulation of the *provability-thesis*. As was shown in section 5, every *truth of reason* can be proved, but principle PROV 2, which maintains that every *truth of fact* is provable, appears very problematic. *Singular* propositions like 'Socrates is snub-nosed' or *particular affirmative* propositions like 'Some women are taller than 2 meters' cannot really be *proved* (at least not by us humans). Only God with his infinitely powerful mind may perhaps analyse Socrates's complete individual concept, $S$, to find out that $S$ contains the predicate 'snub-nosed'. Similarly, only God can perhaps carry out an infinite analysis of the concepts 'women' and 'taller than 2 meters' to find out whether they are *compatible* (and whether there hence exists a *possible individual* which has these properties).

Sentence 130[b] simply repeats the *principle of bivalence*, while in 130[c] two related, but different ideas are mingled in a somewhat unfortunate way. The first idea is to define a »false« or impossible *term* as one which *contains* a contradictory term (like $B$ Not-$B$). The second idea is to define a *proposition* as impossible iff it *entails* a contradictory proposition (like $\beta \wedge \neg \beta$ ). Leibniz, however, uses none of the expressions 'contain' or 'entail' but rather defines a proposition as impossible *into which a contradictory term enters* ("quam ingreditur terminus contradictorius"). This condition would normally be understood to mean that any proposition, in which a contradictory term like $B$ Not-$B$ occurs, is impossible. But this »definition« would clearly be inadequate! E.g., the proposition that $B$ Not-$B$ contains Not-$B$, formally $(B\sim B)c\sim B$, is an instance of law CONJ 2, hence it is necessarily true, but not impossible!

While sentence 130[d] describes the logical relation between the modalities 'possible' and 'impossible' in the usual way, 130[e] comes much as a surprise: Why, for heaven's sake, should anybody believe that, as a consequence of the foregoing definitions, every universal negative proposition is *impossible*? Leibniz's subsequent explanations do not fully solve this puzzle. In sentence 130[*f*] he points out that categorical propositions can alternatively be interpreted "existentially", i.e. referring to *existing* individuals only, or "essentially", i.e. referring to general notions and hence to the larger set of all *possible* individuals. This distinction is illustrated in 130[*g*] by means of the – unorthodox and somewhat surprising – example 'No man is an animal' which Leibniz wants to be understood as dealing not with *actually existing* men only, but rather with all possible men. From this "essential" understanding of the universal proposition Leibniz wants to derive in 130[*h*] that if a singular instance like 'Peter is not an animal', or equivalently 'Peter is a not-animal' is *false* (because 'not-animal' has to be denied of Peter), then this proposition is *necessarily false*, i.e. impossible. And from this observation he tries to derive in 130[*i*] that the hypothesis 130[*e*] is *false*: not every UN is impossible!

Before continuing our commentary on the remaining sentences 130[*j*] and [*k*], let us try to clarify some of these rather obscure thoughts. First of all, the issue of 130[*e*] can't earnestly be the question whether *every UN* is impossible. After all, e.g., 'No man is a *stone*' is *true*, hence this UN is *not impossible*! What is at stake must rather be the question whether every *false UN* is impossible.[79] But what caused Leibniz to raise such a strange question at all?

The most plausible answer consists in the assumption that Leibniz believed that every (false) UN satisfies the criterion mentioned in 130[*c*], i.e. every (false) UN "is one into which a contradictory term enters". In a marginal note to the preceding § 129, Leibniz had considered several formal criteria for the truth of categorical propositions.[80] One such criterion said that the PA 'Some $A$ are $B$' is true iff the conjunctive concept $AB$ is »true« or possible. In view of OPP 2, the UN 'No $A$ is $B$' therefore is true iff $AB$ is »false« or impossible.[81] In the later course of

---

[79] This question in turn is just an instance of the more general question (affirmed in principle PPP 6) whether *every false proposition* is necessarily false.

[80] In particular, he briefly discussed the operation of the "division" or "ablation" of a concept $A$ from a "product" or conjunction like $AB$. This interesting idea must remain out of consideration here.

[81] Cf. AE VI, 4, p. 774, fn. 47: "Videtur optimum ut prius definiamus particulares,

the GI, this criterion will be re-affirmed in various ways. E.g., in § 151 the proposition "tertii adjecti" 'No $A$ is $B$' will be transformed into the proposition "secundi adjecti" '$AB$ is not a thing'. Similarly, in § 151 the "Univ. Neg." will be formalized by the condition "$AB$ is not a thing"[82], and in § 199 Leibniz formulated very succinctly: "*Universal negative*: $AB$ is not [possible]"[83]. Since the UN can alternatively be represented by the formula '$Ac\sim B$', we here have an anticipation of the following counterpart of what M&V call "Leibniz's principle":

Poss 6   $Ac \sim B \leftrightarrow \neg \mathbf{P}(AB).$[84]

Since *every* UN (no matter whether true or false) *affirms* that the corresponding term $AB$ is *impossible*, it may somehow be conceived of as a proposition "into which a contradictory term *enters*". But the mere *occurrence* of such a contradictory *term* in a UN does not make that *proposition* itself contradictory. Hence Leibniz's fear – as expressed in 130[e] – that *every UN* might turn out to be *impossible*, is ungrounded, and his ensuing »refutation« in 130[f]-[i] widely misses the point. Nevertheless, these sentences contain some other important thoughts which are relevant for the main issue raised by PPP 5, 6, *viz*: Are all true propositions necessary, and are all false propositions impossible?

On the one hand, Leibniz's hint to the "essential" vs. "existential" reading of categorical propositions (which was elaborated above by means of *Schemata* 3 and 4) may perhaps be interpreted as evidence for his belief that every false UN, if understood as ranging over all possible

---

nempe $AB$ est notio vera [...] est part. Aff. [...] Cum vero dicimus $AB$ esse falsam notionem, seu negamus part. Aff., fit univ. Neg."

[82]Cf. § 151 GI and similarly § 152: "Univ. Neg. $AB$ non $= AB$ or $AB$ is not a thing". Leibniz's attempt to represent the self-consistency of a term $A$ by means of the equation '$A = A$' is very unapt and was eventually given up in § 155: "All things considered, then, it will perhaps be better for us to say that, in symbols at least, we can always put $A = A$, though nothing is usefully concluded from this when $A$ is not a thing."

[83]Cf. AE VI, 4, 787: "*Universalis Negativa*: $AB$ non est"; Parkinson translates this somewhat misleadingly as "The *universal negative*: $AB$ does not exist" (LLP, p. 87).

[84]This law is later repeated in § 200. Cf. AE VI, 4, 788: "Si dicam $AB$ non est, idem est ac si dicam $A$ continet non-$B$, vel $B$ continet non-$A$, seu $A$ et $B$ sunt inconsistentia."

individuals, will be necessarily false. For, so Leibniz appears to argue, if one considers an arbitrary individual, Peter, and *denies* of him that he has property $B$, then Peter *necessarily* fails to have that property. But the same consideration may be applied also to *affirmative* propositions. If, e.g., 'Every man is an animal' is understood as referring to each *possible individual* $x$, then one might argue as follows. If $x$ falls under the subject 'man', then $x$ *necessarily* falls under the predicate 'animal'. In sum, then, Leibniz may perhaps have meant to defend the following *restricted* thesis:

> PPP 5.1  If a *universal* proposition $\alpha$ is taken "*essentially*" as referring to all possible individuals, then $(\alpha \to \Box\alpha)$.

But this doesn't mean that Leibniz would have accepted the stronger theses PPP 5, 6 according to which *every true* proposition is necessary (and every false proposition impossible). This is evidenced also by the concluding sentences [j], [k] of § 130 where he introduced the following distinction: a UN '$A$ contains Not-$B$' can either be proved by means of a "*perfect* analysis" (which shows in finitely many steps that the predicate Not-$B$ is contained in the subject $A$), or the analysis remains "imperfect" (because it "must be continued *ad infinitum*"). In the latter case, the proposition is "*certain*" (i.e., God can infallibly detect its truth), but it is *not necessary*.

To conclude, let us now consider § 130[bis] which constitutes the main evidence for M&V's claim that, for Leibniz, allegedly every false proposition entails a contradiction:

> (130 [bis]) [a] That is true, therefore, which can be proved, i.e. of which a reason can be given by analysis; [b] otherwise it is false. [c] That is necessary which is reduced by analysis to an identity. [d] That is impossible which is reduced by analysis to a contradiction. [e] A term or a proposition is false if it contains opposites, however this is proved. [f] It is impossible if it contains opposites which are proved by a finite reduction. [g] Therefore $A = AB$, where the proof has been made by a finite analysis, must be distinguished from $A = AB$, where the proof has been made by an infinite analysis. [h] From this there already results what has been said about the necessary, possible, impossible, and contingent.

Sentences 130[bis][a], [b] only repeat the provability-thesis which, as was stressed already several times before, must not be taken at face value. 130[bis][c], [d] similarly repeat the *basic idea*: *necessary* propositions are proved by a reduction to identities; *impossible* propositions are refuted by a reduction to contradictions. In 130[bis][g] two cases for the truth of '$A$ contains $B$' (formally: $A \varepsilon B$, or $A = AB$) are distinguished: perfect reduction (achieved in finitely many steps) vs. imperfect reduction (requiring an infinite analysis). So, the only sentences which need to be discussed here are 130[bis][e] and [f] which M&V interpret as follows:

> Every false propositional term contains a contradiction; an impossible propositional term has the further property that one of the contradictions it contains can be disclosed through a finite as opposed to an infinite process of analysis.

Let us contrast this with Leibniz's own explanation:

> A term or a proposition is false if it contains opposites, however this is proved. It is impossible if it contains opposites which are proved by a finite reduction.

It should first be noted that, unlike M&V, Leibniz doesn't speak of the falsity of a "propositional term" but rather of that of "a term *or* a proposition". He thus tries to summarize the characteristics of »truth« and falsity for entities of two different categories by one and the same comprehensive condition. This shortcut, however, is very dangerous. As was argued in section 1, one should better disentangle the notions of »truth« and falsity with respect to terms and with respect to propositions. In a marginal note to § 2 GI, a term $A$ had been defined as *possible* iff $A$ does *not* contain a contradiction (like $Y$ Not-$Y$)[85]; similarly, in § 32[bis]-34 GI, term $C$ was defined as *impossible* iff $C$ *does* contain a contradiction.[86] Moreover, in § 194 GI, a term $B$ was defined to be »*false*« iff $B$ contains opposites like $A$ Not-$A$ while $B$ is »true« iff $B$ doesn't contain

---

[85] Cf. AE VI, 4, 749, fn. 8: "$A$ non-$A$ *contradictorium* est. *Possibile* est quod non continet contradictorium seu $A$ non-$A$. Possibile est quod non est: $Y$ non-$Y$."

[86] Cf. AE VI, 4, 754: "$B$ non-$B$ est impossibile, seu si $B$ non-$B$ = $C$, erit $C$ impossibile [...] Quod continet $B$, non-$B$ idem est quod *impossibile* [...]".

such a contradiction. Hence the set of all *terms* – and thus in particular the set of all *propositional terms* – can only be split into *two* subsets:

$$\text{Propositional term } B \text{ is } \begin{cases} \text{either »true«, i.e. possible} \\ \text{or »false«, i.e. impossible.} \end{cases}$$

In contrast, according to § 130[bis] GI, the set of all *propositions* is not only dividable into true and false, but both subsets can in turn be divided into propositions which are, or are not, *necessarily* true (or *necessarily* false):

$$\text{Proposition } \alpha \text{ is } \begin{cases} \text{either true, namely } \begin{cases} \text{either necessarily true, i.e.} \\ \text{finitely reducible to an identity} \\ \text{or contingently true, i.e.} \\ \text{not finitely reducible to an identity} \end{cases} \\ \text{or false, namely } \begin{cases} \text{either necessarily false, i.e.} \\ \text{finitely reducible to a contradiction} \\ \text{or contingently false, i.e.} \\ \text{not finitely reducible to a contradiction} \end{cases} \end{cases}$$

For this reason, Leibniz's propositional logic cannot be reduced to a logic of propositional terms à la M&V. Their »reconstruction« surely constitutes a very elegant calculus which shows that the so-called »peripatetic program« can be carried out[87]. But Leibniz would never have accepted such a logic because it leads to a complete collapse of modalities:

COLLAPSE  Everything which is *possible* must be *true*, indeed *necessarily* true; and everything which is *not necessary* must be *false*, indeed *necessarily* false.

# Bibliography

Leibniz's writings are quoted according to the following editions:

---

[87] This issue has been further elaborated in Malink & Vasudevan (2020).

**AE** German Academy of Science (ed.), G. W. Leibniz, *Sämtliche Schriften und Briefe*, esp. Series VI *Philosophische Schriften*, Darmstadt 1930, Berlin 1962 ff.

**C** L. Couturat (ed.), *Opuscules et fragments inédits de Leibniz*, Paris 1903; reprint Hildesheim (Olms) 1960.

**GI** "Generales Inquisitiones de Analysi Notionum et Veritatum", in AE VI, 4, 739-788.

**GP** C. I. Gerhardt (ed.), *G. W. Leibniz Die Philosophischen Schriften*, 7 vol. Berlin 1890, reprint Hildesheim, New York (Olms) 1978.

**LLP** Parkinson, G. H. R. (ed.), *Leibniz Logical Papers – A Selection*, Oxford (Clarendon Press) 1966.

## Secondary Literature

[1] Arnauld, A. & Nicole, P. (1683), *La Logique ou L'Art de Penser*, 5th edition, reprint 1965 Paris (Presses universitaires de France).

[2] Couturat, L. (1901), *La Logique de Leibniz d'après des documents inédits*, Paris, reprint Hildesheim (Olms) 1961.

[3] Ishiguro, H. (1972), *Leibniz's Philosophy of Logic and Language*, London (Duckworth).

[4] Kneale, W. & M. (1962), *The Development of Logic*, Oxford (Oxford University Press).

[5] Lenzen, W. (1984), "Leibniz und die Boolesche Algebra", *Studia Leibnitiana* 16, 187-203.

[6] Lenzen, W. (1986), "'Non est' non est 'est non'", *Studia Leibnitiana* 18, 1-37.

[7] Lenzen, W. (2004), "Logical Criteria for Individual(concept)s", in M. Carrara, A. M. Nunziante & G. Tomasi (eds.), *Individuals, Minds, and Bodies: Themes from Leibniz*, Stuttgart (Steiner Verlag), 87-107.

[8] Lenzen, W. (2019), "Leibniz's Laws of Consistency and the Philosophical Foundations of Connexive Logic", in *Logic and Logical Philosophy* 28, 537-551.

[9] Lenzen, W. (2022), "Rewriting the History of Connexive Logic", in *Journal of Philosophical Logic* 51, 525-553.

[10] Look, B. (2017), "Gottfried Wilhelm Leibniz", in Edward N. Zalta (ed.), *The Stanford Encyclopaedia of Philosophy* (Summer 2017 Edition).

[11] Malink, M. & Vasudevan, A. (2016), "The Logic of Leibniz's *Generales Inquisitiones de Analysi Notionum et Veritatum*", *Review of Symbolic Logic* 9, 686-751.

[12] Malink, M. & Vasudevan, A. (2020), "The Peripatetic Program in Categorical Logic – Leibniz on Propositional Terms", *Review of Symbolic Logic* 13, 141-205.

[13] Mugnai, M. (1992), *Leibniz's Theory of Relations*, Stuttgart (F. Steiner).

[14] Quine, W. V. O. (1953), *From A Logical Point of View*, New York (Harper & Row).

[15] Rescher, N. (1979), *Leibniz – An Introduction to his Philosophy*, Guildford, London & Worcester (Billing & Sons)

[16] Schupp, F. (1982), *Gottfried Wilhelm Leibniz Generales Inquisitiones de Analysi Notionum et Veritatum - Allgemeine Untersuchungen über die Analyse der Begriffe und Wahrheiten*, Hamburg (Meiner).

# Propositional Term Logic & an Intensional Semantics

Robert van Rooij

Elbert Booij

## 1 Introduction

In this paper we will show two things: (i) that syllogistic logic can be extended such that it can also account for propositional reasoning if we allow for complex terms, and (ii) how to provide an intensional semantics for traditional syllogistic reasoning, and how this can be extended to a fact-based semantics for syllogistic-based propositional reasoning.

The paper starts with a quick overview of traditional Aristotelian syllogistics. We will discuss how to extend it with negative, singular, and complex terms. Afterwards we will discuss how propositional logic can be seen (in several ways) as an extension of Aristotelian syllogistics. Thus, in distinction with polish logicians like Lukasiewicz and others, we won't assume that to understand traditional logic we have to presuppose propositional logic, but instead formulate propositional logic by presupposing syllogistic reasoning. There are at least two possible ways to do so. In van Rooij [24] one of those ways was worked out. In this paper, we will show that another way, making use of complex terms, works as well. In the second part of the paper we show that both standard syllogistic logic and its proposition-logical extension can be given an intensional semantics. Interestingly enough, this is exactly in line with Russell's and Wittgenstein's theory of logical atomism.

## 2 Traditional syllogistic reasoning

### 2.1 A proof system for traditional syllogistic logic

Syllogisms are arguments in which a categorical sentence is derived as conclusion from two categorical sentences as premises. Categorical sentences can be of four types: $a$-type (All men are mortal'); $i$-type (Some men are philosophers'; $e$-type (No philosophers are rich), and $o$-type (Some men are not philosophers). A categorical sentence always contains two *terms*. In the $a$-sentence, for instance, the terms are 'men' and 'mortal', while in the $e$-sentence they are 'philosopher' and 'rich'. Thus, the *syntax* of categorical sentences can be formulated as follows: If $T$ and $T'$ are terms, $TaT'$, $TiT'$, $TeT'$, and $ToT'$ are categorical sentences. Because a syllogism has two categorical sentences as premises and one as the conclusion, every syllogism involves only three terms, each of which appears in two of the statements. The term that does not occur in the conclusion is called the *middle term*. The *quality* of a proposition is whether it is *affirmative* (in $a$- and $i$- sentences, the predicate is affirmed of the subject), or *negative* (in $e$ and $o$-sentences, the predicate is denied of the subject). Thus 'every man is a mortal' is affirmative, since 'mortal' is affirmed of 'man'. 'No men are immortal' is negative, since 'immortal' is denied of 'man'. The *quantity* of a proposition is whether it is *universal* ( $a$- and $e$-sentences the predicate is affirmed or denied of "the whole" of the subject) or *particular* (in $i$ and $o$-sentences, the predicate is affirmed or denied of only 'part of' the subject).

A *valid* syllogism is a syllogism that cannot lead from true premises to a false conclusion. Medieval logicians developed a *decision procedure* to determine which syllogisms are valid. This procedure made crucial use of the so-called distribution-value of the terms involved. Whether a term is distributed or not is really a *semantic* question: a term is said to be distributed when it is actually applied to *all* the objects it can refer, and undistributed when it is explicitly applied to only part of the objects to which it can refer. This formulation has been criticised by Geach [10] and other modern logicians, but Sommers [30, 31, 32] and Englebretsen [5] rightly argued that distribution-values are key for understanding many reasoning patterns in syllogistic logic and its possible extensions. As noted by van Benthem [3] and van Eijck [6], distributivity can be redefined in terms of *monotonicity*. A term occurs distributed when it occurs monotone decreasingly/negatively within a sentence, and undistributed when it occurs monotone increasingly/positively within a sentence. Denoting a distributed term by $-$ and an undistributed term by $+$, the following follows at once: $S^-aP^+$, $S^+iP^+$, $S^-eP^-$, and $S^+oP^-$, which we might think of now as a *syntactic* characterisation, just as proposed by Sommers [30, 32]. In terms of the distribution values of terms, we can now state the laws of *quantity* or *distribution*, (R1) and (R2), and of *quality*, (R3). Together, they constitute the rules of the syllogism:[1]

(R1) The middle term must be distributed at least once.

(R2) Every term that is distributed in the conclusion is also distributed in one of the premises.

(R3) The number of negative conclusions must equal the number of negative premises.

The above rules assume existential import. Without this assumption, we have to strengthen (R2) to (R2'):

(R2') Every term that is (un)distributed in the conclusion is (un)distributed in one of the premises.

Medieval logicians and their followers standardly assumed that of all the reasoning schemas stated in syllogistic style, all and only all forms are valid that satisfy those rules. As far as we know, the first one who explicitly *proved* this was Leibniz [?].

The above rules of the syllogism can be reformulated in a more immediate way (following Sommers [30] and Friedman [8]). Suppose we represent the categorical sentences as follows $SaP \equiv -S + P$, $SiP \equiv +S + P$, $SeP \equiv -S - P - 1$, $SoP \equiv +S - P - 1$. Thus, terms occurring distributed are simply represented by '$-$', while undistributed terms are represented by '$+$', and negative sentences are represented by an extra '$-1$'. Then we can say that $\phi_1, \cdots, \phi_n \vdash \psi$ iff $\phi_1 + \cdots + \phi_n = \psi$. From

---
[1]Standardly, more rules are stated, but these can be derived from the rules below. One of the rules normally assumed, for instance, is that at least one of the premiss must be affirmative. But this follows immediately from (R3).

this the above distribution laws immediately follow: in a syllogism the middle term must be once distributed and once undistributed (R1), and the distribution signs of $S$ and $P$ in the conclusion of a syllogism corresponds with their distribution signs in the corresponding premises (R2'). The special use of '−1' for negative sentences accounts for rule (R3).[2]

Aristotle's way to determine which syllogisms are valid was 'proof-theoretic' in nature. He showed that if one starts with one of the first 4 valid syllogisms of the first figure (known as Barbara, Celarent, Darii and Ferio), and use any combination of subalternation, conversion, and the so-called *reductio per impossible*,[3] you end up with a valid syllogism.[4] We won't make use of subalternation — $SaP \vdash SiP$ and $SeP \vdash SoP$ —, but instead implement existential import by taking $SiS$ to be an extra premisse for such inferences. We are also not going to make use of conversion — $SiP \vdash PiS$ and $SeP \vdash PeS$ —, but instead adopt the Law of Identity ($TaT$) and following Leibniz by make heavy use of Aritstotle's *Reductio per impossible*. The rule 'reductio per impossible' can best be formulated in terms of a notion of sentence-negation '¬' which doesn't really exist in traditional logic. However, it can be defined as the contradictory negation featuring in the *square of opposition*: $\neg(SaP) \stackrel{def}{=} SoP$,

---

[2] A problem with this method is that it does not allow for a natural representation of negative terms. Friedman's [8] *multiplicational* method can solve this problem. The multiplicational method represents the categorical sentences as follows: $SaP \equiv \frac{P}{S}$, $SiP \equiv 2SP$, $SeP \equiv \frac{-1}{SP}$, $SoP \equiv \frac{-2S}{P}$. The inference $\phi_1, \cdots, \phi_n \vdash \psi$ is valid iff $\phi_1 \times \cdots \times \phi_n = \psi$. Negative terms $\overline{P}$ are represented as $\frac{1}{P}$.

[3] He required the reductio-rule only to generate the valid syllogisms known as Barocco ($PaM, SoM \vdash SoP$) and Bocardo ($MoP, MaS \vdash SoP$). Because the reductio-rule turns out to be very useful, let us look at one example. Bocardo, for instance, is generated as follows: Suppose that $MaP$ is false, i.e. $MoP$. Given Barbara, this means that either $SaM$ is false, or $SaP$ is false. Assuming $SaM$, this means that $SoP$. Thus, from $MoP$ and $SaM$ we conclude to $SoP$, which is Bocardo. Barocco can be generated in exactly the same way, except that now $SaP$ is assumed, and thus that $SoM$ followed. It is interesting to note that there are exactly these two syllogisms which are declared invalid (or unusable) by Kant in his *Dohna-Wundlacken Logik*. We will see that Leibniz took a quit different route. Although Aristotle's reduction rule is standardly assumed, it is worth remarking that according to Ashworth [1, pp. 244-245], Jungius already suggested an interesting alternative proof of Barocco and Bocardo by making use of contraposition instead of reductio per impossible.

[4] Aristotle was well-aware that any other figure could be taken as the base system as well.

$\neg(SiP) \stackrel{def}{=} SeP$, $\neg(SeP) \stackrel{def}{=} SiP$, and $\neg(SoP) \stackrel{def}{=} SaP$. Now we can state a proof theory for syllogistic reasoning as follows:

1. The first 4 valid Syllogisms of the first figure:

    (a) Barbara: $MaP, SaM \vdash SaP$
    
    (b) Celarent: $MeP, SaM \vdash SeP$
    
    (c) Darii: $MaP, SiM \vdash SiP$
    
    (d) Ferio: $MeP, SiM \vdash SoP$

2. Law of Identity: $\vdash TaT$.

3. *Reductio per impossible*: if the conclusion of a syllogism is false, at least one of the premisses is false as well. $\Gamma, \neg\phi \vdash \psi, \neg\psi \Rightarrow \Gamma \vdash \phi$.

It is useful to see why the extra premisse $SiS$ accounts for subalternation. Look, for instance, at the inference $SaP \vdash SiP$. This follows from $SiS$ and Darii. Similarly, $SeP \vdash SoP$ follows from $SiS$ and Ferio. Following earlier ideas of Ramus, Leibniz [15] showed in his 'Of the Mathematical Determination of Syllogistic Forms' that the rules of conversion are not required, because they can be derived. First, following the practice of (post-)medieval logicians (c.f. [1, Chapter IV], he showed that all the valid syllogisms of the second and third figure can be derived by means of *Reductio per impossible* from Barbara, Celarent, Darii, and Ferio. Next he showed that three rules of conversion follow from some valid syllogisms together with the law of identity. $SiP \vdash PiS$, for instance, follows from Datisi$_3$: $SaS, SiP \vdash PiS$. Similarly, $SeP \vdash PeS$ follows from Cesare$_3$, and $SaP \vdash PiS$ follows from Darapti$_3$. A final rule of conversion, $SeP \vdash PoS$, follows from Festino$_2$ and the law of non-contradiction. These rules of conversion can then be used, finally, to derive the valid syllogisms of the fourth figure.[5]

Medieval logicians simplified the Aristotelian system by instead of taking Barbara, Celarent, Darii, and Ferio to be axioms, they derived those syllogisms by assuming a rule of inference known as the *Dictum de Omni et Nullo* (the maxim of all and none). The dictum is a kind of

---

[5] For much more on Leibniz' method, see Lenzen [19].

*substitution rule*. Although the dictum is not always stated in a clear or general way, Sommers [30, 32] showed that it can be stated very generally and that it makes crucial use of the *distribution* values of terms. We can formulate the *Dictum de Omni et Nullo* as follows:[6]

- **Dictum de Omni** (DDO): $MaP, \Gamma(M)^+ \vdash \Gamma(P)$, where $\Gamma(M)^+$ is a sentence where $M$ occurs positively
- **Dictum de Nullo** (DDN): $MeP, \Gamma_{a/i}(M)^+ \vdash \Gamma_{e/o}(P)$[7]

It is easy to see that the valid syllogisms of the first figure that don't require existential import immediately follow by DDO (Barbara, Darii) or DDN (Celarent, Ferio). For all other valid syllogisms we can make use of these deduction rules plus the law of identity, existential import (by assuming $TiT$ as premisse), and Reductio per impossible.

Negative terms didn't play an important role in Aristotle's theory of syllogisms.[8] Once negative terms are added to the language, however, the proof-system has to change. We will add two things that account for negative terms: a new double negation axiom, and the rule of contraposition. On the other hand, the rule system can now be simplified as well, because the *Dictum de Nullo* is no longer required. The reason is that $SeP$ and $SoP$ can now be written, or defined, as $Sa\overline{P}$ and $Si\overline{P}$, respectively, where $\overline{P}$ is the negative counterpart of $P$.[9,10]

---

[6]The soundness of this rule in the extended syllogistics is proved very elaborately by MacIntosh [22]. But as shown by Sanchez [29], its soundness follows from a much more general relation between monotonicity and positive and negative occurrences of predicates in Lyndon [21].

[7]In our formal terms, medieval logicians formulated the DDN sometimes as $PaM, \Gamma(M)^- \vdash \Gamma(P)$. But this formulation is less general: it doesn't allow one to infer $SoP$ from $MeP$ and $SiM$. A correct and more general formulation of the Dictum de Nullo is the following: $MeP, \Gamma(M)^+ \vdash \Gamma(\overline{P})$. This formulation, however, makes crucial use of negative terms, which we don't have yet. Moreover, once formulated in this way, the Dictum de Nullo can be reduced to the Dictum de Omni: $Ma\overline{P}, \Gamma(M)^+ \vdash \Gamma(\overline{P})$.

[8]In the history of logic, negative terms are also known as *indefinite* or *infinite* terms.

[9]Aristotle allowed for the inference $Sa\overline{P} \vdash SeP$, but not for $SeP \vdash Sa\overline{P}$ (cf. [14, p. 57]. The reason seems to be that 'Socrates is not black' seems to have a 'wider' meaning that 'Socrates is non-black'. Similarly, many traditional logicians didn't allow for the inference $SoP \vdash Si\overline{P}$.

[10]It is also possible to get rid of $SiP$ by defining it as $\neg(Sa\overline{P})$. In that case, we

The proof system **SYL** of syllogistic reasoning consists of the following set of axioms and rules:

(1) $MaP, \Gamma(M)^+ \vdash \Gamma(P)$ \hspace{2em} Dictum de Omni[11]

(2) $\vdash TaT$ \hspace{2em} Law of identity

(3) $\vdash T \equiv \overline{\overline{T}}$[12] \hspace{2em} Double negation

(4) $Sa\overline{P} \vdash Pa\overline{S}$, $SeP \vdash PeS$ \hspace{2em} Contraposition (or conversion,

(5) $\Gamma, \neg\phi \vdash \psi, \neg\psi \;\Rightarrow\; \Gamma \vdash \phi$ \hspace{2em} Reductio per impossible

Having lifted one arbitrary restriction Aristotle put on valid syllogistic reasoning, we might as well slightly extend syllogistic reasoning in another way as well. We could do so by adding a distinguished 'transcendental' term '⊤' to our language, standing for something like 'entity'. Obviously, the sentence $Sa\top$ should always come out true for each term $S$. To reflect this, we will add this sentence as an axiom to **SYL**. But adding ⊤ as an arbitrary term to our language gives rise to a complication once we accept existential import for *all* terms, including negative ones: for negative term $\overline{\top}$ existential import is unacceptable.[13] One way to get rid of this problem is to restrict existential import to *positive* categorical terms only.

We will denote the system consisting of (1), (2), (3), (4), (5) together with the following three rules by **SYL**$^+$.

---

might use instead of the Dictum just Barbara as axiom, but now also use Celarent as axiom. It is interesting to observe that according to some authors the Dictum de Omni et Nullo is indeed just the combination of Barbara and Celarent. Wallis [34], for instance, one of the greatest mathematicians of his time, defined this dictum as follows: 'Whatever way (with truth) be Universally Affirmed or Denied of any Subject; may accordingly be Affirmed or Denied of all things of which that Subject may be truly Affirmed' (Wallis, ch. 5, third part). The axiomatic system that seems to be most standardly assumed was taking Barbara and Celarent (or the Dictum de Omni et Nullo) as axioms, together with the reduction rule, conversion, and subalternation.

[11] Actually, Barbara is enough, because Darii can be derived from the rest.

[12] By this we really mean $\vdash Ta\overline{\overline{T}}$ and $\vdash \overline{\overline{T}}aT$.

[13] Just as it will be unacceptable later for conjunctive terms like 'square circle'.

(6) $\vdash Sa\top$

(7) $\overline{S}aS \vdash \overline{S}aP$.

Rules (6) and (7) tell us what to do with empty and transcendental terms. The use of rule (7) seems to be very limited: $\overline{S}aS$ can be true only if $S$ denotes (extensively) the universe of discourse, and $\overline{S}$ thus the empty set. But, then, the conclusion of (7) follows already from (6), and adding (87 to our system doesn't do much harm either. We will make real use of (6) and (7) only in section 4, where we will think of other sentential connectives as well.[14]

## 2.2 Singular Terms

In many popular logic textbooks examples of valid syllogisms are given that involve singular terms: 'Socrates is a man, every man is rational, thus Socrates is rational'. Still, the problem how to represent and reason with sentences that involve singular terms has bothered traditional logicians for a long time. The inference from 'Socrates is wise' and 'Socrates is old' to 'Some old person is wise' shows that the proposal to treat both premises as particular propositions will not do. The reason is that according to (R1)-(R3) no syllogism is valid with two particular premisses. Most traditional logicians (including Scotus, Ockham and Wallis) concluded that we thus must treat singular propositions as universal ones: thus represent 'Socrates is wise' as something like $SaW$, and 'Socrates is not wise' as $SeW$. One can show that this gives indeed rise to the correct syllogistic inferences. The above inference, for instance, is now a valid syllogism of the form Datisi$_3$. But there is still a problem with this proposal. Normally, $a$ and $e$ propositions are contrary, they can't both be true, but they *can* both be false. Thus, from $\neg(SaP)$ one *cannot* derive $SeP$. However, this inference is possible with singular terms: if 'Socrates is wise' is not true, it follows that 'Socrates is not wise' is true. According to modern logic, singular terms are essentially different from general terms: they are of a different type and cannot be negated. But for traditional logic this would lead to the absurd conclusion that singular terms do not fit into the traditional picture. Leibniz was less pessimistic and proposed that singular terms are terms like any other: they

---

[14]In combination with contraposition, (6) will then just be *ex falso sequitor quilibet*.

can stand in predicate position[15] (and can be negated). Still, perhaps first seen by Sommers [32], Leibniz discovered that there is something special about singular terms like $S$: for them, and only for them, the following holds for all $P$: $SaP$ is true iff $SiP$ is true (and $SeP$ is true iff $SoP$ is true). In terms of the square of opposition this means that for singular propositions, contrary and contradictory negation come down to the same.[16]

Leibniz' logical treatment of proper names was closely related with his metaphysics. 'Real' proper names denote monads, and the truth of any predication about a monad is already contained in the monad itself. We can see a monad as a set of attributes, such that for each attribute, or feature, $f$, it holds that either the monad has it, or not. Thus, a monad can be thought of as a set of maximally consistent/compatible set of attributes. We will make crucial use of this idea in section 4. From an extensional point of view, however, a proper name denotes a set of individuals, just like any other term. The rule that for each proper name $S$ it holds that $SaP$ is true iff $SiP$ is true does not mean that thus the extension of $S$ can have only 1 element. It does mean, however, that all the elements in the extension of $S$ must make the same predicates true, and thus be indistinguishable. By Leibniz' principle of indiscernibles, those elements are thus the same.

As stressed by Sommers [32], if we limit ourselves to singular terms $S_i, S_j$, and $S_k$, one can easily show by syllogistic reasoning that (i) $\vdash S_i a S_i$, (ii) $S_i i S_j \vdash S_j i S_i$, and (iii) $S_1 a S_2, S_2 a S_3 \vdash S_1 a S_3$, and thus that we don't need to introduce a special relation of identity '=' between singular terms that has the properties of an equivalence relation: (i) reflexivity, (ii) symmetry, and (iii) transitivity. Moreover, if $S_i a S_j$ and $S_j a S_i$, it follows immediately with the Dictum de Omni that for every $P$: $S_i a P$ is true iff $S_j a P$ is true. By Leibniz' principle of indiscernebles it follows that $S_1$ en $S_2$ are 'the same': $S_1 \equiv S_2$.

To account for syllogistic reasoning with singular terms, we just separate from all terms those that are called 'singular terms'. Then we add the following rule to our system **SYL**$^+$, and call the result **SYLS**$^+$:

---

[15] For an appealing explanation of why it is *unnatural* for such terms to stand in predicate position, see Sommers [32].

[16] Leibniz' discovery was reinvented by Montague some 250 years later, and became folklore in formal semantics due to Generalized Quantifier Theory.

(9) for all singular terms $I$ and terms $P$: $IiP \dashv \vdash IaP$.

The following derivations, for instance, can now be stated in syllogistic form: 'Every man is rational, Socrates is a man, thus Socrates is rational, $MaR, SaM \vdash_{DDO} SaR$; 'Everybody is rational, thus Socrates is rational' (universal instantiation), $\top^1 aR, Sa\top^1 \vdash_{DDO} SaR$; Madonna is rich and Madonna is famous, so somebody is rich and famous (or better, so somebody rich is famous), $MiR, MaF \vdash_{Conv} MaF, RiM \vdash_{DDO} RiF$.

## 3 Propositional logic

Aristotelian logic doesn't contain propositional logic. But that doesn't mean that we cannot think of propositional logic in traditional terms. In order to do so, we have to treat sentences as terms. Until now we assumed that all terms were 1-ary predicates. To think of propositional logic in traditional terms, we have to allow for 0-ary predicates as well. But once we do so, it is very natural to say that if $S$ and $P$ are 0-ary predicates, the conditional 'If $S$, then $P$' should be represented by the categorical sentence $SaP$. Indeed, this was a quite standard position among traditional logicians. Gassendi (1658) and Wallis (1687) were examples of proponents of this view. Not everybody agreed with this position, though. Kant, for instance, did not.

The idea to represent conditional sentences like 'if $S$, then $P$' by the categorical sentence $SaP$ was explicitly defended by Leibniz and later by Sommers. How should 0-ary predicates be interpreted? This question can be asked for both an extensional and an intensional semantics. We will concentrate on the extensional semantics here. On an intensional semantics $SaP$ is normally taken to be true iff $I(P) \subseteq I(S)$, where $I(S)$ is the set of attributes that it takes to be an $S$ (see section 4). On an extensional semantics, instead, $SaP$ is taken to be true iff $E(S) \subseteq E(P)$, where $E(S)$ is the extension of $S$. Adopting an extensional semantics, we could interpret a 0-ary predicate $T$ as the *set of possible worlds* in which $T$ is true. The conditional $SaP$ is thus true iff all $S$-worlds are also $P$-worlds. This semantics is certainly reasonable (it interprets the conditional as a strict implication), but it doesn't result in the truth-conditional propositional logic that we are used to. To get to standard logic, we have to adopt Frege's idea: 0-ary predicates denote either *the truth* or *the falsity*.[17] Starting from this idea, there are at least two ways to get propositional logic. A first method of obtaining propositional logic is just to use $a$, $e$, $i$ and $o$ as connectives and construct all complex sentences with the help of these 'connectives'. Doing so, we see that $[\phi]a[\psi] \equiv$ '$\phi \to \psi$', $[\phi]i[\psi] \equiv$ '$\phi \wedge \psi$', $[\phi]e[\phi] \equiv$ '$\neg \phi$', and $[[\phi]e[\phi]]a[\psi] \equiv$ '$\phi \vee \psi$'. If we abbreviate '$SaS$' by $\top^0$ we can write the negation of $\phi$ more simply as '$[\top^0]e[\phi]$'. This method was proposed by Sommers [30, 32] and worked out, among others, by Van Rooij [24].

In this paper, instead, we will go for a somewhat different approach, making crucial use of conjunctive terms (or non-traditional two-place connectives), an approach that was in fact proposed by Leibniz, and can be given a more standard-looking analysis as well.

## 3.1 Leibniz's equational analysis

Leibniz had several ways to represent categorical propositions. Most of them making use of conjunctive terms: If $S$ and $T$ are terms, $ST$ is considered to be a (conjunctive) term as well (of the same arity). He represented the four types of categorical sentences by *equations*. The sentence $SaP$ is represented as $S = SP$, $SiP$ as $SP \neq 0$, (where 0 stands for a distinguished element of the algebra) and $SoP$ and $SeP$ thus as $S \neq SP$ and $SP = 0$, respectively.[18] To account for reasoning

---

[17] Casteñada [4] argues that in Leibniz [15] a material implication analysis of conditionals is proposed. Lenzen [19] disagrees.

[18] Sometimes it is easier to make use of another, but equivalent, representation of the categorical sentences that Leibniz also makes extensive use of (see Lenzen, [19] for discussion). According to this representation, $SaP$ should be represented by $\forall Y, \exists Z : YS = ZP$ (with $XP \neq 0$), $SiP$ by $\exists Y, \exists Z : YS = ZP$ (with $YS \neq 0$), $SoP$ by $\exists Y, \forall Z : YS \neq ZP$ (with $YS \neq 0$ and $Z\overline{P} \neq 0$), and $SeP$ by $\forall Y, \forall Z : YS \neq ZP$ (with $YS \neq 0$ and $Z\overline{P} \neq 0$). It is noteworthy that this representation is very close to the representation he adopted in his numerical calculus. It also should be noted that this representation of the categorical sentences is close to the ones adopted by some nominalistic medieval logicians like Ockham who adopted the so-called *identity-theory*. In modern terminology, they gave the following representations by quantifying explicitly over individuals: $SaP \equiv \forall y \in S : \exists z \in P : y = z$; $SiP \equiv \exists y \in S : \exists z \in P : y = z$; $SoP \equiv \exists y \in S : \forall z \in P : y \neq z$, and $SeP \equiv \forall y \in S : \forall z \in P : y \neq z$. Leibniz thought of his system primarily (though certainly not exclusively) from an intensional point of view, and thus of $S, P, Y$, and $Z$ as sets of attributes. Let us now think of the $Y$s and $Z$s as special sets of attributes: as Leibnizian *individual concepts*. Such an individual concept $I_y$ is one for which it holds for any attribute,

he proposed as a rule that identicals can always be substituted for one another — the principle we know as 'Leibniz law' — and the following set of axioms:

(a$_1$) $A = AA$

(a$_2$) $A0 = 0$

(a$_3$) If $A = AB$, then $A \neq A\overline{B}$

(b) $AB = BA$

(c) If $A = B$, then $\overline{B} = \overline{A}$

(d) $A(BC) = (AB)C$[19]

In terms of this system, the identity $A = A$, for instance, can now be derived: $A = AA$ and $AA = A$ are both instantiations of axiom 1, and by substitution it follows that $A = A$. The validity of conversion and contraposition immediately follow from axioms (b) and (c). In fact, one can now check that in terms of the substitution rule and these axioms, all valid syllogisms can be derived (if only consistent terms are used). To illustrate this, let us see that the Dictum de Omni is now redundant, because Barbara and Darii can be easily derived. For Barbara, assume $A = AB$ and $B = BC$. From this we conclude by substitution that $A = A(BC)$. This reduces by (d) and substitution again to $A = AC$, which is what we desired. For Darii, assume $A = AB$ and $CA \neq 0$. By substitution we conclude that $C(AB) \neq 0$. Now, suppose that $CB = BC = 0$. By (a$_2$) and (b) it would follow that also $ABC = CAB = 0$, which is in contradiction with what we concluded from the premises. Thus $CB \neq 0$, as desired.

---

or feature, $f$ that $f \notin I_y$ iff $\overline{f} \in I_y$. Quantifying over individual concepts makes the universal sentence $SaP$ true iff $\forall I_y, \exists I_z : I_y S = I_z P$. But this comes down to $\forall I_y \supseteq S : \exists I_z \supseteq P : I_y = I_z$. Switching from Leibniz' intentional to Ockham's extensional point of view comes down to switching from $I_y \supseteq S$ to $y \in S$. But this means that by making this switch, Leibniz's truth conditions of the categorical sentences come down to Ockham's.

[19] According to Castaneda [4] and Lenzen [19], Leibniz never explicitly stated this associativity axiom, but always assumed it.

The above system is geared towards syllogistic reasoning, and not capable of handing all of propositional logic. But now suppose that we add the following principle that has been explicitly stated by Leibniz to our above system:

(a) $A\overline{B} = C\overline{C}$ iff $A = AB$, for any arbitrary $C$

It is not so difficult to see that by adding this axiom our earlier ($a_1$), ($a_2$), and ($a_3$) become redundant (with $0 = C\overline{C}, \forall C$). But adding this axiom has a much more important impact: it can be shown (cf. Goodstein, [12] that in fact we now have the whole of Boolean algebra. In terms of Boolean algebra we can still account for syllogistic reasoning, of course, but we can also interpret it such that it accounts for propositional logic. The only thing we have to do is to interpret $AB$ as the conjunction of $A$ and $B$ and assume that $A \neq 0$ iff $A = 1$. In this way we restrict ourselves to the two-element Boolean algebra **2** and have the whole of propositional logic.

## 3.2  A more standard analysis using complex terms

Leibniz' analysis of propositional logic is very nice, but looks rather different from the standard way of representing syllogistic sentences. It is possible to account for propositional logic analysis as well, however, in terms of a more standard-looking syllogistic analysis. Again we assume that terms can be interpreted as 0-ary predicates. As mentioned earlier, we will assume that a 0-ary term will denote (from an extensional point of view) either *the* truth, or *the* falsity. To state things in a metaphysically less committing way, let us start again with a domain of individuals $D$. Now we assume that the denotation of any $n$-ary term will be a subset of $D^n$, where $D^2$, for instance, is $\{\langle d_1, d_2 \rangle : d_1, d_2 \in D\}$. Thus, just like any 1-ary term will denote a subset of $D^1 = \{\langle d \rangle : d \in D\} = D$,[20] any 0-ary terms will now denote a subset of $D^0 = \{\langle \rangle\}$. It is clear that $D^0$ has exactly two subsets: $\{\langle \rangle\}$ and $\emptyset$.[21] We say that if a sentence denotes $\{\langle \rangle\}$ it is true, and false otherwise. We can assume that

---

[20] We will assume it is clear how complex 1-ary predicates like $ST$ and $\overline{S}$ should be interpreted extensionally.

[21] At least, if $D$ is not empty.

for primitive 0-ary terms, or sentences, like 'It is raining' the denotation is given by the interpretation function of the model: $V_M$. In terms of this, we define $V_M(SaP) = \{\langle\rangle : V_M(S) \subseteq V_M(P)\}$ and $V_M(SiP) = \{\langle\rangle : V_M(S) \cap V_M(P) \neq \emptyset\}$. Because we have negative terms, we don't need special interpretation rules for $SeP$ and $SoP$. For 0-ary predicte $\phi$ it holds that $V_M(\phi) = \{\langle\rangle\}$ iff $V_M([\top]a[\phi]) = \{\langle\rangle\}$. Notice also that because $[\top]$ always denotes $\{\langle\rangle\}$, in our extensional fragment it holds for every $\phi$ that $V_M([\top]a[\phi]) = V_M([\top]i[\phi])$.[22] Thus, 'the truth' behaves like a proper name. In addition to negative terms, of form $\overline{T}$, we also have conjunctive terms, $ST$ and thus disjunctive terms, $S \vee T = \overline{\overline{S}\overline{T}}$. Negative, conjunctive and disjunctive sentences can now be represented straightforwardly as well. Conjunctive sentences like '$\phi$ and $\psi$', for instance, can be represented both by $[\phi]i[\psi]$ as we did before, as well as by $[\top]i[\phi\psi]$ (or $[\top]a[\phi\psi]$), making use of conjunctive terms. Similarly for disjunctive sentences. Inference can now be stated semantically as follows: $\Gamma \models \phi$ iff $\forall M : \bigcap_{\gamma \in \Gamma} V_M(\gamma) \subseteq V_M(\phi)$.

To reason with sentences involving such complex terms proof-theoretically, we have to add some new inference rules. The first one is needed to account for inferences involving conjunction, while the second one is very standard in generalized quantifier theory (Barwise & Cooper, [2]):[23]

- Conjunction: $SaP$ and $SaQ \dashv \vdash SaPQ$
- Conservativity: $SxP \vdash SxSP$, for any quantifier '$x$' (thus $x \in \{a, i\}$).

---

[22]If we require that all sentences are of subject-predicate structure, a primitive 0-ary term by itself doesn't constitute a sentence. But this is a problem, because then it cannot be used in syllogistic reasoning. To nevertheless reason with such sentences after all, Ramus apparently introduced new inference rules. Wallis and others argued that this is not required. Wallis proposed to use such terms in syllogistic reasoning by assuming that 'it is raining' really means something like 'Now, it is raining'. To account for this we would have to add terms to the language that stand for points of time or place, and add singular terms like 'now', $N$, and 'here' that refers to the present point in time or place. A sentence like 'It is raining' can then be represented by something like $NaR$. A simpler route, which is more in accordance with what Frege did, would be to represent the sentence by $\top iR$ instead.

[23]To make use of the Dictum de Omni, one would also need to know how the distribution values of complex terms determine the distribution values of the (simpler) terms it contains. We leave that to the reader.

One can immediately see that the conjunction rule accounts for conjunction introduction and conjunction elimination at the sentence level. Of course, conjunction elimination is a form of monotonic reasoning, and follows from the Dictum de Omni if we assumed, instead, $\vdash SPaP$ as an axiom ($SaPQ \vdash SPaP, SaPQ \vdash_{DDO} SaP$). The second rule is also stated more generally than required. Indeed, the rule $SaP \vdash SaSP$ can be already derived making use of the laws of identity and conjunction: $SaP \vdash_{identity} SaS, SaP \vdash_{conj} SaSP$. Thus, all conservativy adds is $SiP \vdash SiSP$. As it turns out, this addition still does some important work to prove that the distributivity axioms hold: (a) $x \vee (y \wedge z) = (x \vee y) \wedge (x \vee z)$ and (b) $x \wedge (y \vee z) = (x \wedge y) \vee (x \wedge z)$. We leave to the appendix the details of the proof that with the help of Conjunction and Conservativity our syllogistic proof system is strong enough to account for all axioms of Boolean algebra, and thus can account for propositional logic.

## 3.3 Embedding categorical into propositional sentences

In this section we saw that syllogistics with conjunctive terms can account for propositional logic. This is nice, but, of course, we want something more. We want to be able to *combine* syllogistics and propositional logic. More in particular, we want to know how to represent, interpret and reason with complex propositional sentences that *embed* categorical sentences, such as 'If every woman is happy, every man is happy'. Leibniz had a proposal for how to deal with that: allow categorical sentences to be turned into terms:

> A proposition itself can be conceived as a term: thus, 'Some $A$ is $B$', i.e., '$AB$ is a true term' is a term – namely, '$AB$true'. Again, we have 'Every $A$ is $B$', i.e. '$A$ not-$B$ is false', i.e. '$A$ not-$B$false' is a new term; and again 'No $A$ is $B$', i.e. '$AB$ is false', i.e. '$AB$false' is a new term [15][24]

Although we haven't followed Leibniz' exact idea, we assumed already that a sentence is a term; a 0-ary predicate that has a denotation. Thus, if $S$ and $P$ are 1-ary terms, sentences $SaP$ and $SiP$ are

---
[24]With the substitutions as suggested by Castañeda [4].

0-ary terms. The denotations of categorical sentences are not 1 or 0, as is standardly done, but rather $\{\langle\rangle\}$ (if the sentence is true) or $\emptyset$ ( if the sentence is false). Because we assume this, we can represent conditional sentences like 'If every woman is happy, every man is happy' simply by $[WaH]^0 a[MaH]^0$, and the sentence will be true just in case $WaH$ is false, or $MaH$ is true. To reason with such sentences making use of de Dictum de Omni, we have to know what the distributive value is of each of (the occurrences of) the terms in this sentence. But that has been worked out already by the $+-$ rules of Sommers (1982) ($++ = +, +- = -, -+ = -$ and $-- = +$) together with the distributive marking of universal sentences like $S^- a P^+$. From this it follows, for instance, that because the term $[WaH]^0$ occurs negatively in the whole sentence, term $W$ occurs positively, while the first occurrence of $H$ occurs negatively. Similarly, because the term $[MaH]^0$ occurs positively in the whole sentence, the term $M$ occurs negatively in the whole sentence, while the second occurrence of $H$ occurs positively. By the Dictum de Omni (or monotonicity reasoning more generally) one can then conclude from the truth of the above sentence that also 'If every *human* is happy, every man is happy' and 'If every woman is happy, every man is *happy or sad*' are true, for instance.

## 4 Intensional semantics

### 4.1 Intensional semantics for syllogistics with complex terms

Aritstotle made in his Prior Analytics a distinction between two different types of predicative relations: accidental versus essential predication. 'Animal' is essentially predicated of 'mammal', but 'walking' is not. Although both (1) 'Every man walks' and (2) 'Every man is an animal' are true, the 'reasons' for their respective truths are different. Sentence (1) is true by accident, just because every actual man happens to (be able to) walk. The sentence (2), on the other hand, is true because manhood necessarily involves being animate. In traditional terms it is said that (2) is true *by definition*, although this notion of 'definition' should not be thought of nominalistically: it is the *real* definition. A natural way to account for accidental predication is to say that a sentence of the form

'Every $S$ is $P$' is true just in case every actual $S$-*individual* is also a *P-individual*. A natural way to account for essential predication, on the other hand, is to say that a sentence of the form 'Every $S$ is $P$' is true just in case the *real* definition of $S$ (the set of attributes one needs to have to be an $S$) includes the real definition of $P$ (the set of attributes one needs to have to be a $P$). We will say that the first way to determine whether 'Every $S$ is $P$' is true is *extensional* in nature, the second way *intensional*.

Aristotle clearly assumes that essential predication is stronger than accidental predication, and medieval logicians very much assumed the same. Especially due to the influence of the Port-Royal school of logic, however, it became standard in the 17th century to assume that the two come down to the same thing. Leibniz explicitly endorsed this position. Let's see how we can make sense of it.[25]

It is well-known how to give an *extensional semantics* for syllogis-

---

[25] According to Leibniz, Aristotle, in contrast to a nominalist like Locke, preferred the intensional interpretation:

> *Philalethes* (expressing Locke's view) [...] it appeared to me preferable to reverse the order of the premisses of syllogisms, and to say: *All A is B, all B is C, so all A is C*, rather than saying *All B is C, all A is B, so all A is C*. [...]
>
> *Theophilus* (expressing Leibniz's view) [...] Aristotle may have had a special reason for adopting [what is now] the common arrangement. For rather than saying 'A is B' he usually says 'B is in A' [...]. And with that way of stating it he achieves, through the accepted arrangements, the very connection which you insist upon. For instead of saying 'B is C, A is B, so A is C', Aristotle will express it thus: 'C is in B, B is in A, so C is in A'. For instance, instead of saying 'Rectangles are isogons (i.e. have equal angles), squares are rectangles, so squares are isogons', Aristotle will put the 'middle term' in the middle position without changing the order of the propositions, by stating each of them in a manner which reverses the order of terms, thus: 'Isogon is in rectangle, rectangle is in square, so isogon is in square'. This manner of statement deserves respect; for indeed the predicate is in the subject, or rather the idea of the predicate is included in the idea of the subject. [...] The common manner of statements concerns individuals, whereas Aristotle's refers rather to ideas or universals. [...]
>
> Leibniz, New Essays on Human Understanding, Book 4, chapter 17, sect. 8)

tic logic. Almost all modern textbooks use Venn diagrams to decide whether a syllogistic inference is valid. According to it, interpretation function $E$ assigns to each primitive term $T$ a non-empty subset of the set of objects $D$: $\emptyset \neq E_M(T) \subseteq D$. The sentence $SaP$ is true iff the *extension* of $S$ ($E(S)$) is a subset of the *extension* of $P$, i.e. $E(S) \subseteq E(P)$ iff $E(S) \cap E(P) = E(S)$, and $SiP$ is true iff $E(S) \cap E(P) \neq \emptyset$. $SoP$ and $SeP$ are interpreted as the negations of $SaP$ and $SiP$, respectively. But the representation by means of Venn diagrams assumes the axioms of Boolean algebra, and this much structure is not at all required to model syllogistic reasoning. For the traditional fragment without conjunctive and negative terms, a partially ordered set $\langle U, \leq \rangle$, together with a relation @ which is reflexive and monotonic w.r.t. $\leq$ is already enough (in fact, $\leq$ need only be reflexive and transitive). On the extensional interpretation, $E(T)$ denotes an element of $U$ for any term $T$, and $SaP$ is true iff $E(S) \leq E(P)$ and $SiP$ is true iff $E(S)@E(P)$. The truth-conditions of $SoP$ and $SeP$ are determined as usual in terms of the truth-conditions of $SaP$ and $SiP$. Alternatively, we can assume a special element 0 of $U$ and determine the meet semi-lattice $\langle U, \wedge \rangle$ defined by $\langle U, \leq \rangle$. The special element is such that $\forall T \in Term : 0 \wedge E(T) = 0$. If we start with the semi-lattice, we can define $SaP$ is true iff $E(S) \wedge E(P) = E(S)$ and $SiP$ is true iff $E(S) \wedge E(P) \neq 0$. An argument is valid just in case it preserves truth: $\phi_1, ..., \phi_n \models \psi$ iff $\forall M : V_M(\phi_1) = 1$ and ... and $V_M(\phi_n) = 1$, then $V_M(\psi) = 1$.

For an *intensional semantics*, we also assume a partially ordered set $\langle U, \leq \rangle$, together with a relation @ which is reflexive and monotonic w.r.t. $\leq$. We assume that intensional interpretation function $I$ assigns to each primitive term an element of $U$, just like in the extensional semantics. We demand that $SiP$ is true iff $I(S)@I(P)$ and $SaP$ is true iff $I(S) \geq I(P)$. Note that the 'intensional' order is thus the reverse of the 'extensional' order. We can also start with a semi-order with special element 0. In that case we still say that $SiP$ is true iff $I(S) \wedge I(P) \neq 0$, but now 0 should be thought of as the greatest element: $\forall T \in Term : 0 \wedge I(T) = I(T)$. On this analysis, $SaP$ is true iff $I(S) \wedge I(P) = I(P)$. Validity is always defined as in the extensional analysis.

Thus we see that algebraically speaking, there is not much of a difference between an extensional and an intensional interpretation. This is still the case if we allow for conjunctive and negative terms. Once we have '∧' as an operation between terms, we can also account for conjunctive terms. If we add term-negation to the language, we take $\langle U, \leq \rangle$ to be a distributive lattice (with ∧ and ∨ infimum and supremum), and assume that for all $x, y \in U : \overline{\overline{x}} = x$ and $x \leq y \to \overline{y} \leq \overline{x}$. If we also want empty and universal terms, we assume that we have a whole Boolean algebra.

An algebraic semantics is nice, but a set theoretic analysis is more intuitive. For the extensional semantics, each term then simply denotes a set of individuals. It is straightforward how to proceed with such a semantics. The most straightforward idea for an intensional semantics would be that a term denotes a set of features. Indeed, it is standardly assumed in traditional logic that the *intension* of a term, or concept, consisted of all the essential attributes in it (those that cannot be removed without 'destroying' the concept). Thus, the intension of the term 'triangle' might include the attributes of being polygon, three-sided, three-angled, and so on. On such a semantics, a categorical sentence of the form '(All) men are rational' is true if and only if the *comprehension*, or *intension*, of 'rational' is contained in the *intension* of 'men'. In the whole history of (traditional) logic, this way of interpreting categorical sentences was very popular. Leibniz, for instance, made it his basic principle: a categorical sentence is true iff the predicate is 'part of' the subject.

Straightforward as such an intensional semantics might look, it is not at all obvious how to proceed when the language also contains conjunctive and negative terms. Indeed, Leibniz was never able to give a satisfactory semantics. In fact, he already struggled how to account for categorical sentences of the form $SiP$ (cf. [33, 11]). To solve the difficulties, we propose to *lift* the interpretations: terms should not be interpreted as *sets* of attributes, but rather as *sets of sets* of attributes.[26] The intensional interpretation of terms and sentences is defined as follows:

---

[26]Leibniz solved the problem how to account also for sentences like $SiP$ in quite another way, representing each term, essentially, by an ordered pair of intensions. For modern explanations of this method, see Sotirov [33] and Glashoff [11]. Another way to solve the problem is to represent proporties by a pair of extensions and intensions, as is proposed in formal concept analysis (cf. Willle, [35])

Let $M = \langle A, I, \bot \rangle$ be a model, with $A$ a set of attributes, $I$ an interpretation function which assigns to each primitive term $T$ a subset of $\wp(A)$, $I_M(T) \subseteq \wp(A)$, and $\bot$ a relation between elements of $A$. If for two elements $x, y \in A$ it holds that $x \bot y$, we say that the attributes $x$ and $y$ are incompatible. We assume that for every $f \in A$ there is exactly one $g \in A$: $f \bot g$ (which you might call $\overline{f}$). If $X$ and $Y$ are subsets of $A$, we say that $X \bot Y$ iff $\exists x, y \in X \cup Y : x \bot y$. $Min(X)$ stands for the set $\{x \in X : \neg \exists y \in X : y \subset x\}$. We will assume that for each primitive *categorical* term $T$, each element of $I_M(T)$ is consistent. The interpretation rule for categorical terms in general is defined as follows:

- If $T$ is a primitive term, then $V_M(T) = I_M(T)$
- $V_M(TT') = \{X \cup Y : X \in V_M(T) \ \& \ Y \in V_M(T')\}$
- $V_M(\overline{T}) = Min\{Y \subseteq A : \forall X \in V_M(T) : X \bot Y\}^{27}$

To illustrate, assume that $A = \{x, y, z, v\}$, and $x \bot y$ and $z \bot v$. In that case, the set of compatible sets of features is $\{\{x\}, \{y\}, \{z\}, \{v\}, \{x, z\}, \{x, v\}, \{y, z\}, \{y, v\}\}$. The set of maximally consistent sets of attributes is thus $\{\{x, z\}, \{x, v\}, \{y, z\}, \{y, v\}\}$. Predicates don't denote set of attributes, but rather a *set* of sets of attributes. The sets $\{\{x, z\}\}, \{\{x, v\}\}$, $\{\{y, z\}\}$, and $\{\{y, v\}\}$ can be thought of as the denotations of proper names, which are thus of the same type as ordinary properties. Singleton sets $\{\{x\}\}, \{\{y\}\}, \{\{z\}\}$ and $\{\{v\}\}$ can be thought of as (the denotations of) simple properties. Call them $X, Y, Z$, and $V$, respectively. The compatible conjunctive properties can then be denoted by $XZ, XV, YZ$, and $YV$ (in this simple example, these coincide with the denotations of proper names). Disjunctive properties typically denote non-singleton sets. The disjunctive property consisting of $X$ and $Z$, for instance, denotes $\{\{x\}, \{z\}\} = V_M(X) \cup V_M(Z)$. Whether negative predicates denote singleton sets or not depends on how many incompatibles they have. Because $x$, for instance, is only incompatible with $y$, $\overline{X}$ denotes the singleton sets $\{\{y\}\}$. It is different with conjunctive

---

[27] Notice that now it follows that $I_M(\overline{TT'}) = I_M(\overline{T}) \cup I_M(\overline{T'})$. It follows that a disjunctive term like $T \vee T'$ is interpreted as $I_M(T) \cup I_M(T')$.

properties like $XV$: $V_M(\overline{XV}) = \{\{y\},\{z\}\} = V_M(\overline{X}) \cup V_M(\overline{V})$. Notice that from these we can go back to the original via double negation: $V_M(\overline{\overline{X}}) = \{\{x\}\} = V_M(X)$ and $V_M(\overline{\overline{XV}}) = \{\{x,v\}\} = V_M(XV)$.

Based on this we can, as our first trial, propose the following interpretation rules for sentences:

- $V_M(SaP) = 1$  iff  $\forall X \in I_M(S) : \exists Y \in V_M(P) : X \supseteq Y$,

- $V_M(SiP) = 1$  iff  $\exists X \in V_M(S) : \exists Y \in V_M(P) : \neg(X \bot Y)$.

- $SoP$ and $SeP$ are interpreted as the negations of $SaP$ and $SiP$, respectively.

These interpretation rules for sentences seem straightforward, and in line with the 'predicate-in-subject' analysis. But they are *too intensional*. In particular, although men are not by definition sick, it can still be the case that 'All men are sick' is actually true. Similarly, although the terms *Man* and *Sick* are compatible, it need not be the case that thus the sentence 'Some men are sick' is true. In van Rooij [25] other interpretation rules were used making use of individuals as thought of by Leibniz. According to this proposal, what is wrong with the previous semantic analysis of $SaP$ and $SiP$ is that it is too general, making use of *all possible* individuals. Instead, we should only make use of all *actual* individuals. On a proper intensional analysis, of course, we should not start with individuals, but define these in terms of sets of features, or attributes (cf. [11]). Let $D$ (the possible individuals) be the maximally consistent sets of attributes. For each model (or world) $M$, we will let $D_M$ (the actual individuals) be a subset of $D$, and think of its elements as the set of maximally consistent sets of attributes that 'made it' to $M$ (perhaps due to Leibniz' principle of sufficient reason). We will denote by $F_M(X)$ the elements of $D_M$ that are extensions of $X : \{Y \in D_M : X \subseteq Y\}$, and similarly, $F(X) = X : \{Y \in D : X \subseteq Y\}$ (thus, $F(X)$ is the set of all *possible* individuals that have all features of $X$). We assume that $F_M(I_{(M}(T)) \neq D_M$. Now we define the truth conditions of the categorical propositions as follows:

- $V_M(SaP) = 1$  iff  $\forall X \in V_M(S) : \exists Y \in V_M(P) : F_M(X) \subseteq F_M(Y)$

- $V_M(SiP) = 1$   iff   $\exists X \in V_M(S) : \exists Y \in V_M(P) : F_M(X) \cap F_M(Y) \neq \emptyset$

Notice that $X \supseteq Y$ iff $F(X) \subseteq F(Y)$. In that sense the analysis is thus very similar to Leibniz' natural analysis for sentences of the form $SaP$. But by our use of $F_M(X)$ instead of $F(X)$ it is not the same, because it need not be the case that all maximally consistent set of properties corresponds with an actual individual. It follows, unfortunately, that for sentences involving complex terms, like $STaP$, we can no longer guarantee that we can conclude that $STiP$. In van Rooij [25] it is shown how to turn this analysis into a Boolean analysis.

## 4.2 Intensional semantics for syllogistic-based propositional logic

So far we have thought of the set $A$ in our models as a set of features. This was needed to account for an intensional interpretation of the standard syllogistic inferences. Interestingly, we can also think of our model as a model for **propositional logic**. It is well known that Russell [27] blamed the traditional logical idea that every sentence is of subject-predicate form and that a sentence is true iff the (meaning of) the predicate is 'part of' the (meaning of the) subject for being responsible for the Hegelian philosophical excesses. Ironically enough, however, our intensional semantics is closely related with Russell's [28] own way of thinking about propositional logic in terms of facts.[28] Indeed, we can think of our set $A$ also as a set of original **state of affairs** and of the relation $\perp$ as an incompatibility relation between state of affairs. State of affairs played an important role in Russell's [28] and Wittgenstein's [36] fact-based ontology. **Facts** can be thought of as sets of state of affairs. Facts can be of two kinds: they are either primitive (including negative) or conjunctive. If $\mathbf{p}$ (and $\overline{\mathbf{p}}$) and $\mathbf{q}$ are state of affairs, $\{\mathbf{p}\}$ (and $\{\overline{\mathbf{p}}\}$) and $\{\mathbf{p}, \mathbf{q}\}$ are facts. Worlds can be thought of as conjunctive facts. The set $W$, the worlds, can then be thought of as the set of maximally consistent sets of state of affairs. Disjunctive and conditional facts do not exist. To account for disjunctions and conditionals we make use of

---

[28]This was at least recognized by Wittgenstein in his Tractatus, 5.1222: if $p$ follows from $q$, then the sense of $p$ is contained in the sense of $q$.

**propositions**. Indeed, starting with $A$ as a set of state of affairs, and of $\wp(A)$ as the set of facts, we can think of the elements of $\wp(\wp(A))$ as the set of propositions. Sentences denote propositions. Atomic sentence '$p$' denotes $\{\{\mathbf{p}\}\}$, conjunctive sentence '$p \wedge q$' denotes $\{\{\mathbf{p},\mathbf{q}\}\}$, while disjunctive setence '$p \vee q$' denotes $\{\{\mathbf{p}\},\{\mathbf{q}\}\}$. This way of modelling propositions was proposed by Van Fraassen [7] and our way to provide an intensional semantics for syllogistic reasoning was indeed modelled after Van Fraassen's modelling of propositional logic.[29]

In section 3 we showed that we can account for standard propositional logic as an extension of syllogistic reasoning. This gives rise to the question whether we can do the same now for our intensional semantics. Thus, if $\phi$ and $\psi$ denote propositions, can we think of $\phi \to \psi$ and $\phi \wedge \psi$, for instance, as being represented by $[\phi]a[\psi]$ and $[\phi]i[\psi]$, respectively?

Indeed, sentences like '$(p \wedge q) \to q$' and '$p \wedge q$' now come out as true on the first proposal as discussed above. This is because $\{\mathbf{q}\}$ is a subset of $\{\mathbf{p},\mathbf{q}\}$ and the latter set is consistent. But this first semantics of sentences is obviously much too stringent for a material implication account of conditionals, while conjunctive sentences are predicted too easily to be true. Again, this is due to the too intensional analysis of our first proposal. Our final intensional analysis, fortunately, is much better. For that to make sense, we think now of $F_M$ to be the set of maximally consistent sets of facts that 'made it' to the actual world. But we assume that there is only one such maximally consistent set, and thus $F_M$ is a singleton set. As a result, if $X$ is a set of state of affairs, $F_M(X)$ is either $F_M$ (if all state of affairs in $X$ hold in $M$), or $\emptyset$. A sentence denotes $F_M$ in case it is true, and $\emptyset$ otherwise. But this means that $[\phi]a[\psi]$ is true just in case either $\phi$ is false, or $\psi$ is true, and we interpret it thus like a material implication. Similarly, $[\phi]i[\psi]$ can only be true just in case both $\phi$ and $\psi$ are true, just like in standard propositional logic. We conclude that our intensional analysis provides a nice semantics for propositional logic as well.

---

[29] In fact, Van Fraassen [7] proposed his analysis to provide a semantics for relevance propositional logic, but shows that it can be used to account for standard propositional logic as well.

## 5 Conclusion

Term logic can account for propositional reasoning. This was already clear from the work of Leibniz and Sommers. The contributions of this paper were to show (i) what is required of the proof-system to account for propositional reasoning in a syllogistic way once we account for complex terms, and (ii) how to provide an intensional semantics for syllogistic reasoning such that it also captures propositional logic. Interestingly enough, the latter semantics is fact-based, a type of semantics that has gained enormous popularity recently in philosophy (and natural language semantics).

Term logic can also account for reasoning involving relations (cf. Sommers, [32]) and Aristotle himself already looked into modal syllogisms. How to provide an intensional semantics for reasoning with relations remains a very open issue. Instead, much progress has been made recently how to account for modal syllogisms (cf. Malink, [23], van Rooij & Xie, [26]). It might seem that our intensional semantics could be relevant here. Indeed, it seems natural that to account for Aristotle's intuitions concerning the (in)validity of some modal syllogisms, we should take his metaphysically-laden real definitions into account. Unfortunately, adopting Leibniz' idea that the intensional semantics should match the extensional semantics, as we have done in this paper, won't reflect Aristotle's assumption that essential predication is stronger than accidental predication.

## References

[1] Ashworth, E.J. (1974), *Language and logic in the post-medieval period*, Boston, Reidel.

[2] Barwise, J. & R. Cooper (1981), 'Generalized quantifiers in natural language', *Linguistics and Philosophy*, 4: 159-219.

[3] Benthem, J. van (1983), 'A linguistic turn: New directions in logic', in R. Marcus et al. (eds.), *Logic, Methodology and Philosophy of Science*, Salzburg, pp. 205-240.

[4] Casteñada, H.N. (1990), 'Leibniz's complete propositional logic', *Topoi*, **9**: 15-28.

[5] Englebretsen, G. (1981), *Three Logicians: Aristotle, Leibniz, Sommers and the Syllogistic*, Van Gorcum, Assen.

[6] Eijck, J. van (1985), *Aspects of Quantification in Natural Language*, Dissertation, Philosophical Institute, University of Groningen.

[7] Fraassen, B. van (1969), 'Facts and tautological entailments', *Journal of Philosophy*, **66**: 477- 487.

[8] Friedman, W. (1980), 'Calculemus', *Notre Dame J. Formal Logic*, **21**: 166-174.

[9] Gassendi, P. (1658), *Institutio Logica*, (part of his *Opera Omnia*, Lyon, Anisson/Devenet. (See also the Critical edition with translation and introduction by H. Jones, (1981), Van Gorcum, Assen.)

[10] Geach, P. (1962), *Reference and Generality. An Examination of Some Medieval and Modern Theories*, Cornell University Press, Oxford.

[11] Glashoff, K. (2010), 'An intensional Leibniz semantics for Aristotelian logic', *The Review of Symbolic Logic*, **3**: 262-272.

[12] Goodstein, R.L. (1963), *Boolean Algebra*, Pergamon Press, London.

[13] Kant, I, (1992), *Lectures in Logic, Part III: Dohna-Wundlacken Logik*, edited by J. Young, Cambridge University Press, Cambridge.

[14] Kneale, W. and M. Kneale, (1962), *The Development of Logic*, Clarendon Press, Oxford.

[15] Leibniz, G. (1966), *Logical Papers*, edited and translated by G. H. R. Parkinson, Clarendon Press, Oxford.

[16] Leibniz, G. (1973), 'The nature of truth', in Parkinson (ed.), *Leibniz. Philosophical Writings*, pp. 93-95.

[17] Leibniz, G. (1996), New Essays on Human Understanding, 2nd ed., translated and edited by Peter Remnant and Jonathan Bennett, New York: Cambridge University Press.

[18] Lenzen, W. (1983), 'Zur extensionalen und "intensionalen" interpretationen der Leibnizschen logic', *Studia Leibnitiana*, 129-148.

[19] Lenzen, W. (1990), *Das System der Leibniz'schen Logik*, De Gruyter, Berlin.

[20] Lukasiewicz, J. (1951), *Aristotle's Syllogistic from the standpoint of modern formal logic*, Clarendon Press, Oxford.

[21] Lyndon, R.C. (1959), 'Properties preserved under homomorphism', *Pacific Journal of Mathematics*, **9**: 142-154.

[22] MacIntosh, C (1982), 'Traditional formal logic', Appendix F of Sommers (1982), pp. 387-425.

[23] Malink, M. (2013), *Aristotle's Modal Syllogistic*, Harvard University Press.

[24] Rooij, R. van (2012), 'The propositional and relational syllogistic', *Logique et Analyse*, **55**: No. 217: 85-108.
[25] Rooij, R. van (2014), 'Leibnizian intensional semantics for syllogistic reasoning', in H. Wansing et al.(eds.), *Recent Trends in Philosophical Logic*, Springer, pp. 179-194.
[26] Rooij, R. van & K. Xie (2020), 'A causal analysis of modal syllogisms', in *Monotonicity in Logic and Language*, Springer, Berlin, pp. 183-206.
[27] Russell, B. (1900), *A critical exposition of the Philosophy of Leibniz*, Cambridge University Press, Oxford.
[28] Russell, B. (1924), 'Logical atomism', in J. Muirhead (Ed.), *Contemporary British Philosophy: First statements, Frst series*, (p. 359-383), London.
[29] Sanchez, V. (1991), *Studies on Natural Logic and Categorial Grammar*, PhD thesis, Universiteit van Amsterdam.
[30] Sommers, F. (1970), 'The calculus of terms', *Mind*, **79**: 1-39.
[31] Sommers, F. (1975), 'Distribution matters', *Mind*, **84**: 27-46.
[32] Sommers, F. (1982), *The Logic of Natural Language*, Oxford, Oxford University Press.
[33] Sotirov, J. (1999), 'Arithmetization of Syllogistic a la Leibniz', *Journal of Applied Non-Classical Logics*, **9**: 387-405.
[34] Wallis, J. (1687), *Institutio Logica*, Oxford.
[35] Wille, Rudolf (1982). 'Restructuring lattice theory: An approach based on hierarchies of concepts'. In Rival, Ivan (ed.). *Ordered Sets. Proceedings of the NATO Advanced Study Institute*, Nato Science Series C. Vol. 83. Springer. pp. 445–470.
[36] Wittgenstein, L. (1933), *Tractatus Logico-Philosophicus*, Kegan Paul, Trench, Trubner & Co, London & New York.

# 6 Appendix

To prove that our proof system **SYLS**$^+$ can account for all of propositional logic when Conjunction and Conservativity are added, we have to show that the new proof system can account for all of the laws of Boolean Algebra below:

1. Idempotence
   (a) $X \vee X = X$  (b) $X \wedge X = X$
2. Commutativity
   (a) $X \vee Y = Y \vee X$  (b) $X \wedge Y = Y \wedge X$
3. Associativity
   (a) $(X \vee Y) \vee Z = X \vee (Y \vee Z)$  (b) $(X \wedge Y) \wedge Z = X \wedge (Y \wedge Z)$
4. Distributivity
   (a) $X \vee (Y \wedge Z) = (X \vee Y) \wedge (X \vee Z)$
   (b) $X \wedge (Y \vee Z) = (X \wedge Y) \vee (X \wedge Z)$
5. Identity
   (a) $X \vee \bot = X$  (c) $X \wedge \bot = \bot$
   (b) $X \vee \top = \top$  (d) $X \wedge \top = X$
6. Complementation
   (a) $X \vee \overline{X} = \top$  (c) $X \wedge \overline{X} = \emptyset$
   (b) $\overline{(\overline{X})} = X$  (d) $X - Y = X \cap \overline{Y}$
7. DeMorgan's laws
   (a) $\overline{(X \vee Y)} = \overline{X} \wedge \overline{Y}$  (b) $\overline{(X \wedge Y)} = \overline{X} \vee \overline{Y}$
8. Consistency
   (a) $XaY$ iff $X \vee Y = Y$.
   (b) $XaY$ iff $X \wedge Y = X$

Before we show that these laws are indeed all valid, let us first prove some preliminaries:
(we omit some obvious variants of the sort $SP \leftrightarrow PS$)

1. $Sa\bar{\bar{P}} \vdash SaP$ and $\bar{\bar{S}}aP \vdash SaP$ (double negation 2)
   $Proof$: $\vdash \bar{\bar{P}}aP, Pa\bar{\bar{P}}$ plus DDO.

2. $\vdash PQaP$ and $\vdash PQaQ$ (elim conjunction)
   $Proof$: $\vdash PQaPQ \stackrel{conj}{\Rightarrow} \vdash PQaP$, etc.

3. $SaP \vdash Sa\overline{\bar{P}\bar{Q}}$ and $SaP \vdash Sa\overline{\bar{Q}\bar{P}}$ (intro disjunction)
   $Proof$: $\vdash \bar{P}\bar{Q}a\bar{P}\bar{Q} \stackrel{conj}{\Rightarrow} \vdash \bar{P}\bar{Q}a\bar{P} \stackrel{cp}{\Rightarrow} \vdash \bar{\bar{P}}a\overline{\bar{P}\bar{Q}} \stackrel{dn}{\Rightarrow} \vdash Pa\overline{\bar{P}\bar{Q}}$, etc.

4. $\vdash Pa\overline{\bar{P}\bar{Q}}$ and $\vdash Pa\overline{\bar{Q}\bar{P}}$ (intro disjunction 2)
   $Proof$: id plus DDO.

5. $SaP \vdash STaP$ and $SaP \vdash TSaP$ (strengthening antecedent)
   $Proof$: $\vdash STaST \stackrel{conj}{\Rightarrow} \vdash STaS$, so $SaP \vdash SaP, STaS \vdash STaP$ (DDO), etc.

6. $SaP \vdash SaSP$ and $SaP \vdash SaPS$ (strengthening consequent)
   Proof: $\vdash SaS$, so $SaP \vdash SaP, SaS \vdash SaSP$ (conj), etc.

7. $STaP \vdash STaSP$ and $TSaP \vdash TSaSP$, etc. (strengthening consequent 2)
   Proof: $\vdash STaS$, (ec), so $STaP \vdash STaP, STaS \vdash STaSP$ (conj), etc.

8. $STiP \vdash STiTP$ and $TSiP \vdash TSiTP$, etc. (Conservativity 2, abbreviated as con2)
   Proof: $STiP \vdash_{cons} STi(ST)P$ and, with associativity (see below, does not make use of conservatity) followed by conjunction and DDO, $STiP \vdash STiTP$.

9. $SiP \vdash PiS$ (conversion)
   Proof: $Sa\bar{P} \dashv\vdash Pa\bar{S}$ plus def of $SiP$.

10. $SiPQ \vdash SiP$ (weakening)
    Proof: Darii with $PQaP$, obtained by sa of $PaP$.

11. $\vdash \bar{S}(\overline{\bar{S}\bar{P}})aP$ (disjunctive syllogism)
    Proof: assume $\vdash \bar{S}(\overline{\bar{S}\bar{P}})i\bar{P} \stackrel{con2}{\Rightarrow} \vdash \bar{S}(\overline{\bar{S}\bar{P}})i\bar{S}\bar{P}$. But this is in contradiction with $\vdash \bar{S}(\overline{\bar{S}\bar{P}})a(\overline{\bar{S}\bar{P}})$ (conj).

12. $\vdash S(\overline{SP})a\bar{P}$ (conjunctive syllogism)
    Proof: assume $\vdash S(\overline{SP})iP \stackrel{con2}{\Rightarrow} S(\overline{SP})iSP$. This is in contradiction with $\vdash S(\overline{SP})a\overline{SP}$ (conj).

Making use of these preliminaries, we can now show that if we assume in addition to **SYLS**$^+$ the rules of conjunction and conversativity, we can prove all rules of Boolean Algebra, and thus can account for propositional logic:

1. Idempotence:

   (a) $\vdash \bar{S}a\bar{S} \stackrel{conj}{\Rightarrow} \vdash \bar{S}a\bar{S}\bar{S} \stackrel{cp}{\Rightarrow} \vdash \overline{\bar{S}\bar{S}}a\bar{S} \stackrel{dn}{\Rightarrow} \vdash \overline{\bar{S}\bar{S}}aS$ and
   $\vdash \bar{S}\bar{S}a\bar{S}\bar{S} \stackrel{conj}{\Rightarrow} \vdash \bar{S}\bar{S}a\bar{S} \stackrel{cp}{\Rightarrow} \vdash \bar{\bar{S}}\overline{\bar{S}\bar{S}} \stackrel{dn}{\Rightarrow} \vdash Sa\overline{\bar{S}\bar{S}}$

   (b) $\vdash SaS \stackrel{sc}{\Rightarrow} \vdash SSaS$ and $\vdash SaS \stackrel{conj}{\Rightarrow} \vdash SaSS$

PROPOSITIONAL TERM LOGIC

2. Commutativity:

   (a) $\vdash \bar{P}\bar{S}aP\bar{S} \stackrel{comm}{\Rightarrow} \vdash \bar{P}\bar{S}a\bar{S}P \stackrel{cp}{\Rightarrow} \vdash \overline{\bar{S}P}a\overline{\bar{P}S}$.
   The inference '$comm$' anticipates (b).

   (b) $\vdash STaST \stackrel{conj}{\Rightarrow} \vdash STaS, STaT \stackrel{conj}{\Rightarrow} \vdash STaTS$

3. Associativity:

   (a) $\vdash (\bar{S}P)\bar{Q}a(\bar{S}P)\bar{Q} \stackrel{ass}{\Rightarrow} (\bar{S}P)\bar{Q}a\bar{S}(\bar{P}Q) \stackrel{dn\,dn}{\Rightarrow} \vdash (\overline{\bar{\bar{S}P}})\bar{Q}a\bar{S}(\overline{\bar{\bar{P}Q}})$
   $\stackrel{cp}{\Rightarrow} \vdash \bar{S}(\overline{\bar{P}Q})a(\overline{\bar{S}P})\bar{Q}$. Again anticipation by '$ass$'. The other direction goes $very$ similarly.

   (b) $\vdash$ $S(PQ)aS(PQ)$ $\stackrel{conj}{\Rightarrow} \vdash$ $S(PQ)aS, S(PQ)aPQ$ $\stackrel{conj}{\Rightarrow} \vdash$
   $S(PQ)aS, S(PQ)aP, S(PQ)aQ$
   $\stackrel{conj\,conj}{\Rightarrow} \vdash S(PQ)a(SP)Q$. The other direction goes very similarly here too.

4. Distributivity:

   (a) $\vdash PaP \stackrel{sa}{\Rightarrow} \vdash PQaP \stackrel{cp}{\Rightarrow} \vdash \bar{P}a\overline{PQ} \stackrel{sa}{\Rightarrow} \bar{S}\bar{P}a\overline{PQ} \stackrel{sc2}{\Rightarrow} \vdash \bar{S}\bar{P}a\bar{S}(\overline{PQ})$.
   Analogously $\vdash \bar{S}\bar{Q}a\bar{S}(\overline{PQ})$, hence,
   $\stackrel{cp}{\Rightarrow}\stackrel{cp}{\Rightarrow} \vdash \overline{\bar{S}(\overline{PQ})}a(\overline{\bar{S}\bar{P}}), \overline{\bar{S}(\overline{PQ})}a(\overline{\bar{S}\bar{Q}}) \stackrel{conj}{\Rightarrow} \vdash \overline{\bar{S}(\overline{PQ})}a(\overline{\bar{S}\bar{P}})(\overline{\bar{S}\bar{Q}})$
   Assume $\vdash \bar{S}(\overline{PQ})i(\overline{\bar{S}\bar{P}})(\overline{\bar{S}\bar{Q}}) \stackrel{cons}{\Rightarrow} \bar{S}(\overline{PQ})i\bar{S}(\overline{\bar{S}\bar{P}})(\overline{\bar{S}\bar{Q}}) \stackrel{conv}{\Rightarrow}$
   $\bar{S}(\overline{\bar{S}\bar{P}})(\overline{\bar{S}\bar{Q}})i\bar{S}(\overline{PQ}) \stackrel{w}{\Rightarrow} \bar{S}(\overline{\bar{S}\bar{P}})(\overline{\bar{S}\bar{Q}})i(\overline{PQ})$.
   Now (ds) $\vdash \bar{S}(\overline{\bar{S}\bar{P}})aP$ and $\vdash \bar{S}(\overline{\bar{S}\bar{Q}})aQ$, hence (2×sa + conj) $\vdash$
   $\bar{S}(\overline{\bar{S}\bar{P}})(\overline{\bar{S}\bar{Q}})aPQ$. Contradiction, so $\vdash \bar{S}(\overline{PQ})a(\overline{\bar{S}\bar{P}})(\overline{\bar{S}\bar{Q}}) \stackrel{cp}{\Rightarrow} \vdash$
   $(\overline{\bar{S}\bar{P}})(\overline{\bar{S}\bar{Q}})a\bar{S}(\overline{PQ})$.

   (b) Assume $\vdash S\overline{PQ}i(\overline{SP})(\overline{SQ}) \stackrel{cons}{\Rightarrow} \vdash S\overline{PQ}iS(\overline{SP})(\overline{SQ}) \stackrel{conv}{\Rightarrow} \vdash$
   $S(\overline{SP})(\overline{SQ})iS\overline{PQ} \stackrel{w}{\Rightarrow} \vdash S(\overline{SP})(\overline{SQ})i\overline{PQ}$.
   Now (cs) $\vdash S(\overline{SP})a\bar{P}$ and $\vdash S(\overline{SQ})a\bar{Q}$, so (2×sa + conj)
   $\vdash S(\overline{SP})(\overline{SQ})a\bar{P}\bar{Q}$. Contradiction, so $\vdash S\overline{PQ}a(\overline{SP})(\overline{SQ})$.
   $\vdash SPaS \stackrel{cp}{\Rightarrow} \bar{S}a\overline{SP}$, analogously $\vdash \bar{S}a\overline{SQ}$, so $\stackrel{conj}{\Rightarrow} \vdash$
   $\bar{S}a(\overline{SP})(\overline{SQ})$. Also $\vdash \bar{P}a\overline{SP}$ and $\vdash \bar{Q}a\overline{SQ}$, so (2×sa + conj)
   $\vdash \bar{P}\bar{Q}a(\overline{SP})(\overline{SQ})$. By cp we have $\vdash \overline{(\overline{SP})(\overline{SQ})}aS$ and $\vdash$
   $\overline{(\overline{SP})(\overline{SQ})}a\bar{P}\bar{Q}$, therefore (conj) $\vdash \overline{(\overline{SP})(\overline{SQ})}aS\bar{P}\bar{Q}$

227

5. Identity:

   (a) As $\vdash Ta\top =\vdash Ta\bar{\bot}$ we have $\vdash \bar{S}a\bar{S}, \bar{S}a\bar{\bot} \stackrel{conj}{\Rightarrow} \vdash \bar{S}a\bar{S}\bar{\bot} \stackrel{cp}{\Rightarrow} \vdash \bar{\bar{\bar{S}}}\bot aS$
   $\vdash \bar{S}a\bar{S}, \stackrel{sa}{\Rightarrow} \vdash \bar{S}\bar{\bot}a\bar{S} \stackrel{cp}{\Rightarrow} \vdash Sa\overline{\bar{S}\bar{\bot}}$

   (b) Generally $\vdash Ta\top$, so in particular $\vdash \overline{\bar{S}\bar{\top}}a\top$
   $\vdash \bar{\top}a\bar{\top}, \stackrel{sa}{\Rightarrow} \vdash \bar{S}\bar{\top}a\bar{\top} \stackrel{cp}{\Rightarrow} \vdash \top a\overline{\bar{S}\bar{\top}}$

   (c) $\vdash Ta\top \stackrel{cp}{\Rightarrow} \vdash \bot a\bot \stackrel{sa}{\Rightarrow} \vdash S\bot a\bot$
   Generally $\vdash Ta\top$, so in particular $\vdash \overline{\bar{S}\bar{\bot}}a\top \stackrel{cp}{\Rightarrow} \vdash \bot aS\bot$

   (d) $\vdash SaS \stackrel{sa}{\Rightarrow} \vdash ST aS$
   $\vdash Sa\top, SaS \vdash SaS\top$ (conj)

6. Complementation:

   (a) $\vdash \overline{\bar{\bar{S}}}\bar{S}a\top$ (cp of (c) plus dn)

   (b) $\vdash \bar{\bar{S}}aS$ (dn)

   (c) Assume $\vdash S\bar{S}i\top \stackrel{bl2}{\Rightarrow} \vdash S\bar{S}i\bar{S}\top \vdash \stackrel{w}{\Rightarrow} S\bar{S}i\bar{S}$. But by conjunction $\vdash S\bar{S}aS$. Contradiction, so $\vdash S\bar{S}a\bot$
   $\vdash S\bar{S}a\top \stackrel{cp}{\Rightarrow} \vdash \bot aS\bar{S}$

   (d) $\vdash S\bar{T}aS\bar{T}$

7. De Morgan's laws:

   (a) $\vdash \overline{\bar{S}\bar{T}}a\bar{S}\bar{T}$ (dn2 in both directions)

   (b) $\vdash \overline{ST}a\overline{\bar{S}\bar{T}}$ (2×dn + conj, followed by cp. Works in both directions)

8. Consistency:

   (a) Suppose $\Delta \vdash SaP$, then (cp) $\Delta \vdash \bar{P}a\bar{S} \stackrel{sc}{\Rightarrow} \Delta \vdash \bar{P}a\bar{S}\bar{P} \stackrel{cp}{\Rightarrow} \Delta \vdash \overline{\bar{S}\bar{P}}aP$
   Suppose $\Delta \vdash SaP$, then (cp, sa) $\Delta \vdash \bar{S}\bar{P}a\bar{P} \stackrel{cp}{\Rightarrow} Pa\overline{\bar{S}\bar{P}}$

   (b) Suppose $\Delta \vdash SaP$, then (sa) $\Delta \vdash SPaP$
   Suppose $\Delta \vdash SaP$, then (sc) $\Delta \vdash SaSP$

# A Term Logic of Justification for Epistemic Attitudes

Fabien Schang

## Introduction: Term vs Modal Logic

Agents are used to thinking. A thought is expressed by a proposition, and thinking consists in a set of related propositions that are taken to be true (accepted) or false (rejected) by the agent. A possible world is such a set of such data (a consistent representation of how the world is), and those data are made of objects and properties; now not every property is considered by an arbitrary agent, due to its limited data. For example, Socrates could not think that Joe Biden would win the 2020 US elections. So if Joe Biden does not belong to the set of objects considered by Socrates, then any proposition about Biden is neither true nor false of Socrates' thoughts.

It might be objected that *it* is the case that Joe Biden won the 2020 US elections, whatever Socrates may be able to think about it. However, what is said of Socrates in these lines matches with what Sommers [16] calls *spanning*: a property is said to span an object if, and only if, it makes sense (i.e. it is not absurd) to predicate it of that object. For example, the property of being odd spans the number 2 because it is not absurd (albeit false) to say of 2 that it is odd; whereas being odd does not span Socrates, because it is absurd to assign an algebraic property to a human being. In other words, there is no model where arithmetic properties span individual objects and, if so, not anything can be truly of falsely predicated of a given object. Semantic relevance is mentioned hereby to delimit the set of properties that are to be assigned to objects, accordingly. The same holds for epistemic agents, in the sense that not every proposition is taken to span their thoughts. That is, no epistemic agent is assumed to consider a proposition whose content has never been considered by her. If so, then a representation of how the world is (a "possible world") is a set of propositions whose content (i.e. objects and their properties) is considered by the agent.

An important consequence of that updating view of epistemic attitudes is that modal logic is not in position to model these in a *normal* way. For, according to the minimal criteria of consistency, an agent who knows that $p$ is expected to know $p \vee q$ in accordance to the meaning of logical connectives and irrespective of whether the content of $q$ is entertained by the agent. The same does for possible world semantics and its various accessibility relations: the S5 frame was taken to be a correct semantics to depict the meaning of privative negation, in García-Cruz [8]. But this means that there is an equivalent relation between any possible worlds, i.e. any world is accessible from any other one. This clearly departs from what we explained in the beginning of that section, so that the entertainment criterion – an agent has an epistemic attitude with respect to an arbitrary proposition $p$ iff she entertains both objects and properties of $p$ – requires a non-normal modal semantics. For this reason, modal logic hardly appears to be relevant for whoever takes account of Sommers' criterion of predicates spanning objects. Objects will be epistemic agents and predicates will be knowledge statements, in the following study, and the issue will be about which statements are true or false of these agents once these statements span them, i.e. are considered by agents.

The ensuing sections will include seven issues, together with two appendices. *Section 1* stresses the semantic difference between ignorance and mere lack of knowledge, assuming that ignorance is a case where the corresponding statement spans the agent (whereas lack of knowledge needn't). *Section 2* returns to the seminal work of Englebretsen [5], where the author promotes term logic as a more fine-grained logical form of knowledge statements. It is also noticed that knowledge can be explained in basic terms of justification. *Section 3* extends the former analysis of knowledge statements into a range of epistemic attitudes that go beyond the sole case of knowledge, namely: ignorance, mere belief (that rules knowledge out), and doubt. A term semantics is introduced to make sense of knowledge statements: Submodel Semantics, according to which the meaning of any statement corresponds to a finite set of properties of submodels. *Section 4* introduces the issue of logical relations with respect to their usual definition (in light of Aristotle's theory of opposition), their presentation into a deduction natural form, and their implementation into the area of epistemic attitudes. *Section 5* presents the so-called "ancient negations", i.e. privative negation and infinite negation. It is shown that these make sense only as predicates and not truthfunctional operators (like sentential negation). *Section 6* deals with more complex expressions of epistemic attitude including iterated ancient negation, e.g. privatively infinite negations or infinitely privative negations. A generalization of their logical relations is exposed, according to the number and ordering of ancient negations applied to epistemic statements. *Section 7* introduces a third kind of negation that applies neither to entire propositions (sentential negation) nor to predicates (predicate term negation), viz. individual negation. Finally, *Appendix 1* lists the finite sets of logical squares that occur from the first knowledge statements of Englebretsen [5] to epistemic attitudes with iterated negations. *Appendix 2* explains the logical relations between any kinds of epistemic statement, according to their characteristic models whose cardinality depends on the number of negations together with their ordering.

# 1 Lack of Knowledge vs Ignorance

The statements (a)–(c) are true :

(a) Donald Trump does not know who will win the 2024 US elections.
(b) Richard Nixon does not know who will win the 2024 US elections.
(c) Donald Duck does not know who will win the 2024 US elections.

Are they true in the same way, however? Donald Trump ignores[1] who will win the 2024 elections, whereas Richard Nixon and Donald Duck neither know nor ignore who will win the 2024 US elections (they cannot even think about or consider it). In other words, the latter don't know it insofar as they are unable to know anything about someone they cannot

---

[1] About ignorance, see e.g. Kubyshkyna & Petrolo [10]

ever entertain. Donald Trump does not know who will win the 2024 US elections ; but he does know when these will take place, and he may believe that he will win himself these future elections. Richard Nixon and Donald Duck don't know who will win the 2024 US elections, because they actually know nothing[2] about it. Both know and believe nothing about these future elections, although one can imagine a world at which Nixon would have existed at the XXIst century and would present himself at the 2024 US elections whereas Donald Duck cannot be arguably conceived to do so *qua* duck. Not every predicate may range over Donald Trump and Donald Duck, accordingly.

To summarize, a lack of knowledge differs from a case of ignorance: the latter entails the former, but the converse need not hold insofar as one can fail to know something without ignoring it. The difference between both situations is not taken into account by the normal modal epistemic logic, where knowledge occurs as a modal operator $K$ whose sole opposite is lack of knowledge $\neg K$. A logic that makes a distinction between lack of knowledge and ignorance is suggested in the following, in the tradition of term logic and by means of an alternative formal semantics that differs from Kripke's relational semantics.

## 2  Knowledge, Negation, and Incompatibility

The logical form of a knowledge statement like "John knows that Nixon won" is taken to include three kinds of terms in Englebretsen [5]:

> We will use 'term' in the following way: any statement is a term; any predicate is a term; any propositional object is a term, and it is only an atomic proposition in the present paper. The three discernible terms in *John knows that Nixon won* are: *John knows that Nixon won*, *knows that Nixon won*, and *that Nixon won*. Such terms will be referred to as the first, second, and third terms, respectively. 'First term' and 'statement' will be used interchangeably.

This means that terms are not atoms and may be composed, since the third term occurs in the other two. A term may be part of another

---
[2] About nothingness, see [14].

one, accordingly. Moreover, a special feature of terms is that they can be either affirmed or denied; Englebretsen [5] assigns a single kind of negation to each of the aforementioned terms in a knowledge statement whose logical form is a two-place relation of knowledge

$$\pm a \pm K \pm p$$

between an epistemic agent $a$ and a sentential content (or 'propositional object', Englebretsen would say) $p$. The symbol $\pm$ points out the fact that such a formal expression is not stated yet, i.e. neither affirmed nor denied as it stands. Then three distinct operators are applied to each term of an epistemic statement: the operator $\sim$ symbolizes the negation applied to the first term, whilst $^-$ and $-$ are negations applied to the second and third terms, respectively. It results in a set of 8 knowledge statements, due to the number of parts $2^n = 2^3 = 8$ in a statement of $n = 3$ elements:

(1) $aKp$ — it is the case that $a$ knows that $p$
(2) $aK - p$ — it is the case that $a$ knows that $-p$
(3) $a\overline{K}p$ — it is the case that $a$ does not know that $p$
(4) $a\overline{K - p}$ — it is the case that $a$ does not know that $-p$
(5) $\sim aKp$ — it is not the case that $a$ knows that $p$
(6) $\sim aK - p$ — it is not the case that $a$ knows that $-p$
(7) $\sim a\overline{K}p$ — it is not the case that $a$ does not knows that $p$
(8) $\sim a\overline{K - p}$ — it is not the case that $a$ does not know that $-p$

It strikingly appears that the way Englebretsen [5] deals with epistemic statements differs from the generation of modal logicians that came after Hintikka's seminal work on epistemic logic [9], according to whom knowledge is not a term but a modal sentential operator embedded into an extended first-order logic with quantifiers and possible worlds.

The purpose of the present paper is to do justice to Englebretsen's reading of epistemic statements whilst translating his work into a first-order language including quantifiers and qualifiers. In other words: Yes, there is more than the four epistemic statements available in the line of epistemic modal logic. For these epistemic logics include two negation operators only (applied to either modal operators of knowledge $K$ and belief

$B$), whereas the above logical form of epistemic statements makes a further difference between

(3) It is the case that $a$ does not know that $p$

and

(5) It is not the case that $a$ knows that $p$.

(3) will be treated as a case of *ignorance*, whereas (5) is mere *lack of knowledge* that can be ignorance as well (but need not). And yes, the logical apparatus of terms and their operators may be reconstructed into a first-order logic plus a special semantics I want to introduce in the next section.[3]

## 3 Epistemic attitudes

The aim of Englebretsen [5] was to establish a set of logical relations between (1)-(8) and, in the vein of Chisholm, proceed it by means of a number of square of opposition to account for these logical relations. However, no axiomatics and corresponding semantics was proposed to justify these logical relations. Let us propose now a semantics for the knowledge statements (1)-(8) in order to make a difference between several concepts of epistemology.

The main gist of the following semantics, *Submodel Semantics* (in symbols: $\mathbb{SMS}$), is to the effect that an adequate model for any statement is like a futher term composed of a finite set of other terms, viz. submodels and countersubmodels: a *submodel* (in symbols: $\mathbb{S}$) is a part of a model that the statement must possess to be true; whereas a *countersubmodel* (in symbols: $\overline{\mathbb{S}}$) is a part of a model that the statement must not possess to be true. By extension, a putative 'countercountersubmodel' amounts to a model insofar as what a statement must not not possess is the same

---

[3]By this way, I follow the line of Sedlár & Šebela [15, p. 266]: "Is it possible to formalise term negation within a framework that is still quite close to classical first-order logic? The present article answers affirmatively by formalising term negation within a simple extension of classical first-order logic."

as what it must possess. In symbols: $\overline{\overline{\mathbb{S}}} = \mathbb{S}$. A countersubmodel is a term that is crucially negated to make sense of the whole statement, accordingly, and this makes a crucial difference with all other models that don't matter for the meaning of a given statement: these are *non-models*, i.e. any model whose contribution to the meaning of a statement is zero. There are four kinds of term negations in the end: the three mentioned above by Englebretsen [5], plus the last one applied to submodels. But the difference between these negations won't matter in the following, however, after updating the logical form of knowledge statements and defining their logical relations in $\mathbb{SMS}$. Three set-theoretical operations between terms are available to construct characteristic models : intersection ∩, union ∪, and Boolean negation or complementation |. These obey laws of set theory (like e.g. commutativity and associativity) and, with respect to negation, De Morgan rules and double negation hold in this semantics.[4]

Let $X = \pm@\pm K_a\pm p$ the logical form of a knowledge statement, which is a reformulation of the previous version with a two-place predicate. Three terms may be either affirmed or denied in $X$, that is: a case-operator @[5], an individual knowledge predicate $K_a$, and a sentential content $p$. This results in a corresponding set of 8 rephrased epistemic statements, with the symbolic difference that there is only one kind of negation operator ¬ that obeys double negation[6] and is applied to any of the three components.

(1) $@K_a p$                                           $Kp$ is true of $a$
(2) $@K_a \neg p$                                      $K\neg p$ is true of $a$
(3) $@\neg K_a p$                                      $\neg Kp$ is true of $a$
(4) $@\neg K_a \neg p$                                  $\neg K\neg p$ is true of $a$
(5) $\neg @K_a p$                                      $Kp$ is not true of $a$

---

[4]Union and de Morgan rules will not be used in the following to model statements, however: Section 4 will show that the logical relations between statements crucially rely on Aristotelian relations and their *composed* negations.

[5]This external operator @ may be interpreted as the Aristotelian concept of 'possession', viz. possessing knowledge in the above case.

[6]That is: for any term $t \in X$, $\neg\neg t = t$. Actually, any negation that is *applied to itself* turns into an affirmation. See Englebretsen [5], especially the formulas $(k),(l),(m)$.

(6) $\neg @ K_a \neg p$     $K\neg p$ is not true of $a$
(7) $\neg @ \neg K_a p$     $\neg K p$ is not true of $a$
(8) $\neg @ \neg K_a \neg p$     $\neg K \neg p$ is not true of $a$

An appropriate semantics for $X$ requires a minimal set of two terms in order to define knowledge, in the vein of Englebretsen [5]. I take the latter to view knowledge as undefeated belief, that is: an agent $a$ knows that $p$ iff $a$ is both justified to believe $p$ and not justified to believe its negation $\neg p$. This means that the criterion of truth is not required to define knowledge in the following, unlike the traditional view of knowledge as a *factive* epistemic view according to which knowing $p$ entails that $p$ is true[7]. This also means that *justification* is the minimal unit of meaning to define knowledge and the following epistemic attitudes. I don't take justification to be a self-obvious notion, either: one can have a justification for (i.e. be justified to believe) something by having a mathematical proof, an empirical evidence, a testimony, and the like. Even by reducing justification to one of the roots of rational belief, there is no general agreement on how an agent is entitled to be justified to believe something.[8]

The reason to choose such a definition of knowledge and the philosophical objections to it do not matter for the present purpose, since I merely want to afford a semantics for the logical relations displayed intuitively, but not established demonstratively, in Englebretsen [5].

A minimal model to explain the meaning of $X$ includes two submodels: a submodel ① at which $a$ is justified to believe $p$, $J_a p$; and a submodel ② at which $a$ is justified to disbelieve $p$, that is, to believe its negation $\neg p$, $J_a \neg p$. A characteristic model for any statement $X$, $\mathcal{M}(X)$, amounts to an intersection of submodels or their countersubmodels. Thus, each of the epistemic statements relies on the semantic terms ①, ②, or their negations. These statements may be distinguished between instances of knowledge, ignorance, mere belief, and doubt, with the additional

---

[7] Truth could be included as a proper submodel for knowledge in SMS, admittedly. But it wouldn't change the coming logical relations, after all.

[8] An eclectic (or relativist) is largely more tolerant about the conditions for being justified than a skeptic whose standards acceptance are much higher. Hence a relativist is in position to claim to know what a skeptic should doubt about. At any rate, the following definitions of epistemic attitudes are the same for any epistemic agent.

## A Term Logic of Justification for Epistemic Attitudes

assumption that any such case of epistemic act refers to a statement preceded by the first term @. For knowing, ignoring, merely believing and doubting entails that $a$ thinks something at any rate, i.e. is justified to believe something about it.

*Knowledge*, first. That $a$ knows (the truth of) a sentence (...) means that she is justified to believe it and, in addition, that she is not justified to believe its negation. Formally:

(K) $K_a(...) =: J_a(...) \wedge \neg J_a \neg (...)$

Knowledge may be affirmative or negative, depending on whether its sentential content is affirmed or denied. Thus, its expression corresponds to the first two epistemic statements (1)–(2):

(1) $K_a p = @K_a p =: J_a p \wedge \neg J_a \neg p$ $\qquad \mathcal{M}(K_a p) = ① \cap \overline{②}$
(2) $K_a \neg p = @K_a \neg p =: \neg J_a p \wedge J_a \neg p$ $\qquad \mathcal{M}(K_a \neg p) = \overline{①} \cap ②$

Failure of knowledge refers to cases in which $a$ does not know something, and that amounts to negate the whole epistemic statement. Then failing to know something is either not being justified to believe it, or being justified not to believe it. In other words, $a$ fails to knows something whenever it is not the case that she is both justified to believe it and not justified not to believe it:

($\neg K$) $\neg K_a(...) =: \neg (J_a(...) \wedge \neg J_a \neg (...))$

Both kinds of failure correspond to the 'Boolean' negations of (1)–(2), i.e. the epistemic statements (5)–(6):

(5) $\neg K_a p = \neg @K_a p =: \neg (J_a p \wedge \neg J_a \neg p)$ $\qquad \mathcal{M}(\neg @K_a p) = \overline{① \cap \overline{②}}$
(6) $\neg K_a \neg p = \neg @K_a \neg p =: \neg (\neg J_a p \wedge J_a \neg p)$ $\qquad \mathcal{M}(\neg @K_a \neg p) = \overline{\overline{①} \cap ②}$

Then *ignorance*. The agent $a$ ignores something whenever it is the case that she doesn't know something, i.e. she considers a thing but is still in position to justify its negation. Formally:

(I) $I_a(...) = @\neg K_a(...) =: J_a\neg(...)$

The above definition departs from what Kubyshkina & Petrolo [10] call the Standard View and the New View of ignorance. According to the Standard View, ignoring means the same as not knowing. For not knowing has a broader meaning that encompasses our concept of ignorance: if someone ignores something then she does not know it, whereas someone may not know something without ignoring it (e.g. Nixon does not ignore who will win the 2024 US elections). And according to the New View, ignoring entails not having a true belief. That second view relies on the fact that ignorance has a "factive" character, that is, ignoring something entails that this something is true. I don't subscribe to that view either, since it presupposes the use of ignorance with a that-clause that has a stronger sense than the if-clause: the statement "I ignore that Biden will win the 2024 US elections" does entail that it is the case that Biden will win the 2024 US elections[9]; whereas the statement "I ignore if Biden will win the 2024 US elections" entails nothing in that way and is merely compatible with a reason to believe that Biden will lose. The coming analysis of ignorance goes beyond the above distinction between a standard and a new view of ignorance, and it is largely indebted to the term logic analysis that has been made more than fifty years ago in Englebretsen [5].

Ignorance may be affirmative or negative as well, i.e. it corresponds to the following statements (3)–(4):

(3) $I_a p = @\neg K_a p =: J_a \neg p$  $\qquad \mathcal{M}(I_a p) = ②$
(4) $I_a \neg p = @\neg K_a \neg p =: J_a p$  $\qquad \mathcal{M}(I_a \neg p) = ①$

Negative ignorance is tantamount to rational belief, it the latter is to be

---

[9]The same "factive" character occurs with knowledge accompanied with a that-clause: "I don't know that $p$" entails that $p$ is true, according to Hintikka [9]; whereas "I don't know whether $p$ does not entail it. That thesis of a factive character is questionable (see e.g. Deutscher [4]: I may well say that I don't know that $p$ *because* $p$ is false, just as I may say that unicorns are not kind because they don't exist. This issue about presuppositions and existential import deserves another entire paper, however. See Englebretsen & Schang [6].

## A Term Logic of Justification for Epistemic Attitudes

read as being justified to believe. Then ignoring (the falsity of) $p$, $I\neg p$, amounts to believing (the truth of) $p$. Once again, failure of ignorance occurs whenever either of the above definitions is negated. This leads to the ultimate two epistemic statements (7)-(8):

(7) $\neg I_a p = \neg @ \neg K_a p =: \neg J_a \neg p$ $\qquad \mathcal{M}(\neg I_a p) = \overline{②}$
(8) $\neg I_a \neg p = \neg @ \neg K_a \neg p =: \neg J_a p$ $\qquad \mathcal{M}(\neg I_a \neg p) = \overline{①}$

*Mere belief*, furthermore. The latter differs from belief *simpliciter* by being incompatible with knowledge, i.e. whoever merely believes something does not know it and conversely. It also differs from ignorance in that ignoring something may be entailed by its being known negatively, whereas no negative knowledge entails any mere belief. In other words, merely believing a sentence is like a contingency or two-sided possibility about believing it (to be either true or false):

(M) $M_a(...) =: J_a(...) \wedge J_a \neg (...)$

Mere belief does not correspond to any of the single epistemic statements (1)–(8) but, rather, a twofold ignorance. A quick look at the above definitions shows that affirmative and negative mere beliefs are one and the same epistemic attitude, however. Thus,

$M_a p = I_a \neg p \wedge I_a p =: J_a p \wedge J_a \neg p$ $\qquad \mathcal{M}(M_a p) = ① \cap ②$
$M_a \neg p = I_a p \wedge I_a \neg p =: J_a \neg p \wedge J_a p$ $\qquad \mathcal{M}(M_a \neg p) = ② \cap ①$

It is now taken for granted that both failures of mere affirmative and negative belief are also equivalent with each other, since they are a negation of two equivalent statements. At the same time, these do not amount to a case of knowledge, whether affirmative or negative, for failing to have an epistemic attitude need not entail having another one. Thus,

($\neg M$) $\neg M_a(...) =: \neg(J_a(...) \wedge J_a \neg(...))$

Hence failures of mere belief are equivalent to each other as well:

$$\neg M_a p = \neg(I_a \neg p \wedge I_a p) =: \neg(J_a p \wedge J_a \neg p) \qquad \mathcal{M}(\neg M_a p) = \overline{①} \cap \overline{②}$$
$$\neg M_a \neg p = \neg(I_a p \wedge I_a \neg p) =: \neg(J_a \neg p \wedge J_a p) \qquad \mathcal{M}_a(\neg M_a \neg p) = \overline{②} \cap \overline{①}$$

An ultimate attitude that does not correspond to either of (1)–(8) either may be constructed by means of the initial two submodels characterizing epistemic attitudes. For what of the combination of their countersubmodels, $\overline{\mathcal{A}}$ and $\overline{\mathcal{B}}$?

These are tantamount to the attitude of *doubt*, i.e. having a justification neither for nor against $p$. Thus,

(D) $D_a(...) =: \neg I_a \neg(...) \wedge \neg I_a(...) = \neg J_a(...) \wedge \neg J_a \neg(...)$

Just like merely believing, doubting amounts to the same attitude whenever the sentential content is affirmed or denied:

$$D_a p = \neg I_a \neg p \wedge \neg I_a p =: \neg J_a p \wedge \neg J_a \neg p \qquad \mathcal{M}(D_a p) = \overline{①} \cap \overline{②}$$
$$D_a \neg p = \neg I_a p \wedge \neg I_a \neg p =: \neg J_a \neg p \wedge \neg J_a p \qquad \mathcal{M}(D_a \neg p) = \overline{②} \cap \overline{①}$$

By opposition, failure of doubt does not mean the same as knowledge (although it may entail it). For whoever does not doubt may be justified to believe both affirmatively and negatively:

($\neg$D) $\neg D_a(...) =: \neg(\neg I_a \neg(...) \wedge \neg I_a(...))$

It is also taken for granted that failures of doubt are equivalent to each other.

$$\neg D_a p =: \neg(\neg I_a p \wedge \neg I_a \neg p) \qquad \mathcal{M}(\neg D_a p) = \overline{\overline{①} \cap \overline{②}}$$
$$\neg D_a \neg p =: \neg(\neg I_a \neg p \wedge I_a p) \qquad \mathcal{M}(\neg D_a \neg p) = \overline{\overline{②} \cap \overline{①}}$$

Once all these epistemic attitudes are defined in light of the common logical form advocated in Englebretsen [5], let us consider their logical relations throughout the mentioned theory of opposition.

## 4 Logical relations

Englebretsen [5] displays the exhaustive set of logical relations between epistemic statements (1)–(8) in three ways: first, into a set of five squares of opposition (I)–(V); second, into a set of entailment relations $(a)$–$(j)$ among the previous ones; third, into a set of seven rules $(i)$–$(vii)$ that establish the validity of the five squares through three rules of double negation for every term. The following affords a semantics to confirm the above results, with the help of 𝕊𝕄𝕊 and the usual truth- and falsity-conditions characterizing the so-called *Aristotelian* relations: contrariety, contradictoriness, subcontrariety, and subalternation (together with its converse, superalternation).

Let us recall these logical relations, to begin with. For any pair of statements $X$ and $Y$, these are said to be *contraries* iff they cannot be true together but can be false together; *contradictories* iff they cannot be true together and cannot be false together; *subcontraries* iff they cannot be false together and can be true together; $X$ is the *superaltern* of $Y$ (and, therefore, $Y$ is the *subaltern* of $X$) iff $Y$ cannot be false whenever $X$ is true and cannot be true whenever $X$ is false. For any of these definitions, bivalence holds in the sense that a statement is true (or false) whenever it cannot be false (or true). An easy way to identify contrariety and subcontrariety is by applying the corresponding laws of contrariety and excluded middle, respectively. Thus, any two paired statements are contrary to each other iff their conjunction leads to an antilogy: $X \wedge Y \to \bot$. And these are subcontrary to each other iff their disjunction is a tautology: $X \vee Y \to \top$. In terms of 𝕊𝕄𝕊, this means in both cases that at least one submodel of $X$ is a countersubmodel of $Y$: if $\mathbb{S} \in \mathcal{M}(X)$, then $\overline{\mathbb{S}} \in \mathcal{M}(Y)$.

All logical relations between the statements (1)–(8) can be easily established thanks to their characteristic models in 𝕊𝕄𝕊. On the one hand, any statement $X$ entails another one $Y$ iff the submodels of $Y$ are submodels of $X$. A variant of natural deduction may account for the meaning of Aristotelian relations in terms of a true first relatum (above the line) and a true second relatum (under the line) between parts of submodels $\mathbb{S}$.

On the one hand, a set of natural deduction rules for contrariety holds in 𝕊𝕄𝕊 in accordance to the aforementioned truth-conditional definition.

These two rules, [ct-1] and [ct-2], are to the effect that any submodel of the entry is a countersubmodel in the conclusion. The symmetry between both rules shows the symmetry of contrariety by reversing the premise and the conclusion:

$$[ct\text{-}1] \quad \frac{\ldots \cap \circledS \cap \ldots}{\overline{\circledS}} \qquad [ct\text{-}2] \quad \frac{\overline{\circledS}}{\ldots \cap \circledS \cap \ldots}$$

On the other hand, the simplest deduction rule to define is contradictoriness: any model is the contradictory of its corresponding countermodel, and conversely.

$$[cd\text{-}1] \quad \frac{\ldots \cap \circledS \cap \ldots}{\ldots \cap \overline{\circledS} \cap \ldots} \qquad [cd\text{-}2] \quad \frac{\ldots \cap \overline{\circledS} \cap \ldots}{\ldots \cap \circledS \cap \ldots}$$

The third kind of Aristotelian relation, subcontrariety, consists in exchanging submodels for countersubmodels into the two rules of contrariety [ct-1]–[ct-2]. This is so because, as noticed by Béziau [2, p. 24], "subcontraries are contradictories of contraries". In other words, there is a functional dependence in the definition of the Aristotelian relations: the two relata of a subcontrary relation are the contradictories of the two relata of a contrariety relation [ct-1]. Thus,

$$[sct\text{-}1] \quad \frac{\ldots \cap \overline{\circledS} \cap \ldots}{\circledS} \qquad [sct\text{-}2] \quad \frac{\circledS}{\ldots \cap \overline{\circledS} \cap \ldots}$$

However, the above definition of subcontrariety mustn't be confused with its singular counterpart, viz. being the contradictory of a contrary. For if any two formulas are mutual subcontraries whenever they are the contradictories of any other two formulas that are mutual contraries, this doesn't yet specify the logical relation between the relata of contraries and subcontraries. This relation is subalternation. Indeed, an account of subalternation also displays a functional dependence[10] with another Aristotelian relation: the subaltern, i.e. the second relatum of a subalternation relation, is the *contradictory of the contrary of* the first relatum. This means that the two rules for subalternation, [sb-1] and [sb-2], are alike those of contrariety, except that the second relatum is the contradictory of the second relatum of contrariety. That is,

$$[sb\text{-}1] \qquad\qquad [sb\text{-}2]$$
$$\frac{\ldots \cap \text{\textcircled{S}} \cap \ldots}{\text{\textcircled{S}}} \qquad\qquad \frac{\text{\textcircled{S}}}{\ldots \cap \text{\textcircled{S}} \cap \ldots}$$

The resulting set of 10 subalternation (or entailment) relations correspond to the set of valid logical relations $(a)$–$(j)$ occurring in Englebretsen [5, p. 584].

And given that superalternation is the converse relation of subalternation, the rules for superalternation, [sp-1] and [sp-2], consist in merely exchanging the ordering relation between the first and second relata of subalternation. Thus,

$$[sp\text{-}1] \qquad\qquad [sp\text{-}2]$$
$$\frac{\text{\textcircled{S}}}{\ldots \cap \text{\textcircled{S}} \cap \ldots} \qquad\qquad \frac{\ldots \cap \text{\textcircled{S}} \cap \ldots}{\text{\textcircled{S}}}$$

Finally, and importantly, for any two formulas $X,Y$ that stand in an Aristotelian relation $X$ may also stand in an Aristotelian relation with a third formula $Z$ whereas $Y$ is not. A case in point is e.g. $X = I_a p$, $Y = M_a p$, and $Z = I_a \neg p$. As will be shown in Appendix 1, $I_a p$ and $I_a \neg p$ stand in a non-Aristotelian logical relation of *independence*. In terms of bivalent semantics, this means that the truth-value of the one doesn't entail anything about the truth-value of the other. In terms of $\mathbb{SMS}$, this means that no submodel of the one is a countersubmodel of the other –both are consistent with each other– and at least one submodel of the one doesn't occur in the characteristic model of the other. This can be

---

[10]This dependence is studied in Schang [11, 12], where opposition relations are turned into opposite-forming operators.

figured out with any two distinct submodels ①,①̄:

$$
\begin{array}{cc}
[ind\text{-}1] & [ind\text{-}2] \\
\ldots \cap ① \cap \ldots & \ldots \cap \overline{①} \cap \ldots \\
\ldots \cap \overline{①} \cap \ldots & \ldots \cap ① \cap \ldots
\end{array}
$$

Subalternation and superalternation also stand between consistent submodels, admittedly. But unlike these, independence is not formed by applying contradiction to one of the related models. For whereas e.g. ① is the superaltern of e.g. $\overline{①} \cap ②$, it is not so with e.g. $\overline{①} \cap ② \cap ③$. A good way to identify a case of independence more easily is by noticing that the model of one such related formula is an expansion of the model of one subaltern.[11]

The number and nature of all logical relations between the epistemic statements $(i)$–$(xii)$ will be treated in Appendix 1.

## 5 Alternative knowledge with ancient negations

Not everything has been said about the various ways of knowing something, thus far. For there may be two different cases of ignorance among the preceding examples (a)–(c): Donald Trump does not know who will win the 2024 US elections at the time when the present paper is written; however, Trump could have known it had he been questioned about it after 2024. Richard Nixon does not know will win the 2024 US elections; however, Nixon could have known it had he travelled from his presidential mandate in a future time later than 2024. At the same time, it is arguably impossible to say of Donald Duck that he could have known the winner of the 2024 US elections. For Donald Duck is a duck who, by definition, is not able to know anything like a propositional object by virtue of his animal nature. Here is what seems to be a distinction between each of the three above subjects: Trump and Nixon ignore who will win the US elections, whereas Donald Duck does not know this and

---

[11] For the model $\overline{①} \cap ②$ is a model characterizing a subaltern of the formula whose model is ①, whereas the formula whose model is $\overline{①} \cap ② \cap ③$ ampliates the model $\overline{①} \cap ②$ whilst being independent from the formula characterized by the model ①.

cannot at all; Nixon does not know this, but he knows something else from his own time.

In order to make a difference between these two failures of knowledge that are not equivalent with each other, let us introduce two kinds of *internal* negations that apply to predicates[12]. The first is *privative* negation, inspired by Aristotle's writings in the Chapter 10 of *De Interpretatione*[13] and meaning that a subject is deprived of a knowledge predicate whenever the latter is not true of it but could be (by virtue of its nature)[14]:

$(\widehat{K})$ $\widehat{K}_a(...) =: \neg K_a(...) \wedge \Diamond K_a(...)$

A usual reading of privative negation is by means of the prefix 'un-', then let us read the privative negation of knowledge, $\widehat{K}_a(...)$, as "... is unknown of $a$". It is clear that "unknown" is not the complement of "known" in our term logic, and it is a pity that such a privative expression is restricted to an impersonal use in the third person. A little license with ordinary language would lead to epistemic expressions like "$a$ unknows $p$", where unknowledge is free from any that- or whether-clause and means more than mere lack of knowledge. Note that unknowing also differs from ignoring, and the logical relation between both is to be established thereafter. The same conjunctive definition may be applied to the other epistemic attitudes of ignorance, mere belief, and doubt, by applying one and the same pattern to any such attitude $\Theta = \{K, I, M, D\}$:

$(\widehat{\Theta})$ $\widehat{\Theta}_a(...) =: \neg \Theta_a(...) \wedge \Diamond \Theta_a(...)$

---

[12] Another way to identify the two internal, i.e. privative and infinite negations is by calling them "ancient" negations, due to their metaphysical roots in Aristotelian texts.

[13] For a historical survey of these ancient negations and the corresponding logical squares in the philosophical literature, see e.g. Correia [3]

[14] The metaphysical criteria by virtue of which something "could" satisfy a property are not the main concern of the present paper, but they need to be recalled: for example, Ammonius said that a child is not unjust (besides not being just) because she is not "in the process of possessing justice" *qua* child. The same issue arises about Nixon, Trump, and Donald Duck: Who is "in the process" of knowing who will win the 2024 US elections?

The second internal negation is *infinite* negation, also introduced in the Aristotelian tradition and meaning that $a$ knows *something else* by not knowing something initially:[15]

$(\widetilde{K})$ $\widetilde{K}_a(...) =: \neg K_a(...) \wedge K_a(—)$

Again, one and the same pattern may be applied to the other epistemic attitudes $\Theta$:

$(\widetilde{\Theta})$ $\widetilde{\Theta}_a(...) =: \neg \Theta_a(...) \wedge \Theta_a(—)$

The infinite negation $\widetilde{K}_a(...)$ may be read "... is not-known of $a$", that is, "— is known of $a$". By this way, infinite negation differs from failure of knowledge by including a kind of knowledge of — in addition to a case of ignorance of ... .

A minimal model for privative and infinite negations of knowledge requires more than the preceding two models ①, ② for normal knowledge $K$. This model includes at least 2 propositional objects $p, q$ and 2 kinds of worlds: one actual world, $w_1$, at which something *is* justified; and one arbitrary possible world, $w_2$, at which something *can be* justified. It results in a finite set of 8 submodels for the so-called ancient knowledge statements (combining ancient internal negations and knowledge statements): ①, at which $a$ is justified to believe $p$ at $w_1$; ②, at which $a$ is justified to believe $\neg p$ at $w_1$; ③, at which $a$ is justified to believe $p$ at $w_2$; ④, at which $a$ is justified to believe $\neg p$ at $w_2$; ⑤, at which $a$ is justified to believe $q$ at $w_1$; ⑥, at which $a$ is justified to believe $\neg q$ at $w_1$; ⑦, at which $a$ is justified to believe $q$ at $w_2$; ⑧, at which $a$ is justified to believe $\neg q$ at $w_2$.

The meaning of epistemic statements with privative negation relies on possible worlds, i.e. worlds at which epistemic agents do justify beliefs they cannot at the actual world. Let us consider the first six statements, $(xiii)$–$(xviii)$, before turning to their corresponding contradictories $(xix)$–$(xxiv)$. In order to assign characteristic submodels to these

---

[15]Although the usual symbol for infinite negation is a bar, the tilde is preferred throughout the present paper in order to avoid any confusion with the symbol of complementation (or Boolean negation) over submodels.

formulas, let us view the predicate of justification for an arbitrary epistemic agent $a$ as a two-place relation, $J_a p w_i$, that stands between a proposition $p$ and a world $w_i$. For example, $J_a p w_1$ means that $a$ is justified to believe $p$ at $w_1$ and corresponds to the submodel $\mathcal{A}$.

$(xiii)$ $\mathcal{M}(\widehat{K}_a p) = \overline{J_a p w_1 \cap \overline{J_a \neg p w_1}} \cap J_a p w_2 \cap \overline{J_a \neg p w_2}$ $\quad ① \cap \overline{②} \cap ③ \cap \overline{④}$
$(xiiv)$ $\mathcal{M}(\widehat{K}_a \neg p) = \overline{J_a \neg p w_1 \cap \overline{J_a p w_1}} \cap J_a p w_2 \cap \overline{J_a \neg p w_2}$ $\quad \overline{①} \cap ② \cap ③ \cap \overline{④}$
$(xv)$ $\mathcal{M}(\widehat{I}_a p) = \overline{J_a \neg p w_1} \cap J_a \neg p w_2$ $\quad \overline{②} \cap ④$
$(xvi)$ $\mathcal{M}(\widehat{I}_a \neg p) = \overline{J_a p w_1} \cap J_a p w_2$ $\quad \overline{①} \cap ③$
$(xvii)$ $\mathcal{M}(\widehat{M}_a p) = \overline{J_a p w_1 \cap \overline{J_a \neg p w_1}} \cap J_a p w_2 \cap J_a \neg p w_2$ $\quad \overline{① \cap \overline{②}} \cap ③ \cap ④$
$(xviii)$ $\mathcal{M}(\widehat{D}_a p) = \overline{J_a p w_1 \cap \overline{J_a \neg p w_1}} \cap \overline{J_a p w_2} \cap \overline{J_a \neg p w_2}$ $\quad \overline{①} \cap \overline{②} \cap \overline{③} \cap \overline{④}$

The further six epistemic statements with privative negation, $(xix)$–$(xxiv)$, are the sentential negations of the above statements and their characteristic models are the Boolean negations or complementaries of their contradictories, accordingly. That is,

$(xix)$ $\mathcal{M}(\neg \widehat{K}_a p) = \overline{① \cap \overline{②} \cap ③ \cap \overline{④}}$
$(xx)$ $\mathcal{M}(\neg \widehat{K}_a \neg p) = \overline{\overline{①} \cap ② \cap ③ \cap \overline{④}}$
$(xxi)$ $\mathcal{M}(\neg \widehat{I}_a p) = \overline{\overline{②} \cap ④}$
$(xxii)$ $\mathcal{M}(\neg \widehat{I}_a \neg p) = \overline{\overline{①} \cap ③}$
$(xxiii)$ $\mathcal{M}(\neg \widehat{M}_a p) = \overline{\overline{① \cap \overline{②}} \cap ③ \cap ④}$
$(xxiv)$ $\mathcal{M}(\neg \widehat{D}_a p) = \overline{\overline{①} \cap \overline{②} \cap \overline{③} \cap \overline{④}}$

On the other hand, infinite negation claims explicitly that the subject-term of a statement (the agent $a$, in epistemic statements) does not satisfy some property (a propositional object, e.g. $p$) whilst alluding that it does satisfy another property (another propositional object, e.g. $q$). In other words, infinite negation means that the subject-term is not nothing by being true of something. This negation helps to distinguish Nixon and Trump from Donald Duck in (a)–(c), for the former do know something after all whereas Donald Duck cannot have any epistemic attitude at all. Hence no infinite negation of epistemic attitude is true of Donald Duck, unlike Nixon and Trump.

The characteristic models of epistemic statements with infinite negation,

$\overline{\Theta}_a$, share with those with privative negation $\widehat{\Theta}_a$ the first conjunct of their submodels, that is, negating the corresponding epistemic attitude $\neg \Theta_a$. They differ by their second submodels, since privative negation means that the negated property $p$ can be true of $a$ whereas another property $q$ is true of $a$. Here are the characteristic models of the infinite negation of the four epistemic attitudes $\Theta_a$, together with their sentential negations.

$(xxv)$ $\mathcal{M}(\widetilde{K}_a p) = \overline{J_a p w_1} \cap \overline{J_a \neg p w_1} \cap J_a q w_1 \cap \overline{J_a \neg q w_1}$  $\overline{① \cap ②} \cap ⑤ \cap \overline{⑥}$

$(xxvi)$ $\mathcal{M}(\widetilde{K}_a \neg p) = \overline{J_a \neg p w_1} \cap \overline{J_a p w_1} \cap J_a \neg q w_1 \cap \overline{J_a q w_1}$  $\overline{① \cap ②} \cap \overline{⑤} \cap ⑥$

$(xxvii)$ $\mathcal{M}(\widetilde{I}_a p) = \overline{J_a \neg p w_1} \cap J_a \neg q w_1$  $\overline{②} \cap ⑥$

$(xxviii)$ $\mathcal{M}(\widetilde{I}_a \neg p) = \overline{J_a p w_1} \cap J_a q w_1$  $\overline{①} \cap ⑤$

$(xxix)$ $\mathcal{M}(\widetilde{M}_a p) = \overline{J_a p w_1} \cap \overline{J_a \neg p w_1} \cap J_a q w_1 \cap \overline{J_a \neg q w_1}$  $\overline{① \cap ②} \cap ⑤ \cap \overline{⑥}$

$(xxx)$ $\mathcal{M}(\widetilde{D}_a p) = \overline{J_a p w_1} \cap \overline{J_a \neg p w_1} \cap \overline{J_a q w_1} \cap \overline{J_a \neg q w_1}$  $\overline{① \cap ②} \cap \overline{⑤} \cap \overline{⑥}$

$(xxxi)$ $\mathcal{M}(\neg \widetilde{K}_a p) = \overline{\overline{① \cap ②} \cap ⑤ \cap \overline{⑥}}$

$(xxxii)$ $\mathcal{M}(\neg \widetilde{K}_a \neg p) = \overline{\overline{① \cap ②} \cap \overline{⑤} \cap ⑥}$

$(xxxiii)$ $\mathcal{M}(\neg \widetilde{I}_a p) = \overline{\overline{②} \cap ⑥}$

$(xxxiv)$ $\mathcal{M}(\neg \widetilde{I}_a \neg p) = \overline{\overline{①} \cap ⑤}$

$(xxxv)$ $\mathcal{M}(\neg \widetilde{M}_a p) = \overline{\overline{① \cap ②} \cap ⑤ \cap \overline{⑥}}$

$(xxxvi)$ $\mathcal{M}(\neg \widetilde{D}_a p) = \overline{\overline{① \cap ②} \cap \overline{⑤} \cap \overline{⑥}}$

An addition of privative and infinite negation extends the initial 8 statements (with $\Theta = \{K\}$) to 36 epistemic statements (with $\Theta = \{K, I, M, D\}$). It results in a total set of $36!/34!2! = 630$ pairs of logical relations, including Aristotelian and independence relations. These can be displayed into 62 logical squares (I)–(LXII) including therefore $62 \times 2 = 124$ relations of contrariety and subcontrariety, $62/2 = 31$ relations of contradiction, and $62 \times 2 = 124$ relations of subalternation. There are $124 + 31 + 124 = 279$ Aristotelian relations and $360 - 279 = 81$ independence relations. A presentation of the 62 logical squares of epistemic attitudes with internal negations, (I)–(LXII), occurs in Appendix 2.

## 6 Iterated ancient negations

One can go even further with negations by iterating them over epistemic attitudes, i.e. by applying one of either privative of infinite negation to another one. Despite the lack of intuition behind the resulting statements, this process may lead to interestingly more fine-grained expressions of epistemic attitudes.

For one thing, the hardly intuitive expression of infinite negation required the use of hyphens to make it different from the initial sentential negation in "Socrates is not just". Thus, the meaning of a statement like "Socrate is not-just" is dubious and needs further explanation. The same holds for privative negation, "Socrates is unjust".

On the other hand, a combination of these ancient negations makes even difficult to have a clear understanding of what the resulting statement means. What is the meaning of, e.g., "Socrates is not-unjust or "Socrates is un-not-just", by distinction from "Socrates is not unjust"? Iterating in this way amounts to iterating predicate operators, and the following will consider the resulting expressions and their logical relations between epistemic statements $\Theta$. I assume that all these make perfectly sense despite their not being lexicalizable. For although e.g. "ununknowing" cannot be viewed as a plausible lexicalization of the iterated private negation of knowledge, the ill-formed aspect of "Socrates is un-not-just" is no more a sufficient reason to treat it as a mere non-sense than with iterating operators like "It is possibly possibly necessary that $p$ is contingent" in alethic modal logic. Our definition of ancient negations makes sense of such iterations through a conjunction of two properties, after all.

If ancient negations are iterated just once, it results in a set of four kinds of epistemic statements for any attitude $\Theta = \{K, I, M, D\}$:

$\hat{\hat{\Theta}}_a p$ $\qquad\qquad\qquad\qquad\qquad$ $\hat{\Theta}p$ is not, but could be, true of $a$
$\hat{\tilde{\Theta}}_a p$ $\qquad\qquad\qquad\qquad\qquad$ $\tilde{\Theta}p$ is not, but could be, true of $a$
$\tilde{\hat{\Theta}}_a p$ $\qquad\qquad\qquad\qquad\qquad$ $\hat{\Theta}p$ is not true of $a$, but $\hat{\Theta}q$ is true of $a$
$\tilde{\tilde{\Theta}}_a p$ $\qquad\qquad\qquad\qquad\qquad$ $\tilde{\Theta}p$ is not true of $a$, but $\tilde{\Theta}q$ is true of $a$

To generalize the process of iteration, it can be shown that a $n$-fold

iteration of the two ancient negations yields a set of $2^n$ whole epistemic statements $X$.[16]
$$\widehat{\widehat{\Theta}}_a(...) =: \neg \widehat{\Theta}_a(...) \wedge \Diamond \widehat{\Theta}_a(...)$$

The above means that the propositional object (...) is not unknown but could be unknown. That is, it is not the case that (...) is known by $a$; but it could be the case that (...) is known by $a$. Thus,

$$\widehat{\widehat{\Theta}}_a(...) =: \neg(J_a\neg p \wedge \Diamond(J_a p \wedge \neg J_a \neg p)) \wedge \Diamond(J_a \neg p \wedge \Diamond(J_a p \wedge \neg J_a \neg p))$$

The more complex logical form afforded by iterated ancient negations leads to expanded models, i.e. additional submodels that extend the preceding ones with single ancient negations. If the ancient negation with the broadest scope is privative negation, then double ancient negations require now a set of 3 worlds: the actual world $w_1$, in addition to two distinct possible worlds $w_2, w_3$. And if the broadest scope is infinite negation, then double ancient negations require a set of 3 propositional objects $p, q, r$.

Given that occurrence of possibilities opens the way to a new possible world, a number $n$ of possibilities lead to $n+1$ possible worlds:

$$\mathcal{M}(\Diamond^n J_a(...)) = J_a(...)w_{n+1}$$

Thus, any case in which possibility is embedded into another one with iterated private negation means that a first possible world is ensued by a third one. Appendix 2 lists the ensuing characteristic models for epistemic attitudes with ancient negations.

# 7 Individual negation

There are three ways of negating statement in term logic, as claimed by Englebretsen [5]. Let us consider a knowledge statement, e.g. $@K_a p$. Given that the logical form of $@K_a p$ includes three terms: $@$, $K_a$, and $p$,

---

[16]For example, a 3-fold iteration of ancient negations yields a set of $2^3 = 8$ new epistemic statements from the basic logical form $X$, namely: $\widehat{\widehat{\widehat{X}}}, \widehat{\widehat{\widehat{X}}}, \widehat{\widetilde{\widehat{X}}}, \widehat{\widetilde{\widehat{X}}}, \widetilde{\widehat{\widehat{X}}}, \widetilde{\widehat{\widehat{X}}}, \widetilde{\widetilde{X}}, \widetilde{\widetilde{X}}$.

negation may apply to each of these. Negation applies to an entire statement by standing in front of @, and it behaves like classical negation in $\neg @K_a p$. Let us call 'sentential negation' this first, mainstream operator of negation. Or negation applies to either the knowledge predicate term $K_a$ as in $@\neg K_a p$, or the propositional object $p$ as in $@K_a \neg p$. Let us call 'term negations' the last two ways of negating inside the entire statement. There seems to be no other way to negate anything, according to the logical structure of such knowledge statements.

There is a fourth term that occurs in the above knowledge statement, actually: the individual variable $a$, i.e. the term that refers to the epistemic agent without being separated from the attitude predicate. And whilst internal negations are always taken to be either modal (applied to $K$) or predicate term negations (applied to $p$), another way to negate $@K_a p$ in the vein of ancient negations if by affirming something else about $@K_a p$ whilst negating it. Thus, the privative negation occurring in $@\widehat{K}_a p$ consists in denying that $a$ knows that $p$ whilst affirming that $a$ is in capacity to know it. Infinite negation claims in $@\widetilde{K}_a p$ that $a$ doesn't know that $p$, again, whilst affirming that $a$ knows that another proposition $q$ is true. Here is a fourth way of negating $@K_a p$, in addition to the classical or sentential negation $\neg @K_a p$: $p$ is known, not by $a$, but by $b$. Formally:

$$@K_{\widetilde{a}} p =: @K_b p$$

In other words, individual negation makes sense by affirming that *someone* else knows, just as infinite negation makes sense by meaning that *something* else is known (or ignored, merely believed, doubted) by a given epistemic agent. Only infinite negation may provide a new information when applied to individual terms, insofar as its sentential and privative counterparts do not make any difference with their predicate versions. That is, there is no semantic difference between individual and predicate privative negations: saying that $p$ is not, but could be, true of $a$ is the same as saying that $p$ is not true of $a$ but could be. And saying that it is not the case that $p$ is not known of $a$ is the same as saying it is not the case of $a$ that $p$ is known. Only infinite individual negation departs from its predicate term version, and the former may be inserted into the following extended logical form of knowledge statement:

251

$$\pm @\pm K_{\pm a}\pm p$$

Term logic may extend the knowledge statements displayed in Englebretsen [5] from 8 to 16, accordingly. A corresponding semantics consists in adding new submodels in $\mathbb{SMS}$, i.e. those at which individual terms occur as a new parameter. Since we extended thus far the cardinality of possible world and propositional objects up to 3 elements (by iterating them, see Section 6), the maximal number of submodels that characterize the epistemic attitudes of only one agent $a$ with iterated ancient negations has been $2 \times 3 \times 3 = 18$. Let us now extend submodels for a second epistemic agent, $b$. It results in a subsequent set of 18 submodels, ⑲ − ㊱, such that every such ordered submodel corresponds to the same property as the ordered submodels that characterize the epistemic attitudes of $a$.[17]

Here are the 16 knowledge statements (without ancient negations of knowledge) that occur by introducing individual negation, together with their characteristic models.

(1) $@K_a p$ $\quad \mathcal{M}(@K_a p) = ① \cap \overline{②}$ $\quad$ (9) $\neg @K_a p$ $\quad \mathcal{M}(\neg @K_a p) = \overline{① \cap \overline{②}}$

(2) $@K_a \neg p$ $\quad \mathcal{M}(@K_a \neg p) = \overline{①} \cap ②$ $\quad$ (10) $\neg @K_a \neg p$ $\quad \mathcal{M}(\neg @K_a \neg p) = \overline{\overline{①} \cap ②}$

(3) $@K_{\widetilde{a}} p$ $\quad \mathcal{M}(@K_{\widetilde{a}} p) = ⑲ \cap \overline{⑳}$ $\quad$ (11) $\neg @K_a p$ $\quad \mathcal{M}(\neg @K_a p) = \overline{⑲ \cap \overline{⑳}}$

(4) $@\neg K_a p$ $\quad \mathcal{M}(@\neg K_a p) = ②$ $\quad$ (12) $\neg @K_a p$ $\quad \mathcal{M}(\neg @K_a p) = \overline{②}$

(5) $@K_{\widetilde{a}} \neg p$ $\quad \mathcal{M}(@K_{\widetilde{a}} \neg p) = \overline{⑲} \cap ⑳$ $\quad$ (13) $\neg @K_a p$ $\quad \mathcal{M}(\neg @K_a p) = \overline{\overline{⑲} \cap ⑳}$

(6) $@\neg K_a \neg p$ $\quad \mathcal{M}(@\neg K_a \neg p) = ①$ $\quad$ (14) $\neg @K_a p$ $\quad \mathcal{M}(\neg @K_a p) = \overline{①}$

(7) $@\neg K_{\widetilde{a}} p$ $\quad \mathcal{M}(@\neg K_{\widetilde{a}} p) = ⑳$ $\quad$ (15) $\neg @K_a p$ $\quad \mathcal{M}(\neg @K_a p) = \overline{⑳}$

(8) $@\neg K_{\widetilde{a}} \neg p$ $\quad \mathcal{M}(@\neg K_{\widetilde{a}} \neg p) = ⑲$ $\quad$ (16) $\neg @K_a p$ $\quad \mathcal{M}(\neg @K_a p) = \overline{⑲}$

The characteristic models of the other epistemic attitudes $I, M, D$ with individual negation may be gathered from the above pattern, i.e. by making a correspondence between epistemic attitudes with and without that negation. Thus, each model characterizing an epistemic attitude without individual negation is isomorphic to the one characterizing a corresponding epistemic attitude with individual negation. Thus, for every epistemic attitude $\Theta_a(...)$ and every integer $n$:

If ⓝ $\in \mathcal{M}(\Theta_a(...))$, then ⓝ $+ 18 \in \mathcal{M}(\Theta_{\widetilde{a}}(...))$

---

[17]That is to say, ① $= J_a p w_1$ corresponds to ⑲ $= J_b p w_1$, ② $= J_a \neg p w_1$ corresponds to ⑳ $= J_b \neg p w_1$, and so on.

## Conclusion: What are Terms?

A term logic for epistemic attitudes has been proposed hereby, where each of these attitudes is to be defined in basic terms of justification. This logic is a term logic insofar as any formulas are made of components, and these are terms properly speaking. Moreover, ancient negations have been defined as predicate term operators instead of the usual intensional functions that do justice to the numerous relational semantics of modal logics. Finally, we resorted to a possible world semantics in order to make sense of the concept of possibility, when dealing with privative negation. But these possible worlds have been assumed to be inaccessible from each other, because a property from a first possible world may not occur in a second possible world whenever it doesn't span the objects of the latter.

In sum, the term logic we introduced in the present paper is a *logic* in the sense that it deals with a set of logical relations, i.e. the Aristotelian relations that come from the theory of opposition together with the non-Aristotelian relation of independence. And it is a *term* logic in the sense that the models that characterize its formulas are made of terms, i.e. submodels.

Some other issues are to explored in order to give a complete survey of our term logic of epistemic attitudes. First, the properties of privative negation have not been defined throughout the present paper. Does it validate e.g. excluded middle, such that $A \vee \tilde{A}$ is a theorem characterizing it? Validity has not even been approached hereby; it will be explained in a further paper that discusses what García-Cruz [8] said about the properties of privative negation and its dual of co-privation negation. Second, it can also be applied to complex statements whereas the present paper tackled only logical relations between single propositional attitudes of form $\Theta_a A$. The way to construct model for complex propositional attitudes like e.g. $\Theta_a(A \vee B)$ will be treated into a further paper, together with the ways to construct other definitions of epistemic attitudes like factive knowledge (such that $K_a p$ entails $p$). Finally, the term semantics 𝕊𝕄𝕊 will be applied to reconstruct a series of modal frames like $K$, $D$, $S4$ or $S5$, together with a multiagent epistemic logic that will make use of individual negation (see Section 7) in order to construct the characteristic models of expressions like $K_a K_b p$.

## Appendix 1: Epistemic oppositions

The number and nature of all logical relations between the epistemic statements $(i)$–$(xii)$ may be determined as follows. First, the exhaustive set of logical relations is an arrangement of $n = 2$ pairs of among $k = 12$ elements without repetition. It yields a number of $n!/k!(n-k)! = 12!/2!10! = 66$ logical relations. Second, the nature of the 66 logical relations may be gathered from the nature of a square of opposition. There are 14 cases of contrariety between the epistemic statements $(i)$–$(xii)$, by virtue of the deduction rules $[ct\text{-}1]$–$[ct\text{-}2]$ (see Section 5). Each of the 2 relata $X, Y$ of a contrariety relation has one subaltern in any of the 14 squares of opposition. Thus, there is a set of $2 \times 14 = 28$ subaltern relations between any term $X, Y$ of contraries and the contradictory of its contrary. Furthermore, that any statement has one and one contradictory entails that there are $12/2 = 6$ contradictories into $(i)$–$(xii)$.

A quick computation shows that there remains a subset of $66 - 14 - 14 - 28 - 6 = 66 - 62 = 4$ logical relations that are not Aristotelian relations. These are the so-called cases of *independence*.

The below table lists the exhaustive set of left-to-top logical relations into $(i)$–$(xii)$, including the above four cases of independence $ind$.

|  | $K_a p$ | $K_a \neg p$ | $I_a p$ | $I_a \neg p$ | $M_a p$ | $D_a p$ | $\neg K_a p$ | $\neg K_a \neg p$ | $\neg I_a p$ | $\neg I_a \neg p$ | $\neg M_a p$ | $\neg D_a p$ |
|---|---|---|---|---|---|---|---|---|---|---|---|---|
| $K_a p$ |  | ct | ct | sp | ct | ct | cd | sp | sp | ct | sp | sp |
| $K_a \neg p$ | ct |  | sp | ct | ct | ct | sp | cd | ct | sp | sp | sp |
| $I_a p$ | ct | sb |  | ind | sb | ct | sp | sct | cd | ind | sct | sp |
| $I_a \neg p$ | sb | ct | ind |  | sb | ct | sct | sp | ind | cd | sct | sp |
| $M_a p$ | ct | ct | sp | sp |  | ct | sp | sp | ct | ct | cd | sp |
| $M_a \neg p$ | ct | ct | ct | ct | ct |  | sp | sp | sp | sp | sp | cd |
| $\neg K_a p$ | cd | sb | sb | sct | sb | sb |  | sct | sct | sb | sct | sct |
| $\neg K_a \neg p$ | sb | cd | sct | sb | sb | sb | sct |  | sb | sct | sct | sct |
| $\neg I_a p$ | sb | ct | cd | ind | ct | sb | sct | sp |  | ind | sp | sct |
| $\neg I_a \neg p$ | ct | sb | ind | cd | ct | sb | sp | sct | ind |  | sp | sct |
| $\neg M_a p$ | sb | sb | sct | sct | cd | sb | sct | sct | sb | sb |  | sct |
| $\neg D_a p$ | sb | sb | sb | sb | sb | cd | sct | sct | sct | sct | sct |  |

The logical relations between more complex epistemic attitudes including ancient negation may be established according to their characteristic models and the deduction natural rules for logical relations (see Section 4). Some of these characteristic models are displayed in the following appendix.

# Appendix 2:
# Generalized iterated ancient negations

Models are expanding sets of submodels in SMS: the more negations there are in a number of finite formulas, the more submodels are included into their characteristic models. This means that the maximal model of a given language, including a finite set of single formulas, depends upon the complexity of their logical forms. For example, the maximal model of a statement like $A \circ B$ (where ∘ stands for an arbitrary logical connective) corresponds to the maximal model of the more complex formula between $A$ and $B$. In the case of epistemic attitudes, the models that characterize epistemic statements may be constructed by taking into account the occurrence of ancient negations in their statements.

Let $m$ stand for the number of propositional objects: $p, q, r, ...$; and let $n$ stand for the number of possible worlds: $w_1, w_2, w_3, ....$. Then, on the one hand, the characteristic model of an arbitrary epistemic attitude $\Theta_a(...)$ includes a set of $2mn$ submodels. On the other hand, the number of submodels with different possible worlds and propositional objects is augmented by the number of iterated privative and infinite negation, respectively. That is: if $k$ stands for the number of iterations of an ancient negation, then

$\mathcal{M}(\widehat{\Theta}_a^k(...))$ includes $n + k$ different possible worlds;
$\mathcal{M}(\widetilde{\Theta}_a^k(...))$ includes $m + k$ different propositional objects.

Finally, the content of these submodels depends on which ancient negation applies to $\Theta_a(...)$, and how these negations are ordered in case of iteration: $\widehat{\widetilde{\Theta}}_a$ corresponds to an instance of 'privately infinite' negation of $\Theta_a(...)$, whereas $\widetilde{\widehat{\Theta}}_a$ is a case of 'infinitely private' negation of $\Theta_a(...)$. The following exemplifies that construction in seven by listing the characteristic models of epistemic statements up to two iterations of privative and infinite negation.

With $\{Card(\widehat{\Theta}_a(...)) = 0, Card(\widetilde{\Theta}_a(...)) = 0\} : m = 1, n = 1$
① $= J_a p w_1$, ② $= J_a \neg p w_1$

(i) $\mathcal{M}(K_a p) =$ ① $\cap \overline{②}$

(ii) $\mathcal{M}(K_a\neg p) = \overline{①} \cap ②$
(iii) $\mathcal{M}(I_a p) = ②$
(iv) $\mathcal{M}(I_a\neg p) = ①$
(v) $\mathcal{M}(M_a p) = ① \cap ②$
(vi) $\mathcal{M}(D_a p) = ① \cap \overline{②}$
(vii) $\mathcal{M}(\neg K_a p) = \overline{① \cap \overline{②}}$
(viii) $\mathcal{M}(\neg K_a\neg p) = \overline{\overline{①} \cap ②}$
(ix) $\mathcal{M}(\neg I_a p) = \overline{②}$
(x) $\mathcal{M}(\neg I_a\neg p) = \overline{①}$
(xi) $\mathcal{M}(\neg M_a p) = \overline{① \cap ②}$
(xii) $\mathcal{M}(\neg D_a p) = \overline{① \cap \overline{②}}$

With $\{Card(\widehat{\Theta}_a(...)) = 1, Card(\widetilde{\Theta}_a(...)) = 0\} : m = 1, n = 2$
$① = J_a p w_1, ② = J_a \neg p w_1, ③ = J_a p w_2, ④ = J_a \neg p w_2$

(xiii) $\mathcal{M}(\widehat{K}_a p) = \overline{① \cap \overline{②}} \cap ③ \cap ④$
(xiv) $\mathcal{M}(\widehat{K}_a\neg p) = \overline{①} \cap ② \cap ③ \cap ④$
(xv) $\mathcal{M}(\widehat{I}_a p) = ② \cap ④$
(xvi) $\mathcal{M}(\widehat{I}_a\neg p) = \overline{①} \cap ③$
(xvii) $\mathcal{M}(\widehat{M}_a p) = ① \cap ② \cap ③ \cap ④$
(xviii) $\mathcal{M}(\widehat{D}_a p) = ① \cap \overline{②} \cap ③ \cap \overline{④}$
(xix) $\mathcal{M}(\neg \widehat{K}_a p) = \overline{① \cap \overline{②} \cap ③ \cap \overline{④}}$
(xx) $\neg \mathcal{M}(\neg \widehat{K}_a\neg p) = \overline{\overline{①} \cap ② \cap \overline{③} \cap ④}$
(xxi) $\mathcal{M}(\neg \widehat{I}_a p) = \overline{② \cap ④}$
(xxii) $\mathcal{M}(\neg \widehat{M}_a p) = \overline{① \cap ② \cap ③ \cap ④}$
(xxiii) $\mathcal{M}(\neg \widehat{D}_a p) = \overline{① \cap \overline{②} \cap ③ \cap \overline{④}}$
(xxiv) $\mathcal{M}(\neg \widehat{I}_a\neg p) = \overline{① \cap ③}$

With $\{Card(\widehat{\Theta}_a(...)) = 0, Card(\widetilde{\Theta}_a(...)) = 1\} : m = 2, n = 1$
$① = J_a p w_1, ② = J_a \neg p w_1; ⑦ = J_a q w_1; ⑧ = J_a \neg q w_1$

(xxv) $\mathcal{M}(\widetilde{K}_a p) = \overline{① \cap \overline{②}} \cap ⑦ \cap \overline{⑧}$
(xxvi) $\mathcal{M}(\widetilde{K}_a\neg p) = \overline{①} \cap ② \cap \overline{⑦} \cap ⑧$
(xxvii) $\mathcal{M}(\widetilde{I}_a p) = ② \cap ⑧$
(xxviii) $\mathcal{M}(\widetilde{I}_a\neg p) = \overline{①} \cap \overline{⑦}$
(xxix) $\mathcal{M}(\widetilde{M}_a p) = ① \cap ② \cap ⑦ \cap ⑧$
(xxx) $\mathcal{M}(\widetilde{D}_a p) = ① \cap \overline{②} \cap ⑦ \cap \overline{⑧}$
(xxxi) $\mathcal{M}(\neg \widetilde{K}_a p) = \overline{① \cap \overline{②} \cap ⑦ \cap \overline{⑧}}$
(xxxii) $\mathcal{M}(\neg \widetilde{K}_a\neg p) = \overline{\overline{①} \cap ② \cap \overline{⑦} \cap ⑧}$
(xxxiii) $\mathcal{M}(\neg \widetilde{I}_a p) = \overline{② \cap ⑧}$
(xxxiv) $\mathcal{M}(\neg \widetilde{I}_a\neg p) = \overline{\overline{①} \cap \overline{⑦}}$

## A Term Logic of Justification for Epistemic Attitudes

(xxxv) $\mathcal{M}(\neg \widetilde{M}_a p) = \overline{\overline{① \cap ②} \cap \overline{⑦ \cap ⑧}}$
(xxxvi) $\mathcal{M}(\neg \widetilde{D}_a p) = \overline{\overline{\overline{① \cap ②} \cap \overline{⑦ \cap ⑧}}}$

With $\{Card(\widehat{\Theta}_a(...)) = 2, Card(\widetilde{\Theta}_a(...)) = 0\} : m = 1, n = 3$
$① = J_a p w_1; ② = J_a \neg p w_1; ③ = J_a p w_2; ④ = J_a \neg p w_2; ⑤ = J_a p w_3; ⑥ = J_a \neg p w_3$

(xxxvii) $\mathcal{M}(\widehat{K}_a p) = \overline{\overline{① \cap \overline{②} \cap ③ \cap \overline{④}} \cap \overline{③ \cap \overline{④}} \cap ⑤ \cap \overline{⑥}}$
(xxxviii) $\mathcal{M}(\widehat{K}_a \neg p) = \overline{\overline{① \cap ② \cap \overline{③}} \cap ④ \cap \overline{③} \cap ④ \cap \overline{⑤} \cap ⑥}$
(xxxix) $\mathcal{M}(\widehat{I}_a p) = \overline{② \cap ④} \cap \overline{④ \cap ⑥}$
(xl) $\mathcal{M}(\widehat{I}_a \neg p) = \overline{① \cap ③} \cap \overline{③ \cap ⑤}$
(xli) $\mathcal{M}(\widehat{M}_a p) = \overline{\overline{① \cap ②} \cap ③ \cap ④ \cap \overline{③ \cap ④} \cap ⑤ \cap ⑥}$
(xlii) $\mathcal{M}(\widehat{D}_a p) = \overline{\overline{\overline{① \cap ②} \cap \overline{③ \cap ④}} \cap \overline{③ \cap ④} \cap \overline{⑤ \cap ⑥}}$
(xliii) $\mathcal{M}(\neg \widehat{K}_a p) = \overline{① \cap \overline{②} \cap ③ \cap \overline{④}} \cap \overline{③ \cap \overline{④} \cap ⑤ \cap \overline{⑥}}$
(xliv) $\mathcal{M}(\neg \widehat{K}_a \neg p) = \overline{① \cap ② \cap \overline{③}} \cap ④ \cap \overline{③} \cap ④ \cap \overline{⑤} \cap ⑥$
(xlv) $\mathcal{M}(\neg \widehat{I}_a p) = \overline{② \cap ④ \cap ④ \cap ⑥}$
(xlvi) $\mathcal{M}(\neg \widehat{I}_a \neg p) = \overline{① \cap ③ \cap ③ \cap ⑤}$
(xlvii) $\mathcal{M}(\neg \widehat{M}_a p) = \overline{\overline{① \cap ②} \cap ③ \cap ④ \cap \overline{③ \cap ④} \cap ⑤ \cap ⑥}$
(xlviii) $\mathcal{M}(\neg \widehat{D}_a p) = \overline{\overline{\overline{① \cap ②} \cap ③ \cap ④} \cap \overline{③ \cap ④} \cap ⑤ \cap ⑥}$

With $\{Card(\widehat{\Theta}_a(...)) = 0, Card(\widetilde{\Theta}_a(...)) = 2\} : m = 3, n = 1$
$① = J_a p w_1; ② = J_a \neg p w_1; ⑦ = J_a q w_1; ⑧ = J_a \neg q w_1; ⑬ = J_a r w_1; ⑭ = J_a \neg r w_1$

(xlix) $\mathcal{M}(\widetilde{K}_a \neg p) = \overline{\overline{① \cap \overline{②} \cap ⑦ \cap \overline{⑧}} \cap \overline{⑦ \cap \overline{⑧}} \cap ⑬ \cap \overline{⑭}}$
(l) $\mathcal{M}(\widetilde{K}_a \neg p) = \overline{\overline{① \cap ② \cap \overline{⑦} \cap ⑧} \cap \overline{⑦ \cap ⑧}} \cap \overline{⑬ \cap ⑭}$
(li) $\mathcal{M}(\widetilde{I}_a p) = \overline{② \cap ⑧} \cap \overline{⑧ \cap ⑭}$
(lii) $\mathcal{M}(\widetilde{I}_a \neg p) = \overline{① \cap ⑦} \cap \overline{⑦ \cap ⑬}$
(liii) $\mathcal{M}(\widetilde{M}_a p) = \overline{① \cap ② \cap ⑦ \cap ⑧} \cap \overline{⑦ \cap ⑧} \cap ⑬ \cap ⑭$
(liv) $\mathcal{M}(\widetilde{D}_a p) = \overline{\overline{\overline{① \cap ②} \cap \overline{⑦ \cap ⑧}} \cap \overline{⑦ \cap ⑧} \cap \overline{⑬ \cap ⑭}}$
(lv) $\mathcal{M}(\neg \widetilde{K}_a p) = \overline{① \cap \overline{②} \cap ⑦ \cap \overline{⑧}} \cap \overline{⑦ \cap \overline{⑧}} \cap ⑬ \cap \overline{⑭}$
(lvi) $\mathcal{M}(\neg \widetilde{K}_a \neg p) = \overline{① \cap ② \cap \overline{⑦}} \cap ⑧ \cap \overline{⑦} \cap ⑧ \cap \overline{⑬} \cap ⑭$
(lvii) $\mathcal{M}(\neg \widetilde{I}_a p) = \overline{② \cap ⑧ \cap ⑧ \cap ⑭}$
(lviii) $(\neg \widetilde{I}_a \neg p) = \overline{① \cap ⑦ \cap ⑦ \cap ⑬}$
(lix) $\mathcal{M}(\neg \widetilde{M}_a p) = \overline{\overline{① \cap ②} \cap ⑦ \cap ⑧ \cap \overline{⑦ \cap ⑧} \cap ⑬ \cap ⑭}$

(lx) $\mathcal{M}(\neg \widetilde{D}_a p) = \overline{\overline{① \cap \overline{②} \cap \overline{⑦} \cap \overline{⑧}} \cap \overline{⑦} \cap \overline{⑧} \cap \overline{⑬} \cap \overline{⑭}}$

With $\langle Card(\widehat{\Theta}_a(...)) = 1, Card(\widetilde{\Theta}_a(...)) = 1 \rangle : m = 2, n = 2$
$① = J_a p w_1; ② = J_a \neg p w_1; ③ = J_a p w_2; ④ = J_a \neg p w_2; ⑦ = J_a q w_1; ⑧ = J_a \neg q w_1; ⑨ = J_a q w_2; ⑩ = J_a \neg q w_2$

(lxi) $\mathcal{M}(\widehat{\widetilde{K}}_a p) = \overline{\overline{① \cap \overline{②} \cap ③ \cap \overline{④}} \cap \overline{⑦} \cap \overline{⑧} \cap ⑨ \cap \overline{⑩}}$
(lxii) $\mathcal{M}(\widehat{\widetilde{K}}_a \neg p) = \overline{① \cap ② \cap ③ \cap ④ \cap \overline{⑦} \cap ⑧ \cap \overline{⑨} \cap ⑩}$
(lxiii) $\mathcal{M}(\widehat{\widetilde{I}}_a p) = \overline{\overline{②} \cap \overline{④} \cap \overline{⑧} \cap \overline{⑩}}$
(lxiv) $\mathcal{M}(\widehat{\widetilde{I}}_a \neg p) = \overline{\overline{①} \cap ③ \cap \overline{⑦} \cap ⑩}$
(lxv) $\mathcal{M}(\widehat{\widetilde{M}}_a p) = \overline{\overline{① \cap \overline{②} \cap ③ \cap \overline{④}} \cap \overline{⑦} \cap \overline{⑧} \cap ⑨ \cap \overline{⑩}}$
(lxvi) $\mathcal{M}(\widehat{\widetilde{D}}_a p) = \overline{\overline{① \cap \overline{②} \cap ③ \cap \overline{④}} \cap \overline{⑦} \cap \overline{⑧} \cap ⑨ \cap \overline{⑩}}$
(lxvii) $\mathcal{M}(\neg \widehat{\widetilde{K}}_a p) = \overline{① \cap \overline{②} \cap ③ \cap \overline{④} \cap \overline{⑦} \cap \overline{⑧} \cap ⑨ \cap \overline{⑩}}$
(lxviii) $\mathcal{M}(\neg \widehat{\widetilde{K}}_a \neg p) = \overline{\overline{① \cap ② \cap ③ \cap ④} \cap \overline{⑦} \cap ⑧ \cap \overline{⑨} \cap ⑩}$
(lxix) $\mathcal{M}(\neg \widehat{\widetilde{I}}_a p) = \overline{\overline{②} \cap \overline{④} \cap \overline{⑧} \cap \overline{⑩}}$
(lxx) $\mathcal{M}(\neg \widehat{\widetilde{I}}_a \neg p) = \overline{\overline{①} \cap ③ \cap \overline{⑦} \cap ⑩}$
(lxxi) $\mathcal{M}(\neg \widehat{\widetilde{M}}_a p) = \overline{\overline{① \cap \overline{②} \cap ③ \cap \overline{④}} \cap \overline{⑦} \cap \overline{⑧} \cap ⑨ \cap \overline{⑩}}$
(lxxii) $\mathcal{M}(\neg \widehat{\widetilde{D}}_a p) = \overline{\overline{① \cap \overline{②} \cap ③ \cap \overline{④}} \cap \overline{⑦} \cap \overline{⑧} \cap ⑨ \cap \overline{⑩}}$

With $\langle Card(\overline{\Theta}_a(...)) = 1, Card(\widetilde{\Theta}_a(...)) = 1 \rangle : m = 2, n = 2$
$① = J_a p w_1; ② = J_a \neg p w_1; ③ = J_a p w_2; ④ = J_a \neg p w_2; ⑦ = J_a q w_1; ⑧ = J_a \neg q w_1; ⑨ = J_a q w_2; ⑩ = J_a \neg q w_2$

(lxxiii) $\mathcal{M}(\widehat{\widetilde{K}}_a p) = \overline{\overline{① \cap \overline{②} \cap \overline{⑦} \cap \overline{⑧}} \cap ③ \cap \overline{④} \cap ⑨ \cap \overline{⑩}}$
(lxxiv) $\mathcal{M}(\widehat{\widetilde{K}}_a \neg p) = \overline{\overline{① \cap ② \cap \overline{⑦} \cap ⑧} \cap \overline{③} \cap ④ \cap \overline{⑨} \cap ⑩}$
(lxxv) $\mathcal{M}(\widehat{\widetilde{I}}_a p) = \overline{\overline{②} \cap \overline{⑧} \cap \overline{④} \cap \overline{⑩}}$
(lxxvi) $\mathcal{M}(\widehat{\widetilde{I}}_a \neg p) = \overline{\overline{①} \cap \overline{⑦} \cap ③ \cap ⑨}$
(lxxvii) $\mathcal{M}(\widehat{\widetilde{M}}_a p) = \overline{① \cap \overline{②} \cap \overline{⑦} \cap \overline{⑧} \cap ③ \cap \overline{④} \cap ⑨ \cap \overline{⑩}}$
(lxxviii) $\mathcal{D}(\widehat{\widetilde{K}}_a \neg p) = \overline{① \cap \overline{②} \cap \overline{⑦} \cap \overline{⑧} \cap \overline{③} \cap \overline{④} \cap ⑨ \cap \overline{⑩}}$
(lxxix) $\mathcal{M}(\neg \widehat{\widetilde{K}}_a p) = \overline{① \cap \overline{②} \cap \overline{⑦} \cap \overline{⑧} \cap ③ \cap \overline{④} \cap ⑨ \cap \overline{⑩}}$
(lxxx) $\mathcal{M}(\neg \widehat{\widetilde{K}}_a \neg p) = \overline{\overline{① \cap ② \cap \overline{⑦} \cap ⑧} \cap \overline{③} \cap ④ \cap \overline{⑨} \cap ⑩}$
(lxxxi) $\mathcal{M}(\neg \widehat{\widetilde{I}}_a p) = \overline{\overline{②} \cap \overline{⑧} \cap \overline{④} \cap \overline{⑩}}$
(lxxxii) $\mathcal{M}(\neg \widehat{\widetilde{I}}_a \neg p) = \overline{\overline{①} \cap \overline{⑦} \cap ③ \cap ⑨}$
(lxxxiii) $\mathcal{M}(\neg \widehat{\widetilde{M}}_a p) = \overline{① \cap \overline{②} \cap \overline{⑦} \cap ⑧ \cap ③ \cap \overline{④} \cap ⑨ \cap \overline{⑩}}$

A Term Logic of Justification for Epistemic Attitudes

(lxxxiv) $\mathcal{D}(\neg \widehat{\widehat{K}}_a \neg p) = \overline{\overline{① \cap ② \cap ⑦ \cap ⑧ \cap ③ \cap ④ \cap ⑨ \cap ⑩}}$

For every statement $X$ and $n$ iterations of privative negation, it can be proved three things:

**Proposition 1.** Every formula $X$ and its ancient negation are contraries.
*Proof*: For every formula $X$ such that $Ⓢ \in \mathcal{M}(X)$,
$\widehat{X} =: \neg X \wedge \Diamond X$. Hence $Ⓢ \in \mathcal{M}(X)$ and $\overline{Ⓢ} \in \mathcal{M}(\widehat{X})$. Therefore, $X$ and $\widehat{X}$ are contraries (by virtue of [ct-1]).
$\widetilde{X} =: \neg X \wedge Y$. Hence $Ⓢ \in \mathcal{M}(X)$ and $\overline{Ⓢ} \in \mathcal{M}(\widetilde{X})$. Therefore, $X$ and $\widehat{X}$ are contraries (by virtue of [ct-1]). $\square$

**Proposition 2.** Every formula $X$ and its iterated ancient negations are independent from each other.
*Proof*: By iterating ancient negations of a given formula $X$. Since iterating consists in applying an operator at least twice, there are at least 4 ways of iterating ancient negations with the two operator of privative and infinite negations.
$\widehat{\widehat{X}} =: \neg \widehat{X} \wedge \Diamond \widehat{X} = \neg(\neg X \wedge \Diamond X) \wedge \Diamond(\neg X \wedge \Diamond X)$.
$\widetilde{\widehat{X}} =: \neg(\widehat{X} \wedge \widehat{Y}) = \neg(\neg X \wedge \Diamond Y) \wedge \Diamond(\neg Y \wedge \Diamond Y)$.
$\widehat{\widetilde{X}} =: \neg(\widetilde{X} \wedge \Diamond \widetilde{X}) = \neg(\neg X \wedge Y) \wedge \Diamond(\neg X \wedge Y)$.
$\widetilde{\widetilde{X}} =: \neg(\widetilde{X} \wedge \widetilde{Y}) = \neg(\neg X \wedge Y) \wedge \Diamond(\neg X \wedge Y)$.
Each of these formulas is of logical form $\neg \alpha \wedge \beta$. For each of the above four iterations, $\alpha$ is a contrary of $X$; then the first conjunct $\neg \alpha$ of the entire formula is the contradictory of a contrary of $X$, viz. $\neg \alpha$ is a subaltern of $X$. Now, the second conjunct $\beta$ ampliates the entire formula. Therefore, every iterated ancient negation of $X$ is logically independent from $X$ (by virtue of [ind-1]). $\square$

**Proposition 3.** For any formula $X^n$ including a number $n$ (where $n \geq 0$) of ancient negations and whose characteristic model is $\mathcal{M}(X^n) = \alpha$, its successive ancient negation $X^{n+1}$ is such that $\mathcal{M}(X^{n+1}) = \overline{\alpha} \cap \beta$.[18]

---

[18]There seems to be something in common between that form of iterated negation and what is called *Aufhebung* in Hegel's dialectical logic, insofar as $\alpha$ is a thesis whose negation $\overline{\alpha}$ leads to a synthetic form $\overline{\alpha} \cap \beta$. See Schang [13] about a proposed

*Proof*: For any formula $X^n$, its characteristic model of $X^n$ is of form $\alpha = \overline{①} \cap ⓚ$, and any characteristic model any its successive iteration $X^{n+1}$ is of form $\mathcal{M}(X^{n+1}) = \overline{①} \cap ⓚ \cap \beta$, where $\beta$ is the second conjunct into the definition of either privative or infinite negation. Therefore, $\mathcal{M}(X^{n+1}) = \overline{\alpha} \cap \beta$ whenever $\mathcal{M}^n = \alpha$. □

# References

[1] Berto, F. (2015), "A modality called 'negation'", *Mind* 124(495): 761–793.

[2] Béziau, J.-Y. (2003). "New light on the square of opposition and its nameless corner". *Logical Investigations*, 10: 218-233.

[3] Correia, M. (2017). "Aristotle's Squares of Opposition". *South American Journal of Logic*, 3(1): 313-326.

[4] Deutscher, M. (1969)"Hintikka's conception of epistemic logic". *Australasian Journal of Philosophy*, 47(2): 205-208.

[5] Englebretsen, G. (1969). "Knowledge, Negation and Incompatibility". *Journal of Philosophy*, 66: 580-585.

[6] Englebretsen, G. & Schang, F. (202X). "The Non-Standard Forms of Categorical Propositions". Draft.

[7] García-Cruz, D. & Demey, L. (202X). "Aristotelian diagrams for the Ancient Discussion on Privative and Infinite Negation", forthcoming.

[8] García-Cruz, D. (202X). "Privation, modality, and opposition". Draft.

[9] Hintikka, J. (1962). *Knowledge and Belief: An Introduction to the Logic of the Two Notions*. Ithaca, New York: Cornell University Press.

[10] Kubyshkyna, E. & Petrolo, M. (2019). "A logic for factive ignorance". *Synthese*, 198(6): 5917-5928.

[11] Schang, F. (2013). "Logic in Opposition". *Studia Humana*, 2(3): 31-45.

[12] Schang, F. (2018). "End of the Square?". *South American Journal of Logic*, 4(2): 485-505.

[13] Schang, F. (2020). "Question-Answer Semantics". *Revista de Filosofia Moderna e Contemporeâna*, 8: 73-102.

[14] Schang, F. (202X). "Quantifying statements: Why 'Every thing' is not 'Everything' (among other 'thing's)". Draft (presented at Square of Opposition 2022, Leuven).

[15] Sedlár, I. & Šebela, K. (2018). "Term negation in first-order logic". *Logique et Analyse*, 247: 265-284.

---

formalization of Hegel's *Aufhebung*.

[16] Sommers, F. (1963). "Types and Ontology", *The Philosophical Review* **72**(3): 327-363.

# ON TERMS AND TYPES

LUC SCHNEIDER

## 0 Introduction

### 0.1 Subject matter

There are – according to Van Heijenoort [55] and Cocchiarella [6] – two competing conceptions of logic, namely logic as language and logic as calculus. Viewed as calculus, logic is merely a formal organon of reasoning without any content of its own. Viewed as language, however, logic is deemed to be a *characteristica universalis* that expresses the most basic ontological distinctions. In particular, the logic of modality and tense is thought to provide the basis for the ontology of time and persistence.

The choice of logical paradigm is of the essence. Instead of the common Fregean and Russellian predicate calculus, we will adopt Sommers' [47] Term Functor Logic (TFL) as the framework of our investigations. This decision is based on two reasons: on the one hand, TFL is a formalism close to natural language and everyday reasoning; on the other hand, as we shall argue below, term logic conforms better to the requirements of categorial grammar than predicate logic.

Our first aim is to develop a formal account of (*de re*) modal and temporal categorials within Term Functor Logic, building on the work of Englebretsen [7] [9, pp. 166–176]. Part of this endeavor is the formulation of validity conditions for *de re* modal and temporal categorials.

Our second aim is to develop an ontology based on the cues provided by these semantical considerations as to the furniture of reality, i.e. the fundamental categories. The connotations of terms turn out to be *kinds* which may be identified with sets of possibilia, while the connotations of sentences, i.e., propositions, are equated to sets of worlds or times. Our account of temporal categorials also motivates a revision of our ordinary conception of individuals. Indeed, presumably singular terms that are deemed to uniquely denote individuals need to be re-interpreted as general terms that plurally denote short-lived four-dimensional objects, so-called stages. In other words, ordinary objects turn out to be kinds of stages.

## 0.2 Methodological preliminaries

It may be helpful to outline the methodology that guides the present inquiry. To this end, we must distinguish between two aspects that characterize a method, namely a set of data on the one hand and a mode of explanation on the other hand.

As far as data is concerned, we should start with the observation that not only the reliability, but also the very nature of "intuitions" has been subject to some controversy. Already Popper [30, p. 29 ff] has admonished against methodological essentialism, the view that there are essences to be grasped by some special faculty of a priori intuition. In more recent years, Cappelen has questioned the thesis that present-day analytic philosophers can at all be said to turn to intuitions as a source of data for their theorizing [5, p. 3].

Nor does it seem to be fruitful to invoke "conceptual" or "analytical" truths as a source of a priori knowledge, as argued by Quine [36] and more recently by Williamson [56, chaps. 3, 4]. Indeed, it stands to reason that our beliefs, ranging from every-day observational hunches to the hypotheses of empirical science and the axioms of logic and mathematics, are confronted to experience as a whole, such that no statement may be compared to sensory inputs in isolation (Quine [36, p. 42]). If this holistic view of belief is correct, than it is futile to look for the empirical content of a single assertion, and a fortiori to try to draw a clear line between "synthetic" statements that are true in virtue of their empirical content and "analytic" ones that are true no matter what the observed facts are (Quine [36, p. 43]). Indeed, we can always turn a statement into an "analytic" truth by isolating it from empirical refutations through adjustments of the web of belief as a whole (ibid.). The relative closeness to or distance from experience of certain statements or groups of statements is merely a function of our greater or lesser willingness to revise them instead of others in the light of recalcitrant observations (ibid.).

However, even under the assumption that all knowledge is fallible, one may be able to single out a particular class of beliefs that are as fallible as any other assumptions but are such that the burden of proof lies not with those who affirm them, but with those who deny them. These beliefs are commonsense assumptions about everyday experiences of mid-sized objects (Musgrave [28, p. 284]), which we may call *Moorean beliefs* (following Armstrong [2, p. 26 ff], in homage to G. E. Moore [27]). The special epistemic status of Moorean beliefs follows from the fact that each of them cannot be refuted by scientific theories without the help of other Moorean beliefs regarding the results of scientific experiments or the findings from observation instruments (Armstrong [2, p. 29]). In fact even telescope arrays and large-scale hadron colliders are ultimately midsized objects of everyday experience.

Now, the language of everyday experience in which Moorean beliefs are expressed is ordinary or natural language. Natural language should have the first word (Austin [4, p. 185]), if only for the reason that questions of reference only make sense with respect to a background language (Quine [33, p. 49]), and the regress of background languages has to stop at our home language in which we take our words at face value (ibid.). The logic of natural language is best described as a term logic according to Sommers [47]; this term logic is not a purely formal system, but an empirical account of how humans actually reason (Sommers [51]). Hence, data about the semantics of the logic of natural language are empirical data, tantamount to Moorean beliefs which can be elicited by invoking our own (fallible) linguistic and reasoning capacities.

Given a basic set of Moorean beliefs, the primary mode of explanation to be used by philosophy in general and ontology in particular is that of abduction, i.e., the inference to the best explanation (in the sense of Williamson [56, pp. 356–358]). The goal of any abductive reasoning is to provide a theory or account that best explains a given a set of data. A theory $T$ constitutes the best explanation of a set of data $D$ if a) $T$ is consistent with $D$ or at least with a greater subset of $D$ than any competing theory and b) $T$ better fulfills a given set of meta-theoretical requirements than any other competitor of $T$. The inference to the best explanation consists in a comparative evaluation of theories and is applicable especially in situations where the available empirical evidence underdetermines the choice between candidate theories.

The main meta-theoretical requirements an explanation must conform to are those of conservativism (Quine and Ullian [34, p. 66–68]) and simplicity ([34, p. 69–73]); in the case of an ontological account, conservativism means a reluctance to abandon Moorean beliefs, while simplicity is tantamount to the smallest possible set of categories of entities presupposed by any such account. Besides simplicity, there may be other reasonable meta-theoretical constraints, notably naturalism, i.e. the (fallible) claim that everything that exists, exists in space and time. As we shall see, the latter constraint may be controversial in the light of a commitment to *prima facie* abstract objects such as sets.

## 0.3 Outline of the paper

First, we will provide a general survey of Term Functor Logic, including a rigorous statement of its syntax and semantics. Then we will sketch a formal account of *de re* modal categorials, both alethic and temporal, and formulate the validity conditions of such statements. Using formal semantics of *de re* modal and temporal categorials as a theoretical background, we will finally present an ontology consisting of stages as individuals, kinds as sets of individuals, relations as sets of tuples of individuals and propositions as sets of worlds or times.

# 1 Term Functor Logic *in nuce*

## 1.1 Old Wine in New Wineskins

According to Categorial Grammar (Ajdukiewicz [1]; Gardies [12, chap. 3]), there are two basic categories corresponding to the linguistic functions of reference and truth, which are names ($N$) and sentences ($S$). Predicates are functors that take names as arguments and return (basic) sentences:

$$N_1, \ldots, N_n \Rightarrow S$$

In principle, as purely syntactical functors, predicates should have no semantic content. However, in the Fregean/Russellian paradigm of modern predicate logic, predicates are considered to be syntactical functors as well as, if nominalized, names of attributes or functions (Russell [40, pp. 49–50]), which gives rise to a categorial amphiboly. Not so in traditional term logic, where the sentence-forming operators, the copulae, are strictly syntactical and have no semantic content. They are syncategoremata in contrast to names (terms) and sentences. In this respect, term logic is more compatible with categorial grammar than modern predicate logic.

However, traditional term logic is generally considered to be fundamentally flawed because of the following reasons:

- its lack of any formalism that allows for symbolic reasoning,

- its inability to deal with relations and arguments involving relations,

- its inadequacy for representing propositional logic, and

- its incapacity to tackle singular terms and arguments involving singular terms.

Nonetheless, Fred Sommers ([46, 47, 48], cf. Englebretsen [8, chaps 3 & 4] as well as Sommers and Englebretsen [50]) has developed an updated version of term logic, called Term Functor Logic (TFL) that addresses all these issues by taking cues from the logical works of Leibniz and of the late-19-th century algebraists. TFL constitutes a viable alternative to predicate logic, especially as far as natural language reasoning is concerned, because it espouses the subject/predicate-structure present in ordinary language (Sommers [51]).

*Prima facie*, Term Functor Logic seems to be looking back to a bygone age in the history of logic. However, Aris Noah [29] has shown that Sommers' TFL has close affinities to Quine's ([31], [39, pp. 283–288]) proposal of a variable-free calculus, i.e., Predicate Functor Logic, though their motivations were quite dissimilar: while Sommers was primarily interested in the logic of natural language, Quine's foremost goal was to eliminate bound variables. Be this as it may, TFL could be considered to participate in a larger trend towards variable-free calculi within modern logic.

## 1.2 A Logical Algebra

The categoremata are terms which, if meaningful, express concepts and denote the things that fall under those concepts (Sommers and Englebretsen [50, pp. 15–16]; Englebretsen [8, p. 175]). Each term has a positive or negative polarity, e.g. "+Human" expresses the concept of being human and "−Dead" the concept of being undead. We shall adopt the convention that the "+"-sign can be dropped and that unsigned terms are of positive polarity.

The reason why terms have positive or negative polarity is the fact that some things may be such that the concepts/attributes expressed by the terms are neither true nor false of these things, i.e. that they can neither be affirmed nor denied of them. Indeed, doing so would result in a category mistake (Sommers [43, 44]), as in Chomsky's famous example of a grammatically correct but nonsensical sentence:

> Colourless green ideas sleep furiously.

Copulae combine terms into sentences. In the following we mainly consider split copulae such as "some...are" or "all/every...aren't/isn't" which correspond to the natural word order in English and in other Indo-European languages (Englebretsen [8, p 156]). Split copulae have two parts or facets: quantity ("some", "all") and quality ("is/isn't, "are/aren't"); though symbolized by separate expressions, quality and quantity form a single operator, namely the copula (Englebretsen [8, p. 157]).

Sommers' fundamental insight, which he has borrowed from Leibniz, is that syncategoremata behave like algebraic operators. The same algebraic treatment can be applied to syncategoremata that constitute the logical form of categorials (i.e., "some","is/are","all" and "isn't/aren't") as well as to those that correspond to connectives (like "(both...) and" and "if...then").

"Some" and "is/are" are represented as "+", while "not", "all", and "isn't/aren't" can be viewed as "−" (Sommers [51, p. 124], [52, pp. 10, 20]). Indeed, "some" and "is/are" combined are commutative like "+" (Sommers [52, pp. 18–19]; Englebretsen [8, pp. 112, 152]):

Some   Greeks   are   Athenians   =   Some   Athenians   are   Greeks
↓              ↓           ↓                  ↓              ↓           ↓
+      Greeks   +   Athenians    =   +      Athenians   +   Greeks

Particular negative categorials can be defined using quality and polarity inversion (Sommers [47, p. 177]):

Some   Greeks   aren't   Sages   =   Some   Greeks   are   Non-Sages
↓              ↓             ↓                      ↓             ↓          ↓
+      Greeks     −       Sages   =   +      Greeks    +     − Sages

Assuming that "not" is a "-"-word, the minus-character of "all" (as well as "isn't/aren't") results of the equivalence between a universal affirmative and the negation of a particular negative categorial (Sommers [52, p. 19], [47, p. 178], Englebretsen [8, p. 152]):

All   Greeks   are wise   =   Not:   Some Greeks   aren't wise
↓             ↓          ↓                  ↓       ↓                   ↓
−    Greeks   + wise    =   − (    + Greeks        − wise )

The universal affirmative copula is reflexive and transitive, but asymmetrical, contrary to the particular affirmative copula.

Finally, universally negative categorials may be defined as negations of particular affirmative categorials (Englebretsen [8, p. 158]):

All   Greeks   aren't   Persians   =   Not:   Some Greeks   are Persians
↓             ↓             ↓                         ↓        ↓                  ↓
-     Greeks     -      Persians    =    -    ( + Greeks      + Persians )

Regarding compound sentences, we observe that the conjunction "both ... and" is commutative like "some ... is/are" and thus is a "+"-word (Sommers [52, p. 16]):

Both   $p$   and   $q$   =   Both   $q$   and   $p$
↓             ↓                        ↓             ↓
+     $p$   +     $q$   =   +     $q$   +     $p$

"If ... then" behaves like "all ... is/are" and thus "if" is a "-"-word (Sommers [52, p. 17]):

If   $p$   then   $q$   =   Not:   (both)   $q$   and   not   $p$
↓        ↓              ↓                          ↓            ↓            ↓
-   $p$    +     $q$   =   - (           +       $q$   +     -   $p$ )

## 1.3 Equivalence and validity

Both logical equivalence and validity can be characterised algebraically, validating the conjecture that logical intuition is fundamentally algebraic.

In order to show that logical equivalence can be defined in an algebraical way, we need to introduce two notions, i.e. that of covalence and that of algebraical equality (Sommers and Englebretsen [50, pp. 61–62], Sommers [52, p. 29]). Categorical sentences are algebraically equal if and only if their algebraical sums are equal. So the following two sentences are algebraically equal:

A) Every senator is a citizen: $\quad +(-S + C)$
B) Not: Some senator is a non-citizen: $\quad -(+S + (-C))$

Sentences are terms, so they have positive or negative polarity; the polarity of a sentence is expressed by a prefixed "+" or "−" that we call the judgement sign of the sentence. The valence of a categorical sentence is negative if the judgement sign and the quantity are different, otherwise its valence is positive (Sommers [47, p. 180], Sommers and Englebretsen [50, p. 51]):

$-(+C + S) \quad \Rightarrow$ negative
$-(-D + F) \quad \Rightarrow$ positive

Two categorical sentences are covalent iff they have the same valence (Sommers and Englebretsen [50, p. 51]). It is easy to see that examples A) and B) are covalent.

Categorical statements are logically equivalent if and only if they are both algebraically equal and covalent (Sommers [47, p. 182], [52, p. 7], Sommers and Englebretsen [50, p. 62]). So, A and B, being covalent and algebraically equal, are logically equivalent. Furthermore, $+B - M$ (some bats aren't moving) is logically equivalent to:

$-(-B + M) \quad$ [not: every bat is moving]
$+B + (-M) \quad$ [some bats are motionless]
$+(-M) + B \quad$ [something non-moving is a bat]

The validity of arguments can also be established algebraically (Sommers [46, p. 19]; Sommers and Englebretsen [50, pp. 109–110], Englebretsen [8, pp. 114, 167]). Beforehand, we need to introduce the notion of regularity: an argument is regular if and only if it has as many particular premisses as particular conclusions. The following principle states the sufficient and necessary conditions for validity:

**REGAL**  An argument is valid if, and only if:

1. it is regular, and

2. the sum of its premises is equal to the sum of its conclusion(s).

Thus, the modus Barbara is valid:

| All | $M$ | are | $P$ ; | All | $S$ | are | $M$, | (thus) | All | $S$ | are | $P$ |
|---|---|---|---|---|---|---|---|---|---|---|---|---|
| ↓ | ↓ | | ↓ | ↓ | ↓ | | | | | ↓ | | ↓ |
| - | $M$ | + | $P$ ; | - | $S$ | + | $M$ | | | - $S$ | + | $P$ |

But the following argument is not:

| Some $A$ is $B$ | $+A + B$ | [e.g., some apes are black] |
| Some $C$ isn't $B$ | $+C - B$ | [e.g., some cats aren't black] |
| Some $A$ is $C$ | $+A + C$ | [e.g., some ape are cats] |

The premises do add up to the conclusion but the number of particular conclusion(s) is not equal to the number of particular premises.

The central derivation rule in term logic is the *dictum de omni et nullo (DDO)*: *What is affirmed/denied of a universally quantified subject is likewise affirmed/denied of what that subject-term is affirmed of.*

$$-X +/- Y$$
$$\ldots +X \ldots$$
$$\overline{\ldots +/-Y \ldots}$$

The DDO is essentially a substitution pattern: given a major premiss that is universally quantified, and a minor premiss where the subject term of the major premiss (i.e. the middle term, X) occurs as a predicate term, replace the predicate term of the major premiss (the major term, Y) for the middle term in the minor premiss. The predicate term of the conclusion is the major term and its subject the minor term, i.e. the subject term of the minor premiss (Sommers [47, pp. 129–130, 184], Englebretsen [8, pp. 116–117, 164–166]). In the example of the modus Barbara above, which shows an application of the DDO, the middle term is M, the major term is P and the minor term is S: M is universally quantified in the major, –M+P, and is the predicate of the minor, –S+M; S is substituted for M in the major which yields –S+P.

Algebraically speaking, the DDO consists in the cancellation of the middle term that appears negatively in the major premiss and positively in the minor premiss . Since the conclusion contains the remaining terms, namely the minor and the major term, it is the result of adding the premisses. Furthermore, the major premiss being universal, and the conclusion resulting from a substitution of the middle term by the major term within the minor premiss, the quantity of the conclusion and that of the minor premiss is the same. So, if the conclusion is particular or universal, the minor premiss is also. This means that the argument is regular, since it has as many particular conclusions as particular premisses. Therefore the DDO is valid in virtue of the REGAL principle.

The DDO is not the only derivation rule in Sommers' term logic; other axioms and inference allow transforming the premises of an argument in such a way that the DDO can be applied. A presentation of a complete set of derivation rules for term logic is beyond the scope of the present study; the reader is referred to Sommers [47, pp. 183–186] as well as to Englebretsen [8, pp 167–170] for further details.

## 1.4 Singular terms

There is certainly a syntactical difference between subjects (i.e., quantity and term, e.g. "All men") and predicates (quality and term: "are mortals"). However, term logic does not need to formally differentiate between singular and general terms, though it may do so for convenience sake.

Singular terms (pronouns or proper names) are *semantically* distinguished by the remarkable fact (already noticed by Leibniz) that a particular statement with a singular subject entails the universal with the same singular term as subject and the same predicate (Sommers [45],

[47, pp. 15–16], Englebretsen [8, p. 160 ff]):

$$+A \pm B \Rightarrow -A \pm B$$

e.g. "Some Socrates is (not) bald" $\Rightarrow$ "Every Socrates is (not) bald". Singular subjects are said to have "wild quantity" (marked with "+/−" or "*"):

$$+/- \text{ Tully} + \text{Orator}.$$

Sommers regards singular reference as systematically anaphorical: it is a definite reference that has an indefinite reference as its antecedent (Sommers [47, chap. 5], Englebretsen [8, pp. 126–127]):

"There is an elephant* in the room. It* is purple. Let's call it* Dumbo*"

In term logic, the basic referring expressions are particular subjects such as "an $S$" or "some $S$" which nonidentifyingly refer to an unspecified subset of the things they denote (Sommers [47, p. 59], Englebretsen [8, p. 126]). Generally, indefinite descriptions of the form "some/a $S$" are *epistemic* in the sense that they merely purport to refer to some thing within the perceptual field of the speaker (Sommers [47, p. 57], [8, p. 126]).

Proper names are special-purpose pronouns: their function is to facilitate context-independent reference to that which is referred to epistemically by their antecedents (Sommers [47, p. 230], Englebretsen [8, p. 131]). Since pronouns are rigid designators, so are proper names (Sommers [47, p. 229], Englebretsen [8, p. 131]).

In TFL, we can form proterms by adding superscripts to any general term. E.g. the proterm $A^i$ in the sentence

$$+[+A^i + B] + [+A^i + C]$$

refers to the same things and the whole sentence reads as: "Some $A$s are $B$s and /they/ are $C$s". Let $T$ be the dummy term that denotes "things" in general. The use of proterms is regulated by the following equivalences (Sommers and Englebretsen [50, p. 254]):

$$+A + B = +[+T^i + A] + [+T^i + B]$$
$$-A + B = -[+T^j + A] + [+T^j + B]$$

*Identity* between singular terms can be *defined* in term logic; it is tantamount to predication of singular terms to singular terms (Sommers [45], [47, p. 123], Englebretsen [8, p. 162]), e.g.

$$\pm \text{Tully} \pm \text{Cicero}.$$

We can prove the characterizing formal properties of identity using the wild quantity of singular terms (Sommers [47, pp. 123–124], Englebretsen [8, p. 162]):

**Reflexivity** −Tully+Tully

**Symmetry** +Tully+Cicero ⇒ +Cicero+Tully

**Transitivity** −Tully+Tullius, −Tullius+Cicero ⇒ −Tully+Cicero

**Indistinguishability of Identicals** −Tully+Cicero, −Cicero+Orator ⇒ −Tully+Orator

**Identity of Indistinguishables** Suppose that the Indistinguishability of Identicals holds: then −Tully+Tully entails −Cicero+Tully. But it is true (because of Reflexivity) that −Tully+Tully.

Interestingly enough, singular terms may be negated, and their negations are general terms: thus the negation of "Cicero", "−Cicero" denotes all beings that are not Cicero.

## 1.5 Relational terms

In order to understand how term logic may be able to reckon with relations (Sommers [47, pp. 137–152], Englebretsen [8, p. 128 ff]), one has to take into account the distinction between predicates and predicate terms. Predicates are always monadic, since every sentence in term logic is a dyad of the form:

$$\pm X \pm Y,$$

$\pm X$ being its subject and $\pm Y$ its predicate (Sommers [48, 110–111], Sommers and Englebretsen [50, 91–92]).

However, some predicate terms are phrases headed by a polyadic term, the arity of which is indicated by subscripts; e.g., "Loves$_{12}$" is binary and "Gives$_{123}$" is ternary. The order of the subscripts allows to differentiate converses from each other, such as "Loves$_{21}$" ("Being loved") from "Loves$_{12}$".

A relational phrase is a complex term that consists of a relational term of any arity $n$, called *head*, which is followed by *exactly* $n-1$ occurrences (including possibly repetitions) of monadic terms, called *objects*. An object (i.e. the occurrence of a monadic term) is attached to the head by

- an index matching exactly one of the indices of the head;

- an unsplit copula preceding the object, where "+" stands for "applies to some" and "−" for "applies to every"; for simplicity's sake, we may read the plus or minus preceding the object as a mere quantity.

Some examples are better than any wordy explanations:

−Loves$_{12}$ −Boy$_1$       "is disliked by every boy"
Gives$_{123}$ +Toy$_2$ −Child$_3$   "gives a toy to every child"

Finally, the subject is attached to the relational phrase in the usual manner, but indexed with a subscript that corresponds to the remaining free index of the head:

+Sailor$_1$ +Gives$_{123}$ +Toy$_2$ −Child$_3$   "A sailor gives a toy to every child"
+Teacher$_2$ +−Loves$_{12}$ −Boy$_1$         "A teacher is disliked by every boy"

We have said above that converses may be distinguished by the order of the subscripts of the respective relational term. The two previous examples show a different way of expressing converses, namely the ordering of the subject and the object(s) within a relational categorial sentence.

It should be emphasized that Sommers ([47, 137–152], [48]) explicitly allows relational phrases in which not every position provided by the head is filled by an object, e.g.

| | |
|---|---|
| Gives$_{123}$ +Toy$_2$ | "gives some toy (to somebody)" |
| | or "is given some toy by somebody" |
| Gives$_{123}$ −Child$_3$ | "gives (something) to every child" |
| | or "is given (by sbd.) to a child" |
| Loves$_{12}$ | "is a lover" or "is beloved" |

While this offers a substantial flexibility (while also creating some ambiguity), we can, nonetheless, view these expressions as abbreviations of the full relational phrases with the "missing" positions filled by the generic term "$T$" which stands for "somebody" or "something":

| | |
|---|---|
| Gives$_{123}$ +Toy$_2$ +T$_3$ | "gives some toy (to somebody)" |
| Gives$_{123}$ −Child$_3$ + $T_1$ | "is given (by sbd.) to a child" |
| Loves$_{12}$ + $T_2$ | "is a lover" or "is loving" |
| Loves$_{12}$ + $T_1$ | "is a (be)loved" |

The loss in elegance is compensated by a gain in clarity, but also by the fact that "our" version of TFL is close to a variant for which both correctness and completeness has been established (see section 1.6 below).

Arguments that involve relational terms pose no greater difficulties than those that only concern monadic terms. However, the Dictum de Omni et Nullo (DDO) needs to be generalised in order to deal with relational sentences, since not only the subject, but also any object of the major premiss may be the middle term. Let E*(–M) be the *donor premiss* that contains a negative (i.e. universal) occurrence of the middle term M, E(M) the *host premiss* which contains a positive occurrence of M; the conclusion results from adding the donor premiss to the host premiss, thus cancelling the middle term M (Sommers and Englebretsen [50, pp. 140–141]). The following example shows the pattern of the argument:

1. Some sailor gives a toy to every child.

2. Every girl is a child.

3. Thus, some sailor gives a toy to every girl.

which is formalised in term logic like this:

$E^*(-M)$ + Sailor$_1$+Gives$_{1,2,3}$+Toy$_2$–Child$_3$

$E(M)$ –Girl+Child

$E(E^*)$ + Sailor$_1$+Gives$_{1,2,3}$+Toy$_2$–Girl$_3$

In the example above,

- M = "Child"
- $E^*$ = "+ Sailor$_1$+Gives$_{1,2,3}$+Toy$_2$"
- E = "–Girl"

In order to represent the whole range of relational expressions and sentences in natural language, term logic needs to recur to proterms where predicate logic makes use of bound variables (Sommers and Englebretsen [50, pp. 100–102, 151 f]). For instance, we need pronouns in order to mark the difference between:

"Some barber shaves some barber" : $+B_1 + S_{1,2} + B_2$

and the reflexive:

"Some barber shaves himself" : $+\text{Barber}_1^i + \text{Shaves}_{1,2} + \text{Barber}_2^i$

This concludes our sketch of Sommers' term logic; its main interest is to provide the reader a good grasp of its expressive power and enough background to understand our use of the formalism in the remainder of this article. What it hopefully has shown is the ease of term logic in accounting for everyday reasoning using natural language sentences; in this sense term logic can be claimed to be the logic of commonsense and a viable alternative to predicate logic as a means of natural language analysis.

## 1.6 A formal description of TFL

The preceding informal considerations can be summed up in a more rigorous presentation of TFL. In providing a syntax and semantics of TFL, we take inspiration from previous work of Clifton McIntosh [24], which helpfully includes correctness and completeness proofs for a version of TFL that is close to that adopted in this paper, minus the additional complication of proterms and compound terms other than relational phrases.

### 1.6.1 Syntax of TFL

**Vocabulary:**

1. infinitely many singular monadic term letters:
   $A^*, \ldots, S^*, A'^*, \ldots, S'^*, A''^*, \ldots, S''^*, \ldots$;

2. infinitely many general monadic term letters:
   $A, \ldots, S, A', \ldots, S', A'', \ldots, S'', \ldots$;

3. a general monadic term letter $T$ (meaning "Thing(s)");

4. for any $n > 1$, any permutation $\langle i_1, \ldots, i_n \rangle$ of the tuple $\langle 1, \ldots, n \rangle$, there are infinitely many $n$-adic relational term letters indexed with $\langle i_1, \ldots, i_n \rangle$:
   $A_{i_1\ldots i_n}, \ldots, S_{i_1\ldots i_n}, A'_{i_1\ldots i_n}, \ldots, S'_{i_1\ldots i_n}, A''_{i_1\ldots i_n}, \ldots, S''_{i_1\ldots i_n}, \ldots$;

5. logical signs: $+, -$ ;

6. parentheses: ( , ).

**Terms:**

1. Every term letter is a term.

2. If $\alpha$ is a term, then so are $+\alpha$ and $-\alpha$.

3. For any $n > 1$, if $\beta_{i_1\ldots i_n}$ is an $n$-adic term and, $\alpha^1, \ldots, \alpha^{n-1}$ are monadic terms, then, for any index $k$ in $\{i_1, \ldots, i_n\}$, with $\{j_1, \ldots, j_{n-1}\} = \{i_1, \ldots, i_n\} - \{k\}$, $\beta_{i_1\ldots i_n} \pm \alpha^1_{j_1} \pm \ldots \pm \alpha^{n-1}_{j_{n-1}}$ is a term called a *relational phrase with free index k*.

4. Nothing is a term unless it falls under clauses 1-3.

**Sentences:**

1. If $\alpha$ and $\beta$ are monadic terms, then $\pm\alpha \pm \beta$ is a sentence.

2. If $\alpha$ is a monadic term and $\beta$ a relational phrase with free index $k$, then $\pm\alpha_k \pm \beta$ is a sentence.

3. If $\phi$ is a sentence, then $+\phi$ and $-\phi$ are sentences.

4. If $\phi$ is a sentence and $\psi$ is a sentence, then $\pm\phi \pm \psi$ is a sentence.

5. Nothing is a sentence unless it falls under clauses 1-4.

### 1.6.2 Semantics of TFL

We suppose familiarity with the basic set-theoretic notions of complement $\mathsf{C}$, union $\cup$, intersection $\cap$ and inclusion $\subseteq$. Let $\mathbf{U}$ be the universe of interpretation: $\mathbf{U}^*$ is the set of singleton subsets of $\mathbf{U}$. For any $n \geq 1$, $\mathbf{U}_1 = \mathbf{U}$ and $\mathbf{U}_{n+1} = \mathbf{U}_n \times \mathbf{U}$, i.e., the set of $n$-tuples of elements of $\mathbf{U}$.

**Interpretation:** We write "$[\alpha]$" for the interpretation of a term $\alpha$.

1. For any monadic singular term letter $\alpha^*, [\alpha^*] \subseteq \mathbf{U}^*$.

2. For any monadic general term letter $\alpha, [\alpha] \subseteq \mathbf{U}_1$.

3. $[T] = \mathbf{U}$.

4. For any $n > 1$, for any $n$-adic relational term letter $\beta, [\beta] \subseteq \mathbf{U}_n$ such that $\langle a_1, \ldots, a_n \rangle$ is a member of $[\beta_{1\ldots n}]$ if, and only if, for any permutation $\langle i_1, \ldots, i_n \rangle$ of $\langle 1, \ldots, n \rangle$, the tuple $\langle a_{i_1}, \ldots, a_{i_n} \rangle$ is a member of $[\beta_{i_1 \ldots i_n}]$.

5. For any term letter $A, [+A] = [A]$.

6. For any term letter $A, [-A] = \mathsf{C}[+A]$.

7. For any relational phrase $\beta_{i_1\ldots i_n} \pm \alpha^1_{j_1} \pm \ldots \pm \alpha^{n-1}_{j_{n-1}}$ with free index $k$, such that $\{j_1, \ldots, j_{n-1}\} = \{i_1, \ldots, i_n\} - \{k\}$, $[\beta_{i_1\ldots i_n} \pm \alpha^1_{j_1} \pm \ldots \pm \alpha^{n-1}_{j_{n-1}}]$ is the set of those $y$ such that for some/every $x^1$ in $[\alpha^1]$ and ... and for some/every $x_{n-1}$ in $[\alpha_{n-1}]$, the $n$-tuple $\langle x_1, \ldots, y, \ldots, x_{n-1} \rangle$ is in $[\beta_{i_1\ldots i_n}]$.

8. For any relational phrase $\beta, [+\beta] = [\beta]$.

9. For any relational phrase $\beta, [-\beta] = \mathsf{C}[+\beta]$.

**Truth:**

1. Let $\alpha, \beta$ be any terms:
   (a) $\text{val}(+\alpha + \beta) = 1$ iff $[\alpha] \cap [\beta] \neq \varnothing$;
   (b) $\text{val}(+\alpha - \beta) = 1$ iff $[\alpha] \cap \mathsf{C}[\beta] \neq \varnothing$;
   (c) $\text{val}(-\alpha + \beta) = 1$ iff $[\alpha] \subseteq [\beta]$ ;
   (d) $\text{val}(-\alpha - \beta) = 1$ iff $[\alpha] \subseteq \mathsf{C}[\beta]$.

2. Let $\phi, \psi$ any sentences:
   (a) $\text{val}(+\phi + \psi) = 1$ iff $\text{val}(\phi) = 1$ and $\text{val}(\psi) = 1$;
   (b) $\text{val}(-\phi + \psi) = 1$ iff $\text{val}(\phi) = 0$ and $\text{val}(\psi) = 1$;
   (c) $\text{val}(+\phi - \psi) = 1$ iff $\text{val}(\phi) = 1$ and $\text{val}(\psi) = 0$;
   (d) $\text{val}(-\phi - \psi) = 1$ iff $\text{val}(\phi) = 0$ and $\text{val}(\phi) = 0$;
   (e) $\text{val}(+\phi) = 1$ iff $\text{val}(-\phi) = 0$.

The interpretation as well as the validity conditions provide the background for the more semantically oriented discussions of modalities in TFL that follow.

## 2 De re modalities in Term Functor Logic

### 2.1 Modal categorials

In order to prepare the discussion of temporal reasoning in TFL, we shall first address modal categorials in general, taking cues from Englebretsen's [7] seminal article on modal syllogistics as well from the helpful clarifications and corrections in Englebretsen and Sayward [9, pp. 166–176]

Let us start with two observations. First, we hold that the primary use of modal discourse is contrafactual reasoning about ways in which denizens of the actual world could or could not instantiate other properties or relations than they actually do. It is in this sense that we say that some/all $S$'s are accidentally or essentially $P$'s or non-$P$'s.

Second, the grammatical function of modal words such as "accidentally" or "essentially" in natural language is that of adverbs. Adverbs are basically modifiers; the question is, which parts of speech do modal words modify? A popular view is that modal adverbs modify predicates or predicate terms. However, as McGinn [23, pp. 75–76] has pointed out, the predicate modifier view has two disadvantages, a minor and a major one. The minor disadvantage of the predicate modifier view is that it commits us to strange modal properties such as "being essentially human" or "being accidentally President of the European Union" (McGinn [23, p. 75]). The major drawback of the predicate modifier view is that it doesn't do justice at all to phrases such as "All Europeans are essentially human" or "Some Europeans are accidentally politicians" which seem to convey a modality of ascription rather than of what is being ascribed (McGinn [23, p. 76]).

Indeed, modalities are ways in which properties or relations are (co-) instantiated:

Some/all $S$'s are/aren't essentially/accidentally $P$'s

which would suggest that modal adverbs should be analyzed as modifiers of the copula rather than of the predicate or predicate term. Thus, we follow McGinn [23, p. 77] in adopting the copula modifier view of modalities. Consequently, all modalities are basically *de re*. What seem to be *de dicto* modal statements are actually *de re* statements as to propositions being essentially or accidentally true (McGinn [23, p. 79]). Therefore, *de dicto* statements are fundamentally *de re* metalinguistic statements.

One important lesson that can be drawn from these reflections is that copulae are not modally neutral (McGinn [23, p, 77]). Thus, even the copulae of basic TFL have a modal character; indeed, they are implicitly biased towards actuality: for instance, "Some horses are white" really means "Some horses are actually white". Essential and accidental copulae are marked by the indices "$e$" respectively "$a$"; in the case of split copulae, these indices are attached to the quantity for convenience, though it should be remembered that modalities pertain to (split or unsplit) copulae as a whole. So, for example,

- "Some $S$'s are essentially $P$'s" becomes "$+S +_e P$" and
- "All $S$'s accidentally aren't $P$'s" becomes "$-S -_a P$".

We shall take essentiality as *de re* necessity as well as de re possibility (marked with the index "∘") as primitive notions. Indeed, accidentality can be defined with the help of *de re* possibility and *de re* necessity (essentiality) as follows:

$$\pm S +_a P =_{df} +(\pm S +_\circ P) + -(\pm S +_e P)$$

i.e., all/some $S$'s are accidentally $P$s iff all/some $S$'s are possibly $P$s *and* it is not the case that all/some $S$'s are essentially $P$s.

In order to state the validity conditions of modal categorials, we assume, following Englebretsen and Sayward [9, 169–170], a multiplicity of domains or worlds $d$ such (1) every $d$ is an element of the powerset of the universe of discourse **U** encompassing all possibilia and (2) the intersection of these worlds, $\cap d$, is not empty and contains the actual world. The latter assumption emphasizes that we are reasoning about actual things.

Contrary to standard practice in modal semantics, Englebretsen does not distinguish worlds as points in logical space at which statements are evaluated from (sub)domains as collections of the things that exist at given worlds. Note, however, that this distinction is not drawn in another alternative semantics for modal logics either, namely Lewis' [17] counterpart theory. The conflation of worlds and domains requires that domains are either disjoint (as in Lewis modal realism) or at least not completely overlapping (as in Sommers's semantics of modal categorials).

For any term "$\alpha$", "$[\alpha]$" refers to the interpretation of "$\alpha$" in **U**, and "$[\alpha]_d$" to the set of $\alpha$'s that are among things that are in the domain $d$. The validity conditions of *de re* modal categorials of TFL are as follows (cf. [9, 171]):

$$\mathrm{val}(+S +_o P) = 1 \Leftrightarrow \text{ for some } d : [S]_d \cap [P]_d \neq \varnothing$$
$$\mathrm{val}(+S -_o P) = 1 \Leftrightarrow \text{ for some } d : [S]_d \cap \mathsf{C}[P]_d \neq \varnothing$$
$$\mathrm{val}(-S +_o P) = 1 \Leftrightarrow \text{ for some } d : [S]_d \subseteq [P]_d$$
$$\mathrm{val}(-S -_o P) = 1 \Leftrightarrow \text{ for some } d : [S]_d \subseteq \mathsf{C}[P]_d$$

$$\mathrm{val}(+S +_e P) = 1 \Leftrightarrow \text{ for all } d : [S]_d \cap [P]_d \neq \varnothing$$
$$\mathrm{val}(+S -_e P) = 1 \Leftrightarrow \text{ for all } d : [S]_d \cap \mathsf{C}[P]_d \neq \varnothing$$
$$\mathrm{val}(-S +_e P) = 1 \Leftrightarrow \text{ for all } d : [S]_d \subseteq [P]_d$$
$$\mathrm{val}(-S -_e P) = 1 \Leftrightarrow \text{ for all } d : [S]_d \subseteq \mathsf{C}[P]_d$$

Note that, since all modalities are *de re*, they only pertain to categorials. Complex sentences such as conjunctions and conditionals are to be interpreted in the sole actual world. Moreover, conforming to ordinary language usage, the de re modalities of essentiality and accidentality cannot be meaningfully cumulated: so, for example, "all Europeans are essentially humans" makes sense, but not "all Europeans are essentially essentially humans" or "all Europeans are accidentally essentially humans". Formally, though, such a cumulation of copula modifiers is not a huge problem, as we shall discuss below.

## 2.2 Temporal categorials

Applying the previous considerations to temporal reasoning is straightforward. Again, we begin with the observation that the primary object of temporal discourse is reasoning about change in present things, ways in which present things (sometimes or always) have instantiated or are going to instantiate properties and relations that are the same as, or other than those they currently exemplify. Furthermore, temporal adverbs should be analyzed as copula modifiers which implies that temporal modalities are *de re*. *De dicto* temporal statements are in fact *de re* statements as to propositions being sometimes or always having been or going to be true. In other words, *de dicto* temporal statements are *de re* metalinguistic statements.

Copulae are not temporally neutral. Thus, even the copulae of basic TFL have a temporal character; indeed, they are implicitly biased towards presence: for instance, "Some horses are white" really means "Some horses are presently white". Temporal copulae are marked by four indices ("$p$", "$h$", "$f$" and "$g$") corresponding to the basic modal operators in Priors temporal logic TL; these indices are attached to the quantity of split copulae, e.g.

1. $\pm S \pm_p P$ "some/all $S$'s are/aren't $P$'s always in the past"
2. $\pm S \pm_h P$ " some/all $S$'s are/aren't $P$'s sometimes in the past"
3. $\pm S \pm_f P$ " some/all $S$'s are/aren't $P$'s always in the future"
4. $\pm S \pm_g P$ " some/all $S$'s are/aren't $P$'s sometimes in the future"

For the moment we shall focus on non-iterated *de re* temporal modalities. The time of speech $\sigma$ i.e., the time at which temporal sentences are evaluated, is implicit; by default, it is identical to the present, but may be shifted back and forth along the time line(s) as needed for the analysis of tense in ordinary language. We assume a multiplicity of times, which are a special kind of domains ordered by the relation of temporal precedence $\preccurlyeq$ that is assumed to be transitive and antisymmetric, which is sufficient for the purposes of the weak logic of time envisaged here for the sake of illustration. Furthermore, since de re temporal reasoning is about things that are of the present or of the time of speech $\sigma$, these domains overlap and are such as to contain the things comprised in the present time or in the time of speech $\sigma$.

Below are the validity conditions for the various temporal categorials with non-iterated *de re* temporal modalities; the observing reader will notice that these mirror the validity conditions for de re possibility and necessity, namely for the past and for the future:

$\text{val}(+S +_h P) = 1 \Leftrightarrow$ for some $t$ such that $t \preccurlyeq \sigma : [S]_t \cap [P]t \neq \varnothing$
$\text{val}(+S -_h P) = 1 \Leftrightarrow$ for some $t$ such that $t \preccurlyeq \sigma : [S]_t \cap \mathsf{C}[P]t \neq \varnothing$
$\text{val}(-S +_h P) = 1 \Leftrightarrow$ for some $t$ such that $t \preccurlyeq \sigma : [S]_t \subseteq [P]t$
$\text{val}(-S -_h P) = 1 \Leftrightarrow$ for some $t$ such that $t \preccurlyeq \sigma : [S]_t \subseteq \mathsf{C}[P]t$

$\text{val}(+S +_p P) = 1 \Leftrightarrow$ for all $t$ such that $t \preccurlyeq \sigma : [S]_t \cap [P]_t \neq \varnothing$
$\text{val}(+S -_p P) = 1 \Leftrightarrow$ for all $t$ such that $t \preccurlyeq \sigma : [S]_t \cap \mathsf{C}[P]_t \neq \varnothing$
$\text{val}(-S +_p P) = 1 \Leftrightarrow$ for all $t$ such that $t \preccurlyeq \sigma : [S]_t \subseteq [P]_t$
$\text{val}(-S -_p P) = 1 \Leftrightarrow$ for all $t$ such that $t \preccurlyeq \sigma : [S]_t \subseteq \mathsf{C}[P]_t$

$\text{val}(+S +_g P) = 1 \Leftrightarrow$ for some $t$ such that $\sigma \preccurlyeq t : [S]_t \cap [P]t \neq \varnothing$
$\text{val}(+S -_g P) = 1 \Leftrightarrow$ for some $t$ such that $\sigma \preccurlyeq t : [S]_t \cap \mathsf{C}[P]_t \neq \varnothing$
$\text{val}(-S +_g P) = 1 \Leftrightarrow$ for some $t$ such that $\sigma \preccurlyeq t : [S]_t \subseteq [P]_t$
$\text{val}(-S -_g P) = 1 \Leftrightarrow$ for some $t$ such that $\sigma \preccurlyeq t : [S]_t \subseteq \mathsf{C}[P]_t$

$\text{val}(+S +_f P) = 1 \Leftrightarrow$ for all $t$ such that $\sigma \preccurlyeq t : [S]_t \cap [P]_t \neq \varnothing$
$\text{val}(+S -_f P) = 1 \Leftrightarrow$ for all $t$ such that $\sigma \preccurlyeq t : [S]_t \cap \mathsf{C}[P]_t \neq \varnothing$
$\text{val}(-S +_f P) = 1 \Leftrightarrow$ for all $t$ such that $\sigma \preccurlyeq t : [S]_t \subseteq [P]_t$
$\text{val}(-S -_f P) = 1 \Leftrightarrow$ for all $t$ such that $\sigma \preccurlyeq t : [S]_t \subseteq \mathsf{C}[P]_t$

If the time of speech $\sigma$ is allowed to be fixed by the context, these non-iterated *de re* temporal modalities are sufficient for the purposes of a very rough analysis of tense in ordinary language sentences, though by no means of temporal aspect. If shifts of the time of evaluation are to be explicit, the cumulation of modalities is required: in this case, the copula is marked not by a single, but by a string of indices as in:

$\pm S \pm_{pg} P$, i.e. "some/all $S$'s were always going (not) to be $P$'s at some time in the future"

The current time of speech needs to be explicit in the validity conditions of categorials with iterated de re temporal modalities, since it must be tracked while being shifted during the evaluation process. At the start of the evaluation, the time of speech is the present time by default. The validity conditions for temporal categorials with copulae that have single indices are mutatis mutandis the same as those stated above, e.g.

$\text{val}(+S +_p P, t) = 1 \Leftrightarrow$ for all $t'$ such that $t' \preccurlyeq t : [S]t' \cap [P]t' \neq \varnothing$

Let $\pm S \pm_{i\vec{i}} P$ be a categorial with a copula marked by a list of indices starting with $i$, followed by the reminder list $\vec{i}$. The validity conditions for $\pm S \pm_{i\vec{i}} P$ have to reflect the dependence of the validity of $\pm S \pm_{i\vec{i}} P$ on that of $\pm S \pm_{\vec{i}} P$ as illustrated in the two examples below:

$\text{val}(+S +_{h\vec{i}} P, t) = 1 \Leftrightarrow$ for some $t' \preccurlyeq \sigma : val(+S +_{\vec{i}} P, t') = 1$
$\text{val}(+S +_{f\vec{i}} P, t) = 1 \Leftrightarrow$ for all $t'$ so that $\sigma \preccurlyeq t' : \text{val}(+S +_{\vec{i}} P, t') = 1$

The evaluation eventually terminates in claims about relations between interpretations of terms for a given time.

It should be noted that for the ontological considerations as to the nature of denotata it is indifferent whether we opt for the simple account of non-iterated *de re* temporal modalities or for the complex analysis of temporal categorials with multiple *de re* temporal modalities.

Given that temporal modalities are *de re*, they concern categorials only. Complex sentences such as conjunctions and conditionals are to be interpreted in the sole present moment.

One final remark pertains to the relationship between modal categorials and temporal categorials. The formal properties of the temporal precedence relation are extremely weak: in particular, they do not exclude a given time having more than one predecessor or successor. If so, then branching time is not excluded either. Therefore, it would be safe to *define* being *de re* necessary as being *de re* always in the past and in the future, and being *de re* possible as being *de re* sometimes in the past or in the future:

$$\pm S \pm_\circ P =_{df} +(\pm S \pm_h P) + (\pm S \pm_g -P)$$
$$\pm S \pm_e P =_{df} --(\pm S \pm_p P) --(\pm S \pm_f -P)$$

These definitions would not commit us to necessitarianism given that the time line is not assumed to be serial (though our weak theory of time wouldn't be incompatible with seriality of the time line either).

## 3  On sets and stages

### 3.1  On individuals and complexes

In the remainder of this paper an ontological account of the validity conditions of modal and temporal categorials shall be derived using an inference to the best explanation. To prepare the ground, some general clarifications are in order concerning the individuals that are denizens of the intended domain as well as the ways in which they can be grouped or collected together.

**Individuals.** The intended domain of modal and temporal reasoning is that of experienced space-time; thus, individuals as the denizens of that domain are spatio-temporal entities. More precisely, they are *material bodies*. The assumption of material bodies is a basic Moorean belief inasmuch as the possibility of identifying and re-identifying entities in discourse presupposes the existence of an all-encompassing and intersubjectively accessible frame of reference; this global frame of reference is that of spatio-temporal relations the bearers of which are precisely material bodies (Strawson [53, pp. 38–40]). It should be pointed out, however, that by identifying individuals with material bodies, we have yet to say something substantial about their nature, especially the way they persist through time. We shall return to this question below.

**Complexes** or aggregates are ways in which pluralities of individuals can be gathered together. According to Fine [11, p. 573] (see also Lando [15, pp. 79–81]), these ways can be classified according to whether or not they obey the following principles:

**Absorption** For any $x$ and $y$s, an aggregate of multiple occurrences of $x$ and the $y$s is identical to the aggregate of x and the $y$s.

**Collapse** The aggregate of $x$ is identical to $x$.

**Leveling** For any $x$s and $y$s, the aggregate of the aggregate of the $x$s and the aggregate of the $y$s is identical to the aggregate of the $x$s and the $y$s.

**Permutation** For any $x$s, the aggregate of any sequence of $x$s is identical to the aggregate of a permutation of that sequence.

We will require three notions of aggregate: *fusions* or mereological sums comply with all four principles, *sets* only with *Absorption* and *Permutation*, while *ordered tuples* fail to do so with respect to all aforementioned principles (Fine [11, p. 574]).

Fusions of individuals are individuals and thus spatio-temporal entities themselves; by contrast, sets and tuples are commonly deemed to be abstract objects. Yet Penelope Maddy [22, p. 178 ff] has argued that sets (and *a fortiori* also tuples) of spatio-temporal entities may be objects of perception and of perceptual beliefs. If so, they have also a (scattered) spatio-temporal location which is, of course, parasitic upon those of their members. Whether or not that is the case, sets and tuples of individuals are *identification-dependent* (in the sense of Strawson [53, pp. 16–17] on their members. So even if we decide to regard sets and tuples as abstract entities after all, naturalism can be preserved in a weaker form, namely as the thesis that the ontologically fundamental entities are all spatio-temporal objects.

Having made these clarifications, we can chart the course of the remainder of this paper. In the next two sections, we will analyse the connotations of terms and sentences using the notions of set and tuple. Then we will propose a revision of our ordinary conception of material bodies in terms of so-called stages as contents of momentary space-time regions. Finally, we will wrap up the resulting ontology and review some simplifications of that categorial scheme using the notion of fusion or mereological sum.

## 3.2 Kinds as sets of possibilia or temporalia

Sommers and Englebretsen [50, pp. 14–16] propose a threefold account of the meaning of terms. Each meaningful term is assumed to *express* a concept; furthermore, if there are things that fall under the concept expressed by a term $\alpha$, then the term $\alpha$:

1. denotes the $\alpha$'s

2. signifies the attribute of being $\alpha$.

Otherwise, $\alpha$ does neither denote nor signify anything, but, being meaningful, it still does express a concept. Thus, concepts are assumed by Sommers and Englebretsen (loc. cit.) to provide meaning to terms that happen to be vacuous.

Now, for a given domain d or time t, and for any given meaningful term $\alpha$, the sets $[\alpha]_d$ and $[\alpha]_t$ (the intersection of the interpretation of $\alpha$ with $d$, respectively with $t$) can be empty. However, the set $[\alpha]$, the interpretation of $\alpha$ within the unrestricted universe of discourse of possibilia, cannot be empty, provided $\alpha$ is meaningful and thus signifies an exemplifiable attribute. Therefore, a meaningful term always signifies or connotes an attribute. So either concepts are superfluous or may be conflated with attributes; we have a preference for the second alternative.

Moreover, talk about "properties" or "attributes" may be misleading, inasmuch as it is associated with the dichotomy of substances vs. their dependent characteristics. Now, while there is a syntactic distinction between subjects and predicates as phrases in TFL, terms can occur as quantified subject terms as well as qualified predicate terms. This is illustrated by the commutativity of the affirmative particular copula:

$$+S + T \Leftrightarrow +T + S.$$

There is no fundamental categorial distinction between singular subjects and general predicates in TFL, contrary to the Fregean paradigm of logical form. For this reason, talk about "properties" and "attributes" should be avoided in favour of the more neutral terminology of "types". In this section, we discuss types of *individuals* which shall henceforth be called "kinds".

Kinds are commonly opposed to sets: indeed, it is claimed that while sets are extensional, kinds are intensional. No distinct sets have the same members, but kinds may share all instances without being the same. This makes sense provided one only considers kinds at a given world e.g., the actual one, or at a given time e.g., the present, but not so if one assumes that kinds are instantiated within the union of domains/worlds or of times. If the domain of discourse is that of possibilia or temporalia, then kinds can be assimilated to the sets of individuals that possibly or sometimes fall under the terms that signify or connote them. Thus, kinds can be conceived of to be simply sets of possibilia or temporalia.

An interesting distinction to be made in this context is that between *sparse* and *abundant* kinds (Lewis [17, p. 59 ff]). Sparse kinds are believed to cut reality along its joints: they are deemed to belong to the theoretical inventory of ultimate natural science. Abundant kinds, on the contrary, are gerrymandered or disjunctive and draw arbitrary dividing lines among the denizens of reality. It should be pointed out, however, that the distinction between sparse and abundant kinds may reflect our own interests and biases as agents immersed in a specific culture and technology. Perhaps sparse kinds are merely those that happen to be highlighted by our choices that are informed by our culture and technology.

To sum up, we distinguish, following John Stuart Mill [25, book 1, chap. 1, § 5]. between the denotation and connotation of terms. The denotation of a term are one or several things, i.e., a plurality, while its connotation is a kind, that is, ultimately, a set. It should be noted that according to our account the distinction between denotation and connotation is not that between extension and intension, but that between plural and singular reference. A term plurally denotes the individuals that are members of the kind/set which it singularly connotes. This is reflected by the fact that terms can be read plurally as in

   (All) horses are mammals.

or singularly as in

   The Horse is a Mammal.

It is only for convenience that the connotations of terms are chosen as their interpretation in the semantics of TFL, given the availability of set theory as a proven tool of metalogical investigations.

One complication we still need to mention is that of vagueness. Because any distinction drawn among denizens of reality depends on our choices and conventions, it is determined only as far as is necessary for our practical needs (which may include the requirements of scientific research and technological development). Hence, there is obviously always room for some indeterminacy in the terms used to express such a distinction (i.e. "mountain", "life form", "person"). We agree with Hawley [13, p. 103] that indeterminacy or vagueness pertains to our terms only and is exclusively semantic in nature. Vagueness of terms is semantic indecision (ibid.), the fact that there are multiple candidates for the denotation and the connotation of these terms. Now, the minimal requirement for mutual understanding of speakers who use vague terms is that the candidate denotations and candidate connotations are maximally overlapping and only differ in borderline cases, which are the actual crux of vagueness. Thus, applying Lewis's [20] proposal to pluralities and to sets, we can regard the candidate denotations and connotations of vague terms as partially identical, as "many, but almost one" for all intents and purposes.

As far as relations are concerned, the picture would be mutatis mutandis the same, except for the fact that the instances/members of relations are not possibilia but ordered tuples of possibilia. Leaving aside the question whether ordered tuples are to be considered ontologically primitive or reducible to sets (cf. Kuratowski [14, p. 171]), this account implies that all relations, respectively their members, are ordered. Fine [10] has pointed out that this assumption has two drawbacks:

1. A symmetrical relation (e.g. "in the vicinity of") holds between two individuals (e.g. "Tours is in the vicinity of Paris") in virtue of their being members of two pairs for each direction of the relation (⟨Tours, Paris⟩ and ⟨Paris, Tours⟩), which seems to be one "completion" or connection too many.

2. Each ordered binary relation has a converse, e.g. "loving" vs. "being loved" or "child of" vs. "parent of", and, in general, an n-adic ordered relation has n! converses. The question which arises is which of the converses is ontologically basic, while it really seems that none of them has any entitlement to ontological priority with respect to the other.

A detailed discussion of the problem of relations is outside the purview of this paper, but some hints can be given as to how these issues could be handled in our framework. As far as symmetrical relations are concerned, it may be advisable to view the respective predications as monadic predications of plural subjects, as in "London and Paris are in the vicinity of each other" or "Samuel, John and Barbara are siblings". The problem of converses can be resolved by invoking the distinction between abundant and sparse with respect to relations. As in the case of kinds, relations are singled out as sparse with respect to our practical interests and cultural habits. If so, the problem of whether some relation or its converse is among the sparse relations, vanishes as the distinction between sparse and abundant relations does not cut deep enough to slice into the joints of reality.

Relational terms are as fraught with vagueness as monadic ones: often there is no clear-cut answer whether a relation holds or not. At what distance two cities are not anymore in the vicinity of each other? At what moment exactly the wife fell out of love with her husband? As with kinds, vagueness of relational terms consists in semantic indecision which is due to the existence of many candidate denotations and connotations. As with kinds, the solution to the problem of vagueness consists in simply embracing the multitude of admissible denotations and connotations, while taking into consideration the fact that they are largely overlapping and may be regarded as partially identical, as almost one in all relevant practical respects.

### 3.3 Propositions as sets of worlds or times

Sentences are viewed as terms in TFL and they are attributed the same threefold semantics as other terms. Thus, a *meaningful* sentence expresses a concept, called *proposition*; if there are domains (i.e., worlds or times) that fall under the proposition, then the sentence:

1. denotes the domain(s) that fall(s) under the expressed proposition, and

2. signifies a property of domains, called a *fact*.

In this case, one can also say that the proposition *corresponds* to the fact and therefore is true (Sommers [49], Sommers and Englebretsen [50, p. 23]).

In the light of the previous discussions, we will opt for the much simpler scheme of denotation vs. connotation. So, the denotation of a sentence is simply a domain or a plurality of domains, and its connotation a type which is the set of denoted domains (reminiscent of Lewis [17, pp. 53–55]). We call types of domains *propositions*, following common parlance that identifies facts with true propositions. A proposition is true of those domains or times which are its members. Since worlds and times are sets of possibilia (i.e., kinds), propositions are sets of sets (note, however, that at the end of this paper we will envisage a possible simplification that consists in propositions being conceived of as sets of maximal stages).

The connotation of a categorial is called a *categorial proposition*; categorial propositions are *constitutive types* of domains: so $[+\alpha + \beta]$ connotes the positive type that a domain contains some $\alpha$s that are $\beta$s, while $[-\alpha + \beta]$ connotes the negative type of a domain containing no $\alpha$s that aren't $\beta$s.

If we treat sentences as terms, binary connectives can be accounted for as syncategoremata of standard term logic. We shall focus on implication and conjunction (Sommers [49, p. 178]).

1. An *implication* ("$-[\phi] + [\psi]$") connotes the proposition that all domains denoted by the antecedent are domains denoted by the consequent. Implications are thus (affirmative) general categorical sentences the subject and predicate terms of which are sentences.

2. A *conjunction* ("$+[\phi] + [\psi]$") connotes the proposition that some domains denoted by the first conjunct are domains denoted by the second conjunct. Conjunctions are thus (affirmative) particular categorical sentences the subject and predicate terms of which are sentences.

Since complex sentences are uniquely denoting terms, as they denote the single actual world or present time, $+[\phi] + [\psi]$ entails $-[\phi] + [\psi]$. In general, (affirmative) particular sentences do not imply (affirmative) general sentences unless the respective subject term is uniquely denoting. Since there is only one actual world or present time, from "some $\phi$-world is a $\psi$-world" follows "all $\phi$-worlds are $\psi$-worlds". This entailment needs to hold, so that the parallelism between term-logical syncategoremata and connectives may be maintained, and thus propositional logic be reduced to term logic.

Lewis [17, p. 61] mentions, almost in passing, that propositions, too, can be said to be abundant or sparse, without developing that idea any further. The opposition between gerrymandered and "natural" propositions could be useful to understand what theoretical simplicity means: the simplest explanation consisting in a (conjunctive) sentence the proposition of which is the least gerrymandered and circumscribes the most homogeneous regions in logical space. However, this proposal must remain largely metaphorical as long as no objective measure of the relative homogeneity of propositions is provided.

Sentences containing vague terms are vague themselves. Vagueness of sentences consists in the semantic indecision between a multiplicity of candidate denotations (i.e., pluralities of domains) and connotations (i.e., propositions as types of domains). If vagueness of statements does not prevent understanding between speakers except in borderline cases, the various candidate denotations and connotations must be largely overlapping, which allows them to be regarded as many, but almost one.

## 3.4 Stages as individuals

The question that remains to be answered is: What are the individuals singularly or plurally denoted by terms? The obvious response seems to be that individuals are to be conceived of as (Aristotelian) "substances" or as "continuants", i.e. things that exist in time but have no temporal parts (cf. Simons [42, p. 59]). However, this view of individuals is arguably wrong under the assumption of the semantics of *de re* modal and temporal categorials presented above, especially if we turn our attention to singular terms. Indeed, the validity condition of a *de re* categorial is stated as a set-theoretical relation between the interpretation of the subject term at given domains and the interpretation of the object term at the same domains. The idea is that the interpretation of a term may vary from domain to domain, in particular: from time to time. Now in the case of singular terms, variations of their interpretations from world to world or from time to time are out of the question. In fact, a singular term denotes the same individual in every world and at every time it denotes at all. Thus, a certain predicate fails to hold of a singular subject with respect to a given world or time only if (a) it fails to do so in every world or time, or (b) the subject term does not denote anything at all in this world or at this time.

Hence, our account of validity conditions for *de re* modal and temporal categorials would seem to imply a wholescale essentialism regarding predication of singular subjects. Perhaps not completely so: relational predicates may also fail to hold of their singular relata if at least one of their subjects or objects fails to denote. So, for example, the claim that Romeo loves Juliet may be false in virtue of one of the singular terms "Romeo" or "Juliet" being vacuous at the world or time of the claims' evaluation. Nonetheless this would still represent a rigidity of predication that seems to be incompatible with our views on ordinary objects. We call this the *problem of temporary singulars*.

One could take this conundrum as evidence for the fact that our account of the validity conditions of modal and temporal categorials is simply wrong. It could be alleged that the right approach would have been to interpret $n$-adic terms as sets of n+1-tuples, the last elements of which are *domains*, i.e., worlds or times, as is standard practice in the semantics of modal logic. Following Lewis [17, pp. 202–205], however, one could counter that this would mean that there were no non-relational types and that individuals could not be members of types simpliciter. In other words, there could be no *intrinsic*, i.e. irreducibly monadic types that were not permanently, but merely temporarily instantiated.

It is a matter of controversy how conclusive the argument from temporary intrinsics is (Sider [41, pp. 92–98], Hawley [13, p. 17 ff]) – after all, reference to times may be implicitly present in ordinary language statements relative to every-day objects. There is nonetheless something odd in claiming that the love between Romeo and Juliet is a three-place relation between Romeo, Juliet and a certain time; indeed, the statement that Romeo loves Juliet would seem to be about Romeo and Juliet and nothing else. To say that time is an additional, though implicit, relatum gets something wrong: rather, time has something to do with the way how Love holds between Romeo and Juliet.

So, one alternative proposal is to relativize the second-order relations between types and their members to times in some way or another. Yet it could be argued that this alternative approach is ultimately tantamount to turning first-order types into relations to times (Sider [41, p. 96], Lewis [21]). Furthermore, in our account, the second-order relations between types, as well as between types and their members are set-theoretical relations, which, being mathematical ties that are also part of the (meta)logical apparatus, cannot be relativized to times. Temporalizing set membership or set inclusion is just as meaningless as in the case of the conjunction or the disjunction.

The simplest way to deal with the problem of temporary singulars may be to regard names of continuants not as singular, but as general terms the denotations of which are stages, i.e. instantaneous four-dimensional objects (Hawley [13], Sider [41, chap 5, sec. 8]). Let us call "object terms" those terms denoting stages that correspond to names of physical objects in ordinary language; object terms are akin to individual predicates in the sense of Hawley [13, p. 183]. For any object term $\alpha$ and any time t, $[\alpha]_t$ is a singleton whose sole member is a stage. Thus an object term is a general term that is temporarily singular. However, an object term denotes different stages at different times (cf. Hawley [13, p. 62]). This is so because stages are as fine-grained as time (Hawley [13, p. 51]). That material bodies must be as fine-grained as time follows from the very fact that they are ontologically basic, namely that they are spatially and temporally dense enough to provide waypoints within an encompassing spatial and temporal frame of reference.

Note that if domains, i.e. worlds or times, are sets of stages, then they must be disjoint. Focus on things that are present or actual does not need to be abandoned but it cannot be anymore ensured by the requirement that all times overlap. Instead, one is free to postulate that all object terms denote present as well as future or past stages. Whether the focus on the present still makes sense in a four-dimensional ontology is another question altogether.

The connotations of object terms shall be called "individual kinds". An individual kind is a kind the intersection of which with a given time is always a singleton. Individual kinds are akin to *infimae species* in the sense of Leibniz:

> "...ce que $S$. Thomas asseure sur ce point des anges ou intelligences (quod ibi omne individuum sit species infima) est vray de toutes les substances..."  [16, chap ix]

as well as to *haecceities* or *individual notions* as evoked in the following passage from his "Discours de Métaphysique":

> "Dieu voyant la notion individuelle ou hecceïté d'Alexandre, y voit en même temps le fondement et la raison de tous les prédicats qui se peuvent dire de lui véritablement, comme par exemple qu'il vaincrait Darius et Porus, jusqu'à y connaître a priori (et non par expérience) s'il est mort d'une mort naturelle ou par poison, ce que nous ne pouvons savoir que par l'histoire. Aussi, quand on considère bien la connexion des choses, on peut dire qu'il y a de tout temps dans l'âme d'Alexandre des restes de tout ce qui lui est arrivé, et les marques de tout ce qui lui arrivera... "  [16, chap. viii]

At a fundamental level an ordinary object should be regarded as an haecceity, i.e. an individual kind which provides unity to a manifold of stages. Just as we speak of a forest in the singular, but mean a host of individual trees, so we should think of what we singularly refer to as Alexander, the Mont Blanc or a particular horse as sets of snapshot entities. The view of continuants as (individual) kinds may seem perplexing at first, but has some ontological advantages, not the least being the fact that the coincidence of a statue and a lump of clay, or of a person and a human organism is not more paradoxical than the overlapping of kinds or sets. The fact that coinciding continuants, e.g. a statue and "its" lump of clay, have different modal properties (e.g. the lump may survive a sudden squeeze, but not the statue) can be accounted for in terms of individual kinds sharing some stages as members, without having all members in common, i.e. without being identical (cf. also Hawley [13, p. 183 ff] for further discussion).

Object terms are doubly afflicted by vagueness. On the one hand, they have multiple candidate denotations and connotations, as is best illustrated by names of organisms such as "Bucephalus": our conventions as to the beginnings and endings of lives leave it undecided which is the first, respectively the last stage that is denoted by that name. On the other hand, there is strictly speaking not even a single stage that is denoted by an object term at a given time. In other words, there are many singletons that satisfy the description $[\alpha]t$ for any object term $\alpha$ and time $t$. This is again exemplified by names of organisms. Take the name "Bucephalus" and consider an arbitrary time t when it refers at all: at t, there will be many overlapping horse-shaped stages that differ only in a different hair being removed that are legitimate referents of the name "Bucephalus". Thus, object terms show both a diachronic and a synchronic vagueness. The solution is the same in both cases: accept the plurality of admissible denotations and connotations, and, if enough overlap between these candidate pluralities and sets is given, count them as many, but almost one (Lewis [20]). The idea is that for practical purposes partial identity in the sense of almost complete overlap is good enough to pass for identity.

The dichotomy "abundant vs. sparse" can be also invoked with respect to individual kinds: thus, sparse individual kinds would be less gerrymandered and more homogeneous than abundant ones. As in the other cases where this distinction has been mentioned, the greater or lesser degree of arbitrariness may largely reflect our interests and practical objectives, our cultural and technological biases. It may seem that "Bucephalus" and "Alexander" connote individual kinds that are more natural than "Brangelina", "Orient Express" or "Crown Jewels of the

United Kingdom", but cultural reflexes have elevated the last three examples to the status of first-class denizens of reality.

Finally, one may wonder why we do not follow Quine [38] to the bitter end, and do without object-talk altogether: first replace talk about stages with talk about (instantaneous) space-time regions and then substitute the latter with quadruples of numbers that represent coordinates within a arbitrarily chosen reference system. Indeed, the alternative account would be empirically equivalent to a basic ontology of stages. Nonetheless there are two reasons for resisting these further steps in revising our everyday conceptual scheme. First, the alternative theory would run counter the Moorean assumption that material bodies enjoy ontological priority with respect to all other categories of entities, including space-time regions. However, someone may be comfortable with abandoning this basic belief in view of a gain in conceptual economy, as the resulting ontology would be limited to sets, tuples and numbers. Second, this economy constitutes nonetheless an even stronger reason for not reducing stages to quadruples of coordinates, because it conflicts with the metaontological commitment to naturalism, even with its weaker form, namely the stance that ontologically fundamental entities are all spatio-temporal. Since the proposed revision would result in an ontology of abstract entities, naturalism would have to be abandoned, which would certainly be a greater cost than the rejection of a commonsense belief.

## 3.5 Wrapping up and simplifying the proposed ontology

What emerges from the foregoing discussions can be summed up as a five-fold categorical scheme:

1. at the base level, *stages* as contents of instantaneous space-time regions,

2. *ordered* tuples of items in (1),

3. *kinds* as sets of items in (1); among them also domains, i.e., worlds or times,

4. *relations* as sets of items in (2), and

5. *propositions* as sets of domains, which are items in (3).

This scheme can be substantially rebuilt by introducing the already mentioned notion of *mereological sum*, namely of an aggregate that conforms to the principles of Absorption, Collapse, Leveling and Permutation (Fine [11, p. 574]; cf. Section 3.1). In other words, *composition*, i.e. the operation that takes any things and returns their sum or fusion, is blind or neutral with respect to structure. We adopt the metaphilosophical position of *mereological monism*, i.e. the stance that the mereological theory *par excellence* is *Classical Extensional Mereology (CEM)* (Lando [15, pp. 4–5]; Lewis [18, pp. 72–75]).

CEM consists in the thesis that for any plurality of things, there is (a) at least and (b) at most one sum or fusion of those things (Lando [15, p. 4]). Sub-thesis (a), namely that there is at most one sum of any given things, is called *extensionality of composition*. It can be motivated by the fact that, since composition is blind with respect to structure, the latter does not play any role in the identity conditions of sums (Lando [15, p. 83]), which implies that the identity or diversity of sums is wholly determined by the fact whether or not they share all their parts. Sub-thesis (b), the view that there is at least one sum of any given things, is called the thesis of *unrestricted composition*. This principle can be motivated by the fact that composition is coextensional with existence: whatever exists is the fusion of itself (Lando [15, p. 179]). Existence is purely formal and topic-neutral (according to Quinean metaphilosophy, cf. Lando [15, ibid.]) and is expressed by the quantity-aspect of the copula in term logic. So if existence does not represent a contentful distinction, neither does composition which therefore has to be unrestricted.

Not only is CEM the simplest account of mereological composition, it also allows two substantial simplifications of the categorial scheme above. First of all, in virtue of unrestricted composition, there exists, for every (possible) time, the sum of all stages that are located at that time; let this sum be called *maximal stage at t* or simply *maximal stage*. Since times are disjoint, there is a one-to-one correspondence between times and maximal stages; thus, maximal stages can act as proxies of times in the semantics of temporal categorials. This revision would not require any changes in the formulation of the validity conditions of temporal categorials, the only exception being the definition of the term $[\alpha]_t$, which would designate the set of those $\alpha$'s that are part of the time / maximal stage $t$. Finally, propositions turn out to be kinds of maximal stages, which relieves us from the necessity to assume sets of sets.

One may ask whether maximal stages are denoted by an object term and if so, by which one. The individual kind connoted by that object term obviously represents the totality of spatiotemporal reality; hence the object term "Nature" seems to be appropriate. Note that neither the object term "Nature" nor the individual kind it connotes are vague; they may be the only ones not affected by vagueness. Each maximal stage is a way Nature as the maximal continuant could possibly sometimes be; that each way Nature could possibly sometimes be is an existing maximal stage shall be referred to as *Plenitude* (Lewis [17, p. 86 ff]). Plenitude minimizes what would otherwise be the arbitrariness of Nature only consisting in its present stage which corresponds to the way Nature is contingently now (cf. Unger [54]).

The second simplification made possible by the adoption of CEM relates to the reduction of set theory to mereology as suggested by Lewis's [18, 19] seminal work. The basic idea is that set inclusion can be viewed as parthood between sets ([18, p. 3 ff]): one set is part of another if and only if the former is a subset of the latter ([19, p. 206]). Therefore, any set can be assimilated to the sum or fusion of singletons ([19, p. 212]). We do not require the full power of set theory for this ontological account; indeed, we need only consider sums of singletons of individuals or stages (including maximal stages or times), as well as singletons of ordered tuples.

All these considerations allow us a reduction of our basic ontology to four formal categories:

1. individuals as stages, including times as maximal stages,

2. ordered tuples of items in (1),

3. (fusions of) singletons of items in (1), i.e. kinds and propositions,

4. (fusions of) singletons of items in (2), i.e. relations.

Further simplifications would of course be conceivable, but may be resisted for good reasons. Ordered tuples could be reduced to ordered pairs, the latter being amenable to several alternative set-theoretical definitions, the most famous being of course that of Kuratowski [14, p. 171]. The fact, however, that there is no unique set-theoretical analysis of ordered pairs, while a high-level characterisation of ordered tuples that distinguishes them from sets and sums is available (cf. Section 3.1), is a sufficient rationale to reject the elimination of ordered tuples from the set of basic categories detailed above.

## 4 Conclusion

This paper has had two main objectives. First, we have provided a formal account of de re modal and temporal categorials within Term Functor Logic. Second, reflecting on the semantical considerations arising from that account, we have derived an ontology that is ultimately centered around four primitives: stages as contents of instantaneous slices of spatio-temporal regions, singletons, ordered pairs and fusions. It may be surprising that, contrary to Fregean predicate logic, a neo-Aristotelian term logic does not necessarily lead to a Aristotelian metaphysics but seems far more amenable to a purely extensional post-Quinean ontology of sets, tuples and four-dimensional objects than to a traditional intensional account of substances and their attributes.

# References

[1] Ajdukiewicz, K. (1935). "Die syntaktische Konnexität", in *Studia philosophica* 1, pp. 1–27.
[2] Armstrong, D. M. (2004). *Truth and Truthmakers*. Cambridge: Cambridge University Press.
[3] Austin, J. L. (1979). *Philosophical Papers*. Oxford: Oxford University Press.
[4] Austin, J. L. (1979a). "A plea for excuses", in [3, pp. 175–204].
[5] Cappelen, H. (2012). *Philosophy Without Intuitions*, Oxford: Oxford University Press. 25
[6] Cocchiarella, N. (2001). "Logic and Ontology", in *Axiomathes* 12, pp. 117–150.
[7] Englebretsen, G. (1988). "Preliminary Notes on a New Modal Syllogistic", in *Notre Dame Journal of Formal Logic* 29(3), pp. 381–395.
[8] Englebretsen, G. (1996). *Something to Reckon With: The Logic of Terms*. Ottawa: University of Ottawa Press.
[9] Englebretsen, G., Sayward, C. (2011). *Philosophical Logic: An Introduction to Advanced Topics*. London: Continuum.
[10] Fine, K. (2000). "Neutral Relations", in *Philosophical Review* 199, pp. 1–33.
[11] Fine, K. (2010). "Towards a Theory of Part", in *The Journal of Philosophy* 107 (11), pp. 559–589.
[12] Gardies, J.-L. (1975). *Esquisse d'une grammaire pure*. Paris : Vrin.
[13] Hawley, K. (2001). *How Things Persist*. Oxford: Clarendon Press.
[14] Kuratowski, C. (1921). "Sur la notion de l'ordre dans la Théorie des Ensembles", in *Fundamenta Mathematicae* 2 (1), pp. 161–171.
[15] Lando, G. (2017). *Mereology. A Philosophical Introduction*. London: Bloomsbury.
[16] Leibniz, G. W. (1962). *Discours de métaphysique*. Édition Lestienne. Paris : Vrin.
[17] Lewis, D. K. (1986). *On the Plurality of Worlds*. London: Blackwell.
[18] Lewis, D. K. (1991). *Parts of Classes*. Oxford: Blackwell.
[19] Lewis, D.K. (1993a). "Mathematics as megethology", in *Philosophia Mathematica* 3, pp. 3–23.
[20] Lewis, D. (1993b). "Many, But Almost One", in Keith Campbell, John Bacon & Lloyd Reinhardt (eds.), *Ontology, Causality and Mind: Essays on the Philosophy of D. M. Armstrong*, pp. 23-38. Cambridge: Cambridge University Press.

[21] Lewis, D. K. (2002). "Tensing the Copula", in *Mind* 111, pp. 1–19.
[22] Maddy, P. (1980). "Perception and Mathematical Intuition", in *The Philosophical Review* 89, pp. 163–196.
[23] McGinn, C. (2001). *Logical Properties*. Oxford: Oxford University Press.
[24] McIntosh, C. (1982). "Appendix F", in [47, pp. 387–425].
[25] Mill, J. S. (1843/2012). *A System of Logic, ratiocinative and inductive: being a connected view of the principles of evidence, and the methods of scientific investigation.* 2 vols.. Cambridge: Cambridge University Press.
[26] Moore G. E. (2013). *Philosophical Papers*. London: Routledge.
[27] Moore G. E. (1925). "A Defense of Common Sense", in [26, pp. 32–59].
[28] Musgrave, A. (1993). *Common sense, Science and Scepticism: a historical introduction to the theory of knowledge.* Cambridge: Cambridge University Press.
[29] Noah, A. (1982). "Quine's Version of Term Logic and its Relation to TFL", in [47, pp. 372–385].
[30] Popper, K. R. (1945/2011). *The Open Society and Its Enemies.* London: Routledge.
[31] Quine, W. V. O. (1960). "Variables Explained Away", in *Proceedings of the American Philosophical Society* 104(3), pp. 343–347.
[32] Quine, W. V. O. (1969). *Ontological Relativity and Other Essays.* New York, NY: Columbia University Press.
[33] Quine, W. V. O. (1969a). "Ontological Relativity", in [32, pp. 26–68].
[34] Quine, W. V. O. and Ullian J.S. (1978). *The Web of Belief.* Second Edition. New York, NY: Random House.
[35] Quine, W. V. O. (1980). *From a Logical Point of View. Nine Logico-Philosophical Essays.* Second Edition, revised. Cambridge, MA: Harvard University Press.
[36] Quine, W. V. O. (1980a). "Two Dogmas of Empiricism", in [35, pp. 20–46].
[37] Quine, W. V. O. (1981). *Theories and Things.* Cambridge, MA: Harvard University Press.
[38] Quine, W. V. O. (1981a). "Things and Their Place in Theories", in [37, pp. 1–23].
[39] Quine, W. V. O. (1982). *Methods of Logic.* Fourth Edition. Cambridge, MA: Harvard University Press.
[40] Russell, B. (1903). *The Principles of Mathematics.* Second edition. London: Allen and Unwin.
[41] Sider, T. (2001). *Four-Dimensionalism: an Ontology of Persistence and Time.* Oxford: Oxford University Press.

[42] Simons, P. (2000). "Continuants and Occurrents", in *Proceedings of the Aristotelian Society, Suppl. Vol.* 74, pp. 59–75.
[43] Sommers, F. (1959). "The Ordinary Language Tree", in *Mind* 68, pp. 160–185.
[44] Sommers, F. (1963). "Types and Ontology", in *Philosophical Review* 72, pp. 327–363.
[45] Sommers, F. (1969). "Do We Need Identity ?", in *Journal of Philosophy* 66, pp. 499-504.
[46] Sommers, F. (1970). "The Calculus of Terms", in *Mind* 79, pp. 1–39.
[47] Sommers, F. (1982). *The Logic of Natural Language*. Oxford: Clarendon Press.
[48] Sommers, F. (1990). "Predication in the Logic of Terms", in *Notre Dame Journal of Formal Logic* 31, pp. 106–126.
[49] Sommers, F. (1993). "The World, the Facts and Primary Logic", in *Notre Dame Journal of Formal Logic* 34(2), pp. 169–182.
[50] Sommers, F., Englebretsen, G. (2000). *An Invitation to Formal Reasoning*. Aldershot: Ashgate.
[51] Sommers, F. (2008). "Ratiocination: An Empirical Account", in *Ratio* 21(2), 115–133.
[52] Sommers, F. (2012). "How We Naturally Reason", unpublished manuscript downloaded on April 8, 2022 at https://philarchive.org/archive/SOMHWIv3.
[53] Strawson, P. F. (1959). *Individuals*. London: Methuen.
[54] Unger, P. (1984). "Mimimizing Arbitrariness: Toward a Metaphysics of Infinitely Many Isolated Concrete Worlds", in *Midwest Studies in Philosophy* 9, pp. 29–51.
[55] Van Heijenoort, J. (1967). "Logic as Language and Logic as Calculus", in *Synthese* 17, pp. 324–330.
[56] Williamson, T. (2022). *The Philosophy of Philosophy*. Second Edition. Hoboken/NJ: Wiley Blackwell.

# THE RANGE OF ALGEBRAIC DECISION IN TERM LOGIC

MILES RIND

Algebraic term logic (ATL), also known as term-functor logic (TFL), is a system of logic developed by Fred Sommers to represent the logical forms of statements and arguments in natural language. Sommers offers a rule for deciding logical equivalence between formulas of ATL and a rule for deciding the validity of arguments, both by means of algebraic addition. It is easily shown that neither principle applies to all formulas of ATL without restriction, but the proper range of application of the principles, and thus the extent of decidability by algebraic addition in term logic, has not yet been defined. I identify ways of expanding the range of application of algebraic addition beyond formulas of just two terms, but the question of exactly where the limits of that range lie remains unanswered.

A *term* logic is a system of logic in which all logically material elements belong to the same syntactic substitution class, that of the term, and in which every formula is a dyad composed of a pair of terms combined by one or more logically formative elements. Aristotle's system and its descendants all belong to term logic. *Algebraic* term logic (ATL), also known as *term-functor* logic (TFL), is a system of term logic invented by Fred Sommers in which the formative elements are plus and minus signs.[1] Sommers makes three principal claims for ATL over the now-standard first-order predicate calculus, or modern predicate logic (MPL), as he calls it:

---

[1] Sommers introduces the designation 'term-functor logic' in Sommers [21]. But in Sommers [22], he calls his system 'algebraic term logic' (ATL). The designation is also used in [4].

(1) ATL's powers of expression and deduction extend as far as those of MPL, and indeed farther.

(2) The syntax of ATL for the most part represents the actual logical forms of statements in natural language while that of MPL does not.

(3) Unlike MPL, ATL affords simple algebraic procedures for determining the logical equivalence or non-equivalence of formulas (within a certain range) and the validity or invalidity of arguments (within a certain range).

Part 1 of this paper contains an exposition of ATL that will, I hope, give some idea of the grounds of claims (1) and (2); but my chief concern here is with claim (3), which is the subject of part 2. Sommers offers a rule for determining the logical equivalence or non-equivalence of pairs of formulas of ATL and a rule for determining the validity or invalidity of arguments in ATL. Each rule purports to state necessary and sufficient conditions of the target status in terms of the equality of two algebraic sums derived from the set of formulas under consideration. Neither rule holds good without restriction; but the proper range of application of each has yet to be determined.

Before I take up that task, I need to say something about the choice of a name for the system of logic under consideration here. Sommers seems to have chosen the designation 'term-functor logic' for two reasons: because it facilitated a comparison with Quine's system of so-called predicate-functor logic,[2] and because it allowed him to use the same initialism for it, 'TFL', that he had introduced to abbreviate 'traditional formal logic'.[3] Both of these considerations seem to me parochial, while the two together fail to outweigh the contrary consideration that the designation 'term-functor logic' fails to describe what distinguishes Sommers's system of term logic from other systems of term logic.[4] The

---

[2]See, e.g., [21, p. 124, n. 1].

[3]In [19].

[4]E.g., in [21, p. 107], Sommers notes that "the term/functor style of analysis may be said to go back to Aristotle"; thus, the designation 'term-functor logic', however proprietary, fails to describe what distinguishes Sommers's system of logic from its Aristotelian and scholastic predecessors.

designation 'algebraic term logic' seems to me superior in this respect, and for that reason it is the designation that I employ here.[5]

## Part 1: Exposition of ATL

## 1  Essential Characteristics of ATL

The conception of logical syntax in ATL may best be understood by means of a contrast with the conception of logical syntax in MPL. The latter may be stated as follows:[6]

(1) All logically material elements fall into one or the other of two syntactic categories: singular constants and predicate constants. Singular constants occur only as subjects, and besides variables of quantification they are the only expressions that can do so. Predicate constants are the only material elements that are general in character.

(2) The combination of an $n$-adic predicate constant with just $n$ occurrences of singular constants constitutes a complete formula. Such a formula is logically "atomic": more complex formulas may be composed of it, but it cannot be decomposed into other formulas, and it contains no logically formative elements.

(3) A formula can contain more than one predicate only if it contains a sentential operator.

(4) Negation can only be expressed by a sentential operator.

(5) Logical syntax is properly an aspect only of a purpose-built logical language: a natural language may reflect logical syntax to some degree, but only by luck.

---

[5]Against this, one could also argue that, as the term logics of Boole, Jevons, and others all employ algebraic notation, the term 'algebraic term logic' does not distinguish Sommers's logic from theirs. So perhaps I have to fall back upon the point that in this paper I happen to be concerned specifically with the role of algebraic addition in Sommers's term logic, whatever it be called.

[6]Sommers provides his own point-by-point statements of the opposition between the two views of logical syntax in [20, pp. 181–84], and in [19, p. 17].

The conception of logical syntax in ATL may be stated by opposition to the five points above, thus:

(1) There is just one syntactic category of non-logical element, the elementary term. An elementary term may, according to the interpretation given to it, be singular or general, relational or non-relational, nominal or pronominal, and sub-sentential or sentential; and it may occur in subject position or in predicate position.

(2) Every formula is a dyad consisting of a pair of terms combined by a functor (which may take the form of a separate quantifier and qualifier). None is devoid of logical elements.[7]

(3) The same functors that combine terms to form simple sentences combine them to form compound terms and also combine formulas to form compound formulas. Sentence formulas are just compound terms of a particular sort, and sentential logic is a special branch of term logic.

(4) Negation is an operation upon terms, whether simple or compound. Sentential negation is an instance of term negation.

(5) Logical syntax is an aspect of natural language. A logical language that represents logical form in natural language does so by the use of pairs of opposed positive and negative signs.

ATL may also be characterized by its way of drawing the distinction between logically formative and logically material expressions in natural language. From the perspective of ATL, the distinguishing mark of logically formative expressions, such as 'some', 'every', 'is', '(is) not', 'and', and 'or', is that they form opposed pairs that exhibit the logical behavior of positive and negative signs; hence the possibility of representing them in an algebraic language of pluses and minuses.[8]

So much for the general conception of logical syntax underlying ATL. It remains to explain the specific workings.

---

[7]This is a simplification, as in the sentential part of term logic, a mere letter representing a statement can occur as a complete formula. See, however, point (3).

[8]See [15].

## 2  The Base Language of ATL

The language of ATL is built upon a comparatively simple base. In this base language, which I shall call *basic ATL* or *BATL*, there are just two logical functors: a unary minus sign and a binary plus sign.[9] BATL is not a separate language but simply a fragment of ATL: its syntactic and semantic rules are also rules of ATL. Everything expressible in fully developed ATL has an equivalent in BATL, but BATL, in contrast to the fully developed language, is not designed to represent logical form in natural language. In particular, one may say of it, as Frege said of his conceptual notation, that the distinction between subject and predicate has no place in it.[10]

In what follows, the verb 'to denote' is used without any implication of uniqueness on the side of what is denoted. In the terminology of Leonard and Goodman, it is multigrade on the side of its grammatical object.[11] For example, the word 'dog' may be said to denote Fido, but it may also be said to denote Spot, Bowser, and every other dog in the domain of interpretation. A term denotes objects in its extension (i.e., it denotes every one of them); it does not denote its extension.

With these points made, the syntactic and semantic rules of BATL may be stated as follows, using the italicized letters '$X$' and '$Y$' as schematic symbols for terms. (The reason for the reversal of the alphabetical order of those two letters in schemas will appear later.)

1. Elementary-term rules:

    (a) Syntactic: Any capital letter is an elementary term of ATL.

    (b) Semantic: To any capital letter any term in English may be assigned as its interpretation.

2. Negative-term rules:

    (a) Syntactic: If $X$ is a term of ATL, $(X)$ is a term of ATL.

---

[9]This manner of exposition follows that of Sommers and Englebretsen [23, pp. 49–66].
[10]Frege [7, p. 12].
[11][9, p. 50].

(b) Semantic: $(-X)$ denotes just those objects in the domain not denoted by $X$.

3. Compound-term rules:

   (a) Syntactic: If $X$ and $Y$ are terms of ATL, then $(Y+X)$ is a term of ATL.[12]

   (b) Semantic: $(Y + X)$ denotes just those objects in the domain that are denoted by $X$ and denoted by $Y$.

4. Formula rules:

   (a) Syntactic: If $X$ and $Y$ are terms of ATL, then $Y + X$ and $-(Y + X)$ are formulas of ATL.

   (b) Semantic: $Y+X$ is true just in case some object in the domain denoted by $X$ is denoted by $Y$; $-(Y+X)$ is true just in case it is not the case that some object in the domain denoted by $X$ is denoted by $Y$.

Every term or formula of BATL is generated by means of these rules alone. Using these rules, one can construct a square of opposition as follows:

| $-((-Y)+X)$ | $-(Y+X)$ |
|---|---|
| $Y+X$ | $(-Y)+X$ |

Figure 1: BATL Square of Opposition

These four schemas may be most straightforwardly read, from left to right by row, as 'Not: some $X$ is non-$Y$', 'Not: some $X$ is $Y$', 'Some $X$ is $Y$', and 'Some $X$ is non-$Y$'. The four may also be read according to the traditional A, E, I, and O forms of statement. But BATL, in contrast to the fully developed language ATL, is not designed to represent natural syntax, and its formulas are not of subject-predicate form. We

---

[12]Sommers uses angle brackets ('⟨ ⟩') to enclose non-relational compound terms and reserves round brackets ('( )') for relational ones. I find this multiplication of grouping symbols superfluous and accordingly, following [6], eschew it.

are therefore not obliged to choose a canonical reading of each schema or to follow the order of the terms in each. For example, the first schema may be read with equal correctness as 'Every $X$ is $Y$', 'No $X$ isn't $Y$', 'No non-$Y$ is $X$', 'Every non-$Y$ is non-$X$', 'Not: some non-$Y$ is $X$', and so on. Nor is it necessary to choose the grammatically singular mode of expression: one could read the same formula as 'All $X$'s are $Y$'s', and so on.

Although the schemas in the left side of the square correspond semantically to the statement-forms traditionally called 'affirmative', and the schemas in the right side to those traditionally called 'negative', I have not so labeled the sides, for two reasons. One reason is that negation is present in all the schemas except the one in the lower left. The other is that even that seemingly negation-free schema could be read, however perversely, as 'Some $X$ is not non-$Y$' or 'Not every $X$ is non-$Y$', for example. Affirmative or negative quality cannot be attributed to any of the schemas in themselves, but only to their English readings, and of those only to some and not to others.

I have also refrained from labeling the schemas in the upper half 'universal' and those in the lower half 'particular', even though, once again, they correspond to the statement-forms traditionally so described. This is because the schemas in the upper half do not, in any syntactic respect, express universal quantity: they are simply the negations of the positive formulas in the opposite corners of the lower row.

On the other hand, the opposition between the upper and lower halves of the square cannot be erased or obscured in any English reading, as the opposition between affirmative and negative qualities can be: there is no way to give one of the negative formulas a particular reading, as one can arbitrarily give readings of affirmative or negative quality to the formulas on either side of the square.

This last point is highly consequential. The overall positive or negative character of a formula, corresponding to the logical quantity of its English reading, is a fixed and unalterable attribute of all logically equivalent transformations of that formula. And although it is a syntactic feature, it has a semantic significance. The formulas in the lower half are all positive, not merely in the syntactic sense that they begin without a minus sign, but in the semantic sense that they *posit*, or affirm the presence of, something in the domain of interpretation. Correspondingly, the formulas in the upper half are all negative, not merely in that they begin with a minus sign, but in that they *deny* the presence of something in the domain of interpretation.[13] The positive formulas say, and say only, what *is* in the domain; the negative formulas say, and say only, what is *not* in the domain. That is why no formula of either character has an equivalent of the opposite character.

This overall positive or negative character of formulas in ATL is called their *valence*. But we need to formulate a precise syntactic criterion of application for this term. With the formulas in the table above, it is easy enough to say that a formula is negative if it begins with a minus sign, and otherwise positive. But since the formation rules of BATL (like those of ATL proper) are recursive, it is possible to construct formulas of either positive or negative valence with minus signs at their beginning.[14] For example, the formula '$-(-(+A(-B)))$' begins with a minus sign, but is positive in valence because it is logically equivalent to the formula '$A + (-B)$'.

Every formula of BATL is either a dyad or a negation of an embedded formula; and the same is true of every embedded formula. If a formula is a dyad, it is positive in valence; if it is the negation of a dyad, it is negative; if it is the negation of a negation of a dyad, it is positive in valence; and so on. The rule for determining the valence of a formula is as follows:

> *Formula valence:* A formula is *negative in valence* if and only if there is an odd number of minus signs operating on the outermost dyad in the formula; otherwise, it is *positive in valence*.

---

[13]Sommers says that formulas of the first sort express states of presence and formulas of the second sort express states of absence: [23, p. 36]. Ideally, the term opposed to 'positive' would be not 'negative' but 'sublative', so as to express the opposition between putting (Latin *ponere*, past participle *positum*) and taking away (Latin *tollere*, past participle *sublatum*), as in the pair of logical terms 'modus ponens' and 'modus tollens'. That is, the proper opposed pairs are 'positive'/'sublative' and 'affirmative'/'negative'. Unfortunately, widespread and long-standing custom compels us to make do with 'negative' as the second term of both pairs.

[14]Sommers's criterion of formula valence misses this point. See Sommers [19, p. 180]; Sommers and Englebretsen [23, p. 57].

The concept of valence may also be applied to occurrences of terms in formulas. Semantically considered, an occurrence of a term is positive if it refers, whether in a definite or an indefinite manner, to at least one thing denoted by the term, and negative otherwise.[15] Thus a positive occurrence of a term is an undistributed occurrence and a negative occurrence of a term is a distributed occurrence.[16] The definition of the distinction, however, is again syntactic:

> *Term valence:* An occurrence of a term in a formula is *negative in valence* if and only if there is an odd number of minus signs operating on it; otherwise, it is *positive in valence*.

Thus, for example, in the formula '$-((-A) + (B + C))$', the terms '$(-A)$', '$(B+C)$', '$B$', '$C$', and '$((-A) + (B+C))$' all occur within the scope of a single minus sign and therefore with negative valence, while the term '$A$' occurs within the scope of just two minus signs and therefore with positive valence.

We shall see that the concept of valence is crucial to the determination of logical equivalence and validity in ATL.

## 3 Compound Terms in BATL.

It is easy enough to see how *conjunctive* compounds of terms are formed. An example has already appeared: the formula '$-((-A) + (B+C))$' contains the conjunctive compound term '$(B+C)$', which may be read as '$B$' and '$C$', '$C$' and '$B$', 'a $B$ that is a $C$', or '$C$'s that are $B$'s', and so on. The whole formula may be read as 'Not: some non-$A$ is $B$ and $C$', 'Every $B$ that is $C$ is $A$', and so on.

The rendering of *disjunctive* compounds of terms, such as 'an $A$ or a $B$', requires more finesse. '$A$ or $B$' denotes everything that '$A$' denotes,

---

[15] For Sommers's defense of the concept of indefinite reference, see [19, pp. 49–66]. Sommers's use of the word 'refer' is closer to the scholastic use of 'supponere' than to the use of 'refer' in anglophone philosophy of language. In Sommers and Englebretsen [23], Sommers appears to give up the opposition between referring and denoting, and uses the verb 'to denote' for both purposes — a disastrous choice, in the opinion of this reader. Contrast its use at, e.g., p. 102 with its use at pp. 13–23.

[16] The equivalence between distribution status and valence is most systematically expounded in [16].

everything that '$B$' denotes, and nothing else: it is thus equivalent to 'not both non-$A$ and non-$B$'. Accordingly, the schema for a disjunctive compound term in BATL is '$(-((-Y)+(-X)))$'.

A cousin of the disjunctive form, which I mention rather for the sake of logical completeness than because of its occurring with much frequency in natural language, is the conditional term-form 'if $A$ then $B$' — for example, in the statement 'Every wombat is, if rabid, then dangerous'. This would be a rather contorted way of saying 'Every rabid wombat is dangerous'. But by reading the term-form 'if $A$ then $B$' as 'not both $A$ and non-$B$', we can represent it by the BATL schema '$(-((-Y)+X))$'.

Finally, we come to a special kind of compound term, the *relational* compound. (We may call non-relational compounds *co-ordinate* compounds when we need to distinguish them from relational ones.)[17] Relational compounds in English are a peculiar case for logical syntax, because they appear to contain a logical subject without a recognizable predicate. For example, the statement

3.1. Some elephant is afraid of a mouse

may be regimented as

3.2. Some elephant is a fearer of some mouse

The paraphrase reveals, or at least imputes, two logical subjects: 'some elephant' and 'some mouse'. The former is both the grammatical and the logical subject of the sentence as a whole. The latter is, in common grammatical terms, not a subject but the object of a relational phrase. Yet it has the appearance of a logical subject, and indeed functions as such in inference. E.g., from 'Some elephant is a fearer of some

---

[17]Sommers has no consistent terminology either for relational terms or for relational compounds. In [19], he flatly denies in one place (p. 46) that relational expressions are terms, while in another place (p. 169) he asserts that they are such. In yet another place, he calls relational *terms* 'relations' (p. 182). In [23], the authors recognize relational words as terms, and introduce a useful distinction between *relational* terms such as envier and *complex* terms such as 'envier of every astronaut' (p. 88); but after observing the distinction for a few pages they flout it for the remainder of the book. I follow Englebretsen and Sayward [6, p. 151] in applying the term 'compound term' to relational as well as non-relational compounds.

mouse' and 'Every mouse is a rodent' one may conclude 'Some elephant is a fearer of some rodent'.[18] We may say that 'some elephant' is the *primary subject* of the sentence, while 'some mouse' is a *secondary subject* in the sentence, though it is the *primary* subject of the relational compound term 'fearer of some mouse'.

But calling 'some mouse' the subject of the term 'fearer of some mouse' seems to dictate that the remaining part of the term, 'fearer of', is the predicate.[19] Such an implication is, to say the least, counterintuitive. It is some elephant, not some mouse, that is said in (3.2) to be a fearer. Yet we shall see that there is a sense in which the relational term is predicated of its subject.

Using '$E$' for 'elephant', '$M$' for 'mouse' and '$F$' for 'fearer', we may, as a first draft, try to formalize (3.2) as

3.3. $(F + M) + E$

But this will not do, for at least three reasons. First, (3.3) gives no indication of its relational character. Given the assumed meanings of the term letters, we could just as well read it as 'Some elephant is a fearer that is a mouse' or 'Some elephant is a fearer and a mouse', which is assuredly not what (3.2) means. Second, although this point assumes rules of derivation that have not yet been introduced, from the formula '$(F + M) + E$' one may derive, by association and simplification, the formula '$M + E$', 'Some elephant is a mouse', which certainly does not follow from (3.2). And third, it is desirable that we be able to derive from our formula the equivalent form

3.4. Some mouse is feared by some elephant

(or 'Some mouse is an object of fear to some elephant'), in which primary and secondary subjects have traded places. We cannot do this by simply writing '$(F+E)+M$', as that could equally well be read as 'Some mouse is a fearer of some elephant'.

---

[18]Sommers explains how this sort of inference falls under the rule of *dictum de omni* in [19, pp. 142–47].

[19]Sommers himself draws this conclusion concerning relational compound terms at [19, pp. 148–49].

The third problem points the way to a solution to all three. In term logic, the two relational expressions 'fearer' (or 'afraid of') and 'object of fear' (or 'feared by') are not treated as two distinct terms but as two occurrences of the same term that are differently inflected according to their respective semantic connections in the context of occurrence. In the context of (3.2), this relational term occurs in connection with 'some elephant' as 'fearer', but in connection with 'some mouse' as 'object of fear'; while in (3.4), these two inflections of the same relational term reverse their roles. It is because of this double semantic aspect that 'fearer' may be said to be the predicate of the compound term 'fearer of some mouse'. It would be more natural to say that 'object of fear' is what is predicated, though only implicitly, of 'some mouse' in that context. But in a term-logical perspective, 'fearer' and 'object of fear' are not two distinct terms but two different inflections of a single relational term.

English does not afford a way to express neutrally what is common to the two different inflections 'fearer' and 'object of fear', so let us, for the nonce, adopt the artificial construction '1-fears-2' to serve that office. With this device, we could rewrite (3.2) and (3.4), respectively, thus:

    3.5. Some elephant-1  1-fears-2 some mouse-2
    3.6. Some mouse-2  1-fears-2 some elephant-1

The numerals here are what are called in ATL *pairing indexes*.[20] The concept of pairing is semantic, although it has syntactic criteria of application. Two terms are paired if and only if either they occur together in a formula as the terms of a dyad or they occur in a set of formulas that logically entails a formula in which they occur as the terms of a dyad. (Since a single formula can be considered as a set of formulas that entails that same formula, the first disjunct of the definition is, strictly considered, superfluous.)

Pairing indexes are unnecessary in non-relational formulas, because in any such formula, all terms are paired with one another. They become necessary only in formulas that contain more than one subject — that is

---

[20]Terence Parsons makes use of small Greek letters to a similar purpose in [12, pp. 82–86]. It is curious that Parsons makes no reference to the work of Sommers in this connection.

to say, in relational formulas — because *no two terms occurring as main terms of two distinct relational subjects in a formula are ever paired with each other.* (By a main term I mean a term on which a binary functor immediately operates.) The function of pairing indexes is not so much to indicate which terms can be logically combined in a dyad as rather to indicate which terms *cannot* be so combined. Thus they block the derivation of invalid consequences that would follow from the use of syntactically analogous non-relational compounds (e.g., the erroneous derivation of 'Some elephant is a mouse' from 'Some elephant is a fearer of a mouse'). They also indicate the sense of a relation — for example, which of two subjects in such relational statements as (3.5) and (3.6) is the fearer and which the object of fear.[21]

The syntactic criterion of term pairing is as follows.

*Term pairing:* Two terms occurring in a formula are *paired* just in case either (i) they occur as the terms of a dyad or (ii) one of them is a compound term and the other is a non-subject main component of that term.[22]

Thus, in our examples (3.2) and (3.4), 'elephant' and 'fearer of some

---

[21] The foregoing account of the semantics of term pairing is entirely of my own devising and is designed to make sense of Sommers's mature *practice* of term indexing rather than of his theoretical statements about it, which I have found unhelpful. For sources, see [21, pp. 110–13] and [23, pp. 88–93]. The methods of relational notation used in Sommers's earlier publications, such as [19], are inconsistent and ineffective. Although Sommers eventually devised an adequate system of notation for relational statements, he never gave an adequate account of its semantics. His would-be explanations are founded on the assertion that two terms are paired just in case they are "co-denoting". This seems to me a non-starter. For example, in an ATL version of the statement 'No man owns a unicorn', the relational term 'owns' must bear two indexes, one shared with 'man' and with the compound 'owns a unicorn', and another shared with 'unicorn' alone. But the statement manifestly does not import that *any* two of these terms "co-denote" anything whatever, because its truth does not require that any of them denote anything in the first place. Nor is the matter mended by reading Sommers's use of the verb 'denote' here as a byword for 'refers to', as used in his account of indefinite reference in [19, pp. 49–66].

[22] I use 'non-subject' rather than 'predicate' here because the main components of a co-ordinate compound term, e.g., '$(A+B)$', are paired with each other and with the compound that they belong to, but that compound does not contain a subject or a predicate, even though it has the same syntactic structure as a relational compound such as '$(A_{1\ 2} + B_2)$', in which '$A_{1\ 2}$' is the predicate and '$+B_2$' the subject.

mouse' are paired, 'elephant' and 'fearer' are paired, 'mouse' and 'object of fear' are paired, and 'mouse' and '(what) some elephant fears' are paired, but 'elephant' and 'mouse', being both the main terms of subjects, are not paired. These pairings are indicated by rendering the two statements into BATL thus:

3.7. $(F_{1\ 2} + M_2)_1 + E_1$  Some elephant fears some mouse

3.8. $(F_{1\ 2} + E_1)_2 + M_2$  Some mouse is feared by some elephant

All the examples of relational constructions considered so far have contained just two subjects. But through iteration of the compounding of relational terms, any number of subjects may be combined in a statement. Here is an example (often used by Sommers)[23] of a statement with three subjects:

3.9. A sailor gave a toy to every child

This may initially be paraphrased for purposes of formalization as

3.10 Some sailor is a giver of some toy to every child

But to represent this statement in BATL requires care with logical scope. In a relational statement of just two subjects, the order of logical dominance is the same as the order of occurrence: e.g., in (3.2), 'some elephant' dominates 'some mouse'. But within a relational compound that itself contains more than one subject, as in this case does the term 'giver of some toy to every child', the order of dominance is the reverse of the order of occurrence: 'every child' dominates 'some toy'. (To reverse this order is to interpret the statement to mean that a sailor gave one and the same toy to every one of the children, whether by making it a collective gift or by repeatedly giving it and taking it back.) 'Giver of some toy to every child' presents the further challenge of rendering a universal subject into a language without quantifiers. To adapt the phrase for such translation, we must paraphrase it as 'not: non-giver of some toy to some child'. Taking '1 gave 2 to 3' as the canonical order of the relational term 'giver', (3.10) goes into BATL as

---

[23]See, e.g., [19, p. 148], and [23, p. 89].

3.11. $(-(-(G_1\ _2\ _3 + T_2)) + C_3) + S_1$

The resources to express singular statements (e.g., 'Fido barks'), compound statements (e.g., 'If Fido barks then some elephant fears a mouse'), and statements with pronominal subjects (e.g., 'A dog is barking. It is a mongrel') can be added to BATL; but as such developments are outside of my present concerns I omit them.

## 4 Developed ATL.

The language expounded in the preceding sections is useful for understanding the semantics of term logic, but inconvenient as a medium of expression and deduction. For these purposes, we require a version of ATL with a syntax closer to that of English and other natural languages. We can do this by introducing four new binary functors, each consisting of a *pair* of signs: minus-plus, minus-minus, plus-minus, and plus-plus. They correspond, respectively, to the universal affirmative, universal negative, particular affirmative, and particular negative forms of traditional term logic.[24] The new forms are algebraically as well as (by definition) logically equivalent to their BATL counterparts, according the following set of syntactic and semantic rules:[25]

5. A-forms ('Every $X$ is $Y$'):

   (a) Syntactic: For any terms $x$ and $Y$, $-X + Y$ is a formula of ATL and $(-X + Y)$ is a term of ATL.

   (b) Semantic: $-X + Y$ is true just in case $-((-Y) + X)$ is true, and $(-X + Y)$ denotes just those things in the domain that $(-((-Y) + X))$ denotes.

---

[24] I fudge one point here, namely that I take the universal negative form to be 'Every $X$ isn't $Y$', while tradition favors 'No $X$ is $Y$'. The latter, however, though logically equivalent to a universal negative, is not itself one: it is rather the negation of a particular affirmative. See [14, p. 7 and p. 10, n. 1] and [19, p. 338].

[25] My version of the rules of ATL omits one functor that Sommers includes, namely the unary plus sign. To include it would require the addition of $(+X)$ as a term schema and $+(+/- X +/- Y)$ as a formula schema. The unary plus sign adds syntactic symmetry but is semantically idle. I have accordingly omitted it for the sake of simplicity.

6. E-forms ('Every $X$ isn't $Y$'):

   (a) Syntactic: For any terms $X$ and $Y$, $-X - Y$ is a formula of ATL and $(-X - Y)$ is a term of ATL.

   (b) Semantic: $-X - Y$ is true just in case $-(Y + X)$ is true, and $(-X - Y)$ ) denotes just those things in the domain that $(-(Y + X))$ denotes.

7. I-forms ('Some $X$ is $Y$'):

   (a) Syntactic: For any terms $X$ and $Y$, $+X + Y$ is a formula of ATL and $) + X + Y)$ is a term of ATL.

   (b) Semantic: $+X + Y$ is true just in case $Y + X$ is true and $(+X + Y)$ denotes just those things in the domain that $(Y + X)$ ) denotes.

8. O-forms ('Some $X$ isn't $Y$'):

   (a) Syntactic: For any terms $X$ and $Y$, $+X - Y$ is a formula of ATL and $(+X - Y)$ is a term of ATL.

   (b) Semantic: $+X - Y$ is true just in case $(-Y) + X$ is true and $(X - Y)$ denotes just those things in the domain that $((-Y) + X)$ denotes.

The results can be presented in the following square of opposition:[26]

---

[26] I proceed, as Sommers does in his textbook [23], as if the universal forms $-X + Y$ and $-X - Y$ are interchangeable with their respective equivalents $-(+X + (-Y))$ and $-(+X + Y)$ in all contexts. To do so is contrary to Sommers's professed semantic view that the two are interchangeable only on condition that the term $X$ denotes something: if it does not, then the universal forms, according to Sommers, are undefined and lack a truth value [19, pp. 201 and 290]. This does not mean that the universal and negative forms are not logically equivalent, for logical equivalence, as Sommers understands it, means only that whenever both formulas have a truth value, they necessarily have the same truth value. But while Sommers professes this view in his moments of semantic theory, he consistently violates it in his practices of formal logic. For the sake of simplicity I follow Sommers's formal-logical practice rather than his semantic declarations.

| | |
|---|---|
| A: $-X+Y$ | E: $-X-Y$ |
| I: $+X+Y$ | O: $+X+Y$ |

Figure 2: Developed ATL Square of Opposition

The conditional compound term form 'if $X$ then $Y$', represented in BATL as $(-(+X+(-Y)))$, can now be more compactly and aptly represented as $(-X+Y)$. Disjunctive compound terms, which were represented in BATL by the schema $(-(+(-X)+(-Y)))$, may be represented in developed ATL by the equivalent E-form schema with negative terms, $(-(-X)-(-Y))$.[27]

Relational compound terms can be represented by the schema $(F+/-X)$, where $F$ is a relational term and the medial plus or minus sign expresses particular or universal logical quantity, respectively. Since $(R+X)$ is already covered by rule 3 of BATL, we require only a rule of formation and interpretation for $)F-X)$.[28]

9. R-forms ('$R$ to every $X$'):

   (a) Syntactic: For any terms $F$ and $X$, $(R-X)$ is a term.

   (b) Semantic: $(R-X)$ denotes just those objects in the domain that $(-((-R)+X))$ denotes.

So the transcription of the relational examples (3.1) and (3.4) into ATL will be, respectively, as follows:

4.1. Some elephant fears some mouse  $\quad +E_1+(F_{1\ 2}+M_2)_1$
4.2. Some mouse is feared by some elephant  $\quad +M_2+(F_{1\ 2}+E_1)_2$

---

[27]Sommers abbreviates this schema to $(--X--Y)$ to avoid clutter (e.g., [19, p 185] and [23, pp. 81–82], but I shall adhere to the more explicit notation.

[28]Pairing indexes do not appear in the rule, as they are not parts of terms but supplementary indicators of actual and potential dyadic combinations of terms. The rule also does not specify that $R$ is a relational term, as the interpretation of elementary terms in ATL is covered by rule 1. As noted earlier, there are no substitution classes of terms in ATL. However, my exposition simply dodges the task of explaining the specific manner in which relational terms denote objects. The topic is eminently worth of attention, but even if I had a satisfactory treatment of it, I could not fit it into this paper.

## 5 Derivation in ATL

A system of rules of derivation for ATL will include such rules of equivalence as the following:[29]

*Commutation:* $+X + Y$ and $+Y + X$ are interchangeable.

*Association:* $+X + (Y + Z)$ and $+(+X + Y) + Z$ are interchangeable.

*Internal obversion:* $+/- X + /- Y$ and $+/- X - /+ (-Y)$ are interchangeable.

*External obversion:* $+/- X + /- Y$ and $-/+ (-/+ X - /+ Y)$ are interchangeable.

*Relational obversion:* $R + /- X$ and $-(-R - /+ X)$ are interchangeable.

*Iteration:* An occurrence of a term $X$ is interchangeable with an occurrence of $(-X + X)$ or $(-(-X) - (-X))$.

And it will include several rules of entailment, most prominent among which is the syllogistic rule of *dictum de omni (DDO)*.

Informally expressed, this is the rule that what is asserted of everything denoted by a certain term (hence 'dictum de omni' — 'what is said of all') may be asserted of anything of which that term is asserted. On its narrowest interpretation, this is a rule for pairs of premises of just two elementary terms each containing a middle term $M$ that occurs as the term of a universal subject in one (the omni premise) and as the term of an affirmative predicate in the other (the matrix premise).[30]

---

[29] For sets of similar rules see [19, pp. 183–84]; [11, pp. 395–98]; [3, pp. 168–70]; [23, pp. 253–54]; and [6, pp. 157–60]. The formulations and designations of the rules of obversion are my own.

[30] I take the designations 'omni premise' and 'matrix premise' from [23, pp. 133–35]. The name of the rule is of scholastic origin, though some, including Sommers, claim to find the rule itself in Aristotle. Sommers also calls it the rule of *dictum de omni et de nullo*, as it concerns both affirmative and negative universal predications. But to describe the predicate of a universal negative statement as 'what is said of none' is a logical confusion worthy of Carroll's White King. Another bad naming practice perpetuated by Sommers is that of calling this rule 'the dictum', as if it were a proverb. No one calls it 'the *dici de omni*', even though that Latin phrase was sometimes used for it, just as no one calls, e.g., the rule of association 'the association'.

The conclusion is then the result of substituting the predicate of the omni premise for the predicate of the matrix premise.

DDO may, however, be understood more broadly to allow any premise containing a positive occurrence of a middle term $M$ to count as a matrix premise and any premise in which $M$ occurs negatively as the main term of a universal subject (not necessarily the primary subject) to serve as an omni premise. The conclusion is then the result of replacing the occurrence of $M$ in the matrix premise with a term formed of whatever remains of the omni premise when the subject $-M$ (meaning 'every $M$') is deleted from it. This remnant may be called the *omni fragment*. Schematically stated:

> *Dictum de omni (DDO)*: From $\Phi(M)$, a formula in which the term $M$ occurs positively, and $\Psi(-M)$, a formula in which $M$ occurs negatively as the term of a universal subject, one may derive $\Phi(\Psi)$, the result of substituting for $M$ in $\Phi(M)$ a term formed of what remains of $\Psi(-M)$ when $-M$ is deleted from it.

The application of the rule in this form may be illustrated by the following argument, which is an instance of DDO in which the middle term is the subject term rather than the predicate term of the matrix premise (first premise) and the omni premise (second premise) is a particular rather than a universal statement:

5.1.  $E_1 + (W_{1\ 2} - M_2)$  Some enthusiast watches every movie
 $+M_2 + B_2$  Some movies are boring
 $\therefore +(+E_1 + W_{1\ 2}) + B_2$  Something that some enthusiast watches is boring

The first premise is the omni line. If we delete from it the universal subject '$-M_2$' and replace the latter with an underscore, we get '$+E_1 + (W_{1\ 2}\_\_)$'. The brackets around '$W$' are superfluous and may be deleted, while the remainder, being a dyadic term, needs to be enclosed in brackets to make a term. The resultant omni fragment is '$+E_1 + W_{1\ 2}$'. Substituting this for '$M$' in the second premise, which is the matrix line, yields the conclusion '$+(+E_1 + W_{1\ 2}) + B_2$'.

# Part 2: Algebraic Addition in ATL

## 6 Determining Validity in ATL.

The principle of validity (PV) and the principle of equivalence (PE) are not rules of derivation but rules of decision. This point is evident from the content of both:[31] neither says that from such and such a formula or set of formulas one may derive such and such another formula, but rather that a set of formulas of a certain sort has a certain semantic status if and only if its members have a certain relational syntactic characteristic.

We begin with PV, which says that an argument, or an argument of a certain sort (to be specified), is valid if and only if it meets two conditions:

(i) *Regularity*. The number of premises of positive valence and the number of conclusions of positive valence are the same, namely either both 1 or both 0. In other words, either the conclusion and all the premises are of negative valence, or the conclusion and just one premise are of positive valence.

(ii) *Equality*. The sum of the premises is algebraically equal to the sum of the conclusion, meaning that the result of adding up all the positive and negative occurrences of elementary terms in the premises, including the cancellation of any terms that occur both positively and negatively, and the result of doing the same with the conclusion, are algebraically equal.

The equality condition will need, and shortly will get, further consideration. For now, the use of PV may be illustrated by application to a couple of examples from Lewis Carroll, which I present together with

---

[31] And yet the point is not made clear in any of publication of Sommers or Englebretsen. Indeed, the authors sometimes obscure or even falsify the point. Sommers and Englebretsen [23] contains two derivations in which PE is cited as the justification of a derived line (pp. 150, 161), even though in each case a specific rule of equivalence is available to justify the line, while Englebretsen and Sayward [6] inaptly lists both PE and PV under the heading of "Rules of Inference for Derivations" (pp. 154–56).

their representations in ATL.[32] The first example is a valid argument:

6.1. No misers are unselfish $\quad -(+M+(-S))$
None but misers save eggshells $\quad -(+(-M)+E)$
∴ No unselfish people save eggshells $\quad -(+(-S)+E)$

All (that is, both) of the premises are negative in valence, as is the conclusion: thus the argument satisfies the regularity condition. The algebraic sum of its premises is $M + S + M + E$. The two opposite occurrences of '$M$' cancel each other out, leaving $+S - E$. The sum of the conclusion is also $+S - E$: thus the argument satisfies the equality condition; and thus the argument satisfies both conditions of PV.

Here is an example of an argument that violates the conditions of PV:

6.2. Some epicures are ungenerous $\quad +E+(-G)$
All my uncles are generous $\quad -U+G$
∴ My uncles are not epicures $\quad -U-E$

That the argument is invalid can be concluded from the fact that it has a premise of positive valence and a conclusion of negative valence, in violation of the regularity condition. One may make the same determination by noting that the term 'epicure' has a positive occurrence in the first premise and a negative occurrence in the conclusion, so that the algebraic sum of the premises, $+E - G - U + G$, which is equal to $+E - U$, is not equal to the conclusion. Either feature by itself is sufficient to determine by PV that the argument is invalid.

It is essential to recognize that the strings of letters with pluses and minuses before them that occurred in the preceding two paragraphs are *not* formulas of ATL, even if some of them happen to be typographically identical to such formulas. For example, in the summation of the premises and the conclusion of example (6.1), the string '$+S - E$' was used. This string, though typographically identical to a formula of ATL, is not such a formula in this instance; for the identically spelled formula of ATL is neither logically equivalent to the conclusion of the argument

---

[32]Carroll [2, p. 155]. As charming as Carroll's exercises are, one must exercise caution in using them as examples in ATL, as in Carroll's view, 'Every $X$ isn't $Y$' and 'No $X$ is $Y$'are not logically equivalent: the former, as he interprets it, entails 'There is an $X$', while the latter does not (pp. 74–76).

nor a logical consequence of the set of premises of the argument. As Sommers says of one of his own examples, a string of characters representing the sum of a formula or a set of formulas "has no interpretation".[33]

This creates a problem; for if such strings have no interpretation, then there is no clear sense in which they represent *sums* at all, or in which one of them can be said to represent something that is *equal* to what another represents. For instance, the string '$+E - G - U + G$' is supposed to represent the "sum" of the premises of example (6.2), and that sum is supposed to be "equal" to a sum represented by the string '$+E - U$'. It is clear enough how we are supposed to proceed in such matters, according to PV: what is not at all clear is the nature of what we are doing. How can we do any kind of algebra with uninterpreted strings of characters?

The question goes to the heart of the sense in which Sommers's system of term logic is *algebraic*. For to be algebraic, it is not enough that a system of logic borrow its syncategorematic signs from algebra: there must be a genuine sense in which the operations performed upon strings of characters that contain occurrences of those signs is in fact arithmetical.

The resolution to this difficulty, it seems to me, is to understand that the strings of characters derived by "adding up" the premises and conclusions of arguments, though not formulas of ATL, are formulas of an algebraic *metalanguage*, in which the symbols '+' and '−' represent the operations of addition and subtraction. One could call this metalanguage 'meta-ATL' (or 'MATL' if one insisted on having an initialism for it; but I prefer not to multiply initialisms beyond necessity). In meta-ATL, a plus or minus sign immediately followed by a term letter represents a positive or a negative occurrence of a term in whatever formula or set of formulas of ATL is under consideration. The operations of addition and subtraction upon these strings of symbols are reckonings of the net total of such term occurrences in the set of formulas under consideration. Thus, Sommers's statement that a string of characters derived from a formula or a set of formulas by algebraic addition "has no interpretation" should be understood only in relation to the object lan-

---

[33]Sommers [19, p. 181]. We shall see in a moment that this statement must be understood with a certain qualification.

guage of ATL. Such a string *has* got an interpretation; but in meta-ATL rather than in ATL itself.

As Sommers notes, the algebraic reckoning exercised upon formulas of ATL deviates from the common laws of arithmetic in certain respects:

> That '$-(-x)+y$' is not logically equal to '$+x+y$' is an implicit constraint of the logical algebra that distinguishes it from the familiar everyday algebra of addition and subtraction. On the other hand, logical algebra has a certain license not found in the everyday algebra: the Law of Iteration permits the replacement of '$x$' by '$x$ and $x$' ($+x+x$) and by '$x$ or $x$' ($--x--x$).[34]

The second point has the implication that the cancellation of terms in meta-ATL does not require that their positive and negative occurrences be equal in number. This fact will prove important later (in section 9).

## 7    The Range of Application of PV

Englebretsen and Sayward offer PV as a statement of the necessary and sufficient conditions of validity for all arguments without restriction.[35] It is easily shown that it is nothing of the sort. That it is not a true statement of the *necessary* conditions of validity for all arguments can be proved by the simple exercise of taking any valid argument that satisfies its conditions and adding another premise. The result will be a valid argument that violates either the valence condition or the equality condition, if not both.

One might try to thwart this facile way of refuting unrestricted PV by disallowing arguments with superfluous premises. To do this would require a syntactic criterion of superfluous premise. Whether that is possible or not, it is easy to think of valid arguments without superfluous premises that violate the conditions of PV. We have already seen such an argument in example (5.1) above:

---

[34] [18, p. 28 f.].
[35] Englebretsen and Sayward [6, p. 16]. See also [5, p. 72].

5.1. $+E_1 + (W_{1\ 2} = M_2)$    Some enthusiast watches every movie
$+M_2 + B_2$    Some movies are boring
$\therefore +(+E_2 + W_{1\ 2}) + B_2$    Something that some enthusiast watches is boring

The argument violates the regularity condition yet is valid. Plainly it has no superfluous premise. Hence regularity and equality are not necessary conditions of validity of arguments.

That the conditions of PV are not jointly *sufficient* conditions of validity for all arguments may be shown by the following example:

7.2. A sailor gave a toy to every child    $+S_1 + ((G_{1\ 2\ 3} + T_2) - C_3)$
$\therefore$ A sailor gave to every child a (certain) toy    $+S_1 + ((G_{1\ 2\ 3} - C_3) + T_2)$

It is consistent with the premise that the sailor in question should have several toys, and should give different toys to different children; indeed, this is how the premise would ordinarily be understood. But in such a case, the conclusion is false. Thus the argument is plainly invalid. Yet it meets the conditions of regularity and equality. Hence those conditions are not sufficient conditions of validity.

So PV as formulated above does not state either necessary or sufficient conditions of validity for all arguments. The unfortunate idea that it does so seems to be original to Englebretsen and Sayward. Sommers, by contrast, offers PV as a principle of validity for *syllogistic* arguments only. A syllogistic argument in Sommers's sense may be defined as follows.[36] (Since the term will be subject to redefinition later, the present definition is marked as definition 1.)

> *Syllogistic argument (def. 1):* A set of formulas in ATL is *syllogistic* just in case, for some $n \geq 2$, it consists of just $n$ formulas in which a total of just $n$ elementary terms occur, each such term occurring just once in each of just two formulas. A *syllogistic argument* is a syllogistic set of formulas, one of which is a conclusion and the rest of which are premises.

---

[36] See Sommers and Englebretsen [23, pp. 109–10], for an equivalent definition.

This definition does a certain violence to language, as it allows application of the term 'syllogism' to arguments of a single premise. An argument from a single premise, etymologically, would be only a "logism" and not a "syl-logism", because no premises are put together to draw the conclusion. Nonetheless, as this definition usefully delineates the range of application of PV, we shall adopt it. We may restate PV to include this restriction, thus:

> *Principle of validity for syllogistic arguments ($PV_s$):* A syllogistic argument is valid just in case it is regular and the sum of its premises is algebraically equal to the sum of its conclusion .

$PV_s$ — hereafter simply 'PV' —applies not only to so-called "syllogistic" arguments of a single premise but also to syllogistic arguments of three or more premises, known as "sorites". The following example, again from Lewis Carroll, will illustrate.[37]

| 7.3. | Babies are illogical | $-B + I$ |
| | Nobody is despised who can manage a crocodile | $-(+M + D)$ |
| | Illogical persons are despised | $-I + D$ |
| | Babies cannot manage crocodiles | $-B - M$ |

All formulas are negative in valence; thus the argument is regular. The algebraic sum of the premises is $-B+I-M-D-I+D = -B-M$, which is equal to — in fact, identical with — the conclusion. Thus the argument satisfies PV.

There is a clear affinity between PV and DDO. In fact, any argument that satisfies PV has a derivation using only rules of equivalence and DDO. Given that the rules of equivalence and DDO are sound,[38] and that the conclusion of any valid syllogistic argument is logically equivalent to any other valid conclusion from the same set of premises,

---

[37] Carroll [2, p. 160] (premises) and 185 (conclusion). The exercise consists in determining the conclusion from the given premises. The same example is used in [23, pp. 116–17].

[38] The rules of equivalence are easily proved sound from the semantic rules given in sections 2 and 4 above. A proof of the soundness of DDO is offered in [11, pp. 405–10].

it follows that PV states necessary and sufficient conditions of validity for all syllogistic arguments, on the present definition of 'syllogistic argument'.

## 8 Extending the range of PV (1): arguments with inert terms

But we should consider whether the definition of 'syllogistic argument', and thereby the range of appllication of PV, can be broadened. As stated above, PV applies only to arguments in which each formula contains just two elementary terms. But it is clear that the principle is capable of wider application. Consider, for example, the following argument:

8.1.  Every dog or cat is a furry animal
Every furry animal is a mammal
∴ Every dog or cat is a mammal

If we use '$D$' for 'dog', '$C$' for 'cat', '$F$' for 'furry', '$A$' for 'animal', and '$M$' for 'mammal', the formalization of this argument in ATL will look like this:

8.2.  $-(-(-D) - (-C)) + (+F + A)$
$-(+F + A) + M$
$\therefore -(-(-D) - (-C)) + M$

The argument contains a total of five elementary terms in three formulas. It is therefore, by the definition that we have adopted, not a syllogistic argument. Yet it is plain that the argument is an instance of the traditional syllogistic form Barbara. This can be brought out by using '$O$' for 'dog or cat', '$U$' for 'furry animal', and '$M$', as before, for 'mammal', in which case the ATL version of the argument is

8.3.  $= O + U$
$-U + M$
$\therefore -O + M$

which satisfies the conditions of PV. As a matter of practical logic, there is no good reason to represent (8.1) as (8.2) rather than as the simpler (8.3). Nonetheless, it should be possible to adjust the definition of 'syllogistic argument' to allow PV to apply to (8.2) rather than to require that (8.2) be rewritten as (8.3).

Although (8.2) contains five elementary terms, only one of them, $M$, ever occurs as the main term of a dyad by itself. The others appear only in compound terms which recur intact. Let us say that '$M$', '$(-(-D) - (-C))$', and '$(+F + A)$' are logically *active* terms in the argument, and that the elementary terms '$D$', '$C$', '$F$', and '$A$', by contrast, are logically *inert*. We may define these terms thus:

> *Inert and active terms:* Any term that occurs more than once in a set of formulas as a component of a certain compound term, and never occurs otherwise than as a component of that same compound term, is *inert* in that set of formulas. Any term occurring in a set of formulas that is not inert in that set of formulas is *active* in that set of formulas.

By distinguishing active from inert terms, we can extend the definition of syllogistic sets of formulas, and thus the application of PV, to arguments containing compound terms.

But we cannot simply replace the term 'elementary term' in the definition of 'syllogistic argument' with 'active term', for doing so will have the unfortunate consequence of counting a term and its negation as two distinct active terms in an argument. To avoid this consequence, let us say, adopting a term from Sommers, that an occurrence of a term and an occurrence of its negation are occurrences of the same *absolute term*. Thus, an occurrence of '$M$' and an occurrence of '$(-M)$' count as occurrences of the absolute term '$|M|$' (in Sommers's notation).[39]

With these terms in hand, we can offer the following revised definition of a syllogistic argument in ATL:

> *Syllogistic argument (def. 2):* A set of formulas in ATL is *syllogistic* just in case, for some $n \geq 2$, it consists of just $n$ formulas in which a total of just $n$ active absolute terms occur, each such term occurring just once in each of just two formulas. Such a set of formulas is a *syllogistic argument* just in case just one formula is a conclusion and the others are premises.

---

[39][13, p. 351].

We can still say that PV applies to all and only syllogistic arguments; but we have widened the range of the term 'syllogistic argument' to include *some* arguments containing more than two elementary terms per formula.

## 9 Extending the range of PV (2)

Certain exercises devised by Lewis Carroll for a planned second part of his *Symbolic Logic* suggest yet a further extension of the range of application of PV. These exercises consist of long series of premises from which the student is to draw a "complete" conclusion (a term whose meaning will be explained shortly). But in contrast to the sorites exercises in the first part, the premises in these exercises are what Carroll calls "multiliteral", meaning that more than two letters are required to represent the terms in each one of them (i.e., all the terms that are logically active in them). As these exercises consist of fifteen to twenty premises each, it is not convenient to quote one of them here; but an extract of just a few premises from one exercise will be useful (the actual exercise contains twelve more premises). I have added transcriptions of the premises into ATL and added bold type to the premises to indicate the correspondence of term letters to English words:[40]

| | |
|---|---|
| 9.1. A **l**ogician who eats **p**ork chops for supper will probably l**o**se money | $-(+L+P)+O$ |
| A **g**ambler whose appetite is not **r**avenous will probably l**o**se money | $-(+G+(-R))+O$ |
| A man who does not **g**amble and whose appetite is not **r**avenous is always l**i**vely | $-(+(-G)+(-R))+I$ |
| A l**i**vely **l**ogician who is really in **e**arnest is in no danger of losing money | $-(+(+L+I)+E)-O$ |

Carroll's own notation has an entirely different appearance from that of ATL, but corresponds to it in representing, in effect, the positive or

---

[40]Carroll [2, pp. 331–32]. I do not use the same letters as Carroll uses in his symbolic representation of the argument.

negative valence of each statement and the positive or negative valence of each occurrence of a term within each statement. Carroll does not use the term 'valence', of course, but his terms 'entity' and 'nullity' correspond to the positive and negative valence of formulas, while his use of subscripts and prime marks corresponds to the positive and negative valence of term occurrences.[41]

Carroll's procedure for deriving a conclusion from a given set of premises also has affinities with the algebraic procedure of ATL. Once one has assigned a unique letter to every active absolute term of the argument, one makes what Carroll calls a 'register' of positive and negative occurrences of terms. The register includes the line number of every such occurrence, for use in the proof of the validity of the conclusion that will follow by a method of *reductio ad absurdum* and the use of tree diagrams, though the latter is a part of Carroll's procedure that does not concern us. For (9.1), a Carrollian register will look like this:[42]

| Term | Positive | Negative |
|---|---|---|
| L |  | 1,4 |
| P |  | 1 |
| O | 1,2 | 4 |
| G | 3 | 2 |
| R | 2,3 |  |
| I |  | 3,4 |
| E |  | 4 |

Any term that occurs both positively and negatively is in Carroll's terminology an "eliminand", meaning a term that may be eliminated from the conclusion. Any term that occurs only positively or only negatively is a "retinend", meaning a term that must occur in the conclusion. A "complete" conclusion is one in which all retinends and no eliminands occur.[43] Thus, if the four statements in (9.1) are taken as a complete set of premises, the eliminands are 'O', 'G', and 'I', and the retinends are 'L', 'P', 'R', and 'E'. 'R' occurs only positively while 'L', 'P' and

---

[41] See [2, p. 119].
[42] "Like this", more or less. Carroll, as noted, does not use the terms 'positive' and 'negative', and the orientation of his registers is perpendicular to my arrangement; but the structures correspond.
[43] [2, p. 322].

'$E$' occur only negatively. So, since all the premises are negative in valence, the conclusion must be a formula of negative valence in which all and only the aforementioned retinends occur, and in which they occur with the same valences as they do in the premises. The following is an example of such a conclusion:

9.2.   A logician who eats pork chops   $-(+L + (+P + (-R))) - E$
for supper and whose appetite is
not ravenous is not in earnest

(9.2) is not the only possible complete conclusion; but any valid complete conclusion must be logically equivalent to it. The particular disposition of terms is dictated by one's taste in English sentences.

A point on which Carroll's procedure may be thought to depart from that of ATL is that it does not require a term to have equal numbers of positive and negative occurrences (namely one of each) to be canceled out: a single positive occurrence of a term cancels any number of negative occurrences of the same term, and conversely. But this is entirely consistent with the algebraic procedure of meta-ATL, in which, as noted at the end of section 6, a pair of covalent[44] occurrences of a term are algebraically equal to a single occurrence of the same term. By mathematical induction it follows that the sum of any number of covalent occurrences of the same term is equal to a single occurrence.

If Carroll's multiliteral sorites are to fall within the range of application of PV, we may accommodate them, not by altering the formulation of PV itself, but by rewriting the definition of 'syllogistic argument' to replace the phrase 'just $n$' with 'at least $n$', thus:

> *Syllogistic argument (def. 3):* A set of formulas in ATL is *syllogistic* just in case, for some $n \geq 2$, it consists of just $n$ formulas in which at least $n$ active absolute terms occur, each such term occurring just once in each of at least two formulas. Such a set of formulas is a *syllogistic argument* just in case just one formula is a conclusion and the others are premises.

---

[44]In Sommers's terms, two occurrences of a term are *covalent* if they have the same valence, *divalent* if they have opposite valences [23, p. 51].

This redefinition will allow such arguments as Carroll's to count as syllogistic, and thus allow PV to be applied to them.

But the question arises whether broadening the range of application of PV in this way will admit counterexamples — either invalid arguments that satisfy the conditions of PV or valid arguments that fail to do so. For the moment, I suspend this question and turn to the other rule of decision in ATL, the principle of equivalence (PE).

## 10  Determining Logical Equivalence in ATL.

It was mentioned earlier that PV can be applied to so-called syllogistic arguments of a single premise, under definition 1 of 'syllogistic argument'. In any such argument that satisfies the regularity and equality conditions of PV, premise and conclusion must be formulas of identical valence containing the same two elementary terms occurring with the same respective valences. It can be proved from the semantic rules governing the functors of ATL (see sections 2 and 4 above) that the premise and the conclusion of any such argument will be logically equivalent. The following pair of arguments will illustrate this point:[45]

10.1. None but misers save eggshells $\quad -(+(-M)+E)$
∴ All who save eggshells are misers $\quad -E+M$

10.2. All who save eggshells are misers $\quad -E+M$
∴ None but misers save eggshells $\quad -(+(-M)+E)$

The principle of equivalence (PE) generalizes this relation to all pairs of formulas in ATL:

---

[45] While the sentence 'None but misers save eggshells' is taken from Carroll [2, p. 155], let it not be thought that Carroll would accept both of these arguments as valid. He would deny the validity of (10.1) on the ground that its conclusion "contains" the proposition that there are persons who save eggshells, while the premise does not (p. 76). Carroll allowed that a logician may adopt whatever definition of such words as 'all' and 'some' he chooses, "provided of course that it is consistent with itself and with the accepted facts of Logic"; but apparently he considered the validity of subalternation to be one of these "accepted facts of logic" (p. 232–23; see esp. Bartley's note 2) — despite the fact that at the time of his writing it had already been rejected by Brentano, Peirce, Frege, and Venn. For a summary of this history see [1].

*Principle of equivalence (PE):* Two formulas of ATL are logically equivalent just in case they are covalent and algebraically equal.

If PE is restricted to pairs of formulas containing just two elementary terms, or even, more broadly, to pairs of formulas containing just two active absolute terms, then PE is a corollary of PV and is a provably correct statement of the necessary and jointly sufficient syntactic conditions of logical equivalence. What needs to be determined is whether it applies more widely.

Let us grant that covalence and algebraic equality are *necessary* conditions for logical equivalence in all cases[46] and proceed to the question whether they are jointly *sufficient* conditions. The examples below prove that they are not. Each example is a one-premise argument in which the premise and the conclusion are covalent and algebraically equal, but in which the conclusion does not follow from the premise, and so *a fortiori* is not logically equivalent to it. (It may be noted that the premise of each example follows from the conclusion, but this fact makes no difference to its status as a counterexample.) I have used the term letters '$A$', '$B$', and '$C$' throughout and have supplied arbitrary English readings.

10.4. $-A + (-(-B) - (-C))$  Every positive integer is odd or even
$\therefore -A + (+B + C)$  Every positive integer is odd and even

10.5. $+A_1 + (B_{1\ 2} - C_3)$  Some key unlocks every door
$\therefore +(+A_2 + B_{1\ 2}) - C_3$  Something that some key unlocks isn't a door

10.6. $-(+A + B) + (+A + C)$  Every counselor who is male is a counselor who is a Rotarian
$\therefore -B + C$  Every male is a Rotarian

---

[46] I say, "Let us grant" this point, because, while I believe it to be true, I have no formal proof of it. The requirements of such a proof would be: (1) a sound and complete system of derivation for ATL; (2) a proof of the soundness and completeness of that system; and (3) a proof that for any two formulas, $X$ and $Y$, such that each can be derived from the other in the system in question, $X$ and $Y$ are covalent and algebraically equal. For (1) and (2) see [11].

To explain why these pairs of covalent and algebraically equal formulas fail to be logically equivalent, it will be helpful to have the concept of an *additive formula*. Suppose that a formula contains no binary functor other than the double plus, and that no dyadic compound occurs within it within the scope of a minus sign that is not the dominant sign of the formula. For example, the formula '$-(+(+A + B) + (-C))$' is of the specified character, while the formula '$+(-(+A + B) + C)$' is not, because in the latter a subordinate minus sign operates on the dyadic compound '$(+A + B)$' while in the former no such thing occurs. Let us call any dyadic compound occurring within the scope of a minus sign that is not the dominant functor of the formula a *subordinate negated dyad*. And let us say that any formula that is logically equivalent to a formula that contains no subordinate negated dyad is *additive*.

One can determine whether a given formula is additive by using semantic rules 5–9 (see section 4 above) to reduce it to an equivalent in BATL. The BATL reduction will contain a subordinate negated dyad if and only if the original formula is non-additive. For example, the premise of argument (10.4), '$-A + (-(-B) - (-C))$', reduces to '$-(((-B) + (-C))+A)$', which is additive. The conclusion of (10.4), '$-A+(+B+C)$', by contrast, reduces to '$-(+A + (-(+B + C)))$', which is non-additive. So we may define additive formula thus:

*Additive formula:* A formula is *additive* just in case its BATL reduction contains no subordinate negated dyad.

I pause to make two incidental points about additivity: First, all formulas of just two elementary term occurrences are additive: non-additivity cannot occur without at least three distinct elementary term occurrences. Second, when a term occurs as a term of a negated dyad, its supposition is, in scholastic terms, "confused": "confused and distributed" if it occurs negatively and "merely confused" if it occurs positively.[47] This fact seems to me effectively to rebut Geach's dismissal of confused supposition as a bogus classification founded only on a "sup-

---

[47]See [12, p. 185]. Note that the negated dyad need not be subordinate: e.g., both terms in any universal statement have confused supposition. The BATL reductions of such statements are themselves negated dyads, as shown in the upper half of the BATL square of opposition (figure 1 above).

posed similarity" of merely confused supposition to confused and distributed supposition.[48] Confused supposition corresponds to occurrence within a negated dyad in a BATL reduction.

With this definition in hand, we can make the observation that *every one of the counterexamples (10.4)–(10.6) contains one additive and one non-additive formula*. This invites the conjecture that the applicability of PE requires that either both formulas are additive or both formulas are non-additive. But it is not difficult to come up with a pair of formulas that are covalent, algebraically equal, and both non-additive, but not logically equivalent. We can do this by strategically inserting another letter into the premise of example (10.6):

10.7. $-(+A + (+B + D))$    Every counselor who is a male lawyer is a counselor who is a Rotarian and a lawyer
$+(+A+(+C+D))$
$\therefore -B + C$    Every male is a Rotarian

Once again, the conclusion does not follow from the premise, and *a fortiori* is not logically equivalent to it. So PE cannot reliably be applied any pair of formulas, *one or both* of which are non-additive.

We must therefore restrict the range of application of PE to pairs of additive formulas.[49] Unfortunately, counterexamples can be generated even within the specified range. Here are two examples of invalid one-premise arguments in which the premise and the conclusion are covalent, algebraically equal, and additive:

---

[48][8, p. 90].

[49]At this point we could apply the reasoning applied to the range of PV in section 7 by defining an *effectively additive* set of formulas as a set of formulas whose BATL equivalent set contains no active absolute term that occurs as a member of a subordinate negated dyad and contains no formula containing divalent occurrences of the same term, and then specifying the range of application of PE as pairs of formulas that are effectively additive.

# The Range of Decision

10.8. $+A + A$      Some dog is a dog (= There are dogs)

$\therefore +(+A + B) + (+A + (-B))$      Some dog that barks is a dog that doesnt bark

10.9. $-(+(+A + B) + (+(-A) + B)))$      No flying pig is a non-flying pig

$\therefore -(+B + B)$      No pig is a pig (= There are no pigs)

The premise of (10.8) is true just in case the term '$A$' has denotation. The conclusion, however, is logically false. Thus, the conclusion is not equivalent to the premise. The premise of (10.9) is logically true, while the conclusion is true just in case the term '$B$' has no denotation. So, once again, the conclusion is not equivalent to the premise.

(10.8) and (10.9) can be excluded from the range of application of PE by excluding all *logically determined* formulas. Logically determined additive formulas may be identified by a syntactic criterion:

> *Logical determinacy:* An additive formula is *logically determined* just in case it contains divalent occurrences of the same term. Such a formula is *logically false* if it is positive in valence and *logically true* if it is negative in valence.[50]

Restricting the range of application of PE to pairs of additive, logically undetermined formulas will exclude all of the counterexamples considered so far. But how do we know that there are not other counterexamples?

## 11 Conclusions

The answer is that we don't know until we have a proof of PE within the specified range. I have no such proof to offer. So, until someone

---

[50] Note that this criterion cannot be extended to non-additive formulas. E.g., the non-additive formula '$+A - (+A + B)$' ('Some $A$ is not $AB$') is a positive formula containing divalent occurrences of the term $A$ but is not logically false, while '$-A + (+A + B)$' ('Every $A$ is $AB$') is a negative formula containing divalent occurrences of the same term but is not logically true.

produces such a proof, all that we know of the range of application of PE is that it extends to pairs of formulas each of which contains just two elementary term occurrences (or, on a broader interpretation, two active absolute terms).

An analogous point applies to PV: the widest range in which we know it to have application is the set of arguments that are syllogistic by definition 2. The same techniques that were used to generate counterexamples to PE in the previous section can be used to generate counterexamples to PV in the field of arguments that are syllogistic by definition 3. Here is an example of an argument that is syllogistic by the latter definition, meets the regularity and equality conditions of PV, and is invalid:

11.1.  $-A + (-(-B) - (-C))$     Every chess piece is black or white
       $-(+B+C) + D$              Everything black and white is gray from a distance
       $\therefore -A + D$        Every chess piece is gray from a distance

PV, therefore, cannot reliably be extended to Carrollian multiliteral sorites except as a rule for generating *probative* complete conclusions, which may or may not be valid ones. This is in effect how Carroll himself uses the rule: when he speaks of "solving" one of his "sorites-problems with multiliteral premisses", he does not mean the finding of a so-called "complete" conclusion, which he seems to consider a fairly easy task, but rather the production of a proof of the validity of that conclusion, which requires a long process of tree construction.[51]

It is clear that decision of logical status by algebraic addition has application beyond the range of two-term formulas. The concept of additivity holds promise that a much wider range of application of such procedures can be delineated. But the maximal range in which such procedures can be used without the possibility of counterexamples remains unknown.[52]

---

[51] See, e.g., the "soliloquy" that begins at Carroll [2, p. 289].

[52] I wish to thank George Englebretsen for the generous guidance and encouragement that he has provided to me over the years of my engagement with ATL.

# References

[1] Buckner, Edward. 2015. Existential Import. At http://www.logicmuseum.com/wiki/Existential_import (accessed 4 March 2024).

[2] Carroll, Lewis. 1977. *Lewis Carroll's Symbolic Logic.* Edited by William Warren Bartley III. New York: Clarkson N. Potter.

[3] Englebretsen, George. 1996. *Something to Reckon With: The Logic of Terms.* Ottawa: University of Ottawa Press.

[4] Englebretsen, George. 2007. Teaching Algebraic Term Logic. *American Philosophical Association Newsletter on Teaching Philosophy* 7: 12–16.

[5] Englebretsen, George. 2015. *Exploring Topics in the History and Philosophy of Logic.* Berlin and Boston: Walter de Gruyter.

[6] Englebretsen, George, and Charles Sayward. 2011. *Philosophical Logic: An Introduction to Advanced Topics.* London and New York: Continuum.

[7] Frege, Gottlob. 1967. *Begriffsschrift*, a Formula Language, Modeled upon That of Arithmetic, for Pure Thought. Translated by Jean van Heijenoort. In Jean van Heijenoort, ed., *From Frege to Gödel: A Source Book in Mathematical Logic, 1879–1931*, pp. 1–82. Cambridge, Mass.: Harvard University Press.

[8] Geach, P. T. 1980. *Reference and Generality: An Examination of Some Medieval and Modern Theories.* 3rd edition. Ithaca, N.Y.: Cornell University Press.

[9] Leonard, Henry S. and Nelson Goodman. 1940. The Calculus of Individuals and Its Uses. *The Journal of Symbolic Logic* 5 (1940): 45–55.

[10] Lockwood, Michael. 1982. The Logical Syntax of TFL. Appendix G in Sommers 1982, pp. 426–56.

[11] McIntosh, Clifton. 1982. Appendix F in Sommers 1982, pp. 387–425.

[12] Parsons, Terence. 2014. *Articulating Medieval Logic.* Oxford: Oxford University Presss.

[13] Sommers, Fred. 1963. Types and Ontology. *Philosophical Review* 72: 327–63.

[14] Sommers, Fred. 1970. The Calculus of Terms. *Mind*, new series, 79: 1–39.

[15] Sommers, Fred. 1974. The Logical and the Extra-logical. In Robert S. Cohen, Marx W. Wartofsky, eds., *Methodological and Historical Essays in the Natural and Social Sciences. Boston Studies in the Philosophy of Science*, volume 14, pp. 235–52. Dordrecht: D. Reidel, 1974.

[16] Sommers, Fred. 1975. Distribution Matters. *Mind*, new series, 84: 27–46.

[17] Sommers, Fred. 1976a. Logical Syntax in Natural Language. In A. F.

MacKay and D. D. Merrill, eds., *Issues in the Philosophy of Language: Proceedings of the 1972 Oberlin Colloquium in Philosophy*, pp.11–41. New Haven and London: Yale University Press.

[18] Sommers, Fred. 1976b. Frege or Leibniz? *Studien zu Frege* 3: 11–33.

[19] Sommers, Fred.1982. *The Logic of Natural Language*. Oxford: Clarendon Press.

[20] Sommers, Fred. 1983. Linguistic Grammar and Logical Grammar. In Leigh S. Cauman, Isaac Levi, Charles Parsons, and Robert Schwartz, eds., *How Many Questions? Essays in Honor of Sidney Morgenbesser*, pp. 180–94. Indianapolis: Hackett.

[21] Sommers, Fred. 1990. Predication in the Logic of Terms. *Notre Dame Journal of Formal Logic* 31: 106–26.

[22] Sommers, Fred. 2002. On the Future of Logic Instruction. *American Philosophical Association Newsletter on Teaching Philosophy* 1: 176–80.

[23] Sommers, Fred, and George Englebretsen. 2000. *An Invitation to Formal Reasoning: The Logic of Terms*. Aldershot and Burlington, Vt.: Ashgate.

# THE ROLE OF COPULAS IN REASONING

PEI WANG
*Temple University, USA*
`pei.wang@temple.edu`

## 1 Introduction

The usage of copulas is one of the defining features of term logic, where a typical sentence has a "subject-copula-predicate" format, and a typical inference rule corresponds to a certain copula combination in the premises, as in Aristotle's Syllogistic [1]. On the contrary, in predicate logic and propositional logic, copulas do not play such central roles [6].

While it is widely believed that predicate logic is more capable than term logic both in representation and inference, in the following I will challenge this conclusion from the perspective of artificial intelligence. I will argue that for certain types of working environments and inference tasks, term logic provides a better framework than predicate logic, and a key contribution is made by copulas.

This paper starts with a description of copulas in various term logic models, as well as how their functions and roles are replaced in predicate logic models. This description also relates to other models of reasoning which are not often considered as "logic". After that, the study of reasoning in artificial intelligence is surveyed and analyzed, which leads to a special requirement for logic. A model designed to meet this requirement is described, in which copulas play a fundamental role. This new logic, "Non-Axiomatic Logic (NAL)", is compared with the traditional models, with its features and capabilities discussed.

Though many of the conclusions of this paper have appeared in my previous publications [29, 30, 32, 33, 35], this writing is my first attempt to give the role of copulas in reasoning a more comprehensive and detailed treatment.

## 2 Copulas in Logic

In linguistics, a "copula" is a verb that binds the subject of a sentence to the predicate (though it is sometimes considered part of the predicate), such as the various forms of "being" (am, are, is, was, were) in English.

In term logic, a *copula* indicates a logical relation between the subject term and the predicate term in a proposition (or statement), so a copula $c$ can form a proposition $ScP$ with two terms $S$ and $P$. The order of these two terms is conventional. Though usually the subject term is put first, with the predicate term indicating its type or property, Aristotle also turned the order around by considering "When one thing is predicated of another" [1].

A logical copula is not ambiguous like many linguistic copulas. For instance, the copula "is" in English can mean identity, subsumption, or something else, but they will each be represented as a different logical copula. When there is only one copula involved, it can be omitted in the representation, though the logical relation remains, which is represented implicitly. On the other hand, additional factors can be included in copulas to keep the $ScP$ format for propositions. For example, such a view can be taken for Aristotle's Syllogistic, by considering the forms $a$, $e$, $i$, and $o$ as copulas in a broad sense.

Based on such a representation format, the most obvious inference rule is a deduction rule that corresponds to the transitivity of a copula. In Aristotle's logic, such a syllogism is known as "Barbara" and can be written as

$$\{MaP, SaM\} \vdash SaP$$

In this inference rule, any terms can serve as $S$, $P$, or $M$, so what distinguishes this rule from the other rules is only the copulas and locations of the shared term $M$ in the premises. It is well-known that when terms are interpreted as sets and the copulas as set relations ($a$, $e$, $i$, and $o$ as *subset*, *disjoint*, *intersect*, and *difference*, respectively), all valid syllogisms in Aristotle's logic can be proved as theorems in set theory [15, 3]. At the same time, the meaning of a copula involved is fully revealed by the inference rules in which it appears.

Since mathematical logic [10, 36] came, the dominance of term logic was taken over by predicate logic, which reserves no special role for copulas. As far as mathematical knowledge is concerned, the "predicate with argument" format, $P(a_1, \ldots, a_n)$, is more suitable than the $ScP$ format. Though both formats have a "predicate" $P$, it is defined and handled very differently in the two. The "subject vs. predicate" distinction in term logic is local and relative, so the same term can be a subject in one proposition and a predicate in another. On the contrary, the "argument vs. predicate" distinction in predicate logic is global and absolute, as these two types of terms belong to disjoint domains.

In predicate logic, it is possible to express a traditional copula as a predicate, which is a reason for many people to consider predicate logic as having a higher expressive capability, as in Aristotle's logic there is no obvious way to express a non-copula relation among terms. However, even if a predicate has the same referential semantics as a logical copula, it is still not directly recognized and processed by the inference rules. Predicate logic mainly depends on the inference rules of propositional logic, which are defined and justified according to the truth-values of the propositions involved, without restrictions and requirements on the syntactic structures of these propositions. For example, the truth-value of an implication proposition $p \rightarrow q$ is fully determined by the truth-values of $p$ and $q$, without considering their contents. Overall, the validity of inference rules is based on the requirement that they can only derive true conclusions from true premises [26, 6]. Since truth functions are taken as fundamental, many conceptual relations in term logic are rewritten as propositional relations in predicate logic. For example, $SaP$ in the former usually becomes $(\forall x)(S(x) \rightarrow P(x))$ in the latter [26], and other copulas are transformed into truth-value relations in similar ways.

For the intended usage of mathematical logic, the advantages of predicate logic over term logic are obvious, and consequently, term logic is often considered as only of historical values. The most noticeable exception of this consensus is "Term Functor Logic" (TFL) [24, 25, 7, 8], which is functionally equivalent to first-order predicate logic while keeping the naturalness of term logic, partly because of the use of copulas. In TFL, "the logical forms of statements that are involved in inferences as premises or conclusions can be construed as the result of connecting pairs of terms by means of a logical copula (functor)" [25].

## 3 Logic and Artificial Intelligence

The logic considered in this paper is evaluated mainly according to the needs of Artificial Intelligence (AI). Generally speaking, the objective of AI research is to construct computer systems that work like the human mind, though there are very different opinions on which aspects the similarity should be [34]. For the researchers who want computers to follow the same "laws of thought" as humans, it is natural to look for inspirations and tools from logic, which "is concerned with the principles of valid inference" from the beginning [14].

It is obvious that except for tasks like theorem proving, *classical logic* (first-order predicate logic) does not fully meet the needs of AI, which often deals with problems and tasks outside mathematics. Nevertheless, many researchers still take classic logic as a proper starting point, with the hope that some extensions and revisions will solve the problems in AI while keeping the virtues of classic logic, such as its certainty and exactness [17, 19]. Among the various *non-classical logics* explored in AI, there are *nonmonotonic logic* for defeasible reasoning [22] and *probabilistic logic* for uncertain reasoning [18].

As for the current discussion, the most relevant work is that of *description logic* [5], where the primary focus is to represent human knowledge in a logical language suitable for computerized reasoning. Among the modifications it makes in classical logic, a major addition is to explicitly support the taxonomic structure among concepts using an "IS-A" relation [4], which is basically a logical copula, as it is directly recognized and processed by the inference rules. Though this relation can be expressed in other ways, many knowledge representation frameworks (such as "ontology", "knowledge graph", etc.) choose to directly express and handle it to achieve higher naturalness and efficiency. The result is a hybrid of term logic and predicate logic, where the former supports a taxonomic hierarchy, and the latter handles additional knowledge besides the hierarchical relations.

Though I share opinions with the above works, my own logic model has been motivated by different considerations. My understanding of "intelligence" is "the capability of working with insufficient knowledge and resources and adapting to the environment" [34]. Based on this understanding, I construct a normative theory and a reasoning model to specify how such a system should work, no matter whether it is naturally evolved (like the human mind) or artificially designed (like a computer system) [32].

In this context, "insufficient knowledge" means that the system's knowledge cannot always contain or derive correct or even satisfactory answers for the questions it faces and that what the system "knows" can be wrong, due to incorrect information, biased data, or changes in the environment. It follows that the system's reasoning process is no longer "from truth to truth", as in classical logic or Aristotelian logic.

On the other hand, "insufficient resources" is mainly about the restrictions on computational time and space. As the system may get a new task at any moment and each task has response-time demands (such as a deadline or "as soon as possible"), it often cannot consult all relevant knowledge when processing a task, nor to be specified as a predetermined procedure with a fixed resource demand. It means the system's reasoning processes in task processing cannot depend on task-level algorithms.

However, it does not mean that in such situations all reasoning activities are equally valid (or invalid). It is arguable that such a working environment is the normal case for the human mind, and our existence shows that the regularities observed in the human reasoning process are not completely irrational, even though not "truth-preserving" in the traditional sense. What I have tried is to re-establish the validity of reasoning and rationality of thinking on the foundation of *adaptation*. For a system to be adaptive in its lifetime, it must predict the future according to its past experience and behave accordingly, even if the future and the past cannot be taken as the same. Furthermore, the system needs to spend its available resources in an efficient manner and also manage them according to its own experience.

This type of adaptation requires a "Concept-Centered Knowledge Representation (CCKR)", where a *concept* is an abstraction of an experience segment. The use of concepts allows the system to recognize the partial similarity between the current situation and various past situations, as well as to selectively organize experience in an efficient manner. Since concepts have different levels of abstractness, the same experience can be represented with various granularity, scope, focus, etc. to meet different needs of the system. Unlike the concepts in the "symbolic AI" tradition, the meaning of a concept in CCKR is not determined by the object or event it refers to in the world, but by its location in experience, as revealed by its relations with other concepts.

According to this "Experience-Grounded Semantics (EGS)", the system's knowledge does not provide a description of the world "as it is" but "as the system knows", and the extent to which a statement is "true" cannot be judged by comparing it with facts or future observations (which are not available to the system) but with evidence collected from (past) experience. Reasoning serves the purpose of transforming knowledge to related situations, and valid inference rules are those whose conclusions correctly summarize the evidence provided by the premises. This type of logic is fundamentally different from traditional logic (no matter whether they are built as predicate logic or term logic), while still keeping the normative and formal nature of logic [30, 35].

## 4 Copulas in NAL

Based on the above considerations, in my logic NAL [33] a *concept* is an internal entity (a data structure in the reasoning system) that corresponds to a recognizable component in the experience of the system. It can be a perceived pattern of stimuli, an executable operation, a word or phrase in a (communication) language, or a combination of them. Each concept is uniquely identified within the system by a *term*, which in the simplest form is a string of characters.

A new term can enter the system from the environment, or be composed by the system itself from some existing terms using a term connector. Each compound term corresponds to a concept, which is semantically related to the concepts identified by the terms that are the components of the compound. Among the infinite number of conceptual relations, a small number is directly recognized and processed by the inference rules. Copulas are among them and indicate various forms of substitutability of terms, as well as the concepts identified by them.

The most fundamental copula in NAL is *inheritance*, which is a reflexive and transitive relation between two terms, and is written as "$\rightarrow$". The statement "$S \rightarrow P$" is intuitively close to the "$SaP$" in Aristotle's syllogism, indicating that $S$ is a specialization of $P$ and that $P$ is a generalization of $S$. By defining the *extension* and *intension* of a term as sets containing its specializations and generalizations, respectively, "$S \rightarrow P$" also indicates "$S$ inherits the intension of $P$" and "$P$ inherits the extension of $S$", which is why this copula is named "inheritance".

The above idealized "complete inheritance" can be extended to cover "incomplete inheritance" between terms (concepts), which is the normal case for systems working with insufficient knowledge and resources. In NAL, the idealized (binary-valued) inheritance copula is used to define evidence for a realistic (multi-valued) *inheritance* copula, also written as "$\rightarrow$". The positive evidence for statement "$S \rightarrow P$" are the terms in the *extensional intersection* and the *intensional intersection* of $S$ and $P$, while the terms in the *extensional difference* of $S$ and $P$ and the *intensional difference* of $P$ and $S$ are negative evidence of the statement. The truth-value of a statement is defined as a pair of values in [0, 1], including a *frequency* that is the proportion of positive evidence among all available evidence, and a *confidence* that is the proportion of currently available evidence among all evidence at a fixed "evidential horizon" after a constant amount of future evidence is collected.

Basic inference rules of NAL are syllogistic, each with a truth-value function calculating the evidential support the premises provide for the conclusion. To establish these functions, all involved values in [0, 1] are treated as extended Boolean values in {0, 1}, as explained in detail in [33]. Therefore, corresponding to the Aristotelian syllogism "Barbara", there is the NAL rule

$$deduction : \{M \rightarrow P, S \rightarrow M\} \vdash S \rightarrow P$$

which specifies the transitivity of the (multi-valued) *inheritance* copula.

Following the insight of Peirce, induction and abduction in NAL are obtained from deduction by exchanging the conclusion in the above rule with each of the two premises, respectively:

$$induction : \{S \rightarrow P, S \rightarrow M\} \vdash M \rightarrow P$$

$$abduction : \{M \rightarrow P, S \rightarrow P\} \vdash S \rightarrow M$$

Though Peirce only suggested the cognitive functions of these non-deductive inference types (induction for generalization, and abduction for explanation), using experience-grounded semantics these rules can be justified in the same manner as deduction since they are exactly how evidence is collected by comparing the extension and intension, respectively, of the two terms in the conclusion, as specified earlier.

The following revision rule pools evidence from distinct sources:

$$revision : \{S \rightarrow P, S \rightarrow P\} \vdash S \rightarrow P$$

These four rules form a minimum NAL that can summarize its experience for adaptation while working with insufficient knowledge and resources.

The other inference rules are mostly designed to handle various types of compound terms. For example, if two terms $A$ and $B$ have a relation $R$ that is not *inheritance* (or any other copulas), then the relation is expressed as a set of ordered pairs including that of $A$ and $B$, that is, $(A \times B) \rightarrow R$, like in set theory. This statement still keeps the "subject-copula-predicate" format, though the subject term is a compound $(A \times B)$, whose meaning is partially determined by its (intrinsic) syntactic relations with its components $A$ and $B$, and partially determined by its (acquired) semantic relation with $R$. Here the meaning of $R$ is completely obtained from the system's experience, which is fundamentally different from how a copula (such as "$\rightarrow$") gets its meaning, though both represent "conceptual relations" and are represented as "predicates" in predicate logic.

Overall, there are four basic copulas in NAL. The *similarity* copula ($\leftrightarrow$) is a symmetric version of the *inheritance* copula ($\rightarrow$), and their isomorphic form between two statements (terms with a truth-value) are the *implication* copula ($\Rightarrow$) and the *equivalence* copula ($\Leftrightarrow$), which intuitively represents "if-then" and "if-and-only-if", respectively. While the first pair of copulas specifies the substitutability of the *meanings* (i.e., extensions and intensions) of the two terms linked, the second pair specifies the substitutability (derivability) of their *truth-values*. The inference rules of NAL correspond to different combinations of the copulas in the premises, just like in the case of Aristotle's syllogisms.

## 5 Comparisons and Discussions

Though the previous description of NAL is brief and highly simplified, it still provides enough materials for the role of copulas to be discussed.

First, the four basic copulas are represented and processed very differently from other conceptual relations, as the copulas are innate to the system and directly processed by the inference rules, with experience-independent (operational) meaning. As constants in NAL, they are at the meta-level, while the other relations are object-level concepts, with experience-grounded (acquired) meaning.

The dependency of copulas in NAL corresponds to the view of seeing "reasoning" as concept-substituting, which allows an adaptive system to see the novel (current) situation as familiar (past) situations. This ability of "seeing something as something else" has been suggested as at the center of cognition and intelligence by Ulam [23] and Hofstadter [12, 13], and Peirce also suggested that "Logic may be considered as the science of identity" [20]. This point of view is subtly different from the traditional view of seeing "reasoning" as demonstrations or derivations of new truth from known truth [1, 36, 14].

The most representative case of the traditional view is the reasoning in axiomatic systems, where the truth of axioms is either self-evident or widely accepted, from which the theorems are derived by truth-preserving (deductive) inference rules with guaranteed truth. This feature is shared by many other reasoning systems, where the initial promises are not called "axioms" but "presumptions", "facts", "ground truth", and so on, from which the conclusions (usually not called "theorems") are derived. Though the truth value of these initial promises can be challenged outside the system, within the system they are nevertheless treated just like axioms, and all of these systems do not consider the insufficiency of knowledge and resources as described previously.

This is exactly where NAL is different. Given its assumption on the insufficiency of knowledge and resources, its relation with traditional logic is analogous to the relation between Non-Euclidean geometry and Euclidean geometry, which is why it was named "Non-Axiomatic Logic" (NAL) [28, 33]. In this "axiomatic vs. non-axiomatic" distinction, the key is not on whether a system has "axioms" (according to the common definition), but whether there is "axiom-like" knowledge serving as the standard of truth within the system.

Under the assumption of insufficient knowledge and resources, a system must abstract experience to various levels to become concepts, and the evaluation of the identity, or substitutability, between concepts becomes a central task. Since such relations are domain-independent, they can be captured at the meta-level with justifiable patterns and rules, in spite of the inevitably uncertain and ever-changing nature of the concepts involved. Consequently, it is necessary to separate copulas from ordinary conceptual relations and to build logic systems around the copulas.

The above conclusion does not mean that the copula-free predicate logic is "wrong", but takes it as a different type of reasoning. In axiomatic systems, the focus is on the truth-value relations among the propositions, evaluated according to a given set of axioms or axiom-like propositions that cannot be challenged during the system's lifetime. When the propositions are symbolic and formal, an *interpretation* is needed to map the symbols within the system to concrete entities outside, usually approximately, for the symbols to become meaningful. This treatment gives the system's conclusions the desired certainty, as well as the possibility of applying them to multiple situations under different interpretations. These features are indeed hoped for in mathematics and other formal theories, like logic, but run into various problems when used outside these domains.

Besides the inability to cover non-deductive inference, another well-known issue of classic logic is the lack of semantic relevance between premises and conclusions. As revealed by the paradoxes of implication, many "logically correct" conclusions are intuitively awkward. The proposed solutions, in the form of various models of "relevance logic" [16] and "paraconsistent logic" [21], still attempt to resolve the issue in predicate logic, though it can be argued that the issue does not even exist in term logic [27]. This is the case exactly because of the use of copulas. In a logic like NAL, within a statement a copula relates two terms in meaning, and in a syllogistic inference rule, the two premises must have a shared term for the inference to happen, and the shared term ("middle term") also relates the other two terms (in the conclusion) together in meaning. The inference rule still determines the truth-value of the conclusion according to those of the premises, but that is under the condition that the terms involved are all semantically related. Unlike in propositional/predicate logic, in term logic it is invalid to replace one premise in an inference rule with another statement that happens to have the same truth-value.

A related feature is that inference in a term logic like NAL is not purely "formal" or "syntactic", but also "semantic" in the sense that all the involved terms have their meaning partially used (in the premises) and generated (in the conclusions). Consequently, NAL provides a model of categorization (conceptualization) that explains the forming and evolving of concepts, which happen together with reasoning as different aspects of the same process, rather than as separate processes carried out by separate mechanisms. This closeness to categorization also explains why a term logic uses a "categorical language" that is closer to natural languages than a predicate logic is [24].

However, the conclusion that "term logic provides a more suitable framework for empirical reasoning" is not a suggestion to return to Aristotle's Syllogisms, which is only a (binary) deductive system. Peirce added induction and abduction in their term logic form in an elegant manner and pointed out their cognitive functions, but still did not justify their validity as inference rules. Later, these two names are usually only used with cognitive functions, while their formalization is moved into the framework of predicate/propositional logic, rather than in term logic [9].

NAL, as a normative model designed for adaptive systems, differs from traditional logic (both in the term logic tradition and the predicate logic tradition) by assuming insufficient knowledge and resources, which leads to its concept-centered representation and experience-grounded semantics. In this context, a copula no longer corresponds to an objective relation between objects/events (or their classes) outside the system, but to substitutability between elements (or their patterns) within the system's experience, and this substitutability can be partial and uncertain. The Aristotelian forms *a*, *e*, *i*, and *o* all correspond to the *inheritance* copula in NAL, with their differences indicated by the truth-value that indicates evidential support. Since traditional logic assumes that all relevant evidence is already available, they only need two quantifiers to distinguish the "for all" and "there exist" situations, while in NAL, the *amount* of evidence needs to be quantitatively measured, and the system can never be sure that no new evidence will be found for a specific statement.

Given this understanding of copulas, in NAL all types of inference can be uniformly justified as using the evidence provided by the premises to support the conclusion, and the difference between deductive and non-deductive inference becomes *quantitative* (as indicated by the confidence value of the conclusion), rather than *qualitative* (as whether truth-preserving). Without this "concept substitution" view of reasoning, it is difficult (if not impossible) to use predicate logic with experience-grounded semantics, which is necessary for adaptive systems.

This is also where the key difference between NAL and TFL exists. Though both models extend and revise Aristotle's logic and increase the expressing capability of term logic (by introducing singular terms, relational terms, unanalyzed statements, compound statements, and so on), they are designed with different objectives in mind. Consequently, they make different assumptions on the sufficiency of knowledge and resources, which lead to other differences in the design, including their usages of copulas.

## 6 Conclusions

Non-Axiomatic Logic (NAL) [28, 33] attempts to provide a normative model for reasoning in an adaptive system that must deal with the insufficiency of knowledge and resources [30, 35]. Unlike traditional logic, NAL cannot achieve absolute certainty by depending on a set of axiom-like propositions and only carries out inference that is truth-preserving as defined in model-theoretic semantics [2]. Instead, it uses concept-centered representation and experience-grounded semantics [31]. According to this point of view, knowledge is mainly about the substitutability between concepts, and reasoning is mainly about the spreading of such substitution relations.

As copulas correspond to types of conceptual substitutability without restricting the content of the concepts involved, they provide the basis for a formal language and inference rules defined on the language for the above purpose. The language allows uncertainty caused by conflicting evidence and changing situations, and the rules cover inconclusive inference based on partial evidence. Furthermore, copula-based representation and inference guarantee the semantic relevance of the premises and the conclusion in each inference step.

Axiomatic reasoning in mathematics (and other formal theories) and non-axiomatic reasoning in everyday life (and empirical theories) require different logic models. The practice of NAL shows that copulas can play a central role in the latter type of logic, as well as provides new materials in the discussions about the relationship between logic and thinking [14, 11].

## Acknowledgments

The author benefits greatly from the insightful comments of George Englebretsen and the fruitful discussions with Kotaro Funakoshi.

## References

[1] Aristotle (1882). *The Organon, or, Logical treatises of Aristotle*. George Bell, London. Translated by O. F. Owen.

[2] Barwise, J. and Etchemendy, J. (1989). Model-theoretic semantics. In Posner, M. I., editor, *Foundations of Cognitive Science*, pages 207–243. MIT Press, Cambridge, Massachusetts.

[3] Beziau, J. (2017). Is modern logic non-Aristotelian? In Zaitsev, D. and Markin, V., editors, *The Logical Legacy of Nikolai Vasiliev and Modern Logic*. Springer Verlag.

[4] Brachman, R. J. (1983). What is-a is and isn't: an analysis of taxonomic links in semantic networks. *IEEE Computer*, 16:30–36.

[5] Brachman, R. J. and Schmolze, J. G. (1985). An overview of the KL-ONE knowledge representation system. *Cognitive Science*, 9:171–216.

[6] Copi, I., Cohen, C., and McMahon, K. (2013). *Introduction to Logic*. Pearson custom library. Pearson, 14th edition.

[7] Englebretsen, G. (1981). *Three Logicians*. Van Gorcum, Assen, The Netherlands.

[8] Englebretsen, G. (1996). *Something to Reckon with: the Logic of Terms*. Ottawa University Press, Ottawa.

[9] Flach, P. A. and Kakas, A. C. (2000). Abductive and inductive reasoning: background and issues. In Flach, P. A. and Kakas, A. C., editors, *Abduction and Induction: Essays on their Relation and Integration*, pages 1–27. Kluwer Academic Publishers, Dordrecht.

[10] Frege, G. (1999). Begriffsschrift, a formula language, modeled upon that of arithmetic, for pure thought. In van Heijenoort, J., editor, *Frege and Gödel: Two Fundamental Texts in Mathematical Logic*, pages 1–82. iUniverse, Lincoln, Nebraska. Originally published in 1879.

[11] Gabbay, D. M. and Woods, J. (2001). The new logic. *Logic Journal of the IGPL*, 9(2):141–174.

[12] Hofstadter, D. R. (1995). On seeing A's and seeing As. *Stanford Humanities Review*, 4:109–121.

[13] Hofstadter, D. R. (2001). Analogy as the core of cognition. In Gentner, D., Holyoak, K. J., and Kokinov, B. N., editors, *The Analogical Mind: Perspectives from Cognitive Science*, pages 499–538. MIT Press, Cambridge, Massachusetts.

[14] Kneale, W. and Kneale, M. (1962). *The development of logic*. Clarendon Press, Oxford.

[15] Lukasiewicz, J. (1951). *Aristotle's Syllogistic: From the Standpoint of Modern Formal Logic*. Oxford University Press, London.

[16] Mares, E. (2020). Relevance Logic. In Zalta, E. N., editor, *The Stanford Encyclopedia of Philosophy*. Metaphysics Research Lab, Stanford University, Winter 2020 edition.

[17] McCarthy, J. (1989). Artificial intelligence, logic and formalizing common sense. In Thomason, R. H., editor, *Philosophical Logic and Artificial Intelligence*, pages 161–190. Kluwer, Dordrecht.

[18] Nilsson, N. J. (1986). Probabilistic logic. *Artificial Intelligence*, 28:71–87.

[19] Nilsson, N. J. (1991). Logic and artificial intelligence. *Artificial Intelligence*, 47:31–56.

[20] Peirce, C. S. (1986). Chap. VIII. Of the Copula. In *Writings of Charles S. Peirce: A Chronological Edition, Volume 3: 1872-1878*, pages 90–92. Indiana University Press.

[21] Priest, G., Routley, R., and Norman, J., editors (1989). *Paraconsistent Logic: Essays on the Inconsistent*. Philosophia Verlag, München.

[22] Reiter, R. (1987). Nonmonotonic reasoning. *Annual Review of Computer Sci-*

*ence*, 2:147–186.
[23] Rota, G.-C. (1989). The barrier of meaning. In memorium: Stanislaw Ulam. *Notices of the American Mathematical Society*, 36(2):141–143.
[24] Sommers, F. (1982). *The Logic of Natural Language*. Clarendon Press, Oxford.
[25] Sommers, F. and Englebretsen, G. (2000). *An invitation to formal reasoning: the logic of terms*. Ashgate, Aldershot.
[26] Stebbing, L. S. (1950). *A Modern Introduction to Logic*. Harper & Row, New York, 7th edition.
[27] Steinkrüger, P. (2015). Aristotle's assertoric syllogistic and modern relevance logic. *Synthese*, 192:1413–1444.
[28] Wang, P. (1995). *Non-Axiomatic Reasoning System: Exploring the Essence of Intelligence*. PhD thesis, Indiana University.
[29] Wang, P. (2000). Unified inference in extended syllogism. In Flach, P. and Kakas, A., editors, *Abduction and Induction: Essays on Their Relation and Integration*, pages 117–129. Kluwer Academic Publishers, Dordrecht.
[30] Wang, P. (2004). Cognitive logic versus mathematical logic. In *Lecture notes of the Third International Seminar on Logic and Cognition*, Guangzhou.
[31] Wang, P. (2005). Experience-grounded semantics: a theory for intelligent systems. *Cognitive Systems Research*, 6(4):282–302.
[32] Wang, P. (2006). *Rigid Flexibility: The Logic of Intelligence*. Springer, Dordrecht.
[33] Wang, P. (2013). *Non-Axiomatic Logic: A Model of Intelligent Reasoning*. World Scientific, Singapore.
[34] Wang, P. (2019a). On defining artificial intelligence. *Journal of Artificial General Intelligence*, 10(2):1–37.
[35] Wang, P. (2019b). Toward a logic of everyday reasoning. In Vallverdú, J. and Müller, V. C., editors, *Blended Cognition: The Robotic Challenge*, pages 275–302. Springer International Publishing, Cham.
[36] Whitehead, A. N. and Russell, B. (1910). *Principia Mathematica*. Cambridge University Press, Cambridge.

# HOW LINEAR ARE ENGLEBRETSEN'S LINE DIAGRAMS?

AMIROUCHE MOKTEFI
*Tallinn University of Technology, Estonia*

## 1 Introduction

It is known that George Englebretsen developed a diagrammatic system for term logic. This scheme was introduced in a 1992 article: "Linear diagrams for syllogisms (with relationals)" that appeared in *Notre Dame Journal of Formal Logic* [10]. The system was amended and expanded in subsequent writings, notably: *Something to Reckon with: The Logic of Terms* [11, pp. 188–236], *Line Diagrams for Logic: Drawing Conclusions* [12], and *Figuring it Out: Logic Diagrams* [13]. The design of this system of diagrams was guided by concerns and inspirations that spring from two primary fields, term logic and logic diagrams, as Englebretsen himself recalls:

> "It is a system of formal logic that was the focus of much of my own research beginning half a century ago. Eventually, in the 1980s, my interest turned to the possibility of devising a system of logic diagrams that would be a graphic analogue to term logic in much the way that Venn meant his system of diagrams to be an analogue for Boole's logic. This led to many discussions with Sommers. He said that he had tried to do this for his new version of term logic but kept coming back to simple Venn, and then Peirce, diagrams. I thought something like Leibniz's linear diagrams might be a better place to start. After a number of failed attempts, it turned out that once I thought more deeply about Sommers' notation, keeping it as a kind of guide, I began to make progress." [13, p. xiii]

As such, this scheme can be regarded from a dual perspective. On the one hand, it is presented as a visual analogue for term logic, and such may be viewed as a diagrammatic contribution to term logic. On the other hand, Englebretsen diagrams are a term-logic-inspired contribution to the field of logic diagrams itself. It is this latter perspective that we consider in this paper. In particular, we attempt to situate Englebretsen's diagrammatic technique among the various methods that one might adopt to represent logical propositions with the aid of line segments.

Line diagrams have a long history. Several authors have argued that Aristotle himself must have used some form of such diagrams, as attested by several passages in his writings (see [[13, pp. 8–25]). Line diagrams are also known to have been used in Arabic Logic in the twelfth century (Hodges [18, 19]) and in Europe in the early seventeenth century (Lemanski [23], [24, pp. 200–202]). In the eighteenth and nineteenth centuries, line diagrams enjoyed some popularity and were promoted by such eminent logicians as Gottfried Wilhelm Leibniz, Johann Heinrich Lambert and John Neville Keynes. Finally, line diagrams went forgotten in the twentieth century till they got revived by Englebretsen. How his diagrams resemble and differ from those of his predecessors will be discussed in the following sections.

However, it must be said that our aim is not to reconstruct a detailed history of line diagrams. Rather, we conceptually consider different techniques that make line segments represent logical propositions, regardless of their actual implementation by the logicians of the past. Also, we will occasionally provide, for the purpose of illustration and suggestion, circle diagrams that are analogous to the line diagrams we discuss. We will essentially explore various methods of representation and will point out for reference some logicians who might have used such methods. But we will not describe in detail their work and we will rather direct the reader to primary and, occasionally, secondary sources for further reading on those logicians and their diagrams (see also [1, 15, 34, 2]). In this sense, our paper is only incidentally historical and such a history still remains to be written. Yet, there is a lesson to be learned from the various names of past logicians that will run through this paper: albeit less popular than their circular counterparts, line diagrams have been

## How Linear are Englebretsen's Line Diagrams?

| $x$ ——————— |
|---|

Figure 1: Term $x$

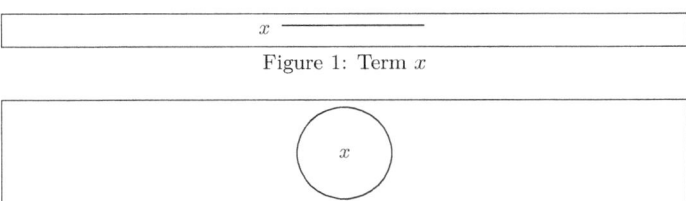

Figure 2: Term $x$

used by logicians throughout the centuries and developed in ways that reflected the interests of logicians at their time.

Since there are many ways in which line diagrams might be said to be linear, it is important to indicate what we take line diagrams to be. Englebretsen himself seems to have used the term in a wide sense, including any "system of diagrams for logic based on the use of points and line segments" [13, p. 16]. In the following, we will narrow our scope to diagrams where line segments are said to represent terms. For instance, in (Fig. 1), the line segment represents the term $x$ (for convenience, the label is inserted on the side, but it might be inserted elsewhere if needed).

Since it is the extension of the term that is represented, one might simply imagine that the individuals that compose that extension form a line[1]. Eric Hammer and Sun-Joo Shin precisely described Lambert's line diagrams as follows:

> "He visualizes individuals as points and imagines all the individuals in a row. Quite naturally, he chose a line to represent a collection, since a line is a collection of points just as a set is a collection of individuals. A concept is assumed to be represented as a line containing the individuals which belong to that concept. For example, the collection of animals can be represented as the line which is drawn connecting those points each of which represents an animal. We can express

---

[1] We assume here, as most traditional logicians did, that terms (and later their opposites) are not empty. This will allow us to focus on the diagrammatic techniques at work rather than the logical disputes on existential import.

Lambert's convention in a way very similar to that we used for Euler's system" [17, p. 4]

As Hammer and Shin rightly observed, this construction of line segments resembles the idea of circle diagrams, as described by Leonhard Euler. Indeed, a circle diagram is conceived as a space in which the individuals that form the extension of a term are gathered. Euler wrote: "As a general notion contains an infinite number of individual objects, we may consider it as a space in which they are all contained" (Euler [14, p. 339]). For instance, the diagram in (Fig. 2) depicts the term $x$ in a manner similar to the line segment in (Fig. 1). We must keep in mind that the shape of the figure is not logically relevant for the representation, as Euler himself pointed out [14, pp. 340–341]. Once terms are represented with line segments, it remains to represent propositions between terms. It is our aim to explore various strategies of representation in the subsequent sections[2]

## 2 Overlapping segments

We take line segments to stand for terms. Then, we wish to express the relations between those terms through the relations between their corresponding segments. An immediate solution consists in making segments (topologically) overlap to express the (logical) overlapping of the terms. Consider two terms $x$ and $y$ and their segments. Suppose we wish to represent the proposition "All $x$ are $y$". Since term $x$ is said to be included in term $y$, we simply depict segment $x$ as part of segment $y$, as shown in (Fig. 3).

This method of representation evidently resembles the general procedure of Euler diagrams. Rememeber that Euler assigned a circle (or a

---

[2] Note that we will not discuss here the solution of logical problems, such as syllogisms, with the various versions of line diagrams. Indeed, despite their diverse approaches of representation, all these methods agree in their treatment of such problems. As Oliver Lemon and Ian Pratt explained: "Inference in [Linear Diagrams] is to be carried out, as is usual with diagrammatic representations, by enumeration of cases. That is, a conclusion follows from a set of premises if all ways of diagramming the premises result in a diagram depicting the conclusion" (Lemon & Pratt [26, p. 575]).

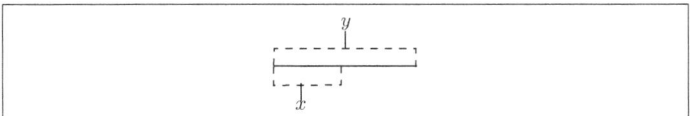

Figure 3: "All $x$ are $y$"

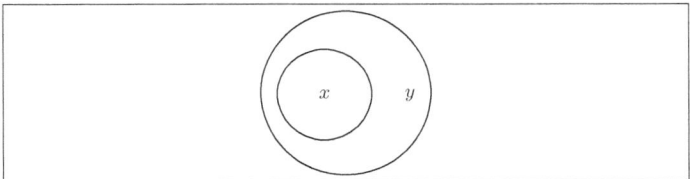

Figure 4: "All $x$ are $y$"

space) to each term, then made those circles (or spaces) interact in such a way as to mimic the logical relations between the terms. Assuming circles $x$ and $y$ to stand for terms $x$ and $y$ respectively, the proposition "All $x$ are $y$" is simply represented by drawing circle $x$ inside circle $y$, as shown in (Fig. 4). In both versions, linear and circular, the figure that stands for term $x$ is depicted as part of the figure that stands for term $y$. As such, the diagram shows directly that the extension of $x$ is part of that of $y$.

In the case of line diagrams, it suffices to imagine that the logical universe (all individuals) stand on a line. Segments of that line gather the individuals who belong to the term (i.e. its extension). Hence, intersections of segments contain shared individuals, and thus, readily exhibit intersections of the term extensions. Also, disjoint segments exhibit sets of individuals which do not intersect. This method readily permits the representation of traditional categorical propositions. For instance, the representation of "No $x$ is $y$" consists in two distinct segments $x$ and $y$, as shown in (Fig. 5). The proposition "Some $x$ are $y$" is represented with two intersecting segments $x$ and $y$, as shown in (Fig. 6).

This first method of using line segments to express logical propositions has a clear merit. The geometrical relations between the segments

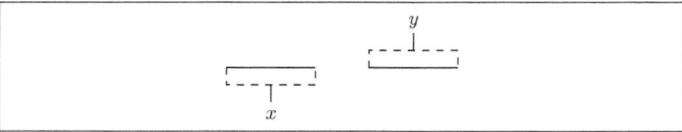

Figure 5: No $x$ is $y$

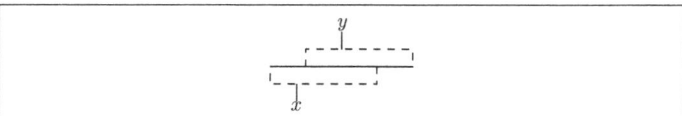

Figure 6: Some $x$ are $y$

truly imitate the logical relations of the terms they represent. As such, diagrams adequately exhibit the relations one wishes to convey. However, this method suffers from an important weakness: it is difficult to visually grasp the extent of a segment without the help of an additional syntactic device. Indeed, except for the case where segments are disjointed (Fig. 5), the overlap of segments makes it complicated to convey to the eye the limit of each segment. Hence, we added braces to exhibit the scope of each segment, as shown in (Fig. 3) and (Fig. 6). Evidently, this addition reduces the readability and immediacy of the diagrams. This limit might explain why this form of diagrams, despite its apparent simplicity, has seldom been used by logicians (except the diagram for "No $x$ is $y$" propositions which does not suffer from this flaw). Still, Leibniz might be said to have used such diagrams, albeit without much explanation, as shown in (Fig. 7).

## 3 Parallel segments

To overcome the ambiguity of representation in the previous method, one might simply draw our segments on distinct parallel lines. This allows a clear visualisation of the extent of each segment, and hence dispense of the braces (or similar devices) which were previously required. Consider again the proposition "All $x$ are $y$". Since $x$ is included in $y$, we

## How Linear are Englebretsen's Line Diagrams?

> B aequilaterum, L regulare, A quadrilaterum.
> Aequilaterum inest seu tribuitur regulari. Ergo
> quadrilaterum aequilaterum inest quadrilatero re-
> gulari sive quadrato perfecto. YS est in RX.
> Ergo RT⊕YS seu RS est in RT⊕RX seu in
> RX.
>
> Scholium. Haec propositio converti non potest, neque enim si A⊕B
> est in A⊕L, sequitur esse B in L.
> Prop. 13. Si L⊕B ∞ L, erit B in L. Si quid adjecto alio
> non fit aliud, adjectum ei inest. Nam B est in L⊕B (per defini-
> tionem inexistentis) et L⊕B ∞ L (ex hyp.), ergo (per prop. 6) B est in L.
>
> RY⊕RX ∞ RX, Ergo RY in RX
> RY in RX, ergo RY⊕RX ∞ RX

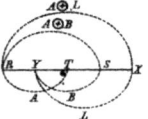
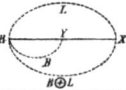

Figure 7: Leibniz diagrams (Gerhardt [16, p. 239])

$$y \quad \text{———}$$
$$x \quad \text{———}$$

Figure 8: All $x$ are $y$

draw segment $x$ strictly under segment $y$, as shown in (Fig. 8). Labels alone unambiguously refer to the segments and their extent. One might rightly object that the relative position of segment $x$ under segment $y$ is not unambiguously conveyed by the diagram. To improve our representation, we introduce vertical dotted lines, as shown in (Fig. 9), to indicate that segment $x$ is identical to the part of $y$ which is delineated by those vertical lines. It is as if we extracted segment $x$ (which is part of segment $y$) and made it slide till its new location on the parallel line. The vertical dotted lines convey this motion to the eye and clearly show that segment $x$ is strictly under, and hence is to be regarded as part of, segment $y$.

This technique can easily be used to represent other categorical propositions. For the proposition "No $x$ is $y$", we obtain (Fig. 10) where the vertical lines clearly show that segments $x$ and $y$ do not in-

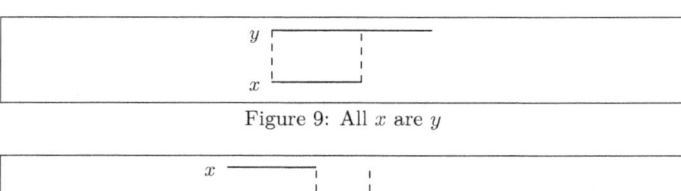

Figure 9: All $x$ are $y$

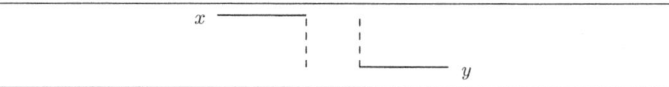

Figure 10: No $x$ is $y$

tersect, and hence, that terms $x$ and $y$ do not share individuals. For the proposition "Some $x$ are $y$", segments $x$ and $y$ are shown to intersect, in (Fig. 11), and hence convey the idea that terms $x$ and $y$ share individuals.

It is not surprising that such a method of representation has been used by many logicians in the past, as it easily and clearly represents the terms and their strict relations. Leibniz, again, is known for using such line diagrams to solve logical problems, as shown in (Fig. 12).

Louis Couturat described, as follows, Leibniz's shift from the method of overlapping segments to that of parallel segments:

> "We know the linear scheme that he uses in the logical papers already published. It consists, in principle, of depicting concepts by segments of one and the same straight line, either contained within one another, or external to one another, or having a common part, etc. However, to distinguish these various segments and mark their unity, it is necessary to use braces, or to designate them by their extreme points, which complicates the writing. Leibniz perfected this system in the following way. He separated the segments previously mixed

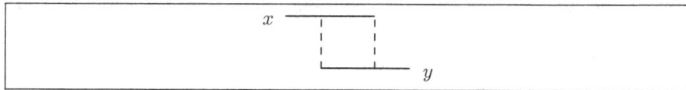

Figure 11: Some $x$ are $y$

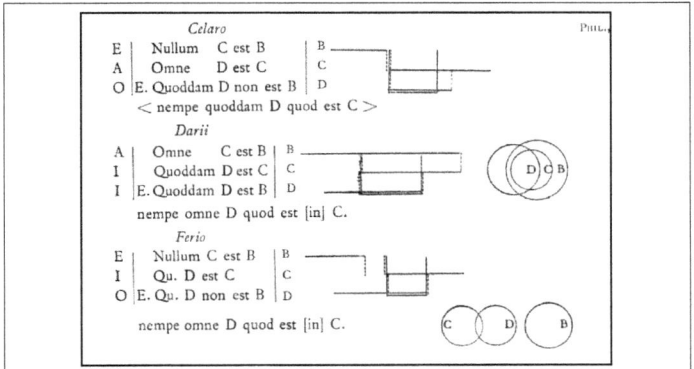

Figure 12: Leibniz diagrams (Couturat [9, p. 295])

on the same straight line, and transported them onto parallel straight lines, perpendicular to their direction; so that it is enough to draw perpendicular (dotted) lines through their ends to note their partial or total inclusion or exclusion relations"[3] [8, pp. 25–26].

As one would expect, such a shift would be unnecessary in the case of circle diagrams where the extent of the circles is generally made clear by its circumference. To make such a shift would produce strange diagrams, such as (Fig. 13) which represents "All $x$ are $y$".

---

[3]Translated from French by the author. Here is the original: "On connait le schématisme linéaire qu'il emploie dans les opuscules logiques déjà publiés. Il consiste, en principe, à figurer les concepts par des segments d'une et même ligne droite, soit contenus l'un dans l'autre, soit extérieurs l'un à l'autre, soit ayant une partie commune, etc. Seulement pour distinguer ces divers segments et marquer leur unité, il faut employer des accolades, ou bien les désigner par leurs points extrêmes, ce qui complique l'écriture. Leibniz a perfectionné ce système de la manière suivante. Il a séparé les segments auparavant confondus sur la même droite, et les a transportés sur des droites parallèles, perpendiculairement à leur direction ; de sorte qu'il suffit de mener par leurs extrémités des lignes (pointillées) perpendiculaires pour constater leurs relations d'inclusion ou d'exclusion partielle ou totale".

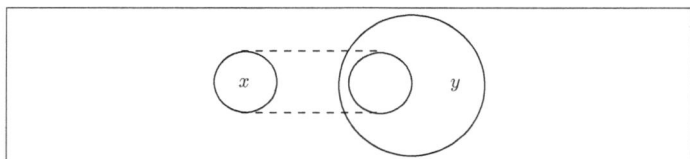

Figure 13: All $x$ are $y$

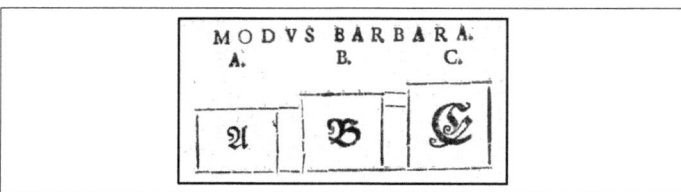

Figure 14: Lange diagrams (Lange [22, p. 354])

Here, circle $x$ is identical to the small circle within $y$. The awkwardness of this representation is that it invites us to 'see' $x$ as part of y while depicting $x$ outside $y$. That's precisely what our parallel line segments do. In (Fig. 9), we are told that $x$ is included in $y$ even though segment $x$ is depicted outside $y$. In (Fig. 11), $x$ and $y$ are said to intersect, even though the segments $x$ and $y$ are displayed on parallel lines and, thus, cannot intersect. We already indicated that this unintended feature results from our modification of the overlapping segments method to overcome its ambiguity. Such a modification is not required in the case of circle diagrams, and thus, it is not surprising that diagrams such as (Fig. 13) are not found among past logic diagrams. However, a resembling scheme has been in use by Johann Christian Lange in his *Nucleus Logicae Weisaniae* (1712). Jens Lemanski already discussed these diagrams and rightly pointed out their similitude with line diagrams (Lemanski [25, pp. 409–412]).

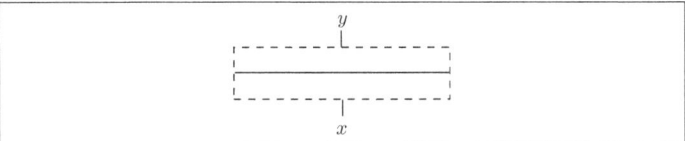

Figure 15: $x$ and $y$ coincide

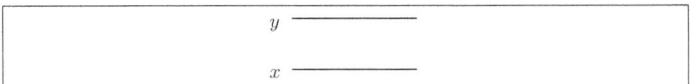

Figure 16: $x$ and $y$ coincide

## 4 Dotted segments

Both previous methods of diagrams (overlapping segments and parallel segments) represent the strict relations between terms with the help of line segments. As such, they fail to represent our imperfect knowledge regarding the scope of the propositions we wish to represent. For instance, the proposition "All $x$ are $y$" indicates that $x$ is part of $y$, but it is unclear whether $x$ is strictly included in $y$ or is identical to it. Yet, the diagrams we used so far to express this proposition, (Fig. 3) and (Fig. 9), represent $x$ as strictly included in $y$. If one wishes to fully represent that proposition with our line diagrams, it is required to add a second diagram depicting the case where $x$ and $y$ coincide, as shown in (Fig. 15) with the overlapping segments method and (Fig. 16) with the parallel segments method.

Naturally, this issue is not proper to line diagrams. The same concern occurs in the case of circle diagrams. For instance, (Fig. 4) shows $x$ strictly inside $y$, even though the proposition it depicts ("All $x$ are $y$") allows $x$ and $y$ to coincide. John Venn famously claimed that the "weak point about [Euler's circles] consists in the fact that they only illustrate in strictness the actual relations of classes to one another. Accordingly they will not fit in with the propositions of common logic" (Venn[38, p. 49]). This concern motivated Venn to invent his celebrated scheme [39] (See [31]). Other logicians, such as Friedrich Ueberweg,

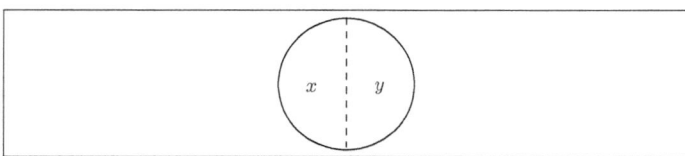

Figure 17: All $x$ are $y$

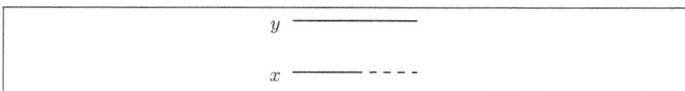

Figure 18: All $x$ are $y$

tackled this issue through the introduction of dotted lines to express uncertainty (Ueberweg [37]) (see [30]). The idea is to mark the borders whose existence is uncertain. Consider again the proposition "All $x$ are $y$". Here, $x$ is either strictly inside or coincides with $y$. This uncertainty is shown, in (Fig. 17), where $y$ covers the full circle while $x$ covers the part of $y$ that is on the left of the dotted line. If that line becomes continuous, $x$ appears strictly within $y$, but if it disappears, then $x$ and $y$ coincide.

The introduction of this technique on line diagrams, using the parallel segments method, is straightforward: dotted lines indicate the portions of segments whose existence is uncertain. Hence, (Fig. 18) represents the proposition "All $x$ are $y$"[4]. If the dotted portion of $x$ exists, then $x$ and $y$ coincide, as in (Fig. 16). But, if the dotted portion disappears, then $x$ is strictly inside $y$, as in (Fig. 8). In this sense, the dotted

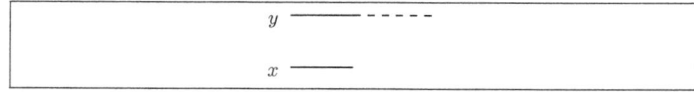

Figure 19: All $x$ are $y$

---

[4]For the ease of the reader, we did not add vertical lines to mark the scope of each segment, as we indicated in the previous section. Their introduction, though possible, would reduce the readability of the diagram.

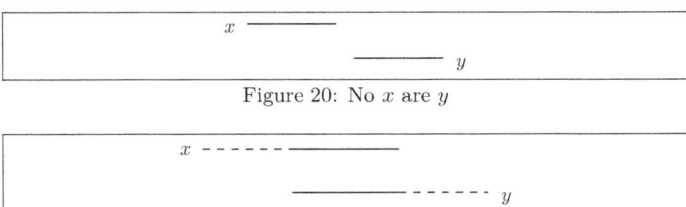

Figure 20: No $x$ are $y$

Figure 21: Some $x$ are $y$

diagram (Fig. 18) combines (Fig. 16) and (Fig. 8), and hence dispenses with drawing multiple diagrams to convey the different relations of $x$ and $y$ permitted by the proposition. Note that the dotted diagram in (Fig. 19), which is slightly different from (Fig. 18), also represents adequately the proposition "All $x$ are $y$".

It is easy to see how other categorical propositions can be represented with this amended version of line diagrams. The proposition "No $x$ is $y$", depicted in (Fig. 20), does not require the usage of dotted lines since there is no uncertainty regarding the precise relation between its terms. The Proposition "Some $x$ are $y$" is represented by (Fig. 21) which, with the help of dotted lines, adequately conveys the different relations permitted between $x$ and $y$. Indeed, the coinciding continuous segments exhibit the portion shared by $x$ and $y$. Then, the dotted lines convey four possible situations: (1) $x$ and $y$ coincide, (2) $x$ is strictly inside $y$, (3) $y$ is strictly inside $x$, (4) $x$ and $y$ partly intersect.

Although such dotted line diagrams were known to Leibniz (Couturat [8, pp. 30–31]), it is through Lambert that they seem to have spread among logicians. Lambert employed such diagrams in his *Neues Organon* (1764), as shown in (Fig. 22).

## 5  Opposite segments

So far, we did not consider negative terms. Yet, their representation might be desirable, especially for propositions and arguments that involve them. Again, this issue is not proper to line diagrams. Modern users of circle diagrams are accustomed to drawing a rectangle to de-

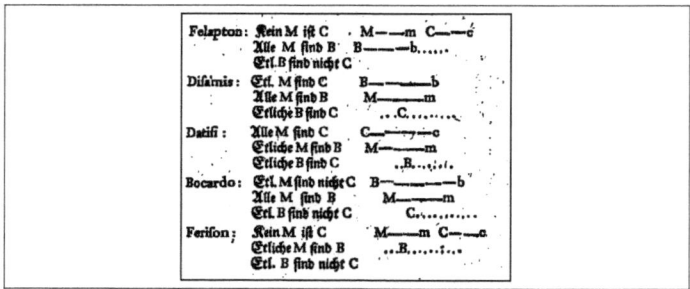

Figure 22: Lambert diagrams (Lambert [21, p. 133])

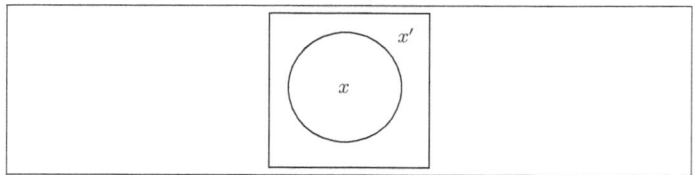

Figure 23: Circle diagram of $x$ and *not-x*

lineate the universe of discourse. This convention advantageously assigns a closed space to the outer region which aggregates the denials of the terms (Bhattacharjee [5]). For instance, in (Fig. 23), $x$ covers the space within the circle while its opposite *not-x* (here $x'$) covers the outer space. Although each term has its space, one might argue that a more symmetrical representation is desirable. This led some logicians to use rectangular diagrams which are equally divided to exhibit the symmetry between terms and their opposites (see, for instance, Lewis Carroll's diagrams [6] and [42, pp. 103–106].

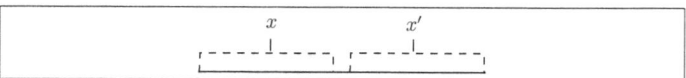

Figure 24: Line diagram of $x$ and *not-x*

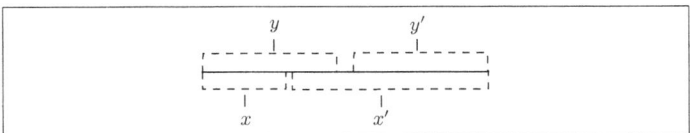

Figure 25: $x$ is strictly included in $y$

To incorporate negative terms in our line diagrams, we need to expand our representation of terms. Indeed, we convene that a large line segment represents the logical universe. Then, we divide that universe into equal (opposite) segments that stand for $x$ and $not\text{-}x$, as shown in (Fig. 24) where we reintroduced the braces to distinguish the extent of the opposite terms[5]. Let us now revisit the representation of categorical propositions with these opposite segments. To ease our exploration, let us consider the representation of the proposition "All $x$ are $y$", in the specific case where $x$ is strictly included in $y$ (and, thus, $not\text{-}y$ is strictly included in $not\text{-}x$). The usage of (expanded) overlapping segments and parallel segments produces (Fig. 25) and (Fig. 26), respectively[6] Both diagrams clearly exhibit the desired relation between $x$, $not\text{-}x$, $y$, and $not\text{-}y$. However, as discussed earlier, if we wish to represent all the relations permitted by the proposition "All $x$ are $y$", we need to use dotted lines that express our imperfect knowledge.

The immediate representation of "All $x$ are $y$" with dotted segments produces (Fig. 27) which shows that $x$ and $y$ have a common portion, as do $not\text{-}x$ and $not\text{-}y$. Uncertainty merely pertains to the dotted portion. However, unlike our practice in the previous section, it is not the existence of this portion that is at issue. Indeed, we know that the large segment (that is made of $x$ and $not\text{-}x$) cannot admit gaps because it

---

[5] We assume that our terms ($x$ and $y$) and their opposites ($not\text{-}x$ and $not\text{-}y$) are not empty.

[6] It is important to keep in mind that the large parallel segments must coincide since both stand for the entire logical universe. Also, parallel segments suggest the idea of using stripes to visually represent propositions. Such devices are not strictly line segments, and hence, do not qualify for our exploration. But much of the ongoing discussion on line diagrams might prove relevant for the logician who, instead, wishes to use stripes. Alexander Macfarlane's *logical spectrum* is an instance of such diagrams (Macfarlane [27]).

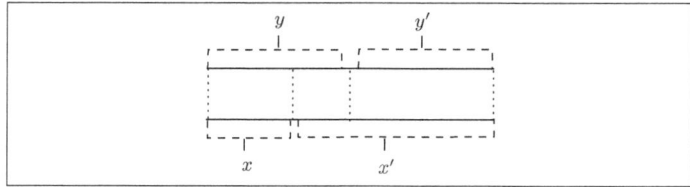

Figure 26: $x$ is strictly included in $y$

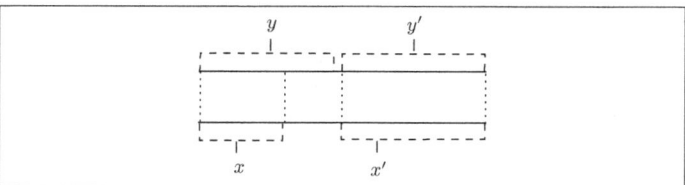

Figure 27: All $x$ are $y$

is an expression of the logical universe. Therefore, each portion of that universe must belong to $x$ or *not-x*. Our uncertainty concerns merely the term to which the dotted segment belongs. If the dotted portion belongs to $x$, the diagram will exhibit the relation where $x$ and $y$ coincide (and, thus, *not-x* and *not-y* coincide as well). If the dotted portion belongs to $x'$, the diagram becomes identical to (Fig. 26) and, hence, exhibits the case where $x$ is strictly in $y$ (and thus, *not-y* is strictly in *not-x*). We observe that (Fig. 27) adequately represents the two relations conveyed by the proposition "All $x$ are $y$". However, this achievement was attained at the cost of a re-interpretation of our dotted segments. We note that (Fig. 27) is presently overloaded. We may simplify it by dropping the braces and adopting different style of line segments for terms and their opposites, as shown in (Fig. 28). We also drop the vertical dotted lines and obtain the simpler (Fig. 29) to represent the proposition "All $x$ are $y$" (it is important that the style of the negative segments is different from the dotted lines used to express uncertainty).

With this simplified version, we can represent other categorical propositions as follows. The proposition "No $x$ is $y$" is depicted in (Fig.

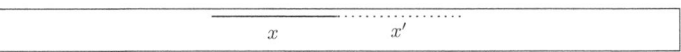

Figure 28: Line diagram of $x$ and *not-x*

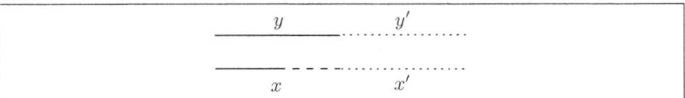

Figure 29: All $x$ are $y$

30) where terms $x$ and $y$ are shown to be disjoint. The uncertainty pertains to the existence of the outer region *not-x not-y*. If the dotted portion belongs to $x'$, then the outer region exists. If the dotted portion belongs to $x$, then the outer region does not exist. In the latter case, $x$ coincides with $y'$ and $x'$ coincides with $y$. The representation of the proposition "Some $x$ are $y$" is shown in (Fig. 31). Here, we see that $x$ and $y$ share a part of the universe. Then, the various combinations of the dotted lines express the different relations between $x$, $y$, *not-x* and *not-y* permitted by the proposition.

With the growing interest in the diagrammatic representation of negative terms at the end of nineteenth century, one finds line diagrams of this sort in the work of Keynes. In his *Studies and Exercises in Formal Logic*, he systematically expanded traditional propositions to incorporate negative terms Keynes [20] (see [33]). For the purpose, he made a thorough use of line diagrams and introduced segments for terms and their opposites, as shown in (Fig. 32).

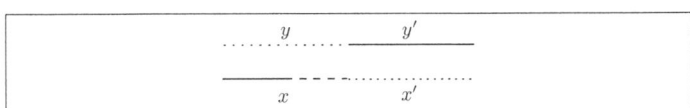

Figure 30: No $x$ are $y$

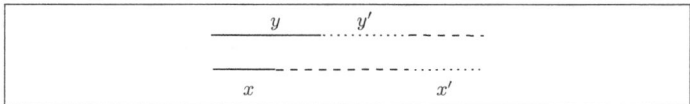

Figure 31: Some $x$ are $y$

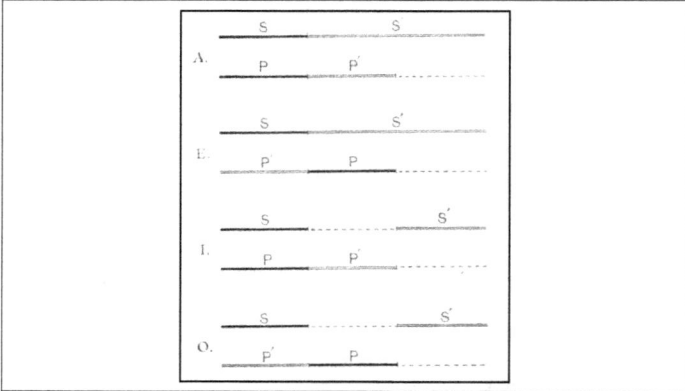

Figure 32: Keynes' line diagrams (Keynes [20, p. 145])

## 6 Compartmented segments

As we saw in the previous sections, our line diagrams become more complex when we attempt to increase their expressive power, by incorporating uncertainty and negative terms. It is known that circle diagrams undertook a similar path. In particular, we alluded earlier to Venn diagrams which aimed at overcoming the shortcomings of earlier circle schemes, generally known as Euler diagrams. Venn's main innovation was to abandon the direct representation of propositions with the relations between the circles. Instead, he first draws a primary diagram exhibiting all the possible combinations of the terms, then marks the resulting compartments to indicate their occupation or emptiness (Englebretsen [13, pp. 34–38]; Moktefi & Lemanski [31]). For illustration,

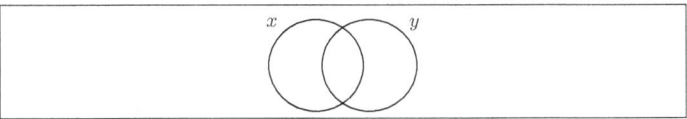

Figure 33: Primary diagram [40, p. 114]

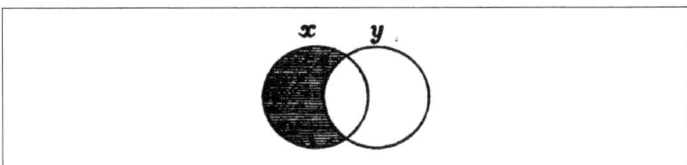

Figure 34: All $x$ are $y$ [40, p. 122]

let us consider the proposition "All $x$ are $y$". To represent this proposition, Venn draws a primary diagram with two intersecting circles, as shown in (Fig. 33). We can see that the diagram has four compartments: $xy$, $x$ not-$y$, not-$xy$ and (the outer region) not-$x$ not-$y$. Since the proposition "All $x$ are $y$" asserts that $x$ not-$y$ is empty, Venn shades that compartment to indicate its emptiness, as shown in (Fig. 34).[7]

A lot may be said about the merits and shortcomings of Venn diagrams. However, it suffices here to point out that they effectively represent categorical propositions, including the uncertainty about their scope and the negative terms they involve. One might wonder what kind of scheme one gets if Venn's method were to be applied on line diagrams. For that purpose, we first draw a large segment that stands for the logical universe, then we divide it into the compartments that are required by the proposition we wish to represent. For instance, if our proposition has two terms $x$ and $y$, we obtain four compartments in the primary diagram, shown in (Fig. 35). The braces exhibit the extent of each compartment. This diagram might be said to be the linear analogue of Venn's primary diagram, shown in (Fig. 33). Although the sequence of the compartments in our line diagram can be altered,

---

[7]In this section, we follow Venn's convention that propositions of the form "All $x$ are $y$" do not entail the existence of their subject.

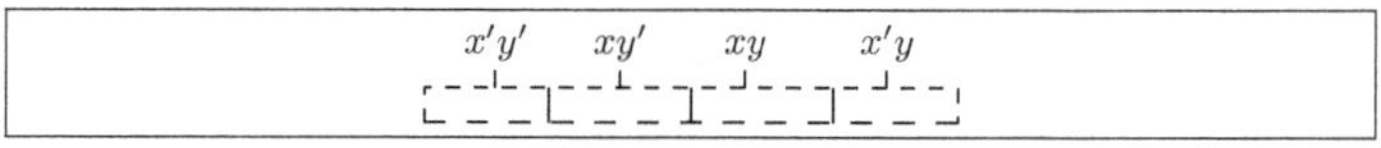

$$x'y' \quad xy' \quad xy \quad x'y$$

Figure 35: Primary line diagram

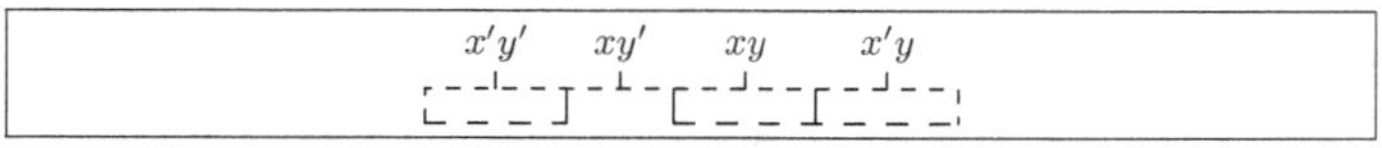

$$x'y' \quad xy' \quad xy \quad x'y$$

Figure 36: All $x$ are $y$

our arrangement conveniently maintains the unity of the segments that represent the terms $x$ and $y$.

It is desired to represent "All $x$ are $y$" on this primary diagram. For the purpose, one might simply remove the segment $x\ y'$, as shown in (Fig. 36). Other propositions can be similarly represented. For "No $x$ is $y$", we simply remove the $xy$ segment, as shown in (Fig. 37). For "Some $x$ are $y$", we need a new graphical convention to indicate that the segment $xy$ is known to exist. For instance, we could simply make it continuous, as shown in (Fig. 38).

These compartmented line diagrams readily and effectively represent our propositions. This merit attests the ingenuity of Venn's method. James Welton, a contemporary of Venn, believed in the potential of such Venn-inspired line diagrams and designed such a scheme, as shown in (Fig. 39). However, such compartmented diagrams, despite their gains in rigour and expressivity, do not possess the apparent naturalness of earlier line diagrams. The search for a balanced scheme is precisely the challenge that the logician faces, and we leave it to the reader to estimate the extent to which Welton succeeded in his endeavour.

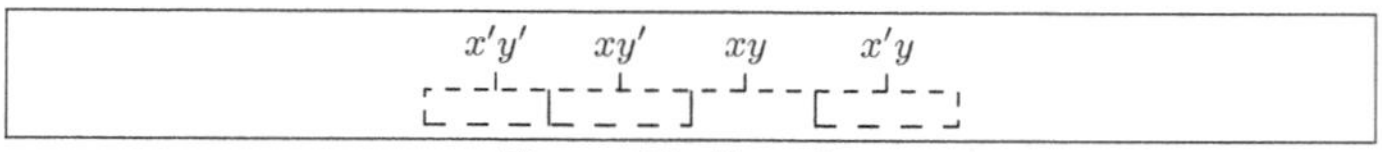

$$x'y' \quad xy' \quad xy \quad x'y$$

Figure 37: No $x$ is $y$

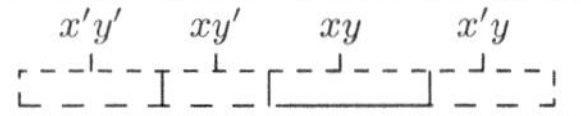

$$x'y' \quad xy' \quad xy \quad x'y$$

Figure 38: Some $x$ are $y$

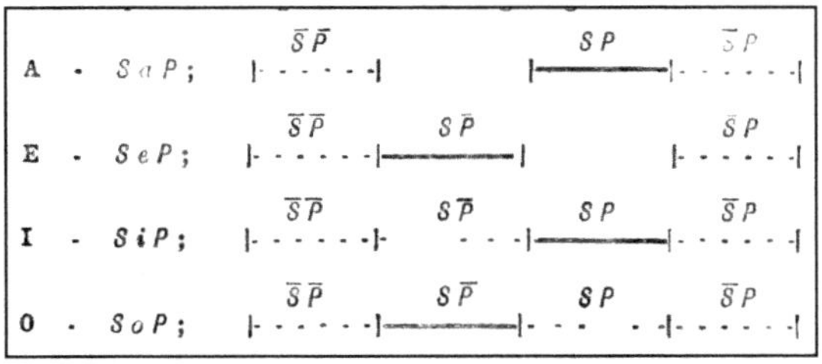

Figure 39: Welton diagrams; Welton [41, p. 223]

## 7 Englebretsen diagrams

So far, we discussed various ways in which segments lines representing terms may be used to express propositions. We now come to Englebretsen's line diagrams and attempt to situate them within this rich set of schemes. First, let us briefly recall some basic features of Englebretsen diagrams, as presented in [13].[8]. In this system, points represent individuals and, similar to previous methods, line segments represent terms. The label is placed next to the (enlarged) right-most endpoint of the segment. For instance, (Fig. 40) represents the term $x$. Then, to represent proposition, Englebretsen's rule is that "the relations that hold between the lines and points of a diagram (inclusion, intersection, exclusion) are meant to mimic the relations that hold between objects represented by those lines and points" [13, p. 79].

Let us consider how Englebretsen's rule is implemented to represent some categorical propositions. We are told that the "inclusion of

---

[8]Note that we will not provide here a detailed exposition of Englebretsen diagrams. We only present the elements of the system that pertain to our purpose, that is the representation of traditional categorical propositions with line diagrams. For a detailed account, readers are invited to consult [13]

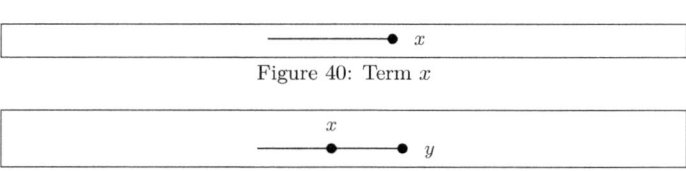

Figure 40: Term $x$

Figure 41: All $x$ are $y$

(the extension of) one term in another is simply represented by the line representing the first term being made a (proper) part of the line representing the other [...] keeping in mind that the label on a term line is meant to apply to the entirety of the line to its left" [13, p. 79]. Hence, the representation of the proposition "All $x$ are $y$" produces (Fig. 41). Englebretsen's representation of the proposition "No $x$ is $y$" produces (Fig. 42) which exhibits "two labelled lines sharing no point" [13, p. 79]. Finally, the proposition "Some $x$ are $y$" is represented with (Fig. 43) which "consists of the two lines representing the two terms sharing at least one common point – i.e., intersect" [13, p. 80].[9]

A comparison between Englebretsen diagrams for categorical propositions with those produced by the previous methods suggests some similitudes but significant differences. First, (Fig. 41) resembles (Fig. 3) which depicts the same proposition with the method of overlapping segments. Both situate segments on the same line and make one segment part of the other. (Fig. 42) rather suggests a connection with the method of parallel segments. However, in the latter method, (Fig. 42) would rather depict the coincidence of $x$ and $y$, not their mutual exclusion. (Fig. 43) attests that Englebretsen's method differs from

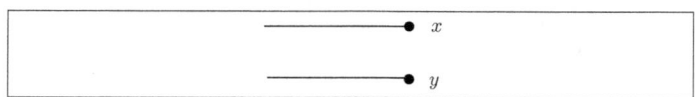

Figure 42: No $x$ is $y$

---

[9]Note that Englebretsen's diagram over-determines the intersection of the segments, however. While the proposition asserts that the terms share *at least* one individual, the segments share *exactly* one point.

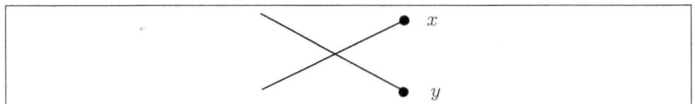

Figure 43: Some $x$ are $y$

both the overlapping and the parallel segments methods. Indeed, here, the two segments intersect, and hence, are neither on the same line nor on parallel lines. Naturally, Englebretsen diagrams also differ from the methods of dotted segments, opposite segments and compartmented segments, as presented earlier, although one might expand them to handle uncertainty and negative terms, the way those methods did.

To fully understand the rationale of Englebretsen diagrams, one needs to conceive their line segments as located within a two dimensional space, not on a line (or set of parallel lines) as the previous linear methods did. This interpretation is supported by Englebretsen's inclusion of a rectangle (that is a spatial object) in his system to "demarcate pertinent domains of discourse" [13, p. 77]. This 'spatial' feature might be overlooked by Englebretsen's readers because these rectangles were "suppressed and left as understood unless context requires otherwise. Nonetheless, it must always be understood that any diagrammatic representation is relative to a specifiable domain" [13, p. 77].

This surprising innovation allows Englebretsen to produce highly iconic diagrams where relations between segments *truly* imitate the relations between the terms. When terms share (do not share) individuals, Englebretsen's segments share (do not share) points, respectively. We saw how the method of parallel segments awkwardly used parallel segments to represent intersecting terms. This anomaly is not found in Englebretsen's diagrams, which share this high iconicity with the method of overlapping segments. In a sense, Englebretsen uses the method of overlapping segments but, instead of keeping his individuals on a (one-dimensional) line, he lets them occupy the (two-dimensional) space. However, Englebretsen diagrams, thanks to their spatial setting and a subtle labelling convention, do not suffer from the ambiguity that overlapping segments have regarding the extent of the segments.

This spatial aspect also suggests that Englebretsen diagrams might have more in common with Euler's circle diagrams than its predecessors had. Indeed, both line diagrams and circle diagrams might be said to be Eulerian in the sense that they represent terms with a geometrical figure, then make the topological relations between the figures stand for the logical relations between the terms. Although exceptions exist[10], logicians commonly did not assign a logical meaning to the shape of the figure and simply used circles for convenience. Many logicians, such as Leibniz and Keynes, actually used to a large extent both circle and line diagrams. These two types of diagrams did not differ in their procedures, but merely in the type of figure they used. However, a hidden feature associated with the figure is the shape of the logical universe itself. Circle diagrams evidently are drawn in a space while line segments stand on a line (or pairs of parallel lines). Englebretsen broke this old (untold) dichotomy and placed his line segments in a space. As such, his diagrams might as well be said to be Euler diagrams where line segments replace circles.

We argued that Englebretsen diagrams, although linear in their usage of line segments to represent terms, share with spatial diagrams (such as Euler's circles), the spatiality of the logical universe. As such, one might regard them as a sort of intermediary type of representation. It is unclear the extent to which Englebretsen would approve such a characterisation. Indeed, his treatment of negative terms shows that he rather maintains his scheme within the tradition of line diagrams. As we saw earlier, negative terms can be represented with line segments in a manner similar to positive terms. It is precisely what Englebretsen does: "Given a specifiable domain, for any term line, P, there will be a (possibly tacit) term line, nonP, such that the two lines share no point in common" [13, p. 82]. This convention was justified in early methods where the universe is represented with a large segment. Hence, whatever is not part of a segment $x$, must belong to its opposite *not-x*, which is a segment (or set of segments) from that universe. However, this practice does not match with Englebretsen's universe which is a space (within a rectangle). As Englebretsen acknowledged (in the case of singular propositions), the negative should be "everything in the domain other than" the term itself

---

[10]See in particular Peirce's version of Euler's diagrams to represent negative terms [32, 3, 35]

[13, p. 100]. Hence, the representation of the opposite of a term should not be a segment, but rather the entire space outside the segment that represents that term. Instead, Englebretsen adopted the old dissonant, albeit workable, convention of having both terms and their opposites represented with line segments [13, pp. 84–85]. Englebretsen shows here a clear inclination towards line diagrams, and it is, thus, reasonable to maintain his scheme within that family.

## 8 Conclusion

Our exploration of the various ways in which line diagrams are said to be linear demonstrated the richness of this family of diagrams and the fruitfulness of its basic principle: to represent a term with a line segment, then to express the logical relations of the terms with the geometrical relations of the segments. The numerous logicians who made use of such diagrams also attest of the attention these schemes have attracted. As such, they played a role, albeit minor, in the competition in which logicians engaged in their search for (better) languages (see [28]).

Englebretsen diagrams are linear in that they use line segments to represent terms. But they differ from older schemes in that they do not adopt a linear delineation of the logical universe. An interesting consequence is that Englebretsen's scheme takes us back closer to our first and simpler method, i.e. the method of overlapping segments. Englebretsen's innovation leads to a wide range of transformations, including an effective and highly iconic representation of categorical propositions, but also some limitations in term of matchedness.

Regardless of *how* it stands among line diagrams, an important merit of Englebretsen's scheme is its very *belonging* to that forgotten family of diagrams. This attention revived interest in this type of representations, especially within the new trend of formal diagrammatic systems. It is known that most of such systems are presently designed by logicians using circle diagrams [36, 4]. (See also [29].) Englebretsen's work is regarded as the first attempt to formalise line diagrams [7, p. 946], and as such, might be said to have opened the way to recent formal line diagrammatic systems.

# References

[1] Baron, M. E. (1969), "A note on the historical development of logic diagrams: Leibniz, Euler and Venn", *Mathematical Gazette*, 53 (384): 113-125.

[2] Bellucci, F., Moktefi, A., & Pietarinen, A.-V. (2014), "Diagrammatic autarchy: linear diagrams in the 17th and 18th centuries", *in* J. Burton & L. Choudhury (eds.), *Diagrams, Logic and Cognition*, CEUR Workshop Proceedings, 1132: 23-30.

[3] Bhattacharjee, R., & Moktefi, A. (2020), "Peirce's inclusion diagrams, with application to syllogisms", *in* A.-V. Pietarinen *et al.* (eds.), *Diagrammatic Representation and Inference*, Cham: Springer, 530-533.

[4] Bhattacharjee, R., Moktefi, A. (2021), "Revisiting Peirce's rules of transformation for Euler-Venn diagrams", *in* A. Basu *et al.* (eds.), *Diagrammatic Representation and Inference*, Cham: Springer, 166-182.

[5] Bhattacharjee, R., Moktefi, A., Pietarinen, A.-V. (2023), "The representation of negative terms with Euler diagrams", *in* J.-Y. Béziau *et al.* (eds.), *Logic in Question*, Cham: Birkhaüser, 43-58.

[6] Carroll, L. (1887), *The Game of Logic*, London: Macmillan.

[7] Chapman, P., Stapleton, G., & Rodgers, P. (2014), "PaL diagrams: A linear diagram-based visual language", *Journal of Visual Languages & Computing*, 25 (6): 945-954.

[8] Couturat, L. (1901), *La Logique de Leibniz*, Paris : Felix Alcan.

[9] Couturat, L. (1903), *Opuscules et Fragments Inédits de Leibniz*, Paris : Felix Alcan.

[10] Englebretsen, G. (1992), "Linear diagrams for syllogisms (with relationals)", *Notre Dame Journal of Formal Logic*, 3: 37-69.

[11] Englebretsen, G. (1996), *Something to Reckon with: The Logic of Terms*, Ottawa: University of Ottawa Press.

[12] Englebretsen, G. (1998), *Line Diagrams for Logic: Drawing Conclusions*, New York: Edwin Mellen Press.

[13] Englebretsen, G. (2020), *Figuring it Out: Logic Diagrams*, Berlin-Boston: De Gruyter.

[14] Euler, L. (1833), *Letters of Euler on Different Subjects in Natural Philosophy Addressed to a German Princess*, New York: J. & J. Harper.

[15] Gardner, M. (1958), *Logic Machines and Diagrams*, New York: McGraw-Hill.

[16] Gerhardt, C. I. (ed.) (1890), *Die Philosophischen Schriften von Gottfried Wilhelm Leibniz*, vol. 7, Berlin: Weidmann.

[17] Hammer, E., & Shin, S.-J. (1998), "Euler's visual logic", *History and*

*Philosophy of Logic*, 19: 1-29.

[18] Hodges, W. (2018), "Two early Arabic applications of model-theoretic consequence", *Logica Universalis*, 12: 37-54.

[19] Hodges, W. (2023), "A correctness proof for Al-Barakat's logical diagrams", *Review of Symbolic Logic*, 16 (2): 369 – 384.

[20] Keynes, J. N. (1894), *Studies and Exercises in Formal Logic*, 3rd ed., London: Macmillan.

[21] Lambert, J. H. (1764), *Neues Organon*, Leibzig: Johann Wendler.

[22] Lange, J. C. (1712), *Nvclevs Logicae Weisianae*, Müller: Gissae-Hassorum.

[23] Lemanski, J. (2017), "Periods in the use of Euler-type diagrams", *Acta Baltica Historiae et Philosophiae Scientiarum*, 5 (1): 50-69.

[24] Lemanski, J. (2021), *World and Logic*, London: College Publications.

[25] Lemanski, J. (2020), "Euler-type diagrams and the quantification of the predicate", *Journal of Philosophical Logic*, 49: 401–416.

[26] Lemon, O., & Pratt, I. (1998), "On the insufficiency of linear diagrams for syllogisms", *Notre Dame Journal of Formal Logic* 39 (4): 573-580.

[27] Macfarlane, A. (1885), "The logical spectrum", *The Philosophical Magazine*, 19: 286-290.

[28] Moktefi, A. (2019a), "The social shaping of modern logic", *in* D. Gabbay et al. (eds.), *Natural Arguments: A Tribute to John Woods*, London: College Publications, 503-520.

[29] Moktefi, A. (2019b), "Diagrammatic reasoning: The end of scepticism?", *in* A. Benedek & K. Nyiri (eds.), *Vision Fulfilled*, Budapest: Hungarian Academy of Sciences, 177-186.

[30] Moktefi, A., Bhattacharjee, R., & Lemanski, J. (2024), "Representing uncertainty with expanded Ueberweg diagrams", *in* J. Lemanski et al. (eds.), *Diagrammatic Representation and Inference*, Cham: Springer, 2024 (submitted).

[31] Moktefi, A., & Lemanski, J. (2022), "On the origin of Venn diagrams", *Axiomathes* 32: 887-900.

[32] Moktefi, A., & Pietarinen, A.-V. (2016), "Negative terms in Euler diagrams: Peirce's solution", *in* M. Jamnik et al. (eds.), *Diagrammatic Representation and Inference*, Berlin: Springer, 286-288.

[33] Moktefi, A., & Schang, F. (2023), "Another side of categorical propositions: the Keynes-Johnson octagon of oppositions", *History and Philosophy of Logic*, 44 (4): 459-475.

[34] Moktefi, A., & Shin, S.-J. (2012), "A history of logic diagrams", *in* D. M. Gabbay et al. (eds.), *Logic: A History of its Central Concepts*, Amsterdam:

North-Holland, 611-682.

[35] Sautter, F. T., & Mendonça, B. R. (2021), "Validity as choiceless unification", *in* A. Basu *et al.* (eds.), *Diagrammatic Representation and Inference*, Cham: Springer, 204–211.

[36] Shin, S.-J. (1994), *The Logical Status of Diagrams*, New York: Cambridge University Press.

[37] Ueberweg, F. (1871), *System of Logic and History of Logical Doctrines*, London: Longmans, Green and Co.

[38] Venn, J. (1880a), "On the employment of geometrical diagrams for the sensible representation of logical propositions", *Proceedings of the Cambridge Philosophical Society*, 4: 47-59.

[39] Venn, J. (1880b), "On the diagrammatic and mechanical representation of propositions and reasonings", *The Philosophical Magazine*, 10: 1-18.

[40] Venn, J. (1894), *Symbolic Logic*, 2$^{nd}$ ed., London: Macmillan.

[41] Welton, J. (1922), *A Manual of Logic*, vol. 1, 2$^{nd}$ ed.. London, W. B. Clive.

[42] Wilson, R., & Moktefi, A. (2019) (eds.), *The Mathematical World of Charles L. Dodgson (Lewis Carroll)*, Oxford: Oxford University Press.

# Names and Definite Descriptions in Natural Logic

Lawrence S. Moss
*Indiana University*

## 1 Introduction

This paper thrusts in two directions. This volume is entitled *New Directions in Term Logic*. I have been involved with an effort that I think of as close to the work on term-functor logic (TFL). (I also use this abbreviation TFL for *traditional formal logic*.) I think of TFL in connection with the influential and inspiring work that Fred Sommers and George Englebretsen did over many years. To keep things straight, I will refer to the area that I have thought of myself as participating in as *natural logic*, here written NL. The first thrust of this paper will be to offer a comparison and contrast of the two efforts. The second will be to illustrate natural logic by obtaining a few new results on names and definite descriptions, a topic which TFL has already treated, and so this will be an opportunity to give an exposition of NL for those involved with the neighboring and overlapping research program of TFL.

Of course it is very hard to characterize either TFL or NL while "standing on one foot". For TFL, I personally have the added difficulty that I am not an expert on the subject. Looking back on textbooks like Sommers [12] and Sommers and Englebretsen [13], collections like Englebretsen [4], and secondary papers in this area, I can offer some opinions. Let me start with some of the many things I like about TFL.

1. To start, the enterprise has a significant educational component. I sense that in some ways the theory was driven by the need to produce good educational materials. I especially like that the example inferences in the books and papers are "real" in ways that

examples of inferences in standard contemporary logic classes are not. I will have more to say about how this component plays out in NL below.

2. The revival of term logic and syllogistic logic and the development of (what I would call) variable-free logical systems are important developments. Since natural language sentences are rendered in the formal system without variables and with few extra symbols, this make it attractive to work in those systems. In the other direction, there has been a lot of work in the TFL area on logical reasoning with relations. This has led to sophisticated machinery which enables one to pay attention to quantifier-scope ambiguities. This aspect of the work stands out because it goes beyond what others have done.

3. The work on TFL has a lot of connection to ontology and epistemology. This is probably the case because much of it was done by people with broader backgrounds as philosophers. Even more, the entire enterprise sees itself as a chapter in the history of logic. For the most part, these connections are missing in the NL area. One exception would be the dissertation Sánchez Valencia [14].

Let me turn now to NL. I mentioned before that NL and TFL are neighboring and overlapping. I would imagine that from the point of view of someone steeped in mathematical logic, they would seem to be very close. To motivate things, here is my personal statement of the goals of natural logic, with some comments on how this relates to TFL:

1. To show that significant parts of natural language inference can be carried out in *algorithmically manageable* logical systems. For me, this suggests that one should not work by translating natural language into first-order logic. (For that matter, I am also wary of "regimenting" into TFL.) I prefer to develop logic in fragments, and then one should try to obtain completeness theorems for those fragments. This is what the rest of this paper will do.

2. Whenever possible, to obtain *complete axiomatizations*, because the resulting logical systems are likely to be interesting. This entails being serious about the semantics of fragments, since sound-

ness and completeness are about the relationship between a syntactically-defined proof system and an independently-defined semantics. Most of this paper illustrates this point as it pertains to names and definite descriptions in several very small logical systems. This point about complete axiomatizations also affects the way I see the educational aspect of NL. I have not tried to produce materials that would teach people how to do basic logical reasoning the way [13] does. But I have been interested in teaching completeness theorems for syllogistic logics *as a way to teach the basics of abstract mathematics*. This is a contrast to teaching logic in order to learn mathematical reasoning in the first place.

3. The topic should be connected to a lot of other areas. One area is natural language semantics itself, studied as a branch of linguistics. In some ways TFL seems to have been developed in isolation from subjects like Montague grammar, generalized quantifier theory, discourse representation theory, and more. This may have been sociological. Be that as it may, I see NL as a meeting point of a large collection of subjects that pursue *inference* from different angles. For a different example, I would hope that natural logic would be be completely mathematical and involve many tools from contemporary mathematical logic and theoretical computer science. For example, one should use computational complexity to separate logical systems. One should build logical systems that run on the computer; there is a lot to learn by doing so. This volume appears at a time when computers are able to do logical reasoning on text at human level (or better), and so this is a watershed moment in the history of logic. The work that we are doing in revitalizing logic should be part of this moment. One should start by asking what current developments in artificial intelligence might teach us about logical reasoning.

4. Finally, a look to what I hope the future would bring. There is a rich literature of psychological studies of inference, also focused on the syllogism. This is a neglected topic in logic, partly for historical reasons and partly because psychologists are often interested in deviations from ideal logical behavior. I would like to

think that eventually the two topics will be reconciled. One would also envision a reworking of natural language semantics based on computational linguistics and on natural logic. But all of this is programmatic at this point.

It should be clear that TFL and NL share a lot of concerns. Throughout the writing of this paper, I have been worried that readers will take it as an un-friendly criticism of TFL rather than the friendly contrast that I intend.

I end this programmatic introduction with a look back at just one small point in Sommers [12], the treatment of "Most" in Appendix D. This short appendix is partly a comment on the logical treatment of "Most", and partly a reply to points on "Most" made by Geach. Sommers holds that the example (Y) below is *not* a syllogism

$$\frac{\text{every M is P}}{\text{most S are M}} \qquad (Y)$$

His reason seems to be that the *Dictum de Omni* (DDO) when "properly construed" does not apply to the minor premise. Even more, he presents an example of (Y) where there are exactly three S's, named *I, *J, and *K. In this situation, the second premise above can be unpacked to read

*I is M and *J is M or *I is M and *K is M or *J is M and *K is M.

(I have simplified the actual text somewhat here.) He then applies DDO to the disjuncts and the first premise, and finally puts things together to reach the conclusion. The point is to argue that (Y) is "no syllogism but a more complicated valid argument in which syllogism plays a part." Finally, he holds that

> To a logician, the idea that 'most' is a quantifier on all fours with 'some' or 'every' ought to be distasteful . . .

My reason for quoting all of this is to make a methodological contrast with the kind of work I want to do in the main body of the paper. To begin, let me note that many (perhaps even "most") of the logicians and linguists who think about quantifiers today *do* regard 'most' is a quantifier on all fours with 'some' or 'every': see Peters and Westerståhl [9] for a reference book on quantifiers. For me, it is especially important to listen to what semanticists say rather than to declare what must be the case based on DDO or anything else logicians tell them; if DDO does not apply to (Y), then maybe we need to widen the scope of DDO, or else look for other valid principles of reasoning. Widening the scope leads to treatments of *monotonicity reasoning* in natural language. Although this is an important topic in NL, this paper will not deal with it. But we do want to look for other valid principles. We shall be concerned with names and definite descriptions, and we will let the inferential phenomena speak for themselves, so to speak, rather than trying to reduce them to principles like DDO alone.

It may also be interesting to note that there are valid reasoning principles concerning "Most" that feel different than (Y) above:

$$\frac{\text{Most M are P} \quad \text{Most M are S}}{\text{Some P are S}} \qquad \frac{\text{All M are P} \quad \text{All P are Q} \quad \text{Most Q are M}}{\text{Most P are M}}$$

On the left, we have another case where, with particular examples, one could reduce particular cases to other kinds of reasoning. This feels unsatisfactory as a general explanation of what is going on: one might apply the pattern above in situations where one doesn't know how many M there are, and where they might not all have names, etc. And as with (Y), the reduction to individual instances is simply not statable in the usual system of logic, first-order logic. For this, see Peters and Westerståhl [9], Part IV. A sadly neglected paper, Purdy [11], also has a proof. Purdy's paper shares many concerns with this one and with natural logic more generally, and its work involving term functor logic to handle relations is something it shares with work in the TFL field.

On the right above, we have a three-premise argument with yet a different feel. It is valid, and to us it is not distasteful but rather interesting. It raises the questions: what is the *complete set of rules* for this kind of reasoning? (See [3] for a partial answer.) Is finiteness of the domain is essential or avoidable? What is the complexity of this reasoning? If we add "Most" to larger fragments, what happens?

**Contributions of this paper** This paper now turns from slogans and research programs to the main "meal": results with motivational points sprinkled in as a kind of "spice."

Perhaps the most famous example of a logical inference is the classic syllogism[1]

> All men are mortal.
> Socrates is a man.
> Socrates is mortal.

However, the simplest forms of logic typically do not include the resources needed to formalize this. We take those forms of logic to be one or another flavor of *syllogistic logic*. We can take this to be the logical system whose only sentences are All p are q, Some p are q, and No p are q. Since this system has no names, it cannot formulate the famous example above. This paper rectifies matters by taking a simple syllogistic logic – not quite the one we just described, but a closely-related system – and adds names to it. After this first addition, we make a second one for *definite descriptions*. This last addition is technically much more challenging, and so it is the substantive goal of the paper. As suggested by our digression into "Most", what we want to do is to obtain completeness theorems for small fragments.

## 2   The Logic $\mathcal{S}^\dagger$

This section reviews a logical system called $\mathcal{S}^\dagger$ in [10].[2] In the syntax, we begin with a set $\mathbb{P}$ of *predicates*; we shall call these *basic nouns*. The set of *literals* is two copies of $\mathbb{P}$ distinguished by using a "bar" symbol:

$$\mathsf{Lit} \quad = \quad \mathbb{P} \cup \{\overline{\mathsf{p}} : \mathsf{p} \in \mathbb{P}\}.$$

In other words, the elements of Lit are either *basic nouns* p, q, etc., or *complemented nouns* $\overline{\mathsf{p}}$, $\overline{\mathsf{q}}$, .... We read $\overline{\mathsf{p}}$ as "non-p". The idea of complementation will be familiar to readers of texts like [12] and [13]. Indeed, those books have a much more thoroughgoing use of the complementarity of assertion and denial.

---
[1] For the origin of this, see [18].

[2] Incidentally, the dagger † refers to the complementation operation. So $\mathcal{S}$ would be the same system without that operation. It will not appear in this paper.

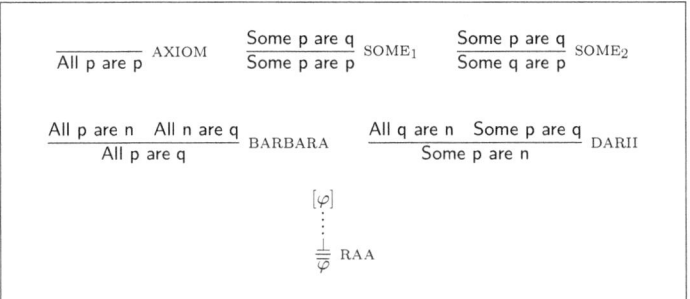

Figure 1: The logic $\mathcal{S}^\dagger$. The letters p and q denote literals (nouns or complemented nouns).

We want to extend the "bar" operation from nouns to all literals, and we do this by stipulating that $\bar{\bar{p}} = p$. That is, we take p and $\bar{\bar{p}}$ to be the same (identical) literal. (This would be like taking "smoker" and "non-nonsmoker" to denote the same concept.)

The logical language under study is called $\mathcal{S}^\dagger$, and it has as sentences

> All p are q
> Some p are q

Here p and q are any literals. They could be basic nouns, and they could be complemented. We use letters like $\varphi$ to denote sentences. Please note that the sentences above are the only sentences in the language; there are no propositional connectives, and the quantifiers All and Some do not occur in other contexts. Indeed, from the point of view of either natural language syntax or formal language theory, this $\mathcal{S}^\dagger$ we really should use scare quotes when calling $\mathcal{S}^\dagger$ a language in the first place. If $\mathbb{P}$ is finite, then the set of sentences here is finite as well.

We turn to the intended semantics for $\mathcal{S}^\dagger$. We start with an arbitrary set $M$, and with interpretations $[\![p]\!]$ of all basic nouns. We require that $[\![p]\!] \subseteq M$ for all p. Then for each complemented noun $\bar{p}$, we define $[\![\bar{p}]\!] = M \setminus [\![p]\!]$. [3] This gives a *model*. In [13], they are called STATES

---

[3] So the semantics here is "classical": we insist that $[\![p]\!]$ and $[\![\bar{p}]\!]$ be complement

OF AFFAIRS, and what I write as $[\![p]\!]$ would be written more simply there as P. We use letters like $\mathcal{M}$ and $\mathcal{N}$ for models. Given a model and a sentence, we define

$$\mathcal{M} \models \text{All p are q} \qquad \text{iff} \qquad [\![p]\!] \subseteq [\![q]\!].$$
$$\mathcal{M} \models \text{Some p are q} \qquad \text{iff} \qquad [\![p]\!] \cap [\![q]\!] \neq \emptyset$$

Note that $\mathcal{M} \models$ All p are $\overline{q}$ iff $[\![p]\!] \cap [\![q]\!] = \emptyset$. We usually read All p are $\overline{q}$ as No p are q. We also usually read No p are p as No p exist, Some p are p as Some p exist, and All $\overline{p}$ are p as Everything is a p.

The $\mathcal{S}$ in $\mathcal{S}^\dagger$ reminds us of "syllogistic", and as we have just observed the classical syllogistic translates into $\mathcal{S}^\dagger$. But we go a little further here: with basic nouns p and q we have sentences of the form Some $\overline{p}$ are q and Some $\overline{p}$ are $\overline{q}$. It is unusual to find negation on the head noun of a sentence like this, and the reader might wonder why we permit it in the syntax. The answer is that it simplifies the syntax to work with a full negation, and it simplifies the analysis of the system considerably.

Given a set $\Gamma \cup \{\varphi\}$, we write $\mathcal{M} \models \Gamma$ to mean that $\mathcal{M}$ satisfies each sentence in $\Gamma$. We write $\Gamma \models \varphi$ to mean that for all models $\mathcal{M}$, $\mathcal{M} \models \Gamma$ implies $\mathcal{M} \models \varphi$. In words, every model of each sentence in $\Gamma$ is also a model of $\varphi$.

We are interested in a proof system for this language. We review the one from [7]. It is given by the rules in Figure 1. Example 2.2 contains examples of formal proofs using those rules. But we need to discuss the the last rule of the system, *reductio ad absurdum* (RAA), before those make sense. Setting this aside for a moment, the rules generate *trees* $\mathcal{T}$. We say that $\Gamma \vdash \varphi$ (without (RAA)) if there is a tree $\mathcal{T}$ whose leaves are labeled with sentences from $\Gamma$, whose root is labeled $\varphi$, and with the property that every non-leaf comes from its parent(s) using one of the rules of the system. We also speak of *proof trees* and *derivations*; these are all the same. A proof system for a semantically-defined logic should be sound (if $\Gamma \vdash \varphi$, then $\Gamma \models \varphi$) and complete (if $\Gamma \models \varphi$, then $\Gamma \vdash \varphi$). We are not going to discuss the soundness here, but we will see the completeness in Theorem 2.12.

---

sets. We could reconsider this, but this paper will not do this. On the other hand, we do move in the direction of partiality in semantics in Section 5.

**Definition 2.1.** Every sentence $\varphi$ in $\mathcal{S}^\dagger$ has a *semantic negation* $\overline{\varphi}$. It is defined as follows:

$$\begin{array}{c|c} \varphi & \overline{\varphi} \\ \hline \text{All } x \text{ are } y & \text{Some } x \text{ are } \overline{y} \\ \text{Some } x \text{ are } y & \text{All } x \text{ are } \overline{y} \end{array} \qquad (1)$$

We call $\overline{\varphi}$ the *semantic negation* of $\varphi$. Please note that the "bar" on sentences is not a symbol in our language, it is an abbreviation "from outside."

The reason that we call this a semantic negation is that for all models $\mathcal{M}$,
$$\mathcal{M} \not\models \varphi \quad \text{iff} \quad \mathcal{M} \models \overline{\varphi}$$
Notice also that $\overline{\overline{\varphi}} = \varphi$ for all $\varphi$.

**Definitions concerning (RAA)** We use $\bot$ ("bottom") as a symbol for a pair of some sentence $\psi$ and its contradictory $\overline{\psi}$ as defined from $\psi$ in (1). The rule (RAA) tells us that if we can prove a contradictory pair, then we may take any sentence $\varphi$, withdraw some or all of the occurrences of $\varphi$ in the leaves of our derivation by putting brackets around them, and then using the rule (RAA) to infer $\overline{\varphi}$ at the root. We obtain a new tree $\mathcal{T}^+$. We allow the case when $\varphi$ does not actually occur in the leaves of the tree $\mathcal{T}$. So in this case, $\mathcal{T}$ and $\mathcal{T}^+$ would have the same set of non-withdrawn leaves. Once again, in the statement of (RAA), we use $\bot$ as an abbreviation for "a sentence and its contradictory."

We write $\Gamma \vdash \varphi$ if there is a proof tree $\mathcal{T}$ whose root is $\varphi$ and all of whose *non-withdrawn* leaves belong to $\Gamma$.

**Example 2.2.** We exhibit two proof trees below. On the left, we have one showing that All p are $\overline{\text{p}}$ $\vdash$ All p are q. That is, $\Gamma$ here is {All p are $\overline{\text{p}}$}, and as is customary with finite sets in this kind of discussion, we drop the set braces. The tree on the right shows that All p are q $\vdash$ All $\overline{\text{q}}$ are $\overline{\text{p}}$.

$$\dfrac{\text{All p are }\overline{\text{p}} \quad \dfrac{[\text{Some p are }\overline{\text{q}}]}{\text{Some p are p}}\,\text{SOME}}{\text{All p are q}}\,\text{RAA} \qquad \dfrac{\dfrac{\text{All p are q} \quad [\text{Some }\overline{\text{q}}\text{ are p}]}{\text{Some }\overline{\text{q}}\text{ are q}}\,\bot}{\text{All }\overline{\text{q}}\text{ are }\overline{\text{p}}}\,\text{RAA}$$

**Example 2.3.** All p are q, All p are $\bar{q}$ ⊢ All p are $\bar{p}$. Here is the reasoning, in brief. From All p are $\bar{q}$, we get All q are $\bar{p}$ using Example 2.2. Then from this and All p are q, we use (BARBARA) to get All p are $\bar{p}$.

We have the following form of *proof by cases*:

**Proposition 2.4.** *[10] If* $\Gamma \cup \{\varphi\} \vdash \psi$, *and also* $\Gamma \cup \{\overline{\varphi}\} \vdash \psi$, *then* $\Gamma \vdash \psi$.

*Proof.* The first hypothesis yields $\Gamma \cup \{\varphi, \overline{\psi}\} \vdash \bot$. Thus we may use (RAA) to withdraw $\varphi$, and we see that $\Gamma \cup \{\overline{\psi}\} \vdash \overline{\varphi}$. This together with the second hypothesis implies that $\Gamma \cup \{\overline{\psi}\} \vdash \varphi$. Obviously we have $\Gamma \cup \{\overline{\psi}\} \vdash \overline{\varphi}$. In other words, $\Gamma \cup \{\overline{\psi}\} \vdash \bot$. So by (RAA) again, $\Gamma \vdash \psi$. □

## 2.1 Orthoposets

One idea behind *algebraic logic* is that some aspects of logic are clarified when one formulates a logical system as family of algebraic objects; one then studies general properties of the algebraic objects and takes information back to the logical system. The best-known example of this is propositional logic, where one introduces boolean algebras as a tool which then gets used in proving the completeness theorem. We do the same thing here, finding *orthoposets* to be the algebraic objects close to the syntax and semantics of $\mathcal{S}^\dagger$.

**Definition 2.5.** An *orthoposet* is a tuple $\mathbb{P} = (P, \leq, 0, ^-)$ such that the following hold:

1. $(P, \leq)$ is a poset: the relation $\leq$ is reflexive, transitive, and antisymmetric.

2. 0 is a *minimum* element: $0 \leq p$ for all $p \in P$.

3. The function $x \mapsto \overline{x}$ is *antitone*: if $x \leq y$, then $\overline{y} \leq \overline{x}$.

4. $x \mapsto \overline{x}$ is *involutive*: $\overline{\overline{x}} = x$.

5. The order $\leq$ and the complement operation $x \mapsto \overline{x}$ are connected by the property of *complement inconsistency*: If $x \leq y$ and $x \leq \overline{y}$, then $x = 0$.

**Example 2.6.** Here are some examples of orthoposets.

1. For all sets $X$ we have a *power set orthoposet* $(\mathcal{P}(X), \subseteq, \emptyset, {}^-)$, where $\mathcal{P}(X)$, the *power set of $X$* is the set of all subsets of $X$; $\subseteq$ is the inclusion relation; $\emptyset$ is the empty set; and $\bar{a} = X \setminus a$ for all subsets $a$ of p.

2. The example below is sometimes called a *Chinese lantern*.

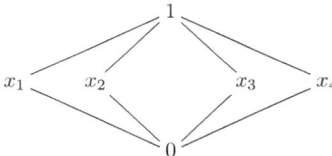

What we mean here is that the poset is the set of six points above, with the evident order. So $0 < x_i < 1$, and for $i \neq j$, $x_i \not\leq x_j$. The 0 is at the bottom. We define the operation $^-$ by: $\bar{0} = 1$, $\bar{1} = 0$, $\bar{x}_1 = x_2$, $\bar{x}_2 = x_1$, $\bar{x}_3 = x_4$, and $\bar{x}_4 = x_3$. Obviously this can be generalized to have any number of complementary pairs in the middle.

3. For every set $\Gamma$ of All-sentences in our current language, we get an orthoposet $\mathbb{P}_\Gamma$ as follows: We first write p $\leq$ q to mean that $\Gamma \vdash$ All p are q. This is a preorder but not a partial order: antisymmetry fails. To rectify this, consider the equivalence relation p $\equiv$ q iff p $\leq$ q and q $\leq$ p. The quotient set Lit/$\equiv$ is a partial order, and this is the universe of our orthoposet. If any p has the property that $\Gamma \vdash$ All p are $\bar{\mathsf{p}}$, then we set $0 = [\mathsf{p}]$. This would imply that $[\mathsf{p}] \leq [\mathsf{q}]$ for all q. Otherwise, we add a fresh 0 to the structure, and also $1 = \bar{0}$. The complement structure has $\overline{[\mathsf{p}]} = [\bar{\mathsf{p}}]$, and of course $\bar{0} = 1$ and $\bar{1} = 0$. The verification of the orthoposet properties comes from the logic. Specifically, the reflexive property of $\leq$ is from (AXIOM), the transitivity is from (BARBARA), and the antisymmetry is from the fact that we are using equivalence classes as above. The fact that 0 is a minimum comes from Example 2.2. The antitonicity of $[\mathsf{p}] \mapsto [\bar{\mathsf{p}}]$ comes from Example 2.2. The involutive property comes from our notational identification

of $\overline{\overline{\mathsf{p}}}$ with p. The complement inconsistency property comes from Example 2.3.

**Definition 2.7.** Let $\mathbb{P}$ be an orthoposet.

1. A set $A \subseteq P$ is *extendible* if for all $p, q \in A$, $p \not\leq \overline{q}$.

2. A set $S \subseteq P$ is a *state* if it is (a) extendible, (b) closed upwards, and (c) contains either $p$ or $\overline{p}$ for each $p \in P$.

**Proposition 2.8.** *A is a state iff it is closed upwards, and for all p, A contains either $p$ or $\overline{p}$ but not both.*

*Proof.* Let $S$ satisfy in (a), (b), and (c) Definition 2.7(2). We show that for all $p$, $S$ does not contains but both $p$ and $\overline{p}$. For if $p, \overline{p} \in S$, then since $p \leq p = \overline{\overline{p}}$, we contradict extendibility.

Conversely, let $A$ be closed upwards and contain exactly one of each pair consisting of an element and its complement. We show that $A$ is extendible. Suppose not. Then we have $p, q \in A$ such that $p \leq \overline{q}$. Since $A$ is closed upwards and contains $p$, it also contains $\overline{q}$. But then $A$ contains both $q$ and $\overline{q}$, contrary to our assumption. $\square$

**Example 2.9.** We revisit Example 2.6 and mention some states.

1. For any set $X$, the power set orthoposet $\mathcal{P}(X)$ contains all of the ultrafilters. But usually there are states which are not ultrafilters. For example, when $X = \{1, 2, 3\}$, the set

$$\{\{1, 2\}, \{1, 3\}, \{2, 3\}, \{1, 2, 3\}\}$$

    is a state of $\mathcal{P}(X)$ but not an ultrafilter (it is not closed under intersection).

2. The Chinese lantern shown in Example 2.6 (2) has exactly four states:

$$\{x_1, x_3, 1\}, \{x_1, x_4, 1\}, \{x_2, x_3, 1\}, \{x_2, x_4, 1\}.$$

3. Let $\mathcal{M} \models \Gamma$. Each point $x \in M$ gives a state $S_x$ of $\mathbb{P}_\Gamma$, where $S_x = \{[\mathsf{p}] : x \in [\![\mathsf{p}]\!]\} \cup \{1\}$. The proof of Theorem 2.12 below implies that every state of $\mathbb{P}_\Gamma$ comes from some point in some fixed *canonical model* $\mathcal{M}_{\mathsf{can}}$.

The verifications here are made easier by Proposition 2.8. The point of these examples will be suggestive to readers familiar with algebraic logic; we are adapting the technique of *completeness via representation* to the logical system at hand.

The following result appears in [2, 5, 19] and in this form in [7].

**Lemma 2.10.** *If $A$ is extendible and $p \in P$ then either $A \cup \{p\}$ or $A \cup \{\overline{p}\}$ is extendible. Every extendible set $A$ is a subset of some state.*

The first assertion here is a direct verification. In the second, we take an extendible state and, well, extend it to a state. If the orthoposet is finite or countably infinite to begin with, then this could be done by a step-by-step process. In the general case, one would need a set-theoretic result like Zorn's Lemma.

**Proposition 2.11.** *Consider a logical system with a semantic negation operation and with a sound proof system that includes (RAA). Assume that there is a fixed contradictory pair $\varphi$, $\overline{\varphi}$ such that $\{\varphi, \overline{\varphi}\}$ is not satisfiable. Also assume that a set $\Gamma$ is consistent in the proof system iff it cannot prove any contradictory pair. Then the following are equivalent:*

1. *The proof system is complete.*

2. *Every consistent set has a model.*

*Proof.* (1)⇒(2): Let $\Gamma$ be consistent. Assume towards a contradiction that $\Gamma$ has no models. Then every model of $\Gamma$ satisfies both sentences in our contradictory pair $\varphi, \overline{\varphi}$. That is, $\Gamma \models \varphi, \overline{\varphi}$. By completeness, $\Gamma \vdash \varphi, \overline{\varphi}$. This means that $\Gamma$ is inconsistent, so we have a contradiction.

(2)⇒(1): Let $\Gamma \cup \{\psi\}$ be a set of sentences, and assume that $\Gamma \models \psi$. We show that $\Gamma \vdash \psi$. If not, we argue that $\Gamma \cup \{\overline{\psi}\}$ is inconsistent. (For if $\Gamma \cup \{\overline{\psi}\}$ were consistent, then by (2), $\Gamma \cup \{\overline{\psi}\}$ has a model, say $\mathcal{M}$. This model satisfies $\Gamma$ and $\overline{\psi}$. But then $\mathcal{M} \not\models \psi$. So we contradict $\Gamma \models \psi$.) Then an application of (RAA) shows that $\Gamma \vdash \overline{\overline{\psi}} = \psi$. □

**Theorem 2.12** ([7]). *The logical system for $\mathcal{S}^\dagger$ is complete.*

*Proof.* (sketch) We use Proposition 2.11 in the direction (2)⇒(1), taking as the fixed contradictory pair to be absolutely any such pair, say All p

are q and Some p are $\overline{q}$. Let $\Gamma$ be consistent. Recall the orthoposet $\mathbb{P}_\Gamma$ from Example 2.6, part (3). Define the *canonical model* $\mathcal{M}_{\mathsf{can}}$ as follows: its universe is the set of states of $\mathbb{P}_\Gamma$, and for every noun x, we take $[\![\mathsf{x}]\!] = \{S : [\mathsf{x}] \in S\}$. This completes the definition of $\mathcal{M}_{\mathsf{can}}$. Since states contain each basic noun x or its complement $\overline{\mathsf{x}}$, we see that for every literal x,
$$[\![\overline{\mathsf{x}}]\!] = \overline{\{S : [\mathsf{x}] \in S\}} = \{S : [\overline{\mathsf{x}}] \in S\}.$$

Let us check that $\mathcal{M}_{\mathsf{can}} \models \Gamma$. For a sentence All u are v in $\Gamma$, let $S \in [\![\mathsf{u}]\!]$. Then $[\mathsf{u}] \in S$. Since $\Gamma$ contains All u are v, $[\mathsf{u}] \leq [\mathsf{v}]$ in $\mathbb{P}_\Gamma$. Since $S$ is closed upward, $[\mathsf{v}] \in S$. So $S \in [\![\mathsf{v}]\!]$, as desired.

For sentences Some u are v in $\Gamma$, we need the following crucial fact: the set $\{\mathsf{u}, \mathsf{v}\}$ must be an extendible subset of $\mathbb{P}_\Gamma$. If not, $\Gamma$ would prove one of the following three sentences: All u are $\overline{\mathsf{u}}$, or All v are $\overline{\mathsf{v}}$, or All u are $\overline{\mathsf{v}}$. In all of these cases, we would get an inconsistency from $\Gamma$. (We omit this verification, but we shall see many similar arguments in the rest of this paper.) Thus $\{\mathsf{u}, \mathsf{v}\}$ is extendible, and so it is a subset of some state, say $S$. This state $S$ belongs to $\mathcal{M}_{\mathsf{can}}$, and indeed $S \in [\![\mathsf{u}]\!] \cap [\![\mathsf{v}]\!]$. Therefore $\mathcal{M}_{\mathsf{can}} \models \Gamma$. □

**Taking stock.** This paper is about adding names and definite descriptions to some rather simple logics. We have not even started on new work, we only mentioned the base system and sketched the completeness. Going forward, one needs the definitions of extendible sets and states in orthoposets (Definition 2.7) Example 2.9(3), and finally the definition of the canonical model in the proof of Theorem 2.12. One should keep in mind that passing from an extendible set to a state which includes it is a form of *model construction*, much like taking a branch in an infinite tree, or extending a filter to an ultrafilter.

I think the main value in the logic of this section might be pedagogical. One can teach students about completeness theorems by examining very weak logical systems like $\mathcal{S}^\dagger$. One would not be teaching anyone how to reason. In fact, one needs some mathematical ability just to read the notation here, to draw the pictures of orthoposets, and to work with the definitions in this section. The next challenge is to take particular finite sets $\Gamma$ and to construct the canonical model $\mathcal{M}_{\mathsf{can}}$ in full detail. In addition, there are many possible good exercises based on the work in this section.

## 3 Adding Names

Recall that this paper has two overall points: one is to lightly comment on natural logic and term-functor logic, and the other is to illustrate natural logic by looking at some new fragments of it and to compare them to [12] and [13]. Those fragments are the additions to $\mathcal{S}^\dagger$ of names and (a very simple form of) definite descriptions. This section adds names to $\mathcal{S}^\dagger$; definite descriptions come later. We start with a set **Names** of *names*, and we use letters like a and b for names. We add to the syntax of $\mathcal{S}^\dagger$ three kinds of sentences

$$\begin{array}{l} \text{a is b} \\ \text{a isn't b} \\ \text{a is a p} \end{array} \quad (2)$$

In this last kind of sentence p may again be a literal (a noun or a complemented noun). We don't need a sentence a isn't a p, since we have a is a $\overline{\text{p}}$. We call this language $\mathcal{S}^\dagger_{\text{names}}$.

**Remark 3.1.** Taking as basic a sentence like a is a p means that we are departing from the TFL treatment of names via wild quantity; see [12] and [13]. In fact, our syntax could be p(a) rather than a is a p; it makes no difference which we write. This would give a syntax close to that of first-order logic (FOL). What we are doing by writing "a is a p" is using the surface form more directly than either TFL or FOL.

**Semantics** In a model $\mathcal{M}$, we interpret names by elements of the universe $M$. (These are sometimes called *points*.) For each name a, $[\![\text{a}]\!] \in M$. Here is the semantics:

$$\begin{array}{lll} \mathcal{M} \models \text{a is b} & \text{iff} & [\![\text{a}]\!] = [\![\text{b}]\!] \\ \mathcal{M} \models \text{a isn't b} & \text{iff} & [\![\text{a}]\!] \neq [\![\text{b}]\!] \\ \mathcal{M} \models \text{a is a p} & \text{iff} & [\![\text{a}]\!] \in [\![\text{p}]\!] \end{array}$$

The last point above is for all basic nouns p. For its complement $\overline{\text{p}}$, it follows that

$$\mathcal{M} \models \text{a is a } \overline{\text{p}} \quad \text{iff} \quad [\![\text{a}]\!] \in [\![\overline{\text{p}}]\!] = M \setminus [\![\text{p}]\!].$$

That is, $\mathcal{M} \models$ a is a $\overline{\text{p}}$ iff $\mathcal{M} \not\models$ a is a p.

$$\frac{}{\text{a is a}}\text{ REF} \qquad \frac{\text{a is b}}{\text{b is a}}\text{ SYM} \qquad \frac{\text{a is b} \quad \text{b is c}}{\text{a is c}}\text{ TRANS}$$

$$\frac{\text{a is a p} \quad \text{a is b}}{\text{b is a p}}\text{ R}_1 \qquad \frac{\text{a is a p} \quad \text{All p are q}}{\text{a is a q}}\text{ R}_2$$

Figure 2: The rules of $\mathsf{S}^\dagger_{\text{names}}$ on top of the rules of $\mathsf{S}^\dagger$. The letters a and b are names, and as before p and q are literals. In addition, we allow (RAA), using as a contradiction sentences a is a p and a is a $\overline{\mathsf{p}}$.

**Remark 3.2.** The denotation $[\![\mathsf{a}]\!]$ of a name a in a model $\mathcal{M}$ is arbitrary: it may be any element of $M$. But this does not mean that every element of $M$ is the denotation of some name or other.

**Definition 3.3.** We extend the notion of a semantic negation (see Definition 2.1) to the new language, as follows:

| $\varphi$ | $\overline{\varphi}$ |
|---|---|
| a is a p | a is a $\overline{\mathsf{p}}$ |
| a is b | a isn't b |
| a isn't b | a is b |

In this, p may be any literal, so we do not need two versions of the first line. That is, the first line covers the case of a complemented noun in the left column.

Recall that the logic $\mathsf{S}^\dagger$ allows us to trigger (RAA) from pairs Some p are q and All p are $\overline{\mathsf{q}}$, where p and q are any literals. Moving to $\mathsf{S}^\dagger_{\text{names}}$, we also allow (RAA) to be triggered from any pair $\varphi$ and $\overline{\varphi}$ as above.

**Example 3.4.** On the left below, we show that
a is a p, a is a q ⊢ Some p are q.

$$\frac{\text{a is a q} \quad \dfrac{\text{a is a p} \quad [\text{All p are }\overline{\mathsf{q}}]}{\text{a is a }\overline{\mathsf{q}}}\text{R}_2}{\text{Some p are q}}\text{RAA} \qquad \frac{\dfrac{\text{a is a p} \quad [\text{a is a }\overline{\mathsf{q}}]}{\text{Some p are }\overline{\mathsf{q}}} \quad \text{All p are q}}{\text{a is a q}}\text{RAA}$$

By the way, we could have taken the example on the left as a rule, call it (R$_{2.5}$), and then derived (R$_2$) from (R$_{2.5}$). This is shown on the right, where the step in the upper-left is an application of (R$_{2.5}$).

**Example 3.5.** Here are some other easy examples:

1. a is a p, b is a p̄ ⊢ a isn't b.
2. All p̄ are p ⊢ a is a p.

We omit the derivations.

## 3.1 Model building and completeness

**Lemma 3.6.** *Let* $\Gamma$ *be consistent. For all names* a, $\{p \in \text{Lit} : \Gamma \vdash$ a is a p$\}$ *is extendible.*

*Proof.* If not, we would have p and q so that $\Gamma$ proves a is a p and a is a q, and also All p are q̄. Using the logic, we get a is a q̄. Thus $\Gamma$ is inconsistent, and this is a contradiction. □

In the next lemma we need some additional notation.

**Definition 3.7.** Fix a set $\Gamma$. Consider the relation $\equiv_\Gamma$ on the set Names given by a $\equiv_\Gamma$ b if $\Gamma \vdash$ a is b. This is an equivalence relation, by the first three rules in the logic. When $\Gamma$ is clear from the context, we drop it from this notation. We also write [a] for the *equivalence class* of a: [a] = $\{b : \Gamma \vdash$ a is b$\}$. Let $N$ be the quotient set Names/$\equiv$.

**Lemma 3.8.** *Let* $\Gamma \subseteq S_{\text{names}}^\dagger$ *consistent. Then there is a model* $\mathcal{N}$ *whose universe* $N$ *is the set in Definition 3.7, such that* $\mathcal{N}$ *satisfies all sentences in* $\Gamma$ *except possibly those of the form* Some x are y.

*Proof.* For each [a] $\in N$, consider

$$T_{[a]} = \{p \in \text{Lit} : \Gamma \vdash \text{a is a p}\}.$$

If [a] = [b] and $\Gamma \vdash$ a is a p, then from $\Gamma$ we have a is b and then b is a p, by ($R_1$). So whether p $\in T_{[a]}$ is independent of the choice of a representative from [a].

Since $\Gamma$ is consistent, the same argument which we saw in Lemma 3.6 tells us that $T_{[a]}$ is extendible as a subset of $\mathbb{P}_\Gamma$. Let $S_{[a]}$ be any state of $\mathbb{P}_\Gamma$ which includes $T_{[a]}$ (Lemma 2.10). We interpret the basic noun p in $\mathcal{N}$ by

$$[\![p]\!] = \{[a] : p \in S_{[a]}\}. \tag{3}$$

For a complemented noun $\bar{p}$, we have $[\![\bar{p}]\!] = \{[a] : \bar{p} \in S_{[a]}\}$. Exactly as before, this is because the state $S_{[a]}$ contains $p$ or else it contains $\bar{p}$.

Let us check that $\mathcal{N}$ satisfies the All-sentences in $\Gamma$. Take such a sentence, say All p are q. Let $[a] \in [\![p]\!]$, so that $p \in S_{[a]}$. Then since states are closed upwards, $q \in T_{[a]} \subseteq U_{[a]}$. So $[a] \in [\![q]\!]$.

Continuing, take a sentence a is b. When this sentence is in $\Gamma$, $[a] = [b]$, and so $\mathcal{N}$ satisfies it. And for a sentence a isn't b, we claim again that $\mathcal{N}$ satisfies it. For suppose not; then $[\![a]\!] = [\![b]\!]$. So $[a] = [b]$. By the definition of $\equiv_\Gamma$, $\Gamma \vdash$ a is b. So $\Gamma$ is inconsistent.

Finally, we check that whenever $\Gamma$ contains a sentence a is a p, then $\mathcal{M} \models$ a is a p. When the literal p is a basic noun, this follows from $p \in T_{[a]} \subseteq S_{[a]}$. When p is complemented, we use the fact noted below (3) that $[\![\bar{p}]\!] = \{[a] : \bar{p} \in S_{[a]}\}$. Thus in this case, $a \in [\![\bar{p}]\!]$, and so $a \notin [\![p]\!]$. □

**Theorem 3.9.** *The logic in Figure 2 is complete for* $\mathcal{S}^\dagger_{\mathsf{names}}$.

*Proof.* We again use Proposition 2.11. Fix a consistent set $\Gamma$.

Forgetting the names, we have a model $\mathcal{M}_{\mathsf{can}}$ defined in the proof of Theorem 2.12. This model satisfies all sentences in $\Gamma$ except those involving names. It cannot satisfy the sentences with names, since the names are not interpreted in this model to begin with.

Let $\mathcal{N}$ be the model from Lemma 3.8. So $\mathcal{N}$ satisfies all sentences in $\Gamma$ except possibly those starting with Some.

Consider $\mathcal{M}_{\mathsf{can}} + \mathcal{N}$. This is the disjoint union of the universes of these models, where we interpret nouns by the union of their interpretations in the two, and interpret the names as in $\mathcal{N}$. Since both models satisfy the All-sentences from $\Gamma$, it is easy to see that $\mathcal{M}_{\mathsf{can}} + \mathcal{N}$ also does. The Some-sentences are satisfied in $\mathcal{M}_{\mathsf{can}}$, and this is unaffected by taking a disjoint union. Similarly, the sentences with names are satisfied in $\mathcal{N}$, hence in $\mathcal{M}_{\mathsf{can}} + \mathcal{N}$. The upshot is that $\mathcal{M}_{\mathsf{can}} + \mathcal{N} \models \Gamma$. □

## 4  The one and only

We make things more interesting by making further additions coming from *definite descriptions*. So if a is a name and p is a literal, then we add sentences a is the p, and a isn't the p. We call the resulting language $S^\dagger_{\text{names,the}}$. And here is the semantics:

$$\begin{array}{llll} \mathcal{M} \models \text{a is the p} & \text{iff} & [\![p]\!] = \{[\![a]\!]\} \\ \mathcal{M} \models \text{a isn't the p} & \text{iff} & [\![p]\!] \neq \{[\![a]\!]\}. \end{array} \quad (4)$$

Please note that $\mathcal{M} \models$ a isn't the p doesn't imply that a is a p, unlike what you would imagine from a real conversation. This is called a *conversational implicature*: nobody would say "a isn't the p" unless it was already known or presupposed that a is a p. We take the negation the way we do so that we have a semantic negation. Given the choice of being more linguistically realistic vs. making our life easier in the logic, we are taking the second choice in this section. However, all of this will be reconsidered in Section 5.

The rules of our logical system are found in Figure 3 together with the rules which we have already seen in Figures 1 and 2. It is probably worth commenting on these rules. For the most part, these are "obvious." This means that the soundness proofs could be easy exercises in a classroom. The rules ($R_{10}$) and ($R_{11}$) are of course more difficult. One might even complain that they are not "rules of inference" at all. See Remark 4.16 for a natural principle which implies these two rules. That principle in effect uses names as variables, and so to be parsimonious we have taken ($R_{10}$) and ($R_{11}$) as rules in our system.

**Proposition 4.1.** *The logical system for* $S^\dagger_{\text{names,the}}$ *is sound.*

*Proof.* We consider the rules in turn. We start each by fixing a model $\mathcal{M}$ of the hypotheses and show that this model satisfies the conclusion.

We begin with ($R_4$). We have a model $\mathcal{M}$ which satisfies the hypothesis, so $[\![p]\!] = \{[\![a]\!]\}$. In particular, $[\![a]\!] \in [\![p]\!]$. Thus, the conclusion holds in $\mathcal{M}$.

In ($R_5$), we have $[\![b]\!] \in [\![p]\!] = \{[\![a]\!]\}$. Thus $[\![a]\!] = [\![b]\!]$.

For ($R_6$), $[\![p]\!] = \{[\![a]\!]\}$ and $[\![p]\!] \cap [\![q]\!] \neq \emptyset$. Let $m \in [\![p]\!] \cap [\![q]\!]$. Since $[\![a]\!]$ is the only element of $[\![p]\!]$, $[\![a]\!] = m$. But then $[\![a]\!] = m \in [\![q]\!]$.

We turn to ($R_7$). We have $[\![b]\!] \in [\![q]\!] \subseteq [\![p]\!] = \{[\![a]\!]\}$. It follows that $[\![b]\!] = [\![a]\!]$. Since $[\![q]\!] \subseteq [\![p]\!]$, $[\![q]\!]$ is either $\emptyset$ or $[\![p]\!]$. It must be $[\![p]\!]$. Thus $[\![q]\!] = [\![p]\!] = \{[\![a]\!]\} = \{[\![b]\!]\}$.

For ($R_8$), $[\![p]\!] = \{[\![a]\!]\}$ and $[\![a]\!] = [\![b]\!]$. So $[\![p]\!] = \{[\![b]\!]\}$.

$$\frac{\text{a is the p}}{\text{a is a p}}\ R_4 \qquad \frac{\text{a is the p} \quad \text{b is a p}}{\text{a is b}}\ R_5$$

$$\frac{\text{a is the p} \quad \text{Some p are q}}{\text{a is a q}}\ R_6 \qquad \frac{\text{a is the p} \quad \text{All q are p} \quad \text{b is a q}}{\text{b is the q}}\ R_7$$

$$\frac{\text{a is the p} \quad \text{a is b}}{\text{b is the p}}\ R_8 \qquad \frac{\text{a is the p} \quad \text{a is the q}}{\text{All p are q}}\ R_9$$

$$\frac{\text{a is the p} \quad \text{b is the } \overline{\text{p}} \quad \text{a is a q} \quad \text{b is a q}}{\text{All } \overline{\text{q}} \text{ are q}}\ R_{10}$$

$$\frac{\text{a is the p} \quad \text{b is the } \overline{\text{p}} \quad \text{a is a } q_1 \quad \text{a is a } q_2 \quad \text{b is a } \overline{q_1}}{\text{All } q_1 \text{ are } q_2}\ R_{11}$$

Figure 3: The rules of the logical form $\mathcal{S}^{\dagger}_{\text{names,the}}$. We take these rules in addition to what we have in Figures 1 and 2.

For ($R_9$), we have $[\![p]\!] = \{[\![a]\!]\} = [\![q]\!]$. Thus $[\![p]\!] = [\![q]\!]$, and so each is a subset of the other.

Turning to ($R_{10}$) and ($R_{11}$), the first two hypotheses imply that $M = \{[\![a]\!], [\![b]\!]\}$, and $[\![a]\!] \neq [\![b]\!]$. Since $[\![a]\!]$ and $[\![b]\!]$ belong to $[\![q]\!]$, we see that *everything* belongs to $[\![q]\!]$. Thus $\mathcal{M} \models$ All $\overline{\text{q}}$ are q.

Similar reasoning applies in ($R_{11}$). Again, $M = \{[\![a]\!], [\![b]\!]\}$. The third and fifth hypotheses imply that $[\![q_1]\!] = \{a\}$, and then by the fourth we have $[\![q_1]\!] \subseteq [\![q_2]\!]$. □

**Example 4.2.** Here are some examples:

1. a is the p, Some p are q ⊢ All p are q.

2. All $\overline{\text{p}}$ are p ⊢ a is a p.

We leave it to the reader to construct the proof trees for these.

**Definition 4.3.** $\Gamma$ is *maximal consistent* if $\Gamma$ is consistent, and if $\Delta$ is a consistent set and $\Gamma \subseteq \Delta$, then $\Delta = \Gamma$.

**Lemma 4.4.** *Let $\Gamma$ be a consistent set.*

1. *$\Gamma$ has a maximal consistent superset.*

2. *If $\Gamma$ is maximal consistent, then $\Gamma \vdash \varphi$ iff $\varphi \in \Gamma$.*

3. *$\Gamma$ is maximal consistent if and only if for each $\varphi$, either $\varphi$ or $\overline{\varphi}$ belongs to $\Gamma$.*

*Proof.* These are all standard for propositional logic, and the main thing is to check that the usual arguments work here even though we do not have the connectives $\wedge$ and $\vee$. Indeed, (1) follows from the derivable rule of cases which we saw in Proposition 2.4. If $\Gamma$ is consistent, then for all $\varphi$ either $\Gamma \cup \{\varphi\}$ or $\Gamma \cup \{\overline{\varphi}\}$ must also be consistent. The other parts are similar. □

**Definition 4.5.** Let $\Gamma \subseteq \mathcal{S}^\dagger_{\text{names,the}}$. The literal p is *a singular literal (for $\Gamma$)* if there is some name a such that $\Gamma \vdash$ a is the p. And we say that the literal a is *singular name (for $\Gamma$)* if for some literal p, $\Gamma \vdash$ a is the p. In other words, if $\Gamma \vdash$ a is the p, then a is a singular name, and p is a singular literal. Note that if p is a singular literal via a and b, then $\Gamma \vdash$ a is b. If a is a singular name via literals p and q, then $p \equiv_\Gamma q$, using ($R_9$).

The main result in this section is Theorem 4.10, showing that every consistent set in $\mathcal{S}^\dagger_{\text{names,the}}$ has a model. Before that, we have three lemmas. These preliminary results are special cases of Theorem 4.10, and we have placed them before the theorem because the theorem calls on them in several places. Also, Lemma 4.8 is noteworthy because the two "corner case" rules of the logic, ($R_{10}$) and ($R_{11}$), are needed exactly in that result.

**Remark 4.6.** Before this we make a convention about our terminology in the rest of this paper. When we have a fixed set $\Gamma$ and a fixed sentence $\varphi$, we often shorten "$\Gamma$ contains $\varphi$" by simply writing $\varphi$.

**Lemma 4.7.** *Let $\Gamma \subseteq \mathcal{S}^\dagger_{\text{names,the}}$ be a maximal consistent set containing All $\overline{p}$ are p and a is the p for some literal p and name a. Then $\Gamma$ has a model $\mathcal{M}$ whose universe $M$ is a singleton set (a set with exactly one element), say $\{*\}$.*

*Proof.* Before we get started, recall from Remark 4.6 that we often omit "Γ contains" from assertions in proofs like this. And also remember that membership in Γ and provability from it are the same, since Γ is maximal consistent.

By Example 3.5, for all names b we have b is a p. Since a is the p, we have b is a.

Let $M = \{*\}$ be a singleton set. The use of the symbol $*$ means that we don't care what object we take here. We are going to build a model $\mathcal{M}$ by interpreting the names and nouns, and then show that $\mathcal{M}$ satisfies all sentences in Γ. Here is how we interpret basic names q and names b.

$$[\![q]\!] = \begin{cases} \{*\} & \text{if a is a q} \\ \emptyset & \text{if a is a } \overline{q} \end{cases}$$

$$[\![b]\!] = *$$

Concerning the names, with a singleton universe we have no other choice. This means that b is c holds in our model, for all names b and c. And indeed, this sentence follows from Γ (and thus belongs to it by maximality). Concerning the basic nouns, notice that $\mathcal{M} \models$ a is a q iff $* \in [\![q]\!]$, and this holds iff a is a q.

Suppose that Γ contains All $q_1$ are $q_2$. To see that this holds in $\mathcal{M}$, we may assume that $[\![q_1]\!] = \{*\}$; otherwise $[\![q_1]\!] = \emptyset$, and our result here is trivial. Then a is a $q_1$. So by $(R_2)$, a is a $q_2$. This implies that $[\![q_2]\!] = \{*\}$. Thus $\mathcal{M}$ satisfies All $q_1$ are $q_2$.

Going the other way, suppose that $\mathcal{M} \models$ All $q_1$ are $q_2$. We break into cases as to whether $* \in [\![q_1]\!] \cap [\![q_2]\!]$, or whether $* \notin [\![q_1]\!]$. In the first case, a is a $q_2$. By $(R_7)$, a is the $q_2$. Then by $(R_9)$, All p are $q_2$. By the first point in Example 2.2, we have All $\overline{p}$ are $\overline{q}_1$, and then All $q_1$ are p by the second. By (BARBARA), we have the desired conclusion All $q_1$ are $q_2$. In the second case, a is a $\overline{q}_1$. We claim that All $q_1$ are $\overline{q}_1$. Otherwise, Some $q_1$ are $q_1$ by (RAA). Our hypothesis in this lemma that All $\overline{p}$ are p (see Example 2.2 again) implies that All $q_1$ are p. By (DARII), Some p are $q_1$. And then a is a $q_1$ by $(R_6)$. This is a contradiction. So again we have the desired conclusion All $q_1$ are $q_2$.

At this point we know that Γ contains All $q_1$ are $q_2$ iff this sentence holds in $\mathcal{M}$. This holds for all literals. It follows that for all literals $q_1$ and $q_2$, Γ contains Some $q_1$ are $q_2$ iff this sentence holds in $\mathcal{M}$.

Finally, if $\Gamma$ contains b is the q, then by ($R_4$), b is a q. We already know that $\mathcal{M} \models$ b is a q. Since the model is a singleton, it satisfies b is the q. Conversely (and this is the most interesting point), suppose that b is the q is true in $\mathcal{M}$. Then we must have $[\![q]\!] = \{*\}$ and $[\![b]\!] = * = [\![a]\!]$. So a is a q. Using ($R_7$), a is the q. Since b is a, by ($R_8$) we have b is the q. $\square$

**Lemma 4.8.** *Let $\Gamma \subseteq S^{\dagger}_{names,the}$ be a maximal consistent set with two singular nouns a and b, and with some literal p such that a is the p and b is the $\overline{p}$. Then there is a model whose universe has exactly two points and such that $\mathcal{M} \models \varphi$ iff $\varphi \in \Gamma$. In particular $\mathcal{M} \models \Gamma$.*

*Proof.* We are going to build a model $\mathcal{M}$ with universe $\{a, b\}$.

We first claim that for every name c, either c is a or else c is b. We know that either c is a p or c is a $\overline{p}$, and using ($R_8$) this determines the interpretation of names: for a name c: we take $[\![c]\!] = $ a if c is a, and otherwise $[\![c]\!] = $ b.

We next interpret the basic nouns q. We take

$$[\![q]\!] = \begin{cases} \{a, b\} & \text{if a is a q and b is a q} \\ \{a\} & \text{if a is a q and b is a } \overline{q} \\ \{b\} & \text{if a is a } \overline{q} \text{ and b is a q} \\ \emptyset & \text{if a is a } \overline{q} \text{ and b is a } \overline{q} \end{cases} \quad (5)$$

Our semantics requires us to take $[\![\overline{q}]\!] = \{a, b\} \setminus [\![q]\!]$. Then we check that (5) holds when we replace q by $\overline{q}$. For example, $[\![\overline{q}]\!] = \{a\}$ if a is a $\overline{q}$ and b is a q; this follows because a is a $\overline{q}$ and b is a q implies that $[\![q]\!] = \{b\}$, and $\{a, b\} \setminus \{b\} = \{a\}$.

We now have our model $\mathcal{M}$, and we check that $\mathcal{M} \models \varphi$ iff $\varphi \in \Gamma$.

Here is the when $\varphi$ is an All-sentence. (This also covers the work for when $\varphi$ is a Some-sentence.) We claim that for all nouns $q_1$ and $q_2$, the following are equivalent:

1. $\Gamma$ contains All $q_1$ are $q_2$.

2. $\mathcal{M} \models$ All $q_1$ are $q_2$.

3. If $[\![a]\!] \in [\![q_1]\!]$, then $[\![a]\!] \in [\![q_2]\!]$; and if $[\![b]\!] \in [\![q_1]\!]$, then $[\![b]\!] \in [\![q_2]\!]$.

(1)⇒(2) is by the soundness of the proof system, and (2)⇒(3) is by the semantics of our language. The important point is to show that (3)⇒(1).

If $[\![a]\!] \in [\![q_2]\!]$ and $[\![b]\!] \in [\![q_2]\!]$, then using ($R_{10}$), we have All $\overline{q}_2$ are $q_2$. From this, All $q_1$ are $q_2$.

If $[\![a]\!] \notin [\![q_1]\!]$, $[\![b]\!] \in [\![q_1]\!]$, and $[\![b]\!] \in [\![q_2]\!]$, then (interchanging a and b and) using ($R_{11}$), All $q_1$ are $q_2$.

If $[\![a]\!] \in [\![q_1]\!]$, $[\![a]\!] \in [\![q_2]\!]$, and $[\![b]\!] \notin [\![q_1]\!]$, then using ($R_{11}$) we have All $q_1$ are $q_2$.

If $[\![a]\!] \notin [\![q_1]\!]$ and $[\![b]\!] \notin [\![q_1]\!]$, then using ($R_{10}$), we have All $q_1$ are $\overline{q}_1$. From this, All $q_1$ are $q_2$.

The four cases above cover all of the ways to have (3). So in all cases, we have All $q_1$ are $q_2$. Thus (1) holds.

We return to our overall verification that $\mathcal{M} \models \varphi$ iff $\varphi \in \Gamma$. When $\varphi$ is a sentence of the form c is a q, this is easy to check based on what we have already seen.

We conclude with sentences $\varphi$ of the form d is the q. Let $\varphi \in \Gamma$. As we know, either d is a or d is b. Without loss of generality it is the first option. So a is the q. It is easy to check that this holds in $\mathcal{M}$, and then so does $\varphi$. In the other direction, suppose that $\mathcal{M} \models$ a is the q. So $\mathcal{M}$ satisfies a is a q and b is a $\overline{q}$. Thus a is a q and b is a $\overline{q}$. By our hypothesis in this result, a is the p. Then by ($R_{11}$), All q are p. By ($R_7$), a is the q. Since d is a, we are done. □

**Lemma 4.9.** *Let* $\Gamma \subseteq \mathcal{S}^\dagger_{\mathsf{names,the}}$ *be a maximal consistent set with two singular nouns* a *and* b, *and with literals* p *and* q *such that* a is the p *and* b is the q, *and* All $\overline{p}$ are q. *Then* $\Gamma$ *has a model whose universe has at most two points.*

*Proof.* We have two cases: (i) b is a p, and (ii) b is a $\overline{p}$. In case (i), then since a is the p and b is a p, we see that a is b. Since b is the q, we have a is the q. By ($R_9$), All p are q. It follows from this and All $\overline{p}$ are q that All $\overline{q}$ are q. Together with the hypothesis b is the q, we are in the situation of Lemma 4.7.

In case (ii), b is a $\overline{p}$ and b is the q, and All $\overline{p}$ are q. Using ($R_7$), b is the q. We are in the situation of Lemma 4.8. □

**Theorem 4.10.** *Every consistent set* $\Gamma \subseteq \mathcal{S}^\dagger_{\mathsf{names,the}}$ *has a model.*

*Proof.* Fix a consistent set $\Gamma$. If $\Gamma$ satisfies the hypothesis of Lemma 4.8, then we are done. This means that in the course of our case-by-case argument, when we reach a case that gives us the hypothesis of Lemma 4.8, then again we are done. The overall plan is to build a model using points along with states of orthoposets. We need three subsidiary constructions and results.

At various points we obtain contradictions to extendibilty (see Claim 4.11 just below, for example). We want to comment on how we shall present these arguments. For example, (6) presents a set $T_a$ defined as the union of three others sets called $(i)$, $(ii)$, and $(iii)$. If $T_a$ is not extendible, then there are two elements of it, say x and y, such that $x \leq_\Gamma \bar{y}$. There are $3 \times 3 = 9$ cases, depending on which of the three sets x is from, and which of the three y is from. However, due to the antitone law, we can drop this from $3^2$ to $\binom{3}{2} = 6$. And we indicate these cases by labeling which sets x and y come from.

For singular names a, let

$$T_a = \quad \{p \in \text{Lit} : \Gamma \vdash \text{a is a p}\} \qquad (i)$$
$$\cup \ \{\bar{q} \in \text{Lit} : \text{for some singular } b \not\equiv a, \Gamma \vdash \text{b is the q}\} \quad (ii)$$
$$\cup \ \{q : \text{for some p, a is the p, and some p are q}\} \qquad (iii)$$
$$(6)$$

**Claim 4.11.** *For all singular* a, $T_a$ *is extendible.*

*Proof.* Suppose not. In case $(i,i)$, see Lemma 3.6. (Here is what we mean by "case $(i,i)$" in this point. Recall the definition of extendible in Definition 2.7. One way for $T_a$ to be non-extendible is that there is p in the set from $(i)$ and also q in the set from $(i)$ such that $p \leq \bar{q}$. In this case, we would contradict Lemma 3.6. These remark apply to similar cases throughout this proof.)

In case $(i, ii)$, we have p, b, and q such that a is a p, b is the q, $a \not\equiv b$, and $p \leq q$. This implies that a is a q, and so a is b. But then $a \equiv b$, and this is a contradiction.

In case $(i, iii)$, we have $p_1$, q, and $p_2$ such that a is a $p_1$, a is the $p_2$, and some $p_2$ are q, and $p_1 \leq \bar{q}$. The first and last give a is a $\bar{q}$. The second and third give a is a q, using ($R_6$). So we have a contradiction: a is a $\bar{q}$ and a is a q.

Let us consider case $(ii, ii)$. We have $q_1$, $q_2$, $b_1 \not\geq a$, and $b_2 \not\geq a$, such that $b_1$ is the $q_1$, and $b_2$ is the $q_2$, and $\overline{q}_1 \leq q_2$. Then $\Gamma$ satisfies the hypothesis of Lemma 4.8.

We continue with case $(ii, iii)$. We have $q_1$, $q_2$ and p and $b \not\geq a$ such that b is the $q_1$, a is the p, some p are $q_2$, and $\overline{q}_1 \leq \overline{q}_2$. This last statement gives $q_2 \leq q_1$. Since a is the p and some p are $q_2$, we have a is a $q_2$. Since $q_2 \leq q_1$, we have a is a $q_1$. Now b is the $q_1$, and so a is b. This contradicts $b \not\geq a$.

In case $(iii, iii)$, we have $p_1$, $p_2$, $q_1$, and $q_2$ so that a is the $p_1$, some $p_1$ are $q_1$, a is the $p_2$, some $p_2$ are $q_2$, and $q_1 \leq \overline{q}_2$. Since a is the $p_1$ and some $p_1$ are $q_1$, we have a is a $q_1$ by ($R_6$). The same argument shows a is a $q_2$. But since a is a $q_1$ and $q_1 \leq \overline{q}_2$, we have a is a $\overline{q}_2$. Thus $\Gamma$ is inconsistent. □

For nonsingular names a, let

$$U_a = \begin{array}{ll} \{p \in \mathsf{Lit} : \Gamma \vdash \text{a is a p}\} & (i) \\ \cup \ \{\overline{q} \in \mathsf{Lit} : \text{for some singular b, } \Gamma \vdash \text{b is the q}\} & (ii) \end{array} \quad (7)$$

**Claim 4.12.** *For all nonsingular* a, $U_a$ *is extendible.*

*Proof.* Suppose not. This time we have three cases. For case $(i, i)$, again see Lemma 3.6. For $(ii, ii)$, see Claim 4.11, case $(ii, ii)$. The only difference is in $(i, ii)$. We would again have p and b such that a is a p, b is the q, and $p \leq q$. This implies that a is a q, and so a is b by ($R_5$). We see that a is singular. This is a contradiction. □

Let $P$ be the set of unordered pairs $\{p, q\}$ such that $\Gamma$ contains some p are q, but neither p nor q is singular. This includes the case $p = q$. For $\{p, q\} \in P$, let

$$V_{\{p,q\}} = \begin{array}{ll} \{p, q\} & (i) \\ \cup \ \{\overline{r} \in \mathsf{Lit} : \text{for some singular b, } \Gamma \vdash \text{b is the r}\} & (ii) \end{array}$$
$$(8)$$

**Claim 4.13.** *For* $\{p, q\} \in P$, $V_{\{p,q\}}$ *is extendible.*

*Proof.* Suppose not. For case $(i, i)$, again see Lemma 3.6. For $(ii, ii)$, see Claim 4.11, case $(ii, ii)$. Case $(i, ii)$ is a little different than what we have seen. We have p (without loss of generality) and also b and r such that b is the r, and $p \leq r$. By ($R_6$), b is an r. Using (DARII), some p are r. By ($R_7$) with b for a, b is the p. And then p is singular; this is a contradiction. □

**The model** Recall that in this proof we began with a maximal consistent set $\Gamma$. We are going to use the claims above to build a model $\mathcal{M}$ and then check that it satisfies $\Gamma$.

For singular a, let $W_a$ be a state including $T_a$. For nonsingular a, let $X_a$ be a state including $U_a$. For $\{p, q\} \in P$, let $Y_{\{p,q\}}$ be a state including $V_{\{p,q\}}$.

Our model $\mathcal{M}_\Gamma$ will have a point $\alpha_{[a]}$ for each equivalence class [a] of a singular name a, and a point $\beta_{[a]}$ for each equivalence class [a] of a nonsingular name a, and two points, $\gamma_{\{p,q\}}$ and $\delta_{\{p,q\}}$ for each $\{p, q\} \in P$. We assume that all of these points are distinct. That is, $M$ is a disjoint union

$$\begin{aligned} M &= \quad \{\alpha_{[a]} : \text{a is singular}\} \\ &\cup \; \{\beta_{[a]} : \text{a is nonsingular}\} \\ &\cup \; \{\gamma_{\{p,q\}}, : \{p, q\} \in P\} \\ &\cup \; \{\delta_{\{p,q\}} : \{p, q\} \in P\} \end{aligned}$$

We shall use letters like $m$ for arbitrary elements of $M$. To interpret the nouns in this model, we say $\alpha_{[a]} \in [\![p]\!]$ iff $p \in W_a$, and similarly for the other sets. We interpret the name a by $\alpha_{[a]}$.

$$\begin{aligned} [\![p]\!] &= \quad \{\alpha_{[a]} : p \in W_a\} \cup \{\beta_{[a]} : p \in X_a\} \\ &\cup \; \{\gamma_{\{x,y\}} : p \leq x \text{ or } p \leq y\} \cup \{\delta_{\{x,y\}} : p \leq x \text{ or } p \leq y\} \\ [\![a]\!] &= \quad \begin{cases} \alpha_{[a]} & \text{if a is singular} \\ \beta_{[a]} & \text{if a is nonsingular} \end{cases} \end{aligned}$$

**Verification** We conclude by checking that $\mathcal{M} \models \Gamma$.

Since all states are closed upwards, we easily see that $\mathcal{M}$ satisfies the All-sentences in $\Gamma$. (See Lemma 3.8 for a very similar argument.)

Consider a Some-sentence in $\Gamma$, say Some p are q. Let us suppose first of all that p is singular. Let a be such that a is the p. Using this and some p are q, we get a is a q by ($R_6$). Then $\alpha_{[a]} \in [\![p]\!] \cap [\![q]\!]$ by ($R_4$)

and the definition of the semantics. So our model satisfies the given sentence in this case. In case q is singular, the same argument works, mutatis mutandis. If neither p nor q is singular, then $\{p, q\} \in P$. In this case, $\{p, q\} \subseteq V_{\{p,q\}} \subseteq Y_{\{p,q\}}$. Therefore $\gamma_{\{p,q\}} \in [\![p]\!] \cap [\![q]\!]$. Again, $\mathcal{M} \models$ Some p are q.

If $\Gamma$ contains a is a p for a literal p, then we check that this sentence holds in $\mathcal{M}$. This is clear for singular a, since $p \in T_a \subseteq W_a$. For nonsingular a, the argument is the same but uses $p \in U_a \subseteq X_a$. We also check the converse. Both of these points apply to $\overline{p}$, since it also is a literal. If $\Gamma$ does not contain a is a p, then it contains a is a $\overline{p}$. Then $\mathcal{M} \models$ a is a $\overline{p}$, by what we have just seen; so $\mathcal{M} \not\models$ a is a p.

Take an identity statement like a is b in $\Gamma$. So $[a] = [b]$. Let us assume first that a is singular. Then so is b. The model has $[\![a]\!] = \alpha_{[a]} = \alpha_{[b]} = [\![a]\!]$. If $\alpha$ is nonsingular, then so is $\beta$, and then we have $[\![a]\!] = \beta_{[a]} = \beta_{[b]} = [\![a]\!]$.

And for a statement in $\Gamma$ like a isn't b, we have $[a] \neq [b]$, and again $\mathcal{M} \models$ a isn't b.

Suppose that $\Gamma$ contains a is the p. Then a is singular. Moreover $\alpha_{[a]}$ will belong to $[\![p]\!]$ because $p \in T_a \subseteq W_a$. We need to be sure that $\alpha_{[a]}$ is the only point in the model which belongs to $[\![p]\!]$. For points of the form $\beta_{[b]}$, $\gamma_{\{x,y\}}$, and $\delta_{\{x,y\}}$, note that $\overline{p} \in U_a$, and also $\overline{p} \in V_{\{x,y\}}$. These all ensure that the points mentioned will not belong to $[\![p]\!]$. For points $\alpha_{[b]}$ such that $\alpha_{[b]} \neq \alpha_{[a]}$, note that $[a] \neq [b]$. That is a $\not\approx$ b. So $T_b$ contains $\overline{p}$. Once again, we have $\alpha_{[b]} \notin [\![p]\!]$.

Finally, suppose that a isn't the p. If we also have a is a $\overline{p}$, then our model arranges that this sentence holds, as we have seen. So we shall assume going forward that a is a p.

We have here three subcases. If p is singular, then there is a name b such that $\Gamma$ contains b is the p. Then a $\not\approx$ b. The name a might be singular or not; without loss of generality, it is singular. Then $\overline{p} \in T_a \subseteq W_a$. In other words, the model arranges that $[\![a]\!] = \alpha_{[a]} \notin [\![p]\!]$. So our model satisfies a isn't the p, as desired.

The second subcase is when p is not singular, and $\Gamma$ contains All p are $\overline{p}$. Intuitively, *no p exist*. In this case, $\overline{p}$ will belong to each set of the form $T_b$, $U_b$, or $V_{\{x,y\}}$. Hence it will belong to each set of the form $W_b$, $X_b$, or $Y_{\{x,y\}}$. This implies that no points in the model will satisfy

p.

The last case is when p is not singular, and yet $\Gamma$ contains some p are p. This time $\{p\} = \{p, p\} \in P$. This time, the model has at least two points in $[\![p]\!]$, namely $\gamma_{\{p\}}$ and $\delta_{\{p\}}$. These points are different. Thus our model satisfies a isn't the p.

This concludes the verification of our model's properties, and hence of Theorem 4.10. □

**Corollary 4.14.** *The logic in Figures 1, 2, and 3 is complete for* $S^\dagger_{names,the}$.

*Proof.* By Lemma 2.10 and Theorem 4.10. □

**Remark 4.15.** Does the proof have to be this complicated? I cannot promise that there is no shorter proof of Theorem 4.10. But I do have a comment that might explain why the proof goes the way it is. One might think that the way to do a proof like this is to *add constants*, in the manner of Henkin's proof of the completeness of first-order logic. The idea would be to add a fresh constant for every literal p such that $\Gamma \vdash$ Some p are p. And then one would build a model whose universe is the set of fresh constants, or rather the set of such constants quotiented by the natural equivalence relation given by the identity relation, is. Indeed, this is close to what we will do in the final theorem of this paper. However, doing this here would mean proving a *lemma on constants*, a result which would tell us that adding a fresh constant on behalf of some literal p above preserves consistency. And for weak logics like this, the lemma on constants is often a lot of work. In effect, we are circumventing it by doing more on the semantic side.

It might be interesting to note that there is a similar point to be made in a result in Mcintosh [6], an appendix to Sommers [12]. McIntosh formulates a logical system called TFL and proves its completeness. We find the following:

> Lemma 4: If $\Gamma$ is a d-consistent set of sentences of TFL, then $\Gamma$ is a d-consistent set of sentences of TFL', where TFL' is the system resulting by adding an infinite list $\mu_1, \mu_2, \ldots$ of new singular term letters to the vocabulary of TL. We assume this without proof.

I didn't want to simply assume this kind of lemma when a proof is available.

**Remark 4.16.** There is a way to simplify the proof system. It would be possible to add a rule of *universal generalization* to the system:

$$\frac{\Gamma \cup \{\text{c is a p}\} \vdash \text{c is a q} \quad \text{c does not occur in } \Gamma}{\Gamma \vdash \text{All p are q}} \text{ UG}$$

Adopting this would make ($R_{10}$) and ($R_{11}$) provable. For that matter, we could also prove (AXIOM), (BARBARA), and ($R_9$). Thus, we could replace these five rules with (UG) and the resulting system would be complete. This would not simplify the completeness proof, and one would need to prove as lemmas the statements of the deleted rules. The change might well complicate the proof search. (But I did not investigate this matter.)

In the other direction, the completeness theorem that we have shown implies that every instance of (UG) is provable.

I do not know whether ($R_{10}$) and ($R_{11}$) can be proved in the logic consisting of Figures 1 and 2 together with ($R_1$)–($R_9$). I conjecture that they are not provable in that system.

## 5  A Proposal for a Logic of Accommodation and Inference in the Spirit of Neutral Free Logic

Our formulation of $\mathcal{S}^{\dagger}_{\text{names,the}}$ has sentences a is b, a isn't b, a is a p, and a is the p, where p is a literal, and a and b are names. But it lacks sentences of several forms that we would want to include:

$$\begin{array}{l}\text{the p is the q}\\ \text{the p isn't the q}\\ \text{the q is a p}\end{array} \quad (9)$$

As before p and q are literals, so the last form of sentence in (9) includes the sentences the q is a $\overline{\text{p}}$. We might want to include these because they are English sentences, and because the presentation of the syntax would be easier and less ad-hoc if we did so.

One way to give the semantics of the sentences in (9) would be to straightforwardly extend what we have done in (4):

$\mathcal{M} \models$ the p is the q      iff     $[\![p]\!] = [\![q]\!]$, and both are singleton sets
$\mathcal{M} \models$ the p isn't the q     iff     $[\![p]\!] \neq [\![q]\!]$, and both are singleton sets
$\mathcal{M} \models$ the q is a p     iff     $[\![q]\!] \in [\![p]\!]$, and $[\![q]\!]$ is a singleton set
(10)

However, there are well-known issues with (10). If $[\![p]\!]$ is not a singleton in a given model, then both of the first two sentences come out false. The second one is not the semantic negation of the first. In a model where $[\![p]\!]$ is not a singleton, we might not want to say that the sentences are false. We might want them to be "undefined" or "anomalous". What we want to do in this final section is to spell out one proposal on how this could be done. So the semantics and proof theory will be *partial* in an appropriate sense.

Incidentally, we are not presenting this material in order to advance a proposal of what sentences like (9) mean. Our view is that this is a complicated matter that one should decide on *before* turning to logic. We are not doing this here. Instead what we are doing is to make a proposal without arguing for it, and then to work out what the logic should be and how the completeness theorem should go.

Our proposal is a *free logic* in that the definite descriptions need not denote. To make life a little simpler, we will take the names a to always denote, and of course this choice could be revised. Our proposal is a *neutral* free logic in that we shall take sentences like those in (9) to be undefined in some models.

The idea of using partiality in connection with presupposition is not new see [1] for one paper in this line of work. That paper contains many more operators and does much more than the system we propose. However, it does not aim for a complete logical system. One place to look at similar logics to ours is Pavlović and Gratzl [8]. However, what we are doing is more 'semantic' somehow in that we build more standard models, and much less substantial in the proof theoretic direction since our base logic is so weak. In addition, our notation is different from theirs.

**Definition 5.1.** We define a logical language $\mathcal{S}^\dagger_{\mathsf{acc}}$ as follows. As before, we begin with a set $\mathbb{P}$ of nouns and a set Names of names.

| | | |
|---|---|---|
| $\mathcal{M} \models \mathsf{a}\downarrow$ | | always |
| $\mathcal{M} \models (\mathsf{the\ p})\downarrow$ | iff | $[\![\mathsf{p}]\!]$ is a singleton set |
| $\mathcal{M} \models (\mathsf{t\ is(n't)\ u})\downarrow$ | iff | $\mathcal{M} \models \mathsf{t}\downarrow$ and $\mathcal{M} \models \mathsf{u}\downarrow$ |
| $\mathcal{M} \models \mathsf{t\ is\ a\ p}\downarrow$ | iff | $\mathcal{M} \models \mathsf{t}\downarrow$ |
| $\mathcal{M} \models (\mathsf{All\ p\ are\ q})\downarrow$ | | always |
| $\mathcal{M} \models (\mathsf{some\ p\ are\ q})\downarrow$ | | always |
| $\mathcal{M} \models \varphi$ | iff | $\mathcal{M} \models \varphi\downarrow$ and $\mathcal{M} \models \varphi$ |

Figure 4: The semantics of $S^\dagger_{\mathsf{acc}}$. Note that we are writing the semantic relation using the symbol $\models$ rather than what we previously used, $\models$, and also that the previous relation $\mathcal{M} \models \varphi$ (see (4) and (10)) figures in to this semantics in the last line.

An *individual term* is either a name a or a definite description the p. We use letters like t and u for individual terms.

We take all of the sentences in the language of Section 4, and together with the sentences in (9). We call these items the *sentences* of $S^\dagger_{\mathsf{acc}}$.

A *definedness assertion*[4] $\mathsf{t}\downarrow$ for t an individual term, and also $\varphi\downarrow$ for $\varphi$ a sentence of this language.

When we speak of *sentences*, we do not include the definedness assertions $\mathsf{t}\downarrow$ and $\varphi\downarrow$.

**Definition 5.2.** Given a model $\mathcal{M}$, and a term t, we define a relation $\mathcal{M} \models \mathsf{t}\downarrow$ in Figure 4. The same figure also defines the overall semantic relation $\mathcal{M} \models \varphi\downarrow$. Note that the previous relation $\mathcal{M} \models \varphi\downarrow$ (defined in (4) with the additions in (10)) figures in to this one. Thus $\mathcal{M} \models \varphi\downarrow$ implies by definition that $\mathcal{M} \models \varphi\downarrow$.

The point of the definition is to formalize the intuition that a definite description the p is only sensible in a model when $[\![\mathsf{p}]\!]$ is a singleton. If this fails, then every sentence that has the p inside will be undefined. (Again, this might or might not be what one wants.)

---

[4] We could also call these "accommodatedness assertions". The connection is that in a given model $\mathcal{M}$, a sentence $\varphi$ is either satisfied ($\mathcal{M} \models \varphi$) or falsfied ($\mathcal{M} \models \overline{\varphi}$) if and only if all of the terms occurring in $\varphi$ are defined (=accommodated).

**Example 5.3.** In a model $\mathcal{M}$ where $[\![p]\!] = \emptyset$, $\mathcal{M} \not\Vdash (\text{the p}) \downarrow$. Thus $\mathcal{M} \not\Vdash$ the p is a p, and we also have $\mathcal{M} \not\Vdash$ the p is a $\bar{\text{p}}$. Also, we have $\mathcal{M} \not\Vdash$ a is the p and $\mathcal{M} \not\Vdash$ a isn't the p. (In contrast, when we return to the semantic notion $\models$ from earlier in the paper, we do have $\mathcal{M} \models$ a isn't the p.) But we do have $\mathcal{M} \Vdash$ All p are p and $\mathcal{M} \Vdash$ All p are $\bar{\text{p}}$. If our model has exactly one element, we would have $\mathcal{M} \Vdash$ a is the $\bar{\text{p}}$. Otherwise, $\mathcal{M} \not\Vdash$ a is the $\bar{\text{p}}$.

**Proposition 5.4.** *For all models $\mathcal{M}$ and all sentences $\varphi$,*

1. $\mathcal{M} \Vdash \varphi \downarrow$ *iff* $\mathcal{M} \Vdash \bar{\varphi} \downarrow$.

2. *If* $\mathcal{M} \Vdash \varphi \downarrow$, *then* $(\mathcal{M} \Vdash \varphi$ *iff it is not the case that* $\mathcal{M} \Vdash \bar{\varphi})$.

3. *If either* $\mathcal{M} \Vdash \varphi$ *or* $\mathcal{M} \Vdash \bar{\varphi}$, *then* $\mathcal{M} \Vdash \varphi \downarrow$.

Naturally, we aim for a proof system and a soundness/completeness result to connect it to the semantics. Since we have changed the notation in the semantic relation, we also do so in the proof system, writing $\Gamma \Vdash \varphi$ in the proof system. We also have assertions $\Gamma \Vdash t \downarrow$ and $\Gamma \Vdash \varphi \downarrow$.

We use the proof system in Figure 5. Perhaps the most interesting rule is (R$_1$), which we would call the rule of *accommodation*. The idea is that when someone reasons with a set $\Gamma$, they tacitly assume, or accommodate, that the definite descriptions like the king are felicitous. On this view, even a sentence like the king isn't bald presupposes a unique king: a hearer would assume that, or accommodate the assertion that, there is exactly one king in whatever situation is under discussion. In this system, one major consequence of accommodation is that if $t \downarrow$, we can infer t is t; this is rule (R$_5$). Without the assumption that $t \downarrow$, we should not make this inference, as Example 5.3 suggests. In other words, the reflexivity rule (REF) is not sound in the semantics of this section, and ($\downarrow_5$) compensates for this rule by adding a hypothesis that the term involved be accommodated.

Rules (R$_2$) and (R$_3$) tell us that if a sentence is accommodated, so is every constituent, and vice-versa. The double lines in the (R$_2$) and (R$_3$) indicate that the inferences could go from the bottom to the top; so from $t \downarrow$ and $u \downarrow$ we may infer upward $(t \text{ is } u) \downarrow$ and also $(t \text{ isn't } u) \downarrow$.

Rule (R$_4$) tells us that names are automatically accommodated. Surely this is controversial. The same goes for (R$_6$) and (R$_7$).

The reductio rule (RAA) in the form from earlier in this paper is not sound for this semantics. For example, if $\Gamma = \emptyset$, and $\varphi$ is the p isn't the p, then from $\Gamma \cup \{\varphi\}$ we have the p$\downarrow$ and so by (R$_5$) we have the p is the p. Thus we have a contradiction. If we had (RAA), we would withdraw $\varphi$ to see that $\Vdash \overline{\varphi}$. That is $\Vdash$ the p isn't the p. But this is exactly what we do not want in this logic. So to prevent this, we add the hypothesis $\varphi\downarrow$ in the statement of (RAA). The example just below shows a usage.

**Example 5.5.** We show

a is the p, b is the q, a isn't b $\Vdash$ the p isn't the q.

Here is our derivation:

$$
\dfrac{\dfrac{\dfrac{\text{a is the p}}{\text{(a is the p)}\downarrow}\downarrow_1}{\text{the p}\downarrow}\downarrow_2 \quad \dfrac{\dfrac{\text{b is the q}}{\text{(b is the q)}\downarrow}\downarrow_1}{\dfrac{\text{the q}\downarrow}{\text{(the p is the q)}\downarrow}\downarrow_2} \quad \dfrac{\dfrac{\text{a is the p} \quad [\text{the p is the q}]}{\text{a is the q}} \quad \dfrac{\text{b is the q}}{\text{b is a q}} R_4}{\text{a is b}} R_5 \quad \text{a isn't b}}{\text{the p isn't the q}} \text{RAA}
$$

Let $\varphi$ be the sentence the p is the q. The subproof on the left establishes that $\varphi\downarrow$. We added $\varphi$ to $\Gamma$, and then derived both $\psi$ and $\overline{\psi}$, where $\psi$ is a is b. We finished the proof with an application of (RAA$\downarrow$), inferring $\overline{\varphi}$ at the root.

**Proposition 5.6.** *If* $\Gamma \vdash \varphi\downarrow$, *then also* $\Gamma \vdash \overline{\varphi}\downarrow$.

**Definition 5.7.** Let $\Gamma$ be a set of sentences in $S^\dagger_{\text{acc}}$. Let

$$C = \{\mathsf{p} : \Gamma \Vdash \text{the p}\downarrow\}.$$

For each $\mathsf{p} \in C$, take a new name new$_\mathsf{p}$ to the language. For different literals p and q in $C$, we want new$_\mathsf{p}$ and new$_\mathsf{q}$ to be different. We only want the new name new$_\mathsf{p}$ when $\mathsf{p} \in C$.

For a term t and a sentence $\varphi$ we define *flattened versions* $\mathsf{t}^\flat$ and $\varphi^\flat$ by replacing each definite description the p inside of it by a new name new$_\mathsf{p}$. (There is no significance to the term "flattened.") Note that $\overline{\varphi}^\flat = \overline{(\varphi^\flat)}$.

## NAMES AND DEFINITE DESCRIPTIONS IN NATURAL LOGIC

$$\frac{\varphi}{\varphi\downarrow}\downarrow_1 \qquad \frac{(\text{t is(n't) u})\downarrow}{\text{t}\downarrow,\text{u}\downarrow}\downarrow_2 \qquad \frac{(\text{t is(n't) a p})\downarrow}{\text{t}\downarrow}\downarrow_3 \qquad \frac{}{\text{a}\downarrow}\downarrow_4$$

$$\frac{\text{t}\downarrow}{\text{t is t}}\downarrow_5 \qquad \frac{}{(\text{All p are q})\downarrow}\downarrow_6 \qquad \frac{}{(\text{some p are q})\downarrow}\downarrow_7$$

All Rules from Figure 1, but change (RAA) to (RAA$_\downarrow$), shown below:

If $\Gamma \Vdash \varphi\downarrow$ and $\Gamma \cup \{\varphi\} \Vdash \bot$, then $\Gamma \Vdash \overline{\varphi}$.

All rules in Figures 2 and 3, with two changes:
First, we generalize names a and b to individual terms t and u.
Second, we drop the reflexivity axiom (R$_1$) in Figure 2.

Figure 5: The proof system of $S^\dagger_{\text{acc}}$.

The map $\varphi \mapsto \varphi^\flat$ is one-to-one. Note also that not every sentence in the target language is of the form $\varphi^\flat$: the ones that are missing are those of the form t is new$_p$ and t isn't new$_p$.

From a fixed set $\Gamma$, we define a *flattened version* $\Gamma^\flat$ as follows:

$$\Gamma^\flat = \{\text{new}_p \text{ is the p} : p \in C\} \cup \{\varphi^\flat : \varphi \in \Gamma \text{ is a sentence}\}.$$

Notice that $\Gamma^\flat$ does not contain any definedness sentences t$\downarrow$ or $\varphi\downarrow$. We may regard $\Gamma^\flat$ as a set of sentences in $S^\dagger_{\text{names,the}}$, the language in our previous section *but over a larger set of names*.

Notice also that for all individual terms u of $S^\dagger_{\text{acc}}$, $\Gamma^\flat \vdash u^\flat$ is u. That is, for a definite description the p occurring in $\Gamma$, we have $\Gamma^\flat \vdash$ new$_p$ is the p. This is by the definition of $\Gamma^\flat$. And for a name a, we have $\Gamma^\flat \vdash$ a is a.

**Example 5.8.** Let $\Gamma = \{(\text{the r})\downarrow, \text{the p isn't the q}\}$. Then we would add new names new$_p$, new$_q$, and new$_r$. The set $\Gamma^\flat$ would have four sentences:

$$\Gamma^\flat = \{\text{new}_p \text{ isn't new}_q, \text{new}_p \text{ is the p}, \text{new}_q \text{ is the q}, \text{new}_r \text{ is the r}\}.$$

**Definition 5.9.** There is an inverse $\varphi \mapsto \varphi^\#$ taking sentences in the new language (with the extra names) back to the original language. It replaces the new name new$_p$ with the corresponding definite description the p.

**Lemma 5.10.** *Let* $\Gamma \subseteq \mathcal{S}^\dagger_{\mathsf{acc}}$, *and let* $\Gamma^\flat \subseteq \mathcal{S}^\dagger_{\mathsf{names,the}}$ *be its flattened version.*

1. *If* $\Gamma$ *is satisfiable using the semantics of this section, so is* $\Gamma^\flat$, *and conversely.*

2. *For all* $\varphi \in \mathcal{S}^\dagger_{\mathsf{acc}}$, *if* $\Gamma \Vdash \varphi$, *then* $\Gamma^\flat \vdash \varphi^\flat$.

3. *For all* $\varphi \in \mathcal{S}^\dagger_{\mathsf{names,the}}$, *if* $\Gamma^\flat \vdash \varphi$, *then* $\Gamma \Vdash \varphi^\#$.

4. *If* $\Gamma$ *is consistent, so is* $\Gamma^\flat$.

*Proof.* The point of (1) is that a model of $\Gamma$ may be regarded as a model of $\Gamma^\flat$. That is, each p which belongs to $C$ (and hence gives rise to a new name $\mathsf{new}_\mathsf{p}$) has to be interpreted by a singleton, and thus the interpretation of $\mathsf{new}_\mathsf{p}$ is clear. The sentences in $\Gamma^\flat$ of the form $\mathsf{new}_\mathsf{p}$ is the p all hold. In the other direction, a model of $\Gamma^\flat$ has to interpret each $\mathsf{p} \in C$ by a singleton set, namely $\{[\![\mathsf{new}_\mathsf{p}]\!]\}$. It follows that all of the definedness assertions in $\Gamma$ are satisfied in the model, as are the sentences from $\Gamma$.

We prove part (2) by induction on the derivations in $\mathcal{S}^\dagger_{\mathsf{acc}}$. This is an easy verification using the fact from before that $\Gamma^\flat \vdash \mathsf{u}^\flat$ is u for all terms u.

We prove part (3) by induction on the derivations in $\mathcal{S}^\dagger_{\mathsf{names,the}}$. Perhaps the most interesting step is for (REF), since that rule is not part of the logic in this section. We need only worry about an application of (REF) with a new name $\mathsf{new}_\mathsf{p}$; that is, $\Gamma \vdash \mathsf{new}_\mathsf{p}$ is $\mathsf{new}_\mathsf{p}$. This name $\mathsf{new}_\mathsf{p}$ comes from some $\mathsf{p} \in C$, and so $\Gamma \Vdash (\text{the } \mathsf{p}) \downarrow$. Using $(\downarrow_5)$, $\Gamma \Vdash$ the p is the p. And, $(\mathsf{new}_\mathsf{p}$ is $\mathsf{new}_\mathsf{p})^\#$ is the p is the p. The rest of the proof here is routine.

Finally, suppose that $\Gamma^\flat$ is inconsistent, say with $\Gamma^\flat \vdash \varphi, \overline{\varphi}$. By the previous part, $\Gamma \Vdash \varphi^\#, \overline{\varphi}^\#$. It is easy to check that $(\overline{\varphi})^\# = \overline{\varphi^\#}$. So indeed, $\Gamma$ is inconsistent. □

**Theorem 5.11.** *The logical system in this section is complete.*

*Proof.* Let $\sigma$ be a syntactic expression in our language $\mathcal{S}^\dagger_{\mathsf{acc}}$, and suppose that $\Gamma \nVdash \sigma$. We have the case that $\sigma$ is a definedness sentence $\mathsf{t}\downarrow$ or $\varphi\downarrow$, and then we also have the case when $\sigma$ is a sentence $\varphi$.

We first consider the case of a definedness sentence. Since $\Gamma \not\Vdash \sigma$, $\sigma$ can only be a term the p for some literal p, or else a sentence t is(n't) u, where either $\Gamma \not\models t\downarrow$ or $\Gamma \not\models u\downarrow$. Let us handle the case when $\sigma$ is the p. (The other cases follow easily from this one.) We move to $\Gamma^\flat$, and we claim that p is not a singular literal. (For if $\Gamma^\flat \vdash$ a is the p, then by $(\downarrow_2)$, $\Gamma \Vdash$ a is the p. And in this case, we would have $\Gamma \Vdash$ the p$\downarrow$, giving a contradiction.) Then by the proof of Theorem 4.10, there is a model $\mathcal{M} \models \Gamma^\flat$ such that $[\![p]\!]$ is not a singleton. This same model can be viewed as a structure for the language of this section. As such $\mathcal{M} \not\models \Gamma$.

For the rest of this proof, we are going to assume that $\sigma$ is a sentence. To match our earlier notation, let's call it $\varphi$. So our assumption is that $\Gamma \not\Vdash \varphi$. We find a model $\mathcal{M} \models \Gamma$ such that $\mathcal{M} \not\models \varphi$.

There are two overall cases, depending on whether $\Gamma \Vdash \varphi\downarrow$ or not.

Let us first assume that $\Gamma \not\Vdash \varphi\downarrow$. In this case, there is some literal p which occurs in $\varphi$ such that $\Gamma \not\Vdash$ the p$\downarrow$. Just as in the first paragraph of this proof, there is a model $\mathcal{M} \models \Gamma^\flat$ such that $[\![p]\!]$ is not a singleton. This same model can be viewed as a structure for the language of this section. As such $\mathcal{M} \models \Gamma$. And since p occurs in $\varphi$, and $[\![p]\!]$ is not a singleton, we see that $\mathcal{M} \not\models \varphi$. So we are done in this case.

Next, we consider the case when $\Gamma \Vdash \varphi\downarrow$. We claim that $\Gamma \cup \{\overline{\varphi}\}$ is consistent. For suppose that $\psi$ is a sentence and $\Gamma \cup \{\overline{\varphi}\} \Vdash \psi, \overline{\psi}$. Now $\Gamma \Vdash \overline{\varphi}\downarrow$ (easily). So we use (RAA)$_\downarrow$ to see that $\Gamma \Vdash \varphi$. This is a contradiction. Our claim is proved.

By Lemma 5.10, $\Gamma^\flat \cup \{\overline{\varphi}^\flat\}$ is consistent in the logical system for $\mathcal{S}^\dagger_{\text{names,the}}$. Thus $\Gamma^\flat \cup \{\overline{\varphi}^\flat\}$ has a model, by Theorem 4.10. This gives a model $\mathcal{M}$ of $\Gamma$, and it also satisfies $(\overline{\varphi}^\flat)^\# = \overline{\varphi}$. So $\mathcal{M} \not\models \varphi$, as desired. □

## 6 Conclusion

This paper suggested additions to a small syllogistic logic intended to handle names and definite descriptions. It did so while commenting on the program of natural logic and how this relates to the overall topic of this book, formal term logic. The technical contribution was a series of completeness theorems. The work on $\mathcal{S}^\dagger$ in Section 2 was not new, and the extension of it to $\mathcal{S}^\dagger_{\text{names}}$ in Section 3 was unsurprising. Section 4 added definite descriptions results in the logic $\mathcal{S}^\dagger_{\text{names,the}}$. The axiomatization and completeness proof were more difficult, by far. We ended with a speculative topic, the neutral free logic treatment of a logic a little larger than $S^\dagger_{\text{names,the}}$. The noteworthy feature of this last logic was that the explicit statements of definedness for the definite descriptions and sentences were first-class citizens in the syntax and proof system.

I mentioned in the Introduction that some of the motivation for these logics was didactic. I do teach the orthoposet-based completeness proof for $S^\dagger$ to students who want to learn mathematical proofs by working with logics. Looking back at the paper, I would add the work in Section 3. The rest of the paper is too involved for beginning students, and the topic overall is for specialists.

All of the work in the paper had to do with fragments of logics. I can see from the collection Englebretsen [4] that several people who were working on TFL in the 1980's were also concerned with fragments. None of them asked about logics like the ones in this paper, logics which started with a weak base system and added logically interesting devices. I would like to think that if they had asked about definite descriptions and done so without casting everything as a subsystem of first-order logic, they might have found logical systems like those in Section 4, or perhaps other ones besides. Be that as it may, I am interested in the axiomatizations and hope that readers of this paper will find the ones here to be elegant and illuminating.

I made a few programmatic points about natural logic (NL) in the Introduction. Here is another set of points about it, from Johan van Benthem's paper [17]. He writes

> In the 1980s, the idea arose that ... natural language is not just a medium for saying and communicating things, but that it also has a 'natural logic', viz. a system of modules for ubiquitous forms of reasoning that can operate directly on natural language surface form. This idea was developed in some detail in [15, 16]. The main proposed ingredients were two [sic] general modules:
>
> (a) Monotonicity Reasoning, i.e., Predicate Replacement,
>
> (b) Conservativity, i.e., Predicate Restriction, and also

(c) Algebraic Laws for inferential features of specific lexical items.

But of course, one can think of many further natural subsystems inside natural language, such as reasoning about collective predication, prepositions, anaphora, tense and temporal perspective, etcetera. Of course, the challenge is then to see how much inference can be done directly on natural language surface form.

Interest in "reasoning that can operate directly on natural language surface form" is the common feature of TFL and NL. Looking back at TFL and Sommers' work, I do feel that the area pursues ingredients (a) and (c), and even sometimes (b), even though the area does not seem to think in terms of "ingredients" and "modules." For just one example, early in this paper I mentioned Appendix D of Sommers [12]. Appendix C of the same book is about 'Any', and Sommers' observations on this word are still worth reading. They are observations that we would now think of as monotonicity facts. Obviously, interest in the algebraic laws of inference are prominent in the whole TFL area. These are also the main interest here. Without mentioning it, the end of this paper is a connection to the language subsystem having to do with *presuppostion*, *accommodation*, and to *pragmatics* more generally. This, too, was in effect considered by Sommers. For his introduction of the notion of "amplitude" is a move in this direction. He writes (see [12], p. 213):

> The amplitude of a term in a statement is determined by my knowledge of the meaning of that statement and this in turn is a matter of my knowing the existential conditions for the truth of the statement.

To get a more adequate logical treatment of the last fragment in this paper, one might look at theories of presupposition and accommodation, at the literature on the *question under discussion* and other related topics. We leave this project to future work.

## Acknowledgements

I am especially grateful to George Englebretsen for encouraging me to write this paper and for his patience with me, and for everything he has done over even more years to advance traditional formal logic.

## References

[1] David Beaver and Emiel Krahmer. A partial account of presupposition projection. *J. Log. Lang. Inf.*, 10(2):147–182, 2001.

[2] C. S. Calude, P. H. Hertling, and K. Svozil. Embedding quantum universes into classical ones. *Foundations of Physics*, 29(3):349–379, 1999.

[3] Jörg Endrullis and Lawrence S. Moss. Syllogistic logic with "most". *Math. Struct. Comput. Sci.*, 29(6):763–782, 2019.

[4] George Englebretsen, editor. *The New Syllogistic*. Peter Lang Publishing, 1987.

[5] F. Katrnoška. On the representation of orthocomplemented posets. *Comment. Math. Univ. Carolinae*, 23:489–498, 1982.

[6] Clifton McIntosh. Appendix F. In *The Logic of Natural Language*, pages 387–425. Clarendon Press, Oxford, 1982.

[7] Lawrence S. Moss. Syllogistic logic with complements. In *Games, Norms and Reasons: Proceedings of the Second Indian Conference on Logic and its Applications*, page 19 pp. Springer Synthese Library Series, Mumbai, 2010.

[8] Edi Pavlović and Norbert Gratzl. Neutral free logic: Motivation, proof theory and models. *J. Philos. Log.*, 52(2):519–554, 2023.

[9] Stanley Peters and Dag Westerståhl. *Quantifiers in Language and Logic*. Clarendon Press, 2006.

[10] Ian Pratt-Hartmann and Lawrence S. Moss. Logics for the relational syllogistic. *Review of Symbolic Logic*, 2(4):647–683, 2009.

[11] William C. Purdy. Surface reasoning. *Australasian Journal of Logic 4*, 33(1):13–36, 2006.

[12] Fred Sommers. *The Logic of Natural Language*. Clarendon Press, Oxford, 1982.

[13] Fred Sommers and George Englebretsen. *An Invitation to Formal Reasoning: the logic of terms*. Routledge, 2000.

[14] V. M. Sánchez Valencia. *Studies on Natural Logic and Categorical Grammar*. PhD thesis, University of Amsterdam, 1991.

[15] Johan van Benthem. *Essays in Logical Semantics*, volume 29 of *Studies in Linguistics and Philosophy*. D. Reidel Publishing Co., Dordrecht, 1986.

[16] Johan van Benthem. *Language in Action*, volume 130 of *Studies in Logic and the Foundations of Mathematics*. North Holland, Amsterdam, 1991.

[17] Johan van Benthem. A brief history of natural logic. In M. Chakraborty, B. Löwe, M. Nath Mitra, and S. Sarukkai, editors, *Logic, Navya-Nyaya and Applications, Homage to Bimal Krishna Matilal*. College Publications, London, 2008.

[18] David A. Wheeler. The origin of All Men are Mortal. https://dwheeler.com/essays/all-men-are-mortal.html, 2023. 2023-02-01 (original 2019-07-07).

[19] N. Zierler and M. Schlessinger. Boolean embeddings of orthomodular sets and quantum logic. *Duke Mathematical Journal*, 32:251–262, 1965.

# Aristotle, Term Logic, and QUARC

Jonas Raab

## 1 Introduction

It is widely agreed that Aristotle is the inventor of *formal* logic. The logic he develops remains the dominant one until Gottlob Frege introduces his logical language in form of the *Begriffsschrift* [24]. As one might expect, their logics and formal languages are strikingly different. Aristotle develops a *term logic*, i.e., a logic which concerns the relation between *terms*. Terms can be affirmed or denied of terms, and can be assigned different quantities.

Fregean languages, on the other hand, distinguish different elements, such as *predicates-symbols*, *individual-constants*, *variables*, and *logical symbols*.[1] This goes beyond the language of term logic in several respects. In particular, terms most closely correspond to certain predicate-symbols, but not every predicate-symbol can easily be considered to be a term. Moreover, the Fregean language knows *quantifiers* which directly indicate something like the quantity in question, whereas Aristotle's term logic does not include them.[2]

Fregean languages are a success-story. Ever since their introduction, they almost completely superseded term languages. The power and flexibility of Fregean languages made the term approach pretty much obsolete—which is also one of the main reasons to prefer Fregean languages. This, however, does not mean that there is *no* competition; and it is the competition that we are interested in here.

---

[1] Frege [24] introduces what we now consider a *second-order* language; however, only the first-order fragment is relevant here.

[2] Note that this does *not* mean that Aristotle and Aristotle's logic do not know *quantification*.

Two of the competitors are Fred Sommers's so-called *Term Functor Logic* (TFL) and Hanoch Ben-Yami's so-called *QUantified ARgument Calculus* (QUARC). Both Sommers and Ben-Yami point towards Aristotelian logic as a potential ally, and as a reason to reject Fregean languages. This is why I consider Aristotle's approach as a base for both TFL and QUARC. Moreover, as both systems attempt to replace the Fregean approach, it is necessary to compare them to it. Overall, we are interested in a somewhat four-fold comparison between Aristotle's logic, the Fregean approach, TFL, and QUARC.

This paper is structured as follows. Section 2 discusses Aristotle's logic. Section 3 provides a generic picture of the Fregean approach as currently understood, but in a form more suitable for our purposes. Section 4 introduces Ben-Yami's QUARC and Section 5 Sommers's TFL; Section 6 compares the systems, though I also compare the approaches within the previous sections. Section 7 concludes the paper.

## 2 Aristotle's Logic

In [40], I reconstruct Aristotle's assertoric logic in a subsystem of QUARC and show that the reconstruction is very close to the original text. The target of the reconstruction is only the first few chapters of Aristotle's *Prior Analytics* (viz., APr A1–6), but I suggest how to introduce complex terms [40, §3.5] which are not to be found in those chapters. This original extension is the relevant one for our purposes, and there is some textual evidence that that's the version Aristotle had in mind (see Section 2.4). We encounter the formalism in Section 2.9.

To arrive there, I don't just consider Aristotle's *Prior Analytics*, but the whole so-called *Organon*. One question to be asked (but, unfortunately, not really answered) is why Aristotle developed a *term* logic. Another question is what counts as a *term* to begin with. In order to answer these questions, I reconstruct parts of the *Organon*, though I cannot discuss every aspect.

In the following, I put the quotations of cited passages—including the Greek text—into footnotes (and I'd suggest ignoring them for the most part).[3]

---

[3]I follow the following translations and Greek texts (though streamline the trans-

## 2.1 Ti Kata Tinos

The general picture is something like this. In *On Interpretation*, Aristotle distinguishes between words (ὄνομα/onoma)[4] and verbs (ῥῆμα/rhêma) (Int 1, 16a1)[5], both of which can then be considered to be *terms* (ὅρος/horos) (Int 3, 16b19f.[6], APr A1, 24b16[7]). Terms on their own do not constitute a sentence and are neither true nor false, yet they are meaningful (Int 1, 16a13–16[8]); a sentence (λόγος/logos)[9] is constituted by the combination of a word and a verb, i.e., by combining appropriate terms. However, not every sentence is significant, i.e., true or false (Int 4, 17a3f.[10]), but every significant sentence must include a verb (Int 5,

---

lation of the technical terms etc.): *Categories* (Cat) and *On Interpretation* (Int): J. L. Ackrill's translation as printed in [5], Greek taken from [1]; *Topics* (Top): R. Smith's translation of Books A and H as printed in [9], all other books by W. A. Pickard-Cambridge as printed in [12], Greek taken from [3]; *Sophistical Refutations* (SE): W. A. Pickard-Cambridge's translation as printed in [12], Greek taken from [3]; *Prior Analytics* (APr): G. Striker's translation of Book A as printed in [10], A. J. Jenkinson's translation of Book B as printed in [12], Greek taken from [4]; *Posterior Analytics* (APo): J. Barnes's translation as printed in [6], Greek taken from [4]; *Metaphysics* (Met): W. D. Ross's translation of Book B as printed in [12], C. Kirwan's translation of Book Γ as printed in [7], D. Bostock's translation of Book Z as printed in [8], Greek taken from [2].

[4]The literal translation is 'name', but what's meant is something like 'word'.

[5]"First we must settle what a word is and what a verb is [Πρῶτον δεῖ θέσθαι τί ὄνομα καὶ τί ῥῆμα]".

[6]"When uttered just by itself a verb is a word and signifies something [αὐτὰ μὲν οὖν καθ' αὑτὰ λεγόμενα τὰ ῥήματα ὀνόματά ἐστι καὶ σημαίνει τι]".

[7]"I call a term that into which a premiss is resolved [Ὅρον δὲ καλῶ εἰς ὃν διαλύεται ἡ πρότασις]".

[8]"Thus words and verbs by themselves—for instance 'man' or 'white' when nothing further is added—are like the thoughts that are without combination and separation; for so far they are neither true nor false [τὰ μὲν οὖν ὀνόματα αὐτὰ καὶ τὰ ῥήματα ἔοικε τῷ ἄνευ συνθέσεως καὶ διαιρέσεως νοήματι, οἷον τὸ ἄνθρωπος ἢ λευκόν, ὅταν μὴ προστεθῇ τι· οὔτε γὰρ ψεῦδος οὔτε ἀληθές πω]".

[9]In APr, Aristotle uses a different word; see below and cf. Kneale and Kneale [26, pp. 34f.].

[10]"There is not truth or falsity in all sentences: a prayer is a sentence but is neither true nor false [οὐκ ἐν ἅπασι δὲ ὑπάρχει, οἷον ἡ εὐχὴ λόγος μέν, ἀλλ' οὔτ' ἀληθὴς οὔτε ψευδής]."

17a9f.[11], Int 10, 19b12[12]).[13]

More importantly, a simple sentence *affirms something of something* (τὶ κατὰ τινός/ti kata tinos) or *denies something of something* (τὶ ἀπὸ τινός/ti apo tinos) (Int 5, 17a20f.[14]; see also, e.g., APo A2, 72a13f.[15])—a structure also appearing in Aristotle's *Metaphysics* (e.g., Met Z17, 1041a20–23[16]).[17] Aristotle also speaks of 'compounded' sentences (Int 5, 17a21f.[18]), though it does not appear that he is concerned with them again throughout the *Organon* (with, maybe, a few exceptions; see below).

A sentence is made up of terms which signify something (Int 4, 16b26f.[19], Int 6, 17a25f.[20]), but, as the 'ti kata/apo tinos' suggests, terms need to be combined (via copula) in order to affirm or deny (Int 4, 16b28ff.[21]). Given this basic structure, we can distinguish between sim-

---

[11]"Every statement-making sentence must contain a verb or an inflexion of a verb [ἀνάγκη δὲ πάντα λόγον ἀποφαντικὸν ἐκ ῥήματος εἶναι ἢ πτώσεως]".

[12]"Without a verb there will be no affirmation or denial [ἄνευ δὲ ῥήματος οὐδεμία κατάφασις οὐδ' ἀπόφασις]".

[13]As has often been noted, Aristotle's writings are ambiguous as to whether claims are about *linguistic expressions* or about *things* expressed by those expressions (see, e.g., Kneale and Kneale [26, §II.2]). That's less of a problem in *De Interpretatione*, but certainly so in the *Categories*; I generally take Aristotle to be interested in *things*, not linguistic items.

[14]"Of these the one is a simple statement, affirming or denying something of something [τούτων δ' ἡ μὲν ἁπλῆ ἐστιν ἀπόφανσις, οἷον τὶ κατὰ τινὸς ἢ τὶ ἀπὸ τινός]".

[15]"The part of a contradictory pair which says something of something is affirmation; the part which takes something from something is a denial [μόριον δ' ἀντιφάσεως τὸ μὲν τὶ κατὰ τινὸς κατάφασις, τὸ δὲ τὶ ἀπὸ τινὸς ἀπόφασις]".

[16]"However, one could ask why a man is such a kind of animal. It is clear that this is not to ask why one who is a man is a man. So what one asks is why it is that one thing is affirmed of another [ζητήσειε δ' ἄν τις διὰ τί ὁ ἄνθρωπός ἐστι ζῷον τοιονδί. τοῦτο μὲν τοίνυν δῆλον, ὅτι οὐ ζητεῖ διὰ τί ὅς ἐστιν ἄνθρωπος ἄνθρωπός ἐστιν· τὶ ἄρα κατά τινος ζητεῖ διὰ τί ὑπάρχει]".

[17]The 'ti kata tinos' is important enough to become the title of [46].

[18]"the other is compounded of simple statements and is a kind of composite sentence [ἡ δ' ἐκ τούτων συγκειμένη, οἷον λόγος τις ἤδη σύνθετος]".

[19]"A sentence is a significant spoken sound some part of which is significant in separation [Λόγος δέ ἐστι φωνὴ σημαντική, ἧς τῶν μερῶν τι σημαντικόν ἐστι κεχωρισμένον]".

[20]"An affirmation is statement affirming something of something, a denial is a statement denying something of something [κατάφασις δέ ἐστιν ἀπόφανσις τινὸς κατὰ τινός, ἀπόφασις δέ ἐστιν ἀπόφανσις τινὸς ἀπὸ τινός]."

[21]"I mean that 'animal', for instance, signifies something, but not that it is or is

ple sentences which affirm or deny something of a subject, and complex sentences which are compounds of simple sentences (Int 5, 17a20ff.[22]). However, as far as I can tell, Aristotle does not mention compounded sentences again, and he does not specify modes of composition (though see Section 2.4).

The basic unit is a sentence which contains two terms, viz., a subject and a verb, where the verb is said/predicated of the subject. Given this basic unit, a few more distinctions are possible. Aristotle distinguishes between things (πράγματα/pragmata) which are universal (καθόλου/katholou) and those which are particular (καθ' ἕκαστον/kath hekaston). Note right away that, in his *Prior Analytics*, Aristotle uses a different expression when referring to a kind of sentence, viz., 'ἐν μέρει' (en merei) (e.g., APr A1, 24a17[23]), which is also translated as 'particular', though a more literal translation would be 'in part'.

The distinction that Aristotle draws is between universal and particular things. He calls 'things' like *human being* 'universal', and 'things' like *Callias* or *Socrates* 'particular'. The distinction is drawn by considering what something can be said of: universal things can be said of several things, particulars cannot (Int 7, 17a38–b1[24])—more on that in Section 2.5.

---

not (though it will be an affirmation or denial if something is added) [λέγω δέ, οἷον ἄνθρωπος σημαίνει τι, ἀλλ' οὐχ ὅτι ἔστιν ἢ οὐκ ἔστιν (ἀλλ' ἔσται κατάφασις ἢ ἀπόφασις ἐάν τι προστεθῇ)]".

[22]"Of these the one is a simple statement, affirming or denying something of something, the other is compounded of simple statements and is a kind of composite sentence [τούτων δ' ἡ μὲν ἁπλῆ ἐστὶν ἀπόφανσις, οἷον τὶ κατὰ τινὸς ἢ τὶ ἀπὸ τινός, ἡ δ' ἐκ τούτων συγκειμένη, οἷον λόγος τις ἤδη σύνθετος]."

[23]"... and this is either universal or particular or indeterminate [οὗτος δὲ ἢ καθόλου ἢ ἐν μέρει ἢ ἀδιόριστος]."

[24]"Now of actual things some are universal, others particular (I call universal that which is by its nature predicated of a number of things, and particular that which is not; man, for instance, is a universal, Callias an particular ['Ἐπεὶ δέ ἐστι τὰ μὲν καθόλου τῶν πραγμάτων τὰ δὲ καθ' ἕκαστον, – λέγω δὲ καθόλου μὲν ὃ ἐπὶ πλειόνων πέφυκε κατηγορεῖσθαι, καθ' ἕκαστον δὲ ὃ μή, οἷον ἄνθρωπος μὲν τῶν καθόλου Καλλίας δὲ τῶν καθ' ἕκαστον]".

## 2.2 Universals and Universally

Both universal and particular things can be the subject of sentences so that things can be said of them (Int 7, 17b1ff.[25])—and, in the case of universal things, that in either of two ways, viz., universally (καθόλου ἀποφαίνηται/katholou apophainêtai) or not (Int 7, 17b3ff.[26]). Examples of something being said universally of a universal are 'every human being is white' and 'no human being is white' (Int 7, 17b5f.[27]). It is of a universal thing, because 'human being' signifies one; and it is said universally, because it is said of every/none of those things.[28] The first of these two sentences counts as affirming something of something (ti kata tinos), whereas the latter as denying something of something (ti apo tinos) as the mode of predication changes, though the latter is not the negation of the former (see Section 2.3).

Something is said of a universal *not universally* when the subject is a universal thing, but the predication is not universally. The examples Aristotle provides are 'human being is white' and 'human being is not white' (Int 7, 17b8ff.[29]). The examples are of universals as 'human being' signifies a universal thing. However, the predications are not universal, because of the quantity of the subject. Regarding this, Aristotle also insists: '"every" does not signify the universal but that it is taken

---

[25]"So it must sometimes be of a universal that one states that something holds or does not, sometimes of an particular [ἀνάγκη δ' ἀποφαίνεσθαι ὡς ὑπάρχει τι ἢ μή, ὁτὲ μὲν τῶν καθόλου τινί, ὁτὲ δὲ τῶν καθ' ἕκαστον]".

[26]"Now if one states universally of a universal that something holds or does not [ἐὰν μὲν οὖν καθόλου ἀποφαίνηται ἐπὶ τοῦ καθόλου ὅτι ὑπάρχει ἢ μή]".

[27]"examples of what I mean by 'stating universally of a universal' are 'every man is white' and 'no man is white' [λέγω δὲ ἐπὶ τοῦ καθόλου ἀποφαίνεσθαι καθόλου, οἷον πᾶς ἄνθρωπος λευκός, οὐδεὶς ἄνθρωπος λευκός]".

[28]Note that the 'no human being is white' can actually be rendered differently, making the universal character explicit: 'every human being is not white'. In this formulation, it is clear that something is predicated universally—and that's the more appropriate way to understand it in the general subject/predicate structure together with the quantity and positive/negative copula involved; in the example sentence, it is *every human being* of whom *white* is *not* said, combining universal quantity with 'negative' predication, i.e., denial (ti apo tinos).

[29]"Examples of what I mean by 'stating of a universal not universally' are 'a human being is white' and 'a human being is not white' [λέγω δὲ τὸ μὴ καθόλου ἀποφαίνεσθαι ἐπὶ τῶν καθόλου, οἷον ἔστι λευκὸς ἄνθρωπος, οὐκ ἔστι λευκὸς ἄνθρωπος]".

universally' (Int 7, 17b11f.[30], cf. also Int 10, 20a9f.[31]). This indicates where the quantity is meant to be applied to. Aristotle rejects that sentences such as 'every human being is every animal' (Int 7, 17b15f.[32]) can ever be true (Int 7, 17b12–15[33]). The quantity is meant to indicate of 'how much' of the subject-term the predicate-term is said.

It is less clear how Aristotle thinks about subjects which are particulars. He affirms that the sentences 'Socrates is white' and 'Socrates is not white' are *contradictories* (Int 7, 17b26–29[34]), but he does not mention anything like a quantity in such cases. Indeed, such sentences only occur very sparingly and do not get a proper discussion (see also Section 2.6).

## 2.3 Affirmation, Denial, and Truth

What Aristotle tells us, though, is how affirmation and denial are related:

> the denial must deny the same things as the affirmation affirmed,
> and of the same thing, whether an individual or a universal (taken
> either universally or not universally). (Int 7, 17b38–18a1[35])

A sentence is only then a denial of another sentence if the terms are the same; the denial of 'every human being is white' is 'not every human being is white',[36] i.e., we keep the terms as they are, and, in a sense, we

---

[30]"τὸ γὰρ πᾶς οὐ τὸ καθόλου σημαίνει ἀλλ' ὅτι καθόλου".

[31]'For "every" does not signify a universal, but that it is taken universally [τὸ γὰρ πᾶς οὐ τὸ καθόλου σημαίνει, ἀλλ' ὅτι καθόλου]'.

[32]"ἔστι πᾶς ἄνθρωπος πᾶν ζῷον".

[33]"It is not true to predicate a universal universally of a subject, for there cannot be an affirmation in which a universal is predicated universally of a subject [ἐπὶ δὲ τοῦ κατηγορουμένου τὸ καθόλου κατηγορεῖν καθόλου οὐκ ἔστιν ἀληθές· οὐδεμία γὰρ κατάφασις ἔσται, ἐν ᾗ τοῦ κατηγορουμένου καθόλου τὸ καθόλου κατηγορηθήσεται]".

[34]'Of contradictory statements about a universal taken universally it is necessary for one or the other to be true or false; similarly if they are about particulars, e.g. "Socrates is white" and "Socrates is not white". [ὅσαι μὲν οὖν ἀντιφάσεις τῶν καθόλου εἰσὶ καθόλου, ἀνάγκη τὴν ἑτέραν ἀληθῆ εἶναι ἢ ψευδῆ, καὶ ὅσαι ἐπὶ τῶν καθ' ἕκαστα, οἷον ἔστι Σωκράτης λευκός – οὐκ ἔστι Σωκράτης λευκός]'.

[35]"τὸ γὰρ αὐτὸ δεῖ ἀποφῆσαι τὴν ἀπόφασιν ὅπερ κατέφησεν ἡ κατάφασις, καὶ ἀπὸ τοῦ αὐτοῦ, ἢ τῶν καθ' ἕκαστά τινος ἢ ἀπὸ τῶν καθόλου τινός, ἢ ὡς καθόλου ἢ ὡς μὴ καθόλου".

[36]Note that that's technically not correct, as Aristotle does not recognize *sentence negation*; nevertheless, for present purposes, I put it like this.

also keep the quantity, though the negation acts on it. Aristotle does not discuss the complex case, but only suggests the following sentences as examples: 'Socrates is white' has as denial 'Socrates is not white' (Int 7, 18a2f.[37]). The correct denial of the more complex sentences is arrived at after further discussion (see, e.g., Int 10, 19b14–18[38]).

The underlying idea is still that of *ti kata tinos*: saying something of something. '*A kata B*' has as its denial '*A apo B*'; the terms remain the same. Aristotle does not specify the denial of '*A apo B*', though we can take the '*A kata B*' as its denial, assuming the only options to be *ti kata tinos* and *ti apo tinos*.

Given this picture, Aristotle suggests when sentences are true and false:

> For it is true to say that it is white or is not white, it is necessary for it to be white or not white; and if it is white or is not white, then it was true to say or deny this. If it is not the case it is false, if it is false it is not the case. (Int 9, 18a39–b3[39])

This understanding of truth is pretty much the same as that in his *Metaphysics* (Met Γ7, 1011b25ff.[40]). The general idea is that if we have a sentence, there are two terms involved, and one term is affirmed/denied of the other. Now, a sentence is true, if what is said actually obtains, and it is false if not. Moreover, under certain conditions, if a sentence is false, its denial is true—since the denial keeps the terms etc. intact,

---

[37] "λέγω δὲ οἷον ἔστι Σωκράτης λευκός – οὐκ ἔστι Σωκράτης λευκός".

[38] "So a first affirmation and denial are: 'a man is', 'a man is not'; then, 'a not-man is', 'a not-man is not'; and again, 'every man is', 'every man is not', 'every not-man is', 'every not-man is not' [ὥστε πρώτη κατάφασις καὶ ἀπόφασις τὸ ἔστιν ἄνθρωπος – οὐκ ἔστιν ἄνθρωπος, εἶτα ἔστιν οὐκ ἄνθρωπος – οὐκ ἔστιν οὐκ ἄνθρωπος, πάλιν ἔστι πᾶς ἄνθρωπος – οὐκ ἔστι πᾶς ἄνθρωπος, ἔστι πᾶς οὐκ ἄνθρωπος – οὐκ ἔστι πᾶς οὐκ ἄνθρωπος]".

[39] "εἰ γὰρ ἀληθὲς εἰπεῖν ὅτι λευκὸν ἢ οὐ λευκόν ἐστιν, ἀνάγκη εἶναι λευκὸν ἢ οὐ λευκόν, καὶ εἰ ἔστι λευκὸν ἢ οὐ λευκόν, ἀληθὲς ἦν φάναι ἢ ἀποφάναι· καὶ εἰ μὴ ὑπάρχει, ψεύδεται, καὶ εἰ ψεύδεται, οὐχ ὑπάρχει".

[40] "This will be plain if we first define what truth and falsehood are: for to say that that which is is not or that which is not is, is a falsehood; and to say that that which is is and that which is not is not, is true; so that, also, he who says that a thing is or not will have the truth or be in error [δῆλον δὲ πρῶτον μὲν ὁρισαμένοις τί τὸ ἀληθὲς καὶ ψεῦδος. τὸ μὲν γὰρ λέγειν τὸ ὂν μὴ εἶναι ἢ τὸ μὴ ὂν εἶναι ψεῦδος, τὸ δὲ τὸ ὂν εἶναι καὶ τὸ μὴ ὂν μὴ εἶναι ἀληθές, ὥστε καὶ ὁ λέγων εἶναι ἢ μὴ ἀληθεύσει ἢ ψεύσεται]".

and similarly the other way around. Also, if $B$ is $A$, it is true to make a corresponding claim ('$A$ kata $B$'), and false to assert the corresponding denial ('$A$ apo $B$'); and if $B$ is not $A$, it is true to deny that $B$ is $A$ ('$A$ apo $B$'), and false to affirm it ('$A$ kata $B$').

## 2.4 Complex Terms

We can also note that, in *On Interpretation*, Aristotle allows *negated terms*, i.e., it is not only sentences which are denials, but we can have affirmations involving negated terms. One of the examples is 'not-human being' (e.g., Int 10, 19b37[41]); another is a negated verb: 'not-just' (Int 10, 19b28[42]). Thus, we can form affirmations out of negated terms: every non-human being is not-just. (Cf. also, e.g., Top E6, 136a33f.[43])

Furthermore, Aristotle also does not exclude the possibility of further *complex terms*. His standard example is 'cloak' (ἱμάτιον/himation) as word for something more complex (an example also occurring at Met Z4, 1029b25–28[44]). For example, Aristotle suggests to introduce the term 'cloak' for the complex 'horse and man', though he denies a certain unity to sentences containing such terms; he rather thinks they are equivalent to compounded sentences (Int 8, 18a19–23[45]).

Aristotle does not say much more about these complexes, though he does say more about the relationship of sentences involving negated terms and denials:

---

[41] "τὸ οὐκ ἄνθρωπος".

[42] "οὐ δίκαιος".

[43] "Thus (e.g.) inasmuch as animate is a property of living creature, animate will not be a property of not-living creature [οἷον ἐπεὶ τοῦ ζῴου ἴδιον τὸ ἔμψυχον, οὐκ ἂν εἴη τοῦ μὴ ζῴου ἴδιον τὸ ἔμψυχον]".

[44] "We must see, therefore, whether there is a formula of what being is for each of these compounds, and whether these too have a what-being-is, e.g. a white man. Suppose 'cloak' to be a word for this [σκεπτέον ἄρ' ἔστι λόγος τοῦ τί ἦν εἶναι ἑκάστῳ αὐτῶν, καὶ ὑπάρχει καὶ τούτοις τὸ τί ἦν εἶναι, οἷον λευκῷ ἀνθρώπῳ [τί ἦν λευκῷ ἀνθρώπῳ]. ἔστω δὴ ὄνομα αὐτῷ ἱμάτιον]".

[45] "Suppose, for example, that one gave the word 'cloak' to horse and man; 'a cloak is white' would not be a single affirmation. For to say this is no different from saying 'a horse and a man is white', and this is no different from saying 'a horse is white and a man is white' [οἷον εἴ τις θεῖτο ὄνομα ἱμάτιον ἵππῳ καὶ ἀνθρώπῳ, τὸ ἔστιν ἱμάτιον λευκόν, αὕτη οὐ μία κατάφασις [οὐδὲ ἀπόφασις μία]· οὐδὲν γὰρ διαφέρει τοῦτο εἰπεῖν ἢ ἔστιν ἵππος καὶ ἄνθρωπος λευκός, τοῦτο δ' οὐδὲν διαφέρει τοῦ εἰπεῖν ἔστιν ἵππος λευκὸς καὶ ἔστιν ἄνθρωπος λευκός]".

> 'No human being is just' follows from 'every human being is not-just', while the contradictory of this, 'not every human being is not-just', follows from 'some human being is just'[.] (Int 10, 20a20–23[46])[47]

Thus, if the predicate-term is negated, the former sentence implies a denial with unnegated predicate-term; and the positive sentence, likewise, implies a denial with negated predicate-term.

Aristotle also suggests the following:

> 'every not-man is not-just' signifies the same as 'no not-man is just'. (Int 10, 20a39f.[48])[49]

This suggests the equivalence of denial and affirmation with negated predicate-terms, though one direction is problematic (see n. 49).

## 2.5 Categories of Terms

Potentially moving away from Aristotle's *formal logic*, let us consider his *Categories* which categorizes the terms. Aristotle notes that terms can be said in or without combination (Cat 2, 1a16f.[50]), and it is the classification of terms without combination—that is: the terms, not sentences resulting from their combination—that he is interested in.

The categorization is based on two concepts:

(i)   being said of a subject, and    (ii)   being in a subject.

Applying these concepts gives rise to a four-fold categorization:

---

[46]"ἀκολουθοῦσι δ' αὗται, τῇ μὲν πᾶς ἐστὶν ἄνθρωπος οὐ δίκαιος ἡ οὐδείς ἐστιν ἄνθρωπος δίκαιος, τῇ δὲ ἔστι τις δίκαιος ἄνθρωπος ἡ ἀντικειμένη ὅτι οὐ πᾶς ἐστὶν ἄνθρωπος οὐ δίκαιος".

[47]These are captured by one of the semantics in Section 2.9; see Theorem 16. The former claim is an instance of (1) (where $\|\overline{\overline{B}}\|_{\mathfrak{M}_A} = \|B\|_{\mathfrak{M}_A}$), the latter of (3).

[48]"τὸ δὲ πᾶς οὐ δίκαιος οὐκ ἄνθρωπος τῷ οὐδεὶς δίκαιος οὐκ ἄνθρωπος ταὐτὸν σημαίνει".

[49]Only one direction holds in one of the semantics of Section 2.9, the other not; see Theorem 16 (2). The other semantics validates both directions, but clashes with different claims of Aristotle; see n. 47.

[50]"Of things that are said, some involve combination while others are said without combination [Τῶν λεγομένων τὰ μὲν κατὰ συμπλοκὴν λέγεται, τὰ δὲ ἄνευ συμπλοκῆς]."

(1) being said of a subject and being in a subject ((i) and (ii)),[51]

(2) being said of a subject, but not being in a subject ((i) and not-(ii)),[52]

(3) not being said of a subject, but being in a subject (not-(i) and (ii)),[53] and

(4) neither being said of a subject nor being in a subject (not-(i) and not-(ii)).[54]

Important for our purposes is what distinguishes (4) from (1)–(3): only *particulars* are neither said of a subject, nor in a subject, i.e., *particulars cannot be predicated* (cf. Met Z3, 1028b33–37[55]). This distinguished feature of particulars is why, in the *Categories*, Aristotle calls them 'substance' *in the strictest and primary sense* (Cat 5, 2a11–14[56]). On the other hand, the kinds and genera of primary substances are secondary substances (Cat 5, 2a14ff.[57]), and, as they are instances of (2), they can be predicated.

---

[51] For example: "knowledge is in a subject, the soul, and is also said of a subject, knowledge-of-grammar [ἡ ἐπιστήμη ἐν ὑποκειμένῳ μέν ἐστι τῇ ψυχῇ, καθ' ὑποκειμένου δὲ λέγεται τῆς γραμματικῆς]" (Cat 2, 1b1ff.).

[52] For example: "human being is said of a subject, the particular human being, but is not in any subject [οἷον ἄνθρωπος καθ' ὑποκειμένου μὲν λέγεται τοῦ τινὸς ἀνθρώπου, ἐν ὑποκειμένῳ δὲ οὐδενί ἐστιν]" (Cat 2, 1a21f.).

[53] For example: "the particular knowledge-of-grammar is in a subject, the soul, but is not said of any subject [ἡ τὶς γραμματικὴ ἐν ὑποκειμένῳ μέν ἐστι τῇ ψυχῇ, καθ' ὑποκειμένου δὲ οὐδενὸς λέγεται]" (Cat 2, 1a25ff.).

[54] For example: "the particular human being or particular horse [ὁ τὶς ἄνθρωπος ἢ ὁ τὶς ἵππος]" (Cat 2, 1b4f.).

[55] "Of the several ways in which substance is spoken of, there are at any rate four which are the most important: the substance of a thing seems to be what being is for that thing, and its universal and its genus, and fourthly the subject. The subject is that of which other things are predicated while it itself is predicated of nothing further [Λέγεται δ' ἡ οὐσία, εἰ μὴ πλεοναχῶς, ἀλλ' ἐν τέτταρσί γε μάλιστα· καὶ γὰρ τὸ τί ἦν εἶναι καὶ τὸ καθόλου καὶ τὸ γένος οὐσία δοκεῖ εἶναι ἑκάστου, καὶ τέταρτον τούτων τὸ ὑποκείμενον. τὸ δ' ὑποκείμενόν ἐστι καθ' οὗ τὰ ἄλλα λέγεται, ἐκεῖνο δὲ αὐτὸ μηκέτι κατ' ἄλλου]".

[56] "A substance—that which is called a substance most strictly, primarily, and most of all—is that which is neither said of a subject nor in a subject, e.g. the particular man or the particular horse [Οὐσία δέ ἐστιν ἡ κυριώτατά τε καὶ πρώτως καὶ μάλιστα λεγομένη, ἣ μήτε καθ' ὑποκειμένου τινὸς λέγεται μήτε ἐν ὑποκειμένῳ τινί ἐστιν, οἷον ὁ τὶς ἄνθρωπος ἢ ὁ τὶς ἵππος]".

[57] "The species in which the things primarily called substances are, are called *sec-*

## 2.6 Particulars and Syllogistic

The immediate relevance for us is that *particulars do not occur as terms in Aristotle's syllogistic*—and particulars are not the only examples of such terms. There is a certain symmetry. In his *Prior Analytics*, Aristotle suggests a term hierarchy. At the bottom of the hierarchy, there are terms—particulars (καθ' ἕκαστα/kath hekasta)—which cannot be predicated:

> That some things are by nature such as to be said of nothing else is clear, for more or less every perceptible thing is such as not to be predicated of anything except accidentally—for we do sometimes say that the white thing there is Socrates, or that what is approaching is Callias. (APr A27, 43a32–36[58])

Aristotle even insists that

> of all the things there are, some are such that they cannot be predicated truly and universally of anything else (for instance, Cleon or Callias, that is, what is particular and perceptible)[.] (APr A27, 43a25ff.[59])

Taken together, it seems as if Aristotle is saying that particulars—and pretty much all perceptible things—cannot be predicated. The latter passage just suggests that they cannot be predicated 'truly and universally', but the former suggests something stronger.

This also suggests that Aristotle does not seem to consider identity statements such as 'Socrates is Callias' or even 'Socrates is Socrates'. Whatever the reason, Aristotle does not consider something like 'is Socrates' or just 'Socrates' as a predicate-term.

---

*ondary substances*, as also are the genera of these species [δεύτεραι δὲ οὐσίαι λέγονται, ἐν οἷς εἴδεσιν αἱ πρώτως οὐσίαι λεγόμεναι ὑπάρχουσιν, ταῦτά τε καὶ τὰ τῶν εἰδῶν τούτων γένη]".

[58]"ὅτι μὲν οὖν ἔνια τῶν ὄντων κατ' οὐδενὸς πέφυκε λέγεσθαι, δῆλον· τῶν γὰρ αἰσθητῶν σχεδὸν ἕκαστόν ἐστι τοιοῦτον ὥστε μὴ κατηγορεῖσθαι κατὰ μηδενός, πλὴν ὡς κατὰ συμβεβηκός· φαμὲν γάρ ποτε τὸ λευκὸν ἐκεῖνο Σωκράτην εἶναι καὶ τὸ προσιὸν Καλλίαν".

[59]"Ἁπάντων δὴ τῶν ὄντων τὰ μέν ἐστι τοιαῦτα ὥστε κατὰ μηδενὸς ἄλλου κατηγορεῖσθαι ἀληθῶς καθόλου (οἷον Κλέων καὶ Καλλίας καὶ τὸ καθ' ἕκαστον καὶ αἰσθητόν)".

This situation is mirrored at the top of the hierarchy. Starting from the particular, we reach another limit:

> But that one also comes to a halt if one goes upwards, we will explain later [at APo A22, 83b24–31[60]]; for the moment let this be assumed. (APr A27, 43a36f.[61])

Both ends of the hierarchy consist of terms which are not the target of Aristotle's syllogistic. Aristotle is explicit (my translation):

> Clearly, the things inbetween admit of both (for they can be predicated of others and others of them). And more or less the arguments and investigations are especially about them. (APr A27, 43a40–43[62])

Note that there are two occurrences of 'σχεδόν' ('schedon'), viz., at APr A27, 43a33 and at APr A27, 43a42, which have been translated as 'more or less'; they suggest the possibility of exceptions. For the former occurrence, the exception is already made explicit. Regarding the latter, Aristotle does not indicate what the exception is meant to be.[63]

---

[60]"Thus one thing will not be said to hold of one thing either in the upward or in the downward direction: the incidentals are said of items in the substance of each thing, and these latter are not infinite; and in the upward direction there are both these items and the incidentals, neither of which are infinite. There must therefore be some term of which something is predicated primitively, and something else of this; and this must come to a stop, and there must be items which are no longer predicated of anything prior and of which nothing else prior is predicated. [οὔτ' εἰς τὸ ἄνω ἄρα ἓν καθ' ἑνὸς οὔτ' εἰς τὸ κάτω ὑπάρχειν λεχθήσεται. καθ' ὧν μὲν γὰρ λέγεται τὰ συμβεβηκότα, ὅσα ἐν τῇ οὐσίᾳ ἑκάστου, ταῦτα δὲ οὐκ ἄπειρα· ἄνω δὲ ταῦτά τε καὶ τὰ συμβεβηκότα, ἀμφότερα οὐκ ἄπειρα. ἀνάγκη ἄρα εἶναί τι οὗ πρῶτόν τι κατηγορεῖται καὶ τούτου ἄλλο, καὶ τοῦτο ἵστασθαι καὶ εἶναί τι ὃ οὐκέτι οὔτε κατ' ἄλλου προτέρου οὔτε κατ' ἐκείνου ἄλλο πρότερον κατηγορεῖται]".

[61]"ὅτι δὲ καὶ ἐπὶ τὸ ἄνω πορευομένοις ἵσταταί ποτε, πάλιν ἐροῦμεν· νῦν δ' ἔστω τοῦτο κείμενον".

[62]"τὰ δὲ μεταξὺ δῆλον ὡς ἀμφοτέρως ἐνδέχεται (καὶ γὰρ αὐτὰ κατ' ἄλλων καὶ ἄλλα κατὰ τούτων λεχθήσεται)· καὶ σχεδὸν οἱ λόγοι καὶ αἱ σκέψεις εἰσὶ μάλιστα περὶ τούτων".

[63]One suggestion here would be the term 'being'. Aristotle does not think that *being* forms a genus (see, e.g., Met B3, 998b22 ["But it is not possible that either unity or being should be a genus of things (οὐχ οἷόν τε δὲ τῶν ὄντων ἓν εἶναι γένος οὔτε τὸ ἓν οὔτε τὸ ὄν)"]), so maybe it can be said of everything else, but nothing of it.

Without the exceptions, the admissible terms for Aristotle's syllogistic are those that (i) can be predicated of other terms *and* (ii) have other terms predicated of them (APr A27, 43a41f.). This also makes sense once we consider the conversion rules (Sections 2.7–2.9). But Aristotle sometimes seemingly uses individuals in syllogisms. As far as I can tell, there are only three passages of this sort (in the *Organon*); let me quote the first in full (the second and third in footnotes 65 and 66, respectively):

> For example, if A is said of B and B of C—one might think that when the terms are so related, there is a syllogism, but in fact nothing necessary comes about, nor a syllogism. For let A designate always being, B, thinkable Aristomenes, and C, Aristomenes. Clearly it is true that A belongs to B, for Aristomenes is always thinkable. And it is also true that B belongs to C, for Aristomenes is a thinkable Aristomenes. But A does not belong to C, since Aristomenes is perishable. For no syllogism resulted from terms related in this way; rather, the premiss AB should have been taken as universal. But this is false—to claim that every thinkable Aristomenes always is, given that Aristomenes is perishable. (APr A33, 47b18–29[64]; see also APr A33, 47b29–37[65] and APr B27, 70a16–20[66])

---

[64] "οἷον εἰ τὸ Α κατὰ τοῦ Β λέγεται καὶ τὸ Β κατὰ τοῦ Γ· δόξειε γὰρ ἂν οὕτως ἐχόντων τῶν ὅρων εἶναι συλλογισμός, οὐ γίνεται δ' οὔτ' ἀναγκαῖον οὐδὲν οὔτε συλλογισμός. ἔστω γὰρ ἐφ' ᾧ Α τὸ ἀεὶ εἶναι, ἐφ' ᾧ δὲ Β διανοητὸς Ἀριστομένης, τὸ δ' ἐφ' ᾧ Γ Ἀριστομένης. ἀληθὲς δὴ τὸ Α τῷ Β ὑπάρχειν· ἀεὶ γάρ ἐστι διανοητὸς Ἀριστομένης. ἀλλὰ καὶ τὸ Β τῷ Γ· ὁ γὰρ Ἀριστομένης ἐστὶ διανοητὸς Ἀριστομένης. τὸ δ' Α τῷ Γ οὐχ ὑπάρχει· φθαρτὸς γάρ ἐστιν ὁ Ἀριστομένης. οὐ γὰρ ἐγίνετο συλλογισμὸς οὕτως ἐχόντων τῶν ὅρων, ἀλλ' ἔδει καθόλου τὴν Α Β ληφθῆναι πρότασιν. τοῦτο δὲ ψεῦδος, τὸ ἀξιοῦν πάντα τὸν διανοητὸν Ἀριστομένην ἀεὶ εἶναι, φθαρτοῦ ὄντος Ἀριστομένους."

[65] "Again, let C designate Miccalus, B educated Miccalus, and A, perishing tomorrow. Clearly it is true to predicate B of C, for Miccalus is an educated Miccalus. And also A of B, for an educated Miccalus might perish tomorrow. But to predicate A of C is false. Indeed, this is the same mistake as before, for it is not universally true that any educated Miccalus will perish tomorrow; but when this was not assumed, there was no syllogism. [πάλιν ἔστω τὸ μὲν ἐφ' ᾧ Γ Μίκκαλος, τὸ δ' ἐφ' ᾧ Β μουσικὸς Μίκκαλος, ἐφ' ᾧ δὲ τὸ Α τὸ φθείρεσθαι αὔριον. ἀληθὲς δὴ τὸ Β τοῦ Γ κατηγορεῖν· ὁ γὰρ Μίκκαλός ἐστι μουσικὸς Μίκκαλος. ἀλλὰ καὶ τὸ Α τοῦ Β· φθείροιτο γὰρ ἂν αὔριον μουσικὸς Μίκκαλος. τὸ δέ γε Α τοῦ Γ ψεῦδος. τοῦτο δὴ ταὐτόν ἐστι τῷ πρότερον· οὐ γὰρ ἀληθὲς καθόλου, Μίκκαλος μουσικὸς ὅτι φθείρεται αὔριον· τούτου δὲ μὴ ληφθέντος οὐκ ἦν συλλογισμός]."

[66] "The proof that wise men are good, since Pittacus is good, comes through the

The general point Aristotle makes in the first two passages is that there is a certain danger when it comes to modal syllogisms involving necessity (APr A33, 47b15–18[67])—hence the modal flavour of these passages. They also involve syllogisms from the first figure, so no conversion occurs. The particular only occurs in subject-term position. It is not clear which mood of the first figure is concerned, and Aristotle's remark that 'the premiss AB should have been taken as universal' does not concern the particular Aristomenes who seems to be chosen just to illustrate the modal point.

The third passage involves a third-figure syllogism whose proofs all rely on conversion. However, the discussion is about *enthymemes* (ἐνθύμημα/enthymêma) which stem from the probable (εἰκός/eikos)—where "the probable is a reputable statement" (APr B27, 70a3f.[68]). Enthymemes are syllogisms involving the probable (APr B27, 70a10[69]), and so might be considered to not entirely fit the discussion of the syllogisms as developed in the first chapters of the *Prior Analytics*.[70]

In his *Posterior Analytics*, Aristotle seems to confirm the point that particulars are not said of anything (APo A1, 71a23f.[71]). Moreover, when he explains what 'in itself' (καθ' αὐτά/kath hauta) means, Aristotle re-iterates that there are things which are not said of anything else (APo

---

last figure. Let A stand for good, B for wise men, C for Pittacus. It is true then to predicate both A and B of C—only men do not say the latter, because they know it, though they state the former. [τὸ δ' ὅτι οἱ σοφοὶ σπουδαῖοι, Πιττακὸς γὰρ σπουδαῖος, διὰ τοῦ ἐσχάτου. ἐφ' ᾧ Α τὸ σπουδαῖον, ἐφ' ᾧ Β οἱ σοφοί, ἐφ' ᾧ Γ Πιττακός. ἀληθὲς δὴ καὶ τὸ Α καὶ τὸ Β τοῦ Γ κατηγορῆσαι· πλὴν τὸ μὲν οὐ λέγουσι διὰ τὸ εἰδέναι, τὸ δὲ λαμβάνουσιν]".

[67]"It often happens that we are deceived about syllogisms because of the necessity, as we said before. But sometimes it is due to the similarity in the position of terms. This must not escape our notice. [Πολλάκις μὲν οὖν ἀπατᾶσθαι συμβαίνει περὶ τοὺς συλλογισμοὺς διὰ τὸ ἀναγκαῖον, ὥσπερ εἴρηται πρότερον, ἐνίοτε δὲ παρὰ τὴν ὁμοιότητα τῆς τῶν ὅρων θέσεως· ὅπερ οὐ χρὴ λανθάνειν ἡμᾶς]."

[68]"τὸ μὲν εἰκός ἐστι πρότασις ἔνδοξος".

[69]"An enthymeme is a syllogism starting from probabilities or signs [Ἐνθύμημα δὲ ἐστι συλλογισμὸς ἐξ εἰκότων ἢ σημείων]". I'm leaving out the 'sign' in the discussion.

[70]In particular, the inference seems rather to be an *induction* than a deduction, inferring from a particular case to a general one.

[71]"this occurs when the items are in fact particulars and are not said of any underlying subject [ὅσα ἤδη τῶν καθ' ἕκαστα τυγχάνει ὄντα καὶ μὴ καθ' ὑποκειμένου τινός]."

A4, 73b5–10[72]), and he insists that "every term is always universal" (APo B13, 97b25[73]). Since terms corresponding to particulars are not universal—as particulars are exactly those things which aren't universal (Int 7, 17a38–b1)—relevant terms are not those of particulars.

## 2.7 The Syllogistic

With these preliminaries out of the way, let's consider the syllogistic. Aristotle develops it in his *Prior Analytics*; our focus is the assertoric part. Aristotle starts by suggesting which notions need to be introduced: *sentence/premiss* (πρότασις/protasis), *term* (ὅρος/horos), *syllogism* (συλλογισμός/syllogismos), *this (not) being in that as in a whole* (τὸ ἐν ὅλῳ (μὴ) εἶναι τόδε τῷδε/to en holô (mê) einai tode tôde), *predicate of all* (κατὰ παντὸς κατηγορεῖσθαι/kata pantos katêgoreisthai), and *predicate of none* (κατὰ μηδενὸς κατηγορεῖσθαι/kata mêdenos katêgoreisthai) (APr A1, 24a11–15[74]).

Syllogisms consist of sentences/premisses, and Aristotle defines sentences/premisses as *affirming or denying something of something* (APr A1, 24a16f.[75])—bringing back the *ti kata/apo tinos* structure. Both the 'ti' and the 'tinos', i.e., the predicate and the subject, respectively, are *terms*, as sentences/premisses are resolved in terms which are combined

---

[72]"Again, certain items are not said of some other underlying subject: e.g. whereas what is walking is something different walking (and similarly for what is white), substances, i.e. whatever means this so-and-so, are not just what they are in virtue of being something different. Well, items which are not said of an underlying subject I call things in themselves, and those which are said of an underlying subject I call incidental. [ἔτι ὃ μὴ καθ' ὑποκειμένου λέγεται ἄλλου τινός, οἷον τὸ βαδίζον ἕτερόν τι ὂν βαδίζον ἐστὶ καὶ τὸ λευκὸν ⟨λευκόν⟩, ἡ δ' οὐσία, καὶ ὅσα τόδε τι σημαίνει, οὐχ ἕτερόν τι ὄντα ἐστὶν ὅπερ ἐστίν. τὰ μὲν δὴ μὴ καθ' ὑποκειμένου καθ' αὑτὰ λέγω, τὰ δὲ καθ' ὑποκειμένου συμβεβηκότα]."

[73]"ἀεὶ δ' ἐστί πᾶς ὅρος καθόλου".

[74]"Then, to define what is a premiss, what is a term, and what a syllogism, and which kind of syllogism is perfect and which imperfect. After that, what it is for this to be or not to be in that as in a whole, and what we mean by 'to be predicated of all' or 'of none' [εἶτα διορίσαι τί ἐστι πρότασις καὶ τί ὅρος καὶ τί συλλογισμός, καὶ ποῖος τέλειος καὶ ποῖος ἀτελής, μετὰ δὲ ταῦτα τί τὸ ἐν ὅλῳ εἶναι ἢ μὴ εἶναι τόδε τῷδε, καὶ τί λέγομεν τὸ κατὰ παντὸς ἢ μηδενὸς κατηγορεῖσθαι]."

[75]"A premiss, then, is a sentence that affirms or denies something of something [Πρότασις μὲν οὖν ἐστι λόγος καταφατικὸς ἢ ἀποφατικός τινος κατά τινος]".

by a positive or negative copula (APr A1, 24b16ff.[76]).

There are different ways of affirming/denying something of something, viz., *universally* (καθόλου/katholou), *particularly* (ἐν μέρει/en merei), and *indeterminately* (ἀδιόριστος/adioristos) (APr A1, 24a17[77]). As noted in Section 2.1, the quantity is put as 'ἐν μέρει' (en merei), which can be translated as 'in part'. This contrasts with the universal predication which does not just predicate 'in part', but universally. Aristotle characterizes these as follows:

> By 'universal' I mean belonging to all or to none of something; by 'particular', belonging to some or not to some, or not to all; by 'indeterminate', belonging without universality or particularity, as in 'of contraries there is a single science' or 'pleasure is not a good'. (APr A1, 24a18–22[78])

The universal affirmation and denial say something of all of the subject; the universal affirmation/denial says of all of the subject that a term applies/does not apply to it. The particular affirmation/denial says only of *part of* the subject (hence, the ἐν μέρει/en merei-phrasing) that a term does/does not apply to it. The 'indeterminate' case just does not indicate whether all or only part of the subject is meant; it doesn't play much of a role for us.

Given that terms built up sentences/premisses which say something of something, a *syllogism* is

> an argument in which, certain things being posited, something other than what was laid down results by necessity because these things are so. (APr A1, 24b18ff.[79])

---

[76] "I call a term that into which a premiss is resolved, that is, what is predicated and what it is predicated of, with the addition of 'to be' or 'not to be' [Ὅρον δὲ καλῶ εἰς ὃν διαλύεται ἡ πρότασις, οἷον τό τε κατηγορούμενον καὶ τὸ καθ' οὗ κατηγορεῖται, προστιθεμένου [ἢ διαιρουμένου] τοῦ εἶναι ἢ μὴ εἶναι]".

[77] "and this is either universal or particular or indeterminate [οὗτος δὲ ἢ καθόλου ἢ ἐν μέρει ἢ ἀδιόριστος]".

[78] "λέγω δὲ καθόλου μὲν τὸ παντὶ ἢ μηδενὶ ὑπάρχειν, ἐν μέρει δὲ τὸ τινὶ ἢ μὴ τινὶ ἢ μὴ παντὶ ὑπάρχειν, ἀδιόριστον δὲ τὸ ὑπάρχειν ἢ μὴ ὑπάρχειν ἄνευ τοῦ καθόλου ἢ κατὰ μέρος, οἷον τὸ τῶν ἐναντίων εἶναι τὴν αὐτὴν ἐπιστήμην ἢ τὸ τὴν ἡδονὴν μὴ εἶναι ἀγαθόν".

[79] "συλλογισμὸς δέ ἐστι λόγος ἐν ᾧ τεθέντων τινῶν ἕτερόν τι τῶν κειμένων ἐξ ἀνάγκης συμβαίνει τῷ ταῦτα εἶναι".

Put differently, a syllogism is a valid argument [42, §1], which is not trivial, i.e., something *new* has to be concluded (cf. SE 1, 164b27–165a2[80]).

Aristotle makes it clear that there must be a logical relationship between the sentences/premisses in order for a syllogism to obtain, a relationship that concerns the terms constituting the premisses (APr A1, 24b20ff.[81]).

The relationship Aristotle singles out is *being in another as in a whole* which he explains as follows: *A is in B as in a whole* iff *B* is predicated of all of *A*. Moreover, he explains: *B is predicated of all of A* iff there is no *A* that is not *B* (i.e., all *A* are *B*). Similarly, *B is predicated of none of A* iff there is no *A* that is *B* (i.e., no *A* are *B*) (APr A1, 24b26–30[82]).

Overall there are four sentence-types, depending on quantity and mode. The quantity can either be universal or particular ('in part'), and the mode can be positive ('kata', affirming) or negative ('apo', denying). These account for the relation of the two terms involved in sentences. Given two terms $A$ and $B$, we get a sentence $AB$ whose predicate-term is $A$ and whose subject-term is $B$. The sentence $AB$ can be either

(**a**) universal-affirmative ("all $B$ are $A$"; $AaB$), or

(**i**) particular-affirmative ("some $B$ are $A$"; $AiB$), or

(**e**) universal-negative ("all $B$ are not $A$"; $AeB$), or

(**o**) particular-negative ("some $B$ are not $A$; $AoB$).

---

[80]"For a syllogism rests on certain statements such that they involve necessarily the assertion of something other than what has been stated, through what has been stated [ὁ μὲν γὰρ συλλογισμὸς ἐκ τινῶν ἐστι τεθέντων ὥστε λέγειν ἕτερον ἐξ ἀνάγκης τι τῶν κειμένων διὰ τῶν κειμένων]".

[81]"By 'because these things are so' I mean that it results through these, and by 'resulting through these' I mean that no term is required from outside for the necessity to come about [λέγω δὲ τῷ ταῦτα εἶναι τὸ διὰ ταῦτα συμβαίνειν, τὸ δὲ διὰ ταῦτα συμβαίνειν τὸ μηδενὸς ἔξωθεν ὅρου προσδεῖν πρὸς τὸ γενέσθαι τὸ ἀναγκαῖον]".

[82]"For one thing to be in another as in a whole is the same as for the other to be predicated of all of the first. We speak of 'being predicated of all' when nothing can be found of the subject of which the other will not be said and the same account holds for 'of none' [τὸ δὲ ἐν ὅλῳ εἶναι ἕτερον ἑτέρῳ καὶ τὸ κατὰ παντὸς κατηγορεῖσθαι θατέρου θάτερον ταὐτόν ἐστιν. λέγομεν δὲ τὸ κατὰ παντὸς κατηγορεῖσθαι ὅταν μηδὲν ᾖ λαβεῖν [τοῦ ὑποκειμένου] καθ' οὗ θάτερον οὐ λεχθήσεται· καὶ τὸ κατὰ μηδενὸς ὡσαύτως]".

Aristotle calls sentence-types **e** and **o** *privative* (στερητικός/sterêtikos); he does not think of them as involving what we would understand as a *negation*. Indeed, he has different ways of referring to the same type. On the one hand, a sentence can be an *affirmation* (κατάφασις/kataphasis) and a *denial* (ἀπόφασις/apophasis) (and, derivatively, sentences can be affirmative (καταφατικός/kataphatikos) and negative (ἀποφατικός/apophatikos)), and he refers to the sub-types as *universal* and *particular*. On the other hand, he refers to the denials as *privative*; e.g., he speaks of the "universal privative premiss" (APr A2, 25a5f.[83]). The 'privative' applies to the copula—it suggests a negative copula—the 'universal' to the subject—indicating the quantity of the subject.

This second way singles out the subject (universally or particularly) and notes the privation, i.e., that a term does not apply to it. For example, some human beings are *not* healthy, i.e., *lack* health, and so health is privative to those human beings. The whole sentence is a denial (ti apo tinos), and the predicate-term is privative (apo), i.e., the subject lacks the corresponding property.

Since both constituents of a sentence are terms, there is a natural question as to their relationship. Aristotle notes that three of the sentence-types *convert* (ἀντιστρέφειν/antistrephein), viz., **a**, **i**, and **e**. Sentence-type **o**, however, does not.

*Converting* a sentence means interchanging the predicate-term and subject-term; the sentence $AB$ converts to $BA$. Sentence-types **i** and **e** convert to the same sentence-type; sentence-type **a**, on the other hand, converts to an **i**-type sentence (APr A2, 25a5–13[84]). The conversions

---

[83] "τὴν μὲν ἐν τῷ ὑπάρχειν καθόλου στερητικήν".

[84] "... it is necessary for the universal privative premiss of belonging to convert with respect to its terms. So, for instance, if no pleasure is a good, then neither will any good be a pleasure. And the positive premiss converts necessarily, though not universally, but to the particular; for instance, if every pleasure is a good, it is necessary that some good be also a pleasure. Of the particular premisses the affirmative necessarily converts to the particular, for if some pleasure is a good, then some good will also be a pleasure; but for the privative premiss this is not necessary. For it is not the case that, if man does not belong to some animal, then animal also does not belong to some man [τὴν μὲν ἐν τῷ ὑπάρχειν καθόλου στερητικὴν ἀνάγκη τοῖς ὅροις ἀντιστρέφειν, οἷον εἰ μηδεμία ἡδονὴ ἀγαθόν, οὐδ' ἀγαθὸν οὐδὲν ἔσται ἡδονή· τὴν δὲ κατηγορικὴν ἀντιστρέφειν μὲν ἀναγκαῖον, οὐ μὴν καθόλου ἀλλ' ἐν μέρει, οἷον εἰ πᾶσα ἡδονὴ ἀγαθόν, καὶ ἀγαθόν τι εἶναι ἡδονήν· τῶν δὲ ἐν μέρει τὴν μὲν καταφατικὴν

can be summarized as follows (symbolizing 'converts to' as '⤳'):

(**a**-**i**-conv) $AaB \leadsto BiA$  (**i**-**i**-conv) $AiB \leadsto BiA$

(**e**-**e**-conv) $AeB \leadsto BeA$  (**o**-**o**-conv) $AoB \not\leadsto BoA$

That Aristotle claims that these sentence-types convert—and that without any suggestion of a restriction in place—suggests that predicate- and subject-terms are on a par as worked out in Section 2.6. Suppose Aristotle allowed *particulars* into his syllogistic. Then sentences with such particulars cannot convert, since particulars can*not* play the role of predicates; the validity of the conversions rules out terms denoting particulars.

With all these preliminaries out of the way, Aristotle goes on to introduce three figures and to establish their syllogisms. The figures come about by considering the different roles three terms, $A$, $B$, $C$, can play. Sentences of the form '$AB$' have '$B$' as their subject-term and '$A$' as their predicate-term. Since the syllogisms come about via the relation of the terms, one term has to occur in two premisses as to establish a relation between the other terms. The three figures encode exactly that.[85]

The first figure has one term—the so-called *middle* term (μέσον/ meson)—occurring as predicate-term in one premise and subject-term in the other, i.e., the premises are $AB$ and $BC$ (APr A4, 25b35f.[86]). The conclusion concerns the other terms $A$ and $C$—the so-called *extremes* (ἄκρα/akra) (APr A4, 25b36f.[87]).

The first-figure syllogisms are the following:

---

ἀντιστρέφειν ἀνάγκη κατὰ μέρος (εἰ γὰρ ἡδονή τις ἀγαθόν, καὶ ἀγαθόν τι ἔσται ἡδονή), τὴν δὲ στερητικὴν οὐκ ἀναγκαῖον· (οὐ γὰρ εἰ ἄνθρωπος μὴ ὑπάρχει τινὶ ζῴῳ, καὶ ζῷον οὐχ ὑπάρχει τινὶ ἀνθρώπῳ]".

[85] I'm ignoring the fourth figure that Aristotle does not mention and which is not necessary to establish all the syllogisms.

[86] "I call 'middle' the term that is itself in another and in which there is also another—the one that also has the middle position [καλῶ δὲ μέσον μὲν ὃ καὶ αὐτὸ ἐν ἄλλῳ καὶ ἄλλο ἐν τούτῳ ἐστίν, ὃ καὶ τῇ θέσει γίνεται μέσον]".

[87] "Extremes are what is in another and that in which there is another [ἄκρα δὲ τὸ αὐτό τε ἐν ἄλλῳ ὂν καὶ ἐν ᾧ ἄλλο ἐστίν]".

ARISTOTLE, TERM LOGIC, AND QUARC

(B<u>a</u>rb<u>a</u>r<u>a</u>) $\dfrac{AaB \quad BaC}{AaC}$  (C<u>e</u>lar<u>e</u>nt) $\dfrac{AeB \quad BaC}{AeC}$

(D<u>a</u>r<u>ii</u>) $\dfrac{AaB \quad BiC}{AiC}$  (F<u>e</u>r<u>io</u>) $\dfrac{AeB \quad BiC}{AoC}$

The second figure has the middle term only as predicate-term (APr A5, 26b34–37[88]), and comprises the following syllogisms:

(C<u>esare</u>) $\dfrac{MeN \quad MaX}{NeX}$  (F<u>e</u>st<u>i</u>n<u>o</u>) $\dfrac{MeN \quad MiX}{NoX}$

(C<u>a</u>m<u>e</u>str<u>e</u>s) $\dfrac{MaN \quad MeX}{NeX}$  (B<u>a</u>r<u>o</u>c<u>o</u>) $\dfrac{MaN \quad MoX}{NoX}$

The third figure has the middle term only as subject-term (APr A6, 28a10–13[89]), and comprises the following syllogisms:

(D<u>a</u>r<u>a</u>pt<u>i</u>) $\dfrac{PaS \quad RaS}{PiR}$  (D<u>a</u>t<u>i</u>s<u>i</u>) $\dfrac{PaS \quad RiS}{PiR}$

(F<u>e</u>l<u>a</u>pt<u>o</u>n) $\dfrac{PeS \quad RaS}{PoR}$  (B<u>o</u>c<u>a</u>rd<u>o</u>) $\dfrac{PoS \quad RaS}{PoR}$

(D<u>i</u>s<u>a</u>m<u>i</u>s) $\dfrac{PiS \quad RaS}{PiR}$  (F<u>e</u>r<u>i</u>s<u>o</u>n) $\dfrac{PeS \quad RiS}{PoR}$

Given the conversion rules, certain further conclusions can be drawn. For example, given (B<u>a</u>rb<u>a</u>r<u>a</u>) with conclusion $AaC$, we can apply (<u>a</u>-<u>i</u>-conv) and infer $CiA$. Moreover, we can apply (<u>i</u>-<u>i</u>-conv) and infer $AiC$.

## 2.8 The Square of Opposition

Aristotle is taken to endorse the *square of opposition*, though he does not state it explicitly [26, p. 56]. The four vertexes of the square are

---

[88]"When the same thing belongs to all of one and none of the other, or to all or none of both other terms, I call this sort of figure the second. And in this figure I call middle the term that is predicated of both ["Ὅταν δὲ τὸ αὐτὸ τῷ μὲν παντὶ τῷ δὲ μηδενὶ ὑπάρχῃ, ἢ ἑκατέρῳ παντὶ ἢ μηδενί, τὸ μὲν σχῆμα τὸ τοιοῦτον καλῶ δεύτερον, μέσον δὲ ἐν αὐτῷ λέγω τὸ κατηγορούμενον ἀμφοῖν]".

[89]"If one term belongs to all, another to none of the same thing, or both to all or both to none, I call this sort of figure the third. And in this figure I call the middle the term of which both the predicated terms are said [Ἐὰν δὲ τῷ αὐτῷ τὸ μὲν παντὶ τὸ δὲ μηδενὶ ὑπάρχῃ ἢ ἄμφω παντὶ ἢ μηδενί, τὸ μὲν σχῆμα τὸ τοιοῦτον καλῶ τρίτον, μέσον δ' ἐν αὐτῷ λέγω καθ' οὗ ἄμφω τὰ κατηγορούμενα]".

labelled by the four sentence-types, and the relations between these types are captured by edges.

The possible relations are *contradictories*, *contraries*, *subcontraries*, and *subalternation*. Consider two sentences $\varphi$ and $\psi$. They are *contradictories* iff exactly one of them is true; they are *contraries* iff they cannot both be true, but can both be false; they are *subcontraries* iff they cannot both be false, but can both be true; and $\psi$ is a subaltern of $\varphi$ iff $\varphi$ implies $\psi$. For example, an **a**-type sentence has the corresponding **o**-type sentence as its contradictory, the corresponding **e**-type sentence as its contrary, and the corresponding **i**-type sentence as its subaltern. Figure 1 pictures the square.

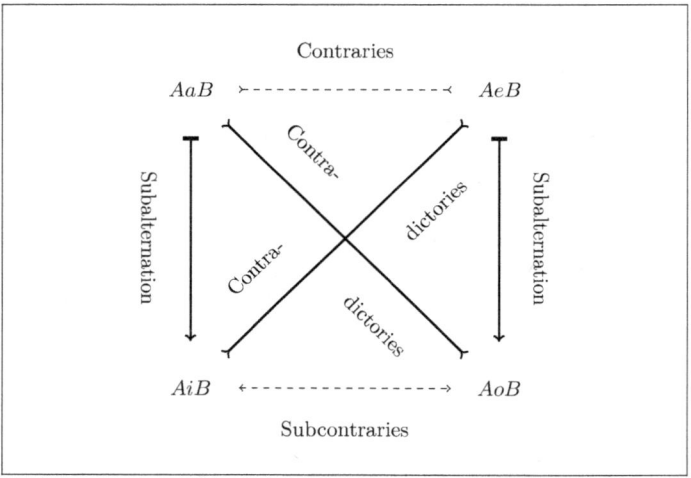

Figure 1: The Square of Opposition

Except for subalternation, the relations are *symmetrical*; only subalternation is directed. Moreover, given (**a**-**i**-conv) and (**i**-**i**-conv), we can account for the sentence-type **i** being the subaltern of sentence-type **a**: by (**a**-**i**-conv), $AaB$ implies (converts to) $BiA$ which, by (**i**-**i**-conv), implies (converts to) $AiB$.

Moreover, given the other relations, we can see that the **e**-type sentence has the **o**-type sentence as its subaltern. If $AeB$ holds, then $AaB$ cannot hold as its contrary. Thus, as exactly one of $AaB$ and $AoB$ has to be true, it follows that $AoB$ must be true.

Aristotle provides the definitions for *contradictories* and *contraries* in *On Interpretation* (Int 7, 17b16–20[90], Int 7, 17b20–23[91], respectively), and he notes that contradictories cannot, but contraries can be true together (Int 7, 17b23–29[92]). Moreover, in his *Topics*, Aristotle suggests the *subalternations* (Top B1, 109a3–6[93]).

Lastly, the *subcontraries* result from the established relations as well. For, in cases where both the **a**- and **e**-type sentence are false—as they can be as contraries—their contradictories must be true, i.e., the **o**-/**i**-type sentence is true as the contradictory of the **a**-/**e**-type sentence. Moreover, one of the **i**- and **o**-type sentence has to be true. Suppose that the **i**/**o**-type sentence is false. Then its contradictory **e**/**a**-type sentence is true which implies the corresponding **o**/**i**-type sentence. For the same reason, one of the **a**- and **e**-type sentence has to be false.

---

[90]"I call an affirmation and a negation contradictory opposites when what one signifies universally the other signifies not universally, e.g. 'every man is white' and 'not every man is white', 'no man is white' and 'some man is white' [Ἀντικεῖσθαι μὲν οὖν κατάφασιν ἀποφάσει λέγω ἀντιφατικῶς τὴν τὸ καθόλου σημαίνουσαν τῷ αὐτῷ ὅτι οὐ καθόλου, οἷον πᾶς ἄνθρωπος λευκός – οὐ πᾶς ἄνθρωπος λευκός, οὐδεὶς ἄνθρωπος λευκός – ἔστι τις ἄνθρωπος λευκός]".

[91]"But I call the universal affirmation and the universal denial contrary opposites, e.g. 'every man is just' and 'no man is just' [ἐναντίως δὲ τὴν τοῦ καθόλου κατάφασιν καὶ τὴν τοῦ καθόλου ἀπόφασιν, οἷον πᾶς ἄνθρωπος δίκαιος – οὐδεὶς ἄνθρωπος δίκαιος]".

[92]"So these cannot be true together, but their opposites may be both true with respect to the same thing, e.g. 'not every man is white' and 'some man is white'. Of contradictory statements about a universal taken universally it is necessary for one or the other to be true or false; similarly if they are about particulars, e.g. 'Socrates is white' and 'Socrates is not white' [διὸ ταύτας μὲν οὐχ οἷόν τε ἅμα ἀληθεῖς εἶναι, τὰς δὲ ἀντικειμένας αὐταῖς ἐνδέχεται ἐπὶ τοῦ αὐτοῦ, οἷον οὐ πᾶς ἄνθρωπος λευκός, καὶ ἔστι τις ἄνθρωπος λευκός. ὅσαι μὲν οὖν ἀντιφάσεις τῶν καθόλου εἰσὶ καθόλου, ἀνάγκη τὴν ἑτέραν ἀληθῆ εἶναι ἢ ψευδῆ, καὶ ὅσαι ἐπὶ τῶν καθ᾽ ἕκαστα, οἷον ἔστι Σωκράτης λευκός – οὐκ ἔστι Σωκράτης λευκός]".

[93]"for when we have proved that a predicate belongs in every case, we shall also have proved that it belongs in some cases. Likewise, also, if we prove that it does not belong in any case, we shall also have proved that it does not belong in every case [δείξαντες γὰρ ὅτι παντὶ ὑπάρχει, καὶ ὅτι τινὶ ὑπάρχει δεδειχότες ἐσόμεθα· ὁμοίως δὲ κἂν ὅτι οὐδενὶ ὑπάρχει δείξωμεν, καὶ ὅτι οὐ παντὶ ὑπάρχει δεδειχότες ἐσομεθα]".

Aristotle is also aware that you can use the square to refute sentences. For, if you want to refute an **a**-type sentence, it suffices to establish the corresponding **o**-type sentence; and similarly with **e**- and **i**-type sentences (Top B3, 110a32–37[94]).

## 2.9 Formal Syllogistic

In order to have a comparison base, let me briefly introduce some formalism capturing Aristotle's syllogistic. The presentation is based on, but also differs from, [40] where more discussion and details can be found.

**Definition 1** *(The Language $\mathcal{L}_A$)*
The *language of Aristotelian Syllogistic* ($\mathcal{L}_A$) consists of the following:

- a countable set $\mathsf{STerm}_{\mathcal{L}_A}$ of *(simple) terms*,
- the set of *logical symbols* including '¬', '—', '∧', '∨', '→', '↔', '∀', and '∃', and
- the set of *auxiliary symbols* including '(' and ')'.

The '—'-symbol is used to distinguish term-negation from a negative copula.[95] The remaining symbols are to be understood as indicated below.

Since we allow complex terms, let us introduce them:

**Definition 2** *(Complex $\mathcal{L}_A$-Terms)*
The *(full) set of terms* ($\mathsf{Term}_{\mathcal{L}_A}$) is recursively defined as follows:

(1) if $A \in \mathsf{STerm}_{\mathcal{L}_A}$, then $A \in \mathsf{Term}_{\mathcal{L}_A}$;

---

[94]"Of course, in refuting a statement there is no need to start the discussion by securing any admission, whether the attribute is said to belong to all or to none of something; for if we prove that in any case whatever the attribute does not belong, we shall have refuted the universal assertion of it, and likewise if we prove that it belongs even in a single case, we shall refute the universal denial of it [πλὴν ἀνασευκάζοντι μὲν οὐδὲν δεῖ ἐξ ὁμολογίας διαλέγεσθαι, οὔτ᾽ εἰ παντὶ οὔτ᾽ εἰ μηδενὶ ὑπάρχειν εἴρηται· ἐὰν γὰρ δείξωμεν ὅτι οὐχ ὑπάρχει ὁτῳοῦν, ἀνῃρηκότες ἐσόμεθα τὸ παντὶ ὑπάρχειν· ὁμοίως δὲ κἂν ἑνὶ δείξωμεν ὑπάρχον, ἀναιρήσομεν τὸ μηδενὶ ὑπάρχειν]".

[95]Including '—' differs from, but is equivalent to, the set-up in [40, §3.5].

(2) if $A \in \mathsf{Term}_{\mathcal{L}_\mathsf{A}}$, then $\overline{A} \in \mathsf{Term}_{\mathcal{L}_\mathsf{A}}$;

(3) if $A, B \in \mathsf{Term}_{\mathcal{L}_\mathsf{A}}$, then, $(A \circ B) \in \mathsf{Term}_{\mathcal{L}_\mathsf{A}}$ ($\circ \in \{\wedge, \vee, \rightarrow, \leftrightarrow\}$).

Given the language and the terms, we can define the formulas:

**Definition 3** *($\mathcal{L}_\mathsf{A}$-Formulas)*
The set of $\mathcal{L}_\mathsf{A}$-*formulas* ($\mathsf{Form}_{\mathcal{L}_\mathsf{A}}$) is defined as follows:

- If $A, B \in \mathsf{Term}_{\mathcal{L}_\mathsf{A}}$, then
    - $(\forall A)B \in \mathsf{Form}_{\mathcal{L}_\mathsf{A}}$
    - $(\forall A)\neg B \in \mathsf{Form}_{\mathcal{L}_\mathsf{A}}$
    - $(\exists A)B \in \mathsf{Form}_{\mathcal{L}_\mathsf{A}}$
    - $(\exists A)\neg B \in \mathsf{Form}_{\mathcal{L}_\mathsf{A}}$.

The formulas are to be read as follows: '$(\forall A)B$' as "all $A$ are $B$" ($BaA$), '$(\exists A)B$' as "some $A$ are $B$" ($BiA$), '$(\forall A)\neg B$' as "all $A$ are not $B$" or "no $A$ is $B$" ($BeA$), and '$(\exists A)\neg B$' as 'some $A$ are not $B$" ($BoA$). Note that, according to Definition 2, complex terms are covered by Definition 3. For example, formulas of the form '$(\forall \overline{(A \wedge B)})\neg \overline{C}$' are allowed, and should correspondingly be read as "all not-($A$-and-$B$) are not not-$C$" or "no not-($A$-and-$B$) is not-$C$".

The absence/occurrence of a negation-symbol '$\neg$' indicates whether the formula is affirming or denying, respectively; it represents the copula. According to Definition 3, at most one negation-symbol occurs in a formula. Negation does not act on sentences, and we need to ensure in a different way how the sentences are related with respect to affirmation and denial; as in [40], this achieved via positive ('+') and negative ('−') semantic clauses (Definitions 6–7).

The quantifier-symbols indicate the quantity, i.e., whether a sentence is universal ($\forall$) or particular ($\exists$) (or, whether the predication is *universally* or *particularly*)—and that's all they are doing: They are just a means to make explicit what kind of sentence is represented; instead of Aristotle's way of suggesting that, e.g., '$BA$' is universal affirming/-denying sentence, we directly depict it as '$(\forall A)B$'/'$(\forall A)\neg B$'.

Given this understanding, let me introduce a model-theoretic semantics. I provide two ways of doing so. One interpretation allows *empty* terms, i.e., the structure can assign the empty extension as interpretation of terms (I refer to it as 'the empty semantics'); the other interpretation forces the simple terms to be non-empty (shown to be inadequate below).

**Definition 4 ($\mathcal{L}_A$-Model)**
Let $\mathcal{L}_A$ be a language of Aristotelian Syllogistic. An $\mathcal{L}_A$-model is a tuple $\mathfrak{M}_A = \langle D, \|\cdot\|_{\mathfrak{M}_A} \rangle$ such that

(1) $D$ is a non-empty set (the *universe*)

(2) $\|\cdot\|_{\mathfrak{M}_A}$ is an *interpretation-function of* $\mathfrak{M}_A$ such that

    (a) if $A \in \mathsf{STerm}_{\mathcal{L}_A}$, $\|A\|_{\mathfrak{M}_A} \subseteq D$;

    (b) if $A \in \mathsf{Term}_{\mathcal{L}_A}$ is a complex term of the form '$\overline{B}$' for some $B \in \mathsf{Term}_{\mathcal{L}_A}$, then $\|A\|_{\mathfrak{M}_A} = D \setminus \|B\|_{\mathfrak{M}_A}$;

    (c) if $A \in \mathsf{Term}_{\mathcal{L}_A}$ is a complex term of the form '$(B \wedge C)$' for some $B, C \in \mathsf{Term}_{\mathcal{L}_A}$, then $\|A\|_{\mathfrak{M}_A} = \|B\|_{\mathfrak{M}_A} \cap \|C\|_{\mathfrak{M}_A}$.

Since Definition 4 allows for empty terms, we have to specify that the domain $D$ is non-empty. Negated terms are interpreted as the set-theoretic difference between the extension of a term and the domain. Complex terms are treated as expected; Definition 4 only specifies the clause for conjunctive terms ('∧'); the others are definable given clauses (2b)–(2c).

**Definition 5 (NE-$\mathcal{L}_A$-Model)**
Let $\mathcal{L}_A$ be a language of Aristotelian Syllogistic. A *non-empty* $\mathcal{L}_A$-*model* is a tuple $\mathfrak{M}_{ne} = \langle D, \|\cdot\|_{\mathfrak{M}_{ne}} \rangle$ such that

(1) $D$ is a set (the *universe*);

(2) $\|\cdot\|_{\mathfrak{M}_{ne}}$ is an *interpretation-function of* $\mathfrak{M}_{ne}$ such that

    (a) if $A \in \mathsf{STerm}_{\mathcal{L}_A}$, $\emptyset \neq \|A\|_{\mathfrak{M}_{ne}} \subseteq D$;

    (b) if $A \in \mathsf{Term}_{\mathcal{L}_A}$ is a complex term of the form '$\overline{B}$' for some $B \in \mathsf{Term}_{\mathcal{L}_A}$, then $\|A\|_{\mathfrak{M}_{ne}} = D \setminus \|B\|_{\mathfrak{M}_{ne}}$;

    (c) if $A \in \mathsf{Term}_{\mathcal{L}_A}$ is a complex term of the form '$(B \wedge C)$' for some $B, C \in \mathsf{Term}_{\mathcal{L}_A}$, then $\|A\|_{\mathfrak{M}_{ne}} = \|B\|_{\mathfrak{M}_{ne}} \cap \|C\|_{\mathfrak{M}_{ne}}$.

In contrast to Definition 4, Definition 5 does not need to enforce the domain to be non-empty, as clause (2a) effectively takes care of it.[96] The remaining clauses are the same as in Definition 4.

Given the different models, we can introduce corresponding satisfaction relations. Since $\mathcal{L}_A$ does not contain sentence-negation, we need to ensure that, for example, an **a**-type sentence has an **o**-type sentence as its contradictory by introducing positive ('+') and negative ('−') clauses. Moreover, since we take the validity of the square of opposition as a condition for any adequate satisfaction relation, we need to interpret the sentences accordingly. This results in different clauses for the **a**- and **o**-type sentences. (Note that, as Lemma 3.6 of [40] shows, the number of clauses is reducible to four.)

**Definition 6** *(Satisfaction $\models_A$)*
Let the *satisfaction-relation* $\mathfrak{M}_A \models_A \varphi$ for $\mathcal{L}_A$-formulas $\varphi$ and $\mathcal{L}_A$-model $\mathfrak{M}_A$ be defined as follows: Let $A, B \in \mathsf{Term}_{\mathcal{L}_A}$, then:

($\mathbf{a}_+$) $\mathfrak{M}_A \models_A (\forall A)B$ iff $\|A\|_{\mathfrak{M}_A} \cap \|B\|_{\mathfrak{M}_A} = \|A\|_{\mathfrak{M}_A}$ and $\|A\|_{\mathfrak{M}_A} \neq \emptyset$;

($\mathbf{a}_-$) $\mathfrak{M}_A \not\models_A (\forall A)B$ iff $\mathfrak{M}_A \models_A (\exists A)\neg B$.

($\mathbf{i}_+$) $\mathfrak{M}_A \models_A (\exists A)B$ iff $\|A\|_{\mathfrak{M}_A} \cap \|B\|_{\mathfrak{M}_A} \neq \emptyset$.

($\mathbf{i}_-$) $\mathfrak{M}_A \not\models_A (\exists A)B$ iff $\mathfrak{M}_A \models_A (\forall A)\neg B$.

($\mathbf{e}_+$) $\mathfrak{M}_A \models_A (\forall A)\neg B$ iff $\|A\|_{\mathfrak{M}_A} \cap \|B\|_{\mathfrak{M}_A} = \emptyset$.

($\mathbf{e}_-$) $\mathfrak{M}_A \not\models_A (\forall A)\neg B$ iff $\mathfrak{M}_A \models_A (\exists A)B$.

($\mathbf{o}_+$) $\mathfrak{M}_A \models_A (\exists A)\neg B$ iff $\|A\|_{\mathfrak{M}_A} \cap \|B\|_{\mathfrak{M}_A} \neq \|A\|_{\mathfrak{M}_A}$ or $\|A\|_{\mathfrak{M}_A} = \emptyset$.

($\mathbf{o}_-$) $\mathfrak{M}_A \not\models_A (\exists A)\neg B$ iff $\mathfrak{M}_A \models_A (\forall A)B$.

In order for the square of opposition to hold, we must ensure that an **a**-type sentences imply **i**-type sentences. The usual way to do so is by only allowing non-empty terms as in Definition 5, but Definition 4 allows for empty terms. Thus, a model $\mathfrak{M}_A$ can only satisfy an **a**-type sentence

---

[96] Note that Definition 1 does not enforce $\mathsf{STerm}_{\mathcal{L}_A}$ to be non-empty. If $\mathsf{STerm}_{\mathcal{L}_A} = \emptyset$, clause (2a) does still not produce a problem as the clause is then vacuous.

if the term happens to be non-empty, i.e., if $\|A\|_{\mathfrak{M}_A} = \emptyset$, no sentence of the form '$(\forall A)B$' can be satisfied. Since **a**-type sentences have **o**-type sentences as their contradictories, the satisfaction-clause ($\mathbf{o}_+$) needs to include the cases in which the subject-term is empty.

As the non-empty $\mathcal{L}_A$-models $\mathfrak{M}_{ne}$ don't allow non-empty terms, the clauses are simpler than those of Definition 6. However, as shown in Theorem 14, there is no fully general formal analogue of (**a**-**i**-conv) and so the square of opposition does not follow.

**Definition 7 (NE-Satisfaction $\models_{ne}$)**
Let the *non-empty satisfaction-relation* $\mathfrak{M}_{ne} \models_{ne} \varphi$ for $\mathcal{L}_A$-formulas $\varphi$ and non-empty $\mathcal{L}_A$-model $\mathfrak{M}_{ne}$ be defined as follows: Let $A, B \in \text{Term}_{\mathcal{L}_A}$, then:

($\mathbf{a}_+^{ne}$) $\mathfrak{M}_{ne} \models_{ne} (\forall A)B$ iff $\|A\|_{\mathfrak{M}_{ne}} \cap \|B\|_{\mathfrak{M}_{ne}} = \|A\|_{\mathfrak{M}_{ne}}$.

($\mathbf{a}_-^{ne}$) $\mathfrak{M}_{ne} \not\models_{ne} (\forall A)B$ iff $\mathfrak{M}_{ne} \models_{ne} (\exists A)\neg B$.

($\mathbf{i}_+^{ne}$) $\mathfrak{M}_{ne} \models_{ne} (\exists A)B$ iff $\|A\|_{\mathfrak{M}_{ne}} \cap \|B\|_{\mathfrak{M}_{ne}} \neq \emptyset$.

($\mathbf{i}_-^{ne}$) $\mathfrak{M}_{ne} \not\models_{ne} (\exists A)B$ iff $\mathfrak{M}_{ne} \models_{ne} (\forall A)\neg B$.

($\mathbf{e}_+^{ne}$) $\mathfrak{M}_{ne} \models_{ne} (\forall A)\neg B$ iff $\|A\|_{\mathfrak{M}_{ne}} \cap \|B\|_{\mathfrak{M}_{ne}} = \emptyset$.

($\mathbf{e}_-^{ne}$) $\mathfrak{M}_{ne} \not\models_{ne} (\forall A)\neg B$ iff $\mathfrak{M}_{ne} \models_{ne} (\exists A)B$.

($\mathbf{o}_+^{ne}$) $\mathfrak{M}_{ne} \models_{ne} (\exists A)\neg B$ iff $\|A\|_{\mathfrak{M}_{ne}} \cap \|B\|_{\mathfrak{M}_{ne}} \neq \|A\|_{\mathfrak{M}_{ne}}$.

($\mathbf{o}_-^{ne}$) $\mathfrak{M}_{ne} \not\models_{ne} (\exists A)\neg B$ iff $\mathfrak{M}_{ne} \models_{ne} (\forall A)B$.

Given a notion of satisfaction, we can introduce the usual notions:

**Definition 8**
Let $T \subseteq \text{Form}_{\mathcal{L}_A}$, $\Vdash \in \{\models_A, \models_{ne}\}$, and $\mathfrak{M}_{\Vdash} = \left\{ \begin{array}{ll} \mathfrak{M}_A & \text{if } \Vdash \text{ is } \models_A \\ \mathfrak{M}_{ne} & \text{if } \Vdash \text{ is } \models_{ne} \end{array} \right\}$.

(1) $\varphi$ *is a logical consequence of* $T$ $(T \Vdash \varphi)$ iff for all $\mathcal{L}_A$- models $\mathfrak{M}_{\Vdash}$, if $\mathfrak{M}_{\Vdash} \Vdash \psi$ for all $\psi \in T$, then $\mathfrak{M}_{\Vdash} \Vdash \varphi$.

If $T = \{\varphi_1, \ldots, \varphi_n\}$, we write '$\varphi_1, \ldots, \varphi_n \Vdash \varphi$' for '$\{\varphi_1, \ldots, \varphi_n\} \Vdash \varphi$'.

(2) $\varphi$ is logically valid iff $\emptyset \Vdash \varphi$ ($\Vdash \varphi$).

(3) $T$ is satisfiable iff there is an $\mathcal{L}_\mathsf{A}$-model $\mathfrak{M}_{\Vdash}$ such that $\mathfrak{M}_{\Vdash} \Vdash \varphi$ for all $\varphi \in T$.

Given these definitions, we can formulate some results. First, we can note that the empty $\mathcal{L}_\mathsf{A}$-models see sentence-types **a** and **e** as contraries:

**Lemma 9** *(Contraries)*
$(\forall A)B$ and $(\forall A)\neg B$ are contraries in $\mathcal{L}_\mathsf{A}$-models $\mathfrak{M}_\mathsf{A}$:

(1) $\{(\forall A)B, (\forall A)\neg B\}$ is *not* satisfiable;

(2) there are $\mathcal{L}_\mathsf{A}$-models $\mathfrak{M}_\mathsf{A}$ such that $\mathfrak{M}_\mathsf{A} \not\models_\mathsf{A} (\forall A)B$ and $\mathfrak{M}_\mathsf{A} \not\models_\mathsf{A} (\forall A)\neg B$.

*Proof.* Let $\mathfrak{M}_\mathsf{A}$ be an $\mathcal{L}_\mathsf{A}$-model.

**(1):** Suppose that $\mathfrak{M}_\mathsf{A} \models_\mathsf{A} (\forall A)B$ and $\mathfrak{M}_\mathsf{A} \models_\mathsf{A} (\forall A)\neg B$. Then, by $(\mathbf{a}_+)$, $\|A\|_{\mathfrak{M}_\mathsf{A}} \cap \|B\|_{\mathfrak{M}_\mathsf{A}} = \|A\|_{\mathfrak{M}_\mathsf{A}} \neq \emptyset$, and, by $(\mathbf{e}_+)$, $\|A\|_{\mathfrak{M}_\mathsf{A}} \cap \|B\|_{\mathfrak{M}_\mathsf{A}} = \emptyset$, a contradiction.

**(2):** Let $\|A\|_{\mathfrak{M}_\mathsf{A}} = \{a, b\}$ and $\|B\|_{\mathfrak{M}_\mathsf{A}} = \{a\}$. Then, $\|A\|_{\mathfrak{M}_\mathsf{A}} \cap \|B\|_{\mathfrak{M}_\mathsf{A}} \neq \emptyset$, i.e., by $(\mathbf{i}_+)$, $\mathfrak{M}_\mathsf{A} \models_\mathsf{A} (\exists A)B$. And, since $\|A\|_{\mathfrak{M}_\mathsf{A}} \cap \|B\|_{\mathfrak{M}_\mathsf{A}} \neq \|A\|_{\mathfrak{M}_\mathsf{A}}$, by $(\mathbf{o}_+)$, $\mathfrak{M}_\mathsf{A} \models_\mathsf{A} (\exists A)\neg B$. By $(\mathbf{e}_-)$ and $(\mathbf{a}_-)$, respectively, the result follows.

$\square$

Two characteristics of the semantics are the following:

**Theorem 10**
The following hold.

(1) $\not\models_\mathsf{A} (\exists A)A$ 

(2) $(\forall A)B \not\models_\mathsf{A} (\exists A)B$

(3) $\not\models_\mathsf{ne} (\exists A)A$

(4) $(\forall A)B \not\models_\mathsf{ne} (\exists A)B$

*Proof.* **(1):** Let $\mathfrak{M}_\mathsf{A}$ be an $\mathcal{L}_\mathsf{A}$-model such that $\|A\|_{\mathfrak{M}_\mathsf{A}} = \emptyset$. Then, $\|A\|_{\mathfrak{M}_\mathsf{A}} \cap \|A\|_{\mathfrak{M}_\mathsf{A}} = \emptyset$, i.e., by $(\mathbf{e}_+)$, $\mathfrak{M}_\mathsf{A} \models_\mathsf{A} (\forall A)\neg A$. Thus, by $(\mathbf{i}_-)$, $\mathfrak{M}_\mathsf{A} \not\models_\mathsf{A} (\exists A)A$.

**(2):** Let $\mathfrak{M}_A \models_A (\forall A)B$. Then, by $(\mathbf{a}_+)$, $\|A\|_{\mathfrak{M}_A} \cap \|B\|_{\mathfrak{M}_A} = \|A\|_{\mathfrak{M}_A} \neq \emptyset$. Therefore, by $(\mathbf{i}_+)$, $\mathfrak{M}_A \models_A (\exists A)B$.

**(3):** Let $\mathfrak{M}_{ne}$ be a non-empty $\mathcal{L}_A$-model such that $\|A\|_{\mathfrak{M}_{ne}} = D$. Then, $\|\overline{A}\|_{\mathfrak{M}_{ne}} = \emptyset$. Thus, $\|\overline{A}\|_{\mathfrak{M}_{ne}} \cap \|\overline{A}\|_{\mathfrak{M}_{ne}} = \emptyset$, so, by $(\mathbf{e}_+^{ne})$, $\mathfrak{M}_{ne} \models_{ne} (\forall \overline{A})\neg \overline{A}$. By $(\mathbf{i}_-^{ne})$, $\mathfrak{M}_{ne} \not\models_{ne} (\exists \overline{A})\overline{A}$.

**(4):** Consider the model in (3). Since $\|\overline{A}\|_{\mathfrak{M}_{ne}} \cap \|B\|_{\mathfrak{M}_{ne}} = \|\overline{A}\|_{\mathfrak{M}_{ne}}$, by $(\mathbf{a}_+^{ne})$, $\mathfrak{M}_{ne} \models_{ne} (\forall \overline{A})B$. However, since $\|\overline{A}\|_{\mathfrak{M}_{ne}} \cap \|B\|_{\mathfrak{M}_{ne}} = \emptyset$, by $(\mathbf{e}_+^{ne})$, $\mathfrak{M}_{ne} \models_{ne} (\forall \overline{A})\neg B$, and so, by $(\mathbf{i}_-^{ne})$, $\mathfrak{M}_{ne} \not\models_{ne} (\exists \overline{A})B$. □

Theorem 10 (3)–(4) imply that the non-empty semantics does not validate the square of oppostion; for example, sentence-types **a** and **e** fail to be contraries:

**Corollary 11**
In the non-empty semantics, $(\forall A)B$ and $(\forall A)\neg B$ are *not* contraries. In general, given a non-empty $\mathcal{L}_A$-model $\mathfrak{M}_{ne}$, if $\|A\|_{\mathfrak{M}_{ne}} = \emptyset$, then $\{(\forall A)B, (\forall A)\neg B\}$ is satisfiable in $\mathfrak{M}_{ne}$.

*Proof.* Consider the proof of Theorem 10 (4). The model $\mathfrak{M}_{ne}$ is such that both $\mathfrak{M}_{ne} \models_{ne} (\forall \overline{A})B$ and $\mathfrak{M}_{ne} \models_{ne} (\forall \overline{A})\neg B$. □

The empty semantics has formal analogues of the conversions:

**Theorem 12** *(Conversion)*
The following conversions hold:

(**a**-**i**-conv$^{\models_A}$) $(\forall A)B \models_A (\exists B)A$

(**i**-**i**-conv$^{\models_A}$) $(\exists A)B \models_A (\exists B)A$

(**e**-**e**-conv$^{\models_A}$) $(\forall A)\neg B \models_A (\forall B)\neg A$

The non-empty semantics only validates two such conversions:

**Theorem 13** *(NE-Conversion)*
The following conversions hold:

($\underline{\mathbf{i}}$-$\underline{\mathbf{i}}$-conv$^{\models_{\text{ne}}}$) ($\exists A)B \models_{\text{ne}} (\exists B)A$

($\underline{\mathbf{e}}$-$\underline{\mathbf{e}}$-conv$^{\models_{\text{ne}}}$) ($\forall A)\neg B \models_{\text{ne}} (\forall B)\neg A$

*Proof.* Let $\mathfrak{M}_{\text{A}}$ be an $\mathcal{L}_{\text{A}}$-model.

($\underline{\mathbf{a}}$-$\underline{\mathbf{i}}$-conv$^{\models_{\text{A}}}$) Suppose that $\mathfrak{M}_{\text{A}} \models_{\text{A}} (\forall A)B$. Then, by $(\mathbf{a}_+)$, $\|A\|_{\mathfrak{M}_{\text{A}}} \cap \|B\|_{\mathfrak{M}_{\text{A}}} = \|A\|_{\mathfrak{M}_{\text{A}}}$ and $\|A\|_{\mathfrak{M}_{\text{A}}} \neq \emptyset$. Thus, $\|B\|_{\mathfrak{M}_{\text{A}}} \cap \|A\|_{\mathfrak{M}_{\text{A}}} = \|A\|_{\mathfrak{M}_{\text{A}}} \cap \|B\|_{\mathfrak{M}_{\text{A}}} = \|A\|_{\mathfrak{M}_{\text{A}}} \neq \emptyset$. So, by $(\mathbf{i}_+)$, $\mathfrak{M}_{\text{A}} \models_{\text{A}} (\exists B)A$.

($\underline{\mathbf{i}}$-$\underline{\mathbf{i}}$-conv$^{\models_{\text{A}}}$) Suppose that $\mathfrak{M}_{\text{A}} \models_{\text{A}} (\exists A)B$. Then, by $(\mathbf{i}_+)$, $\|A\|_{\mathfrak{M}_{\text{A}}} \cap \|B\|_{\mathfrak{M}_{\text{A}}} \neq \emptyset$, i.e., $\|B\|_{\mathfrak{M}_{\text{A}}} \cap \|A\|_{\mathfrak{M}_{\text{A}}} \neq \emptyset$ and so, by $(\mathbf{i}_+)$, $\mathfrak{M}_{\text{A}} \models_{\text{A}} (\exists B)A$.

($\underline{\mathbf{e}}$-$\underline{\mathbf{e}}$-conv$^{\models_{\text{A}}}$) Suppose that $\mathfrak{M}_{\text{A}} \models_{\text{A}} (\forall A)\neg B$. Then, by $(\mathbf{e}_+)$, $\|A\|_{\mathfrak{M}_{\text{A}}} \cap \|B\|_{\mathfrak{M}_{\text{A}}} = \emptyset$, so also $\|B\|_{\mathfrak{M}_{\text{A}}} \cap \|A\|_{\mathfrak{M}_{\text{A}}} = \emptyset$, i.e., by $(\mathbf{e}_+)$, $\mathfrak{M}_{\text{A}} \models_{\text{A}} (\forall B)\neg A$.

($\underline{\mathbf{i}}$-$\underline{\mathbf{i}}$-conv$^{\models_{\text{ne}}}$) and ($\underline{\mathbf{e}}$-$\underline{\mathbf{e}}$-conv$^{\models_{\text{ne}}}$) are shown in the same way. □

The non-empty semantics does not validate the third conversion.

**Theorem 14** *(NE-Conversion-Failure)*
The formal analogue of ($\underline{\mathbf{a}}$-$\underline{\mathbf{i}}$-conv) fails for the non-empty semantics:

($\underline{\mathbf{a}}$-$\underline{\mathbf{i}}$$\not\!\!\times$conv$^{\models_{\text{ne}}}$) $(\forall A)B \not\models_{\text{ne}} (\exists B)A$

*Proof.* Let $\mathfrak{M}_{\text{ne}}$ be a non-empty $\mathcal{L}_{\text{A}}$-model such that $\|C\|_{\mathfrak{M}_{\text{ne}}} = D$. Then, by Definition 5 (2b), $\|\overline{C}\|_{\mathfrak{M}_{\text{ne}}} = D \setminus \|C\|_{\mathfrak{M}_{\text{ne}}} = D \setminus D = \emptyset$.

Now, let $\|A\|_{\mathfrak{M}_{\text{ne}}} = \|\overline{C}\|_{\mathfrak{M}_{\text{ne}}}$, and suppose that $\mathfrak{M}_{\text{ne}} \models_{\text{ne}} (\forall A)B$. Then, by $(\mathbf{a}^{\text{ne}}_+)$, $\|A\|_{\mathfrak{M}_{\text{ne}}} \cap \|B\|_{\mathfrak{M}_{\text{ne}}} = \|A\|_{\mathfrak{M}_{\text{ne}}}$, i.e., $\|A\|_{\mathfrak{M}_{\text{ne}}} \cap \|B\|_{\mathfrak{M}_{\text{ne}}} = \emptyset$, so also $\|B\|_{\mathfrak{M}_{\text{ne}}} \cap \|A\|_{\mathfrak{M}_{\text{ne}}} = \emptyset$. By $(\mathbf{e}^{\text{ne}}_+)$, $\mathfrak{M}_{\text{ne}} \models_{\text{ne}} (\forall B)\neg A$. Thus, by $(\mathbf{i}^{\text{ne}}_-)$, $\mathfrak{M}_{\text{ne}} \not\models_{\text{ne}} (\exists B)A$.

Therefore, $(\forall A)B \not\models_{\text{ne}} (\exists B)A$. □

All we get is a restricted version:

**Theorem 15** *(Restricted NE-a-i-Conversion)*
Let $\mathfrak{M}_{\text{ne}}$ be a non-empty $\mathcal{L}_{\text{A}}$-model. Then:

($\underline{\mathbf{a}}$-$\underline{\mathbf{i}}$-conv$^{\models_{\mathsf{ne}}}$ ↾ ne)  $(\exists A)A, (\forall A)B \models_{\mathsf{ne}} (\exists B)A$

I take this as evidence that Definitions 4 and 6 are the correct ones since the problem arises already with negative terms—which Aristotle explicitly discusses in his *Organon*, even though not in his *Prior Analytics*. Other complex terms might end up empty, too, even though the simpler terms are not. Let $\mathfrak{M}_{\mathsf{ne}}$ be a non-empty $\mathcal{L}_\mathsf{A}$-structure. Suppose that $\emptyset \subsetneq \|A\|_{\mathfrak{M}_{\mathsf{ne}}} \subsetneq D$. Then, $\emptyset \subsetneq \|\overline{A}\|_{\mathfrak{M}_{\mathsf{ne}}} \subsetneq D$. However, $\|(A \wedge \overline{A})\|_{\mathfrak{M}_{\mathsf{ne}}} = \emptyset$. Therefore, $\|(A \wedge \overline{A})\|_{\mathfrak{M}_{\mathsf{ne}}} \cap \|B\|_{\mathfrak{M}_{\mathsf{ne}}} = \|(A \wedge \overline{A})\|_{\mathfrak{M}_{\mathsf{ne}}}$ since $\emptyset \cap \|B\|_{\mathfrak{M}_{\mathsf{ne}}} = \emptyset$. Thus, by $(\mathbf{a}^{\mathsf{ne}}_+)$, $\mathfrak{M}_{\mathsf{ne}} \models_{\mathsf{ne}} (\forall (A \wedge \overline{A}))B$, but $\mathfrak{M}_{\mathsf{ne}} \not\models_{\mathsf{ne}} (\exists B)(A \wedge \overline{A})$. This also means again that $(\forall C)D \not\models_{\mathsf{ne}} (\exists C)D$.

Of course, one could still insist on the non-emptiness of terms. One option, though I don't take it to be particularly plausible, is to only allow terms which don't lead to empty ones. That, of course, rules out simultaneously having negated and conjunctive terms.[97] Another option is to do the same as in Definition 6, though then there is no reason to assume that terms are non-empty to begin with. There might be different options available, but I take Definitions 4 and 6 to be the correct ones. Nevertheless, the semantics to be developed in the following sections are more like the non-empty one from Definitions 5 and 7.

Regarding the empty semantics, there is a difference between denials and affirmations modulo negated terms:

**Theorem 16** *(Negation)*
The following hold:

(1) $(\forall A)B \models_\mathsf{A} (\forall A)\neg\overline{B}$;  (2) $(\forall A)\neg\overline{B} \not\models_\mathsf{A} (\forall A)B$;

(3) $(\exists A)B \models_\mathsf{A} (\exists A)\neg\overline{B}$;  (4) $(\exists A)\neg\overline{B} \not\models_\mathsf{A} (\exists A)B$.

---

[97]If we only consider negated terms, the option has some plausibility. Given the discussion of Section 2.6, assigning the whole domain as interpretation of a term might push us to the top of the term-hierarchy and, thus, to terms that Aristotle dismisses as relevant for his syllogistic. Thus, if the only complex terms are negated terms, we might change Definition 5 (2b) to

(2) (b*)  if $A \in \mathsf{STerm}_{\mathcal{L}_\mathsf{A}}$, $\emptyset \subsetneq \|A\|_{\mathfrak{M}_{\mathsf{ne}}} \subsetneq D$

which resolves the problem as for any $A \in \mathsf{Term}_{\mathcal{L}_\mathsf{A}}$, $\|\overline{A}\|_{\mathfrak{M}_{\mathsf{ne}}} \neq \emptyset$.

*Proof.* Let $\mathfrak{M}_A$ be an $\mathcal{L}_A$-model.

**(1):** Let $\mathfrak{M}_A \models_A (\forall A)B$. By $(\mathbf{a}_+)$, $\|A\|_{\mathfrak{M}_A} \cap \|B\|_{\mathfrak{M}_A} = \|A\|_{\mathfrak{M}_A} \neq \emptyset$. By Definition 4 (2b) $\|\overline{B}\|_{\mathfrak{M}_A} = D \setminus \|B\|_{\mathfrak{M}_A}$, i.e., $\emptyset = \|B\|_{\mathfrak{M}_A} \cap \|\overline{B}\|_{\mathfrak{M}_A}$. Therefore, $\emptyset = \|A\|_{\mathfrak{M}_A} \cap \|B\|_{\mathfrak{M}_A} \cap \|\overline{B}\|_{\mathfrak{M}_A} = \|A\|_{\mathfrak{M}_A} \cap \|\overline{B}\|_{\mathfrak{M}_A}$. Thus, by $(\mathbf{e}_+)$, $\mathfrak{M}_A \models_A (\forall A)\neg \overline{B}$.

**(2):** Let $\|A\|_{\mathfrak{M}_A} = \emptyset$. Then, $\|A\|_{\mathfrak{M}_A} \cap \|\overline{B}\|_{\mathfrak{M}_A} = \emptyset$, i.e., by $(\mathbf{e}_+)$, $\mathfrak{M}_A \models_A (\forall A)\neg \overline{B}$. Also, as $\|A\|_{\mathfrak{M}_A} = \emptyset$, by $(\mathbf{o}_+)$, $\mathfrak{M}_A \models_A (\exists A)\neg B$. Therefore, by $(\mathbf{a}_-)$, $\mathfrak{M}_A \not\models_A (\forall A)B$.

**(3):** Let $\mathfrak{M}_A \models_A (\exists A)B$. By $(\mathbf{i}_+)$, $\|A\|_{\mathfrak{M}_A} \cap \|B\|_{\mathfrak{M}_A} \neq \emptyset$. By Definition 4 (2b), $\|\overline{B}\|_{\mathfrak{M}_A} = D \setminus \|B\|_{\mathfrak{M}_A}$, so $\|A\|_{\mathfrak{M}_A} \cap \|\overline{B}\|_{\mathfrak{M}_A} \neq \|A\|_{\mathfrak{M}_A}$. Therefore, by $(\mathbf{o}_+)$, $\mathfrak{M}_A \models_A (\exists A)\neg \overline{B}$.

**(4):** Let $\|A\|_{\mathfrak{M}_A} = \emptyset$. Then, by $(\mathbf{o}_+)$, $\mathfrak{M}_A \models_A (\exists A)\neg \overline{B}$. Also, $\|A\|_{\mathfrak{M}_A} \cap \|B\|_{\mathfrak{M}_A} = \emptyset$, so, by $(\mathbf{e}_+)$, $\mathfrak{M}_A \models_A (\forall A)\neg B$. Thus, by $(\mathbf{i}_-)$, $\mathfrak{M}_A \not\models_A (\exists A)B$. $\square$

As shown in Lemma 3.29 of [40], the non-empty-semantics validates that sentence-types **e/i** imply corresponding sentence-types **a/i**, i.e., $\mathfrak{M}_{ne} \models_{ne} (qA)\neg B$ iff $\mathfrak{M}_{ne} \models_{ne} (qA)\overline{B}$ ($q \in \{\forall, \exists\}$) whereas the empty semantics only validates the direction from sentence-type **a/i** to sentence-type **e/o**, i.e., $(qA)B \models_A (qA)\neg \overline{B}$, but $(qA)\neg B \not\models_A (qA)\overline{B}$.

## 2.10 Identity

The syllogistic is lacking any treatment of *identity*. As suggested in Section 2.6, Aristotle does not consider something like 'is Socrates' to be a term; the *Organon* does not seem to include any identity claims.

Yet, Aristotle formulates some principles to test for the (non-)identity of terms. Given two terms $A$ and $B$, we can compare them with respect to other terms $C$.[98] One principle Aristotle suggests is that if $A$ and $B$ are identical, then if $A$ is identical to $C$, $B$ must also be identical

---

[98]Note that Aristotle does not speak about *terms*, but I take it to apply to them.

to $C$ (Top H1, 152a31f.[99]). Whereas finding differences breaks identities (cf. Top A18, 108b2ff.[100]), being identical to something else suffices for identity, i.e., if $A$ is identical to $C$, and $B$ is identical to $C$, then $A$ and $B$ are identical too (SE 6, 168b31f.[101]).

Aristotle does not say much more about this, and its not entirely clear of what sort of things he claims identity, though he seems to formulate a (more general) version of Leibniz's law:

> Speaking generally, one ought to be on the look-out for any discrepancy anywhere in any sort of predicate of each term, and in the things of which they are predicated. For all that is predicated of the one should be predicated also of the other, and of whatever the one is a predicate, the other should be a predicate as well. (Top H1, 152b25–29[102])

I put it in terms of *terms* above. It should be clear that there are no (explicit) principles to establish identities between terms, but the principles allow us to break some (see also Top H1, 152b34f.[103]). Suppose that we establish that $(\forall A)C$ and $(\exists B)\neg C$. Then $A$ and $B$ cannot be identical. Also, if we establish that $(qC)A$ and $(qC)\neg B$ ($q \in \{\forall, \exists\}$), then $A$ and $B$ cannot be identical. However, if $A$ and $B$ are identical, we can conclude from $(qC)A$ that $(qC)B$, as well as $(qB)C$ from $(qA)C$. As $A$ and $B$ are terms, they can occur both in subject- and predicate-position, and the principle Aristotle suggests is meant to check both options. $A$ and $B$ are only then identical if the same terms are predicated of them

---

[99] "Again, look and see if, supposing the one to be the same as something, the other also is the same as it; for if they are not both the same as the same thing, clearly neither are they the same as one another [Πάλιν σκοπεῖν εἰ ᾧ θάτερον ταὐτόν, καὶ θάτερον· εἰ γὰρ μὴ ἀμφότερα τῷ αὐτῷ ταὐτά, δῆλον ὅτι οὐδ' ἀλλήλοις]."

[100] "for when we have found any difference whatever between the things proposed, we shall have shown that they are not the same thing [εὑρόντες γὰρ διαφορὰν τῶν προκειμένων ὁποιανοῦν δεδειχότες ἐσόμεθα ὅτι οὐ ταὐτόν]".

[101] "for we claim that things that are the same as one and the same thing are also the same as one another [τὰ γὰρ ἑνὶ καὶ ταὐτῷ ταὐτὰ καὶ ἀλλήλοις ἀξιοῦμεν εἶναι ταὐτά·]".

[102] "Καθόλου δ' εἰπεῖν ἐκ τῶν ὁπωσοῦν ἑκατέρου κατηγορουμένων καὶ ὧν ταῦτα κατηγορεῖται σκοπεῖν εἴ που διαφωνεῖ· ὅσα γὰρ θατέρου κατηγορεῖται, καὶ θατέρου κατηγορεῖσθαι δεῖ, καὶ ὧν θάτερον κατηγορεῖται, καὶ θάτερον κατηγορεῖσθαι δεῖ."

[103] "Moreover, see whether the one can exist without the other; for, if so, they will not be the same [Ἔτι εἰ δυνατὸν θάτερον ἄνευ θατέρου εἶναι· οὐ γὰρ ἂν εἴη ταὐτόν]".

$((qA)C$ and $(qB)C)$ *and* they are predicated of the same terms $((qC)A$ and $(qC)B)$.

## 3 A Fregean Approach

### 3.1 Background

Aristotle's logical system remained the dominant system until Gottlob Frege developed his *Begriffsschrift* [24]. That does not mean, though, that the syllogistic did not undergo *any* changes at all. One notable change is the treatment of particulars as terms in a way analogous to other terms (see Parkinson's introduction in [29]). Given the formalism from Section 2.9, a model $\mathfrak{M}_A$ interprets such terms $A$ as $\|A\|_{\mathfrak{M}_A} = \{a\}$ for an $a \in D$. Thus, for any term $B$, $(\exists A)B \models_A (\forall A)B$, i.e., the **i**-type sentence implies the **a**-type sentence. And, as already encoded in the square of opposition (Section 2.8), the latter also implies the former. For example, if $a \in D$ is Socrates and we blur the line between predicates and individuals, then 'some Socrates is human' implies 'every Socrates is human', and vice versa.[104]

However, Aristotelian syllogistic is limited in its expressive power. In particular, two limitations are generally pointed out, viz., Aristotelian syllogistic does not know *relational* terms, and, based on this, cannot deal with several quantificational phrases (see, e.g., Frege [24], Carnap [21], Russell [43, ch. 22], Kneale and Kneale [26, pp. 31, 487], and Link [30, p. 10]).

The main limitation is the syllogistic's restriction to terms which we can take to correspond to unary predicates so that it cannot account for *relations*. According to the *ti kata tinos*, the basic structure of sentences is subject-predicate. This means that the syllogistic cannot—in its current form—account for relational statements such as 'point $a$ lies between point $b$ and point $c$'. Frege overcomes this limitation by replacing the "concepts *subject* and *predicate* by *argument* and *function*" [24, p. 7, his emphases]. Of course, the most basic structure is still that of subject-predicate and is captured by a function applying

---

[104]As I've mostly treated terms as plural, it would be better to say 'some/every Socrates *are* human'.

to an argument—something that presupposes individual-constants that are not included in the syllogistic as presented above—but that immediately generalizes once the function is allowed to take more than one argument. Moreover, the subject-predicate structure is broken up once we consider sentences within the range of application of the syllogistic; for example, a sentence like 'all human beings are mortal' is not taken to have 'all human beings' or 'human beings' as its subject (depending on how one understands the quantity indicated by 'all'), but is analysed in terms for *quantifiers, variables, connectives,* and *functions* applying to *arguments*.

The other limitation is what might be called *nested quantification* (aka *multiply general propositions*). The subject-predicate structure does not rely on quantifiers, but the quantity of its subject is somehow indicated; Aristotle specifies it explicitly by saying, e.g., 'let $AB$ be a universal affirmative sentence', and the proposed formalism from Section 2.9 captures it by including a quantifier-symbol in front of the subject-term. Thus, the syllogistic can capture sentences like 'all human beings are mortal', but it lacks the means to express 'all human beings have someone they like' or 'some human beings like all human beings'. What's lacking is another way to even attach a quantity, and, as we have seen in Section 2.2, Aristotle does not think that sentences can be true if a quantity is assigned to more than the subject-term.

The Fregean approach with its function-argument analysis, on the other hand, has the means to assign quantities to several parts of sentences. Indeed, the *quantifiers* are treated as proper constituents of sentences. Only considering the first-order fragment, we can see that given arguments $a_1, \ldots, a_n$ and an $n$-ary function-symbol $f$, we can form a sentence '$f(a_1, \ldots, a_n)$' in which every argument-place allows to be quantified in. For example, '$\forall x_1 \exists x_2 f(x_1, x_2)$' is a sentence with nested quantifiers which can capture a sentence properly outside of Aristotelian syllogistic.

Overall, Frege captures a sentence like 'all human beings are mortal' as consisting of a *quantifier* ('$\forall$') binding a *variable* ('$x$') and acting on a *complex formula* with a *conditional* ('$\rightarrow$') as its main connective whose antecedent and consequent are *functions* applied to an *argument* ('$H(x)$', '$M(x)$'). None of these explicitly appears in the original sentence, and

Frege is well aware that his formal language departs from ordinary language [24, p. 6]; he thinks he introduces a tool for "certain scientific purposes" [24, p. 6], comparing it to the introduction of a microscope to better the human eye. Similarly, Carnap compares natural language to a "crude, primitive pocketknife" which is "useful for a hundred different purposes" [22, p. 938], but not so much for specific purposes requiring greater precision. In this sense, we can—and will—understand the formal languages and their formalisms as *explications*.

One strength of the Fregean explication is that it allows for fairly simple solutions to the aforementioned limitations of Aristotelian syllogistic. For, as sentences are not forced to have subject-predicate structure, it is possible to allow relational predicates like '$x$ lies between $y$ and $z$' ('$B(x, y, z)$'), and nest quantifiers. For example, we can render a sentence like 'every point $a$ lies between some points $b$ and $c$' as '$\forall x \exists y \exists z (B(x, y, z))$'.

## 3.2 A Formalism

To make the approach formally precise and to have a basis for comparison, let me introduce a convenient formalism (which more or less follows [39, ch. 3]). It should be clear that the following exposition is not following *Frege* in any detail, and is geared toward better comparison between the different formalisms to be introduced in the following and the one introduced in Section 2.9. Nevertheless, the following can rightly be claimed to expose, or explicate, a *Fregean* formalism.

The following exposition is not entirely standard, though it does not deviate much from a standard exposition. Insofar as it deviates, it is geared towards running in parallel with the exposition of the QUARC (Section 4.2). As I explain much of what's going on here already, I can present the QUARC-formalism succinctly while just pointing out the QUARC-specific features.

We start by specifying the vocabulary of a Fregean language.

**Definition 17** *(Fregean Language)*
A *Fregean language* ($\mathcal{L}_F$) consists of the following:

- a countably infinite set $\mathsf{Var}_{\mathcal{L}_F} = \{v_0, v_1, v_2, \ldots\}$ of (*individual-*) *variables*,

- a countable set $\mathsf{Const}_{\mathcal{L}_\mathsf{F}} = \{c_0, c_1, c_2, \ldots\}$ of (*individual-*)*constants*,
- for every $n > 0$, a countable set $\mathsf{Pred}^n_{\mathcal{L}_\mathsf{F}} = \{P^n_0, P^n_1, P^n_2, \ldots\}$ of *n-ary predicate-symbols*,
- the set of *logical symbols* including '$\neg$', '$\wedge$', '$\vee$', '$\rightarrow$', '$\leftrightarrow$', '$=$', '$\forall$', and '$\exists$', and
- the set of *auxiliary symbols* including '(', ')', and ','.

The sets are assumed to be disjoint. Let $\mathsf{Pred}_{\mathcal{L}_\mathsf{F}} = \bigcup_{n>0} \mathsf{Pred}^n_{\mathcal{L}_\mathsf{F}}$.

The basic vocabulary of a Fregean language $\mathcal{L}_\mathsf{F}$ is fairly standard and extends the language of Aristotelian Syllogistic $\mathcal{L}_\mathsf{A}$ in several ways. Firstly, $\mathcal{L}_\mathsf{F}$ contains what can be taken to correspond to $\mathcal{L}_\mathsf{A}$-terms, but also predicate-symbols of any arity. It also contains individual-constants and individual-variables, making it a first-order language. Lastly, $\mathcal{L}_\mathsf{F}$ has an additional logical symbol, viz., '$=$', and it lacks the term-negation '$-$'. Even though the languages overlap significantly (as we can consider $\mathcal{L}_\mathsf{A}$ to be a sublanguage of $\mathcal{L}_\mathsf{F}$), the formation rules for $\mathcal{L}_\mathsf{F}$ are significantly different from those of $\mathcal{L}_\mathsf{A}$.

**Definition 18** *($\mathcal{L}_\mathsf{F}$-Formula)*
Let $\mathcal{L}_\mathsf{F}$ be a Fregean language. The set of $\mathcal{L}_\mathsf{F}$-*formulas* ($\mathsf{Form}_{\mathcal{L}_\mathsf{F}}$) is recursively defined by:

(1) given $n > 0$ $\mathcal{L}_\mathsf{F}$-constants $c_1, \ldots, c_n$ and $P \in \mathsf{Pred}^n_{\mathcal{L}_\mathsf{F}}$, $P(c_1, \ldots, c_n) \in \mathsf{Form}_{\mathcal{L}_\mathsf{F}}$;

(2) given $\mathcal{L}_\mathsf{F}$-constants $c_1$ and $c_2$, $(c_1 = c_2) \in \mathsf{Form}_{\mathcal{L}_\mathsf{F}}$;

(3) if $\varphi \in \mathsf{Form}_{\mathcal{L}_\mathsf{F}}$, then $\neg\varphi \in \mathsf{Form}_{\mathcal{L}_\mathsf{F}}$;

(4) if $\varphi, \psi \in \mathsf{Form}_{\mathcal{L}_\mathsf{F}}$ and $\circ \in \{\wedge, \vee, \rightarrow, \leftrightarrow\}$, then $(\varphi \circ \psi) \in \mathsf{Form}_{\mathcal{L}_\mathsf{F}}$;

(5) if $\varphi(c) \in \mathsf{Form}_{\mathcal{L}_\mathsf{F}}$, $x \in \mathsf{Var}_{\mathcal{L}_\mathsf{F}}$, and $q \in \{\forall, \exists\}$, then $qx\varphi[x/c] \in \mathsf{Form}_{\mathcal{L}_\mathsf{F}}$.

Definition 18, in contrast to Definition 3, introduces a recursion to generate all the formulas. Clause (1) captures the function-argument structure, viz., the elements of $\mathsf{Const}_{\mathcal{L}_\mathsf{F}}$ are the arguments to the functions contained in $\mathsf{Pred}_\mathsf{F}$. Additionally, the symbol '=' figures as binary function/predicate. As usual, we can consider the formulas obtained by clauses (1)–(2) to be *atomic* and the constituents of *complex* formulas arrived by the remaining clauses.

Clause (5) allows for nested quantification for which clauses (1)–(2) provide the places. For example, for $P \in \mathsf{Pred}^2_{\mathcal{L}_\mathsf{F}}$ and $c_1, c_2 \in \mathsf{Const}_{\mathcal{L}_\mathsf{F}}$, clause (1) guarantees that $P(c_1, c_2) \in \mathsf{Form}_{\mathcal{L}_\mathsf{F}}$. Applying clause (5) to it, $x \in \mathsf{Var}_{\mathcal{L}_\mathsf{F}}$ and $\forall$, leads to $\forall x \varphi(x, c_2) \in \mathsf{Form}_{\mathcal{L}_\mathsf{F}}$. Applying it again to this, $y \in \mathsf{Var}_{\mathcal{L}_\mathsf{F}}$ and $\exists$, we get $\exists y \forall x P(x, y) \in \mathsf{Form}_{\mathcal{L}_\mathsf{F}}$.

Definition 18 is non-standard insofar as it does not allow for *open* formulas. Clause (5) is the only clause introducing *variables*, and those variables are bound. Because of this and in order to have a better comparable formalism, we understand quantification as *substitutional* and treat it accordingly below.

Given this increase in complexity, the interpretations of such Fregean languages have to be more complex, too, though the underlying model remains the same; we just make more use of it.

**Definition 19 ($\mathcal{L}_\mathsf{F}$-Model)**
Let $\mathcal{L}_\mathsf{F}$ be a Fregean language. A *model for* $\mathcal{L}_\mathsf{F}$ ($\mathcal{L}_\mathsf{F}$-model) is an ordered pair $\mathfrak{M}_\mathsf{F} = \langle D, \|\cdot\|_{\mathfrak{M}_\mathsf{F}} \rangle$ such that

(1) $D$ is a set (the *domain* of $\mathfrak{M}_\mathsf{F}$);

(2) $\|\cdot\|_{\mathfrak{M}_\mathsf{F}}$ is an *interpretation-function of* $\mathfrak{M}_\mathsf{F}$ such that

   (a) if $c \in \mathsf{Const}_{\mathcal{L}_\mathsf{F}}$, then $\|c\|_{\mathfrak{M}_\mathsf{F}} \in D$;
   
   (b) if $P \in \mathsf{Pred}^1_{\mathcal{L}_\mathsf{F}}$, then $\emptyset \neq \|P\|_{\mathfrak{M}_\mathsf{F}} \subseteq D$;
   
   (c) if $n > 1$ and $P \in \mathsf{Pred}^n_{\mathcal{L}_\mathsf{F}}$, then $\|P\|_{\mathfrak{M}_\mathsf{F}} \subseteq D^n$.

Clause (2b) forces unary predicates to be assigned a non-empty extension. This has been done in order for a smoother comparison with QUARC. Moreover, in this way we also generate better comparability to (non-empty) $\mathcal{L}_\mathsf{A}$-models.

Moreover, as QUARC relies on *substitutional quantification*, we understand it similarly here. Hence, in order to correctly interpret the formulas involving quantification, we need to make sure that the interpretation does not rely on the specific choice of $\mathsf{Const}_{\mathcal{L}_\mathsf{F}}$. In order to do so, we first expand the underlying language (Definition 20), enriching it with further individual-constants, and then making sure that the interpretation keeps up (Definition 21). With these in place, we can specify when a model satisfies a formula (Definition 22).

**Definition 20 ($\mathcal{L}_\mathsf{F}$-$A$-*Expansion*)**
Let $\mathcal{L}_\mathsf{F}$ be a Fregean language. Let $\mathfrak{M}_\mathsf{F} = \langle D, \|\cdot\|_{\mathfrak{M}_\mathsf{F}} \rangle$ be an $\mathcal{L}_\mathsf{F}$-model. Let $A \subseteq D$. The $\mathcal{L}_\mathsf{F}$-$A$-*expansion of* $\mathcal{L}_\mathsf{F}$ is the language $\mathcal{L}'_\mathsf{F} := \mathcal{L}_\mathsf{F} \cup \{c_a | a \in A\}$ where the $c_a$ are new (individual-)constants not contained in $\mathcal{L}_\mathsf{F}$.

If $A = \{a\}$, we call $\mathcal{L}'_\mathsf{F}$ an $\mathcal{L}_\mathsf{F}$-$a$-expansion.

The idea is that we consider part of the domain of a model $\mathfrak{M}_\mathsf{F}$ and introduce new names for the elements of the chosen part. The new symbols need to be interpreted in the correct way too, which cannot be done in the original model $\mathfrak{M}_\mathsf{F}$ so that we have to expand it to $\mathfrak{M}'_\mathsf{F}$ in the following way.

**Definition 21 ($\mathcal{L}_\mathsf{F}$-*Model Expansion*)**
Let $\mathcal{L}_\mathsf{F}$ be a Fregean language and $\mathfrak{M}_\mathsf{F} = \langle D, \|\cdot\|_{\mathfrak{M}_\mathsf{F}} \rangle$ an $\mathcal{L}_\mathsf{F}$-model. Let $A \subseteq D$, and $\mathcal{L}'_\mathsf{F}$ be an $\mathcal{L}_\mathsf{F}$-$A$-expansion. The $A$-*expansion of* $\mathfrak{M}_\mathsf{F}$ *to* $\mathcal{L}'_\mathsf{F}$ is the model $\mathfrak{M}'_\mathsf{F} = \langle D', \|\cdot\|_{\mathfrak{M}'_\mathsf{F}} \rangle$ such that

(1) $D' = D$;   (2) $\|\cdot\|_{\mathfrak{M}_\mathsf{F}} \subseteq \|\cdot\|_{\mathfrak{M}'_\mathsf{F}}$;

(3) $\|c_a\|_{\mathfrak{M}'_\mathsf{F}} = a \in A$ for every new symbol $c_a$.

The domains of the model $\mathfrak{M}_\mathsf{F}$ and its expansion $\mathfrak{M}'_\mathsf{F}$ are the same. The new constants are interpreted according to how they have been introduced. Since $A \subseteq D$ and $D' = D$, $A \subseteq D'$, and the new symbols $c_a$ for $a \in A$ just provide names for the elements $a \in D$.

Lastly, the expanded interpretation-function $\|\cdot\|_{\mathfrak{M}'_\mathsf{F}}$ extends the interpretation-function $\|\cdot\|_{\mathfrak{M}_\mathsf{F}}$, i.e., it leaves unaltered the interpretations of the original model $\mathfrak{M}_\mathsf{F}$. In particular, suppose that $\|P\|_{\mathfrak{M}_\mathsf{F}} = \{a\}$ for $a \in D$, but there is no $c \in \mathsf{Const}_{\mathcal{L}_\mathsf{F}}$ such that $\|c\|_{\mathfrak{M}_\mathsf{F}} = a$. We can then expand the language to $\mathcal{L}'_\mathsf{F}$ to include $c_a \in \mathsf{Const}_{\mathcal{L}'_\mathsf{F}}$ without altering

$\|P\|_{\mathfrak{M}_F}$; all that the expansion does is give a name to a (potentially) unnamed object without altering the interpretation of the $P \in \mathsf{Pred}_{\mathcal{L}_F}$.

With this machinery, we can define the corresponding satisfaction-relation. It also suffices to expand the language by *one* individual-constant at a time as we quantify over *all* such expansions so that no element of $D$ gets missed.

**Definition 22** *(Satisfaction $\models_F$)*
Let the *Fregean satisfaction-relation* $\mathfrak{M}_F \models_F \varphi$ for $\varphi \in \mathsf{Form}_{\mathcal{L}_F}$ and $\mathcal{L}_F$-model $\mathfrak{M}_F = \langle D, \|\cdot\|_{\mathfrak{M}_F}\rangle$ be recursively defined as follows:

(1) $\mathfrak{M}_F \models_F P(c_1, \ldots, c_n)$ iff $\langle \|c_1\|_{\mathfrak{M}_F}, \ldots, \|c_n\|_{\mathfrak{M}_F}\rangle \in \|P\|_{\mathfrak{M}_F}$;

(2) $\mathfrak{M}_F \models_F c_1 = c_2$ iff $\|c_1\|_{\mathfrak{M}_F} = \|c_2\|_{\mathfrak{M}_F}$;

(3) $\mathfrak{M}_F \models_F \neg\varphi$ iff it is not the case that $\mathfrak{M}_F \models_F \varphi$ ($\mathfrak{M}_F \not\models_F \varphi$);

(4) $\mathfrak{M}_F \models_F \varphi \wedge \psi$ iff $\mathfrak{M}_F \models_F \varphi$ and $\mathfrak{M}_F \models_F \psi$;

(5) $\mathfrak{M}_F \models_F \exists x \varphi[x]$ iff for some $a$-expansion $\mathfrak{M}'_F$ of $\mathfrak{M}_F$, $\mathfrak{M}'_F \models_F \varphi[c_a/x]$;

(6) $\mathfrak{M}_F \models_F \forall x \varphi[x]$ iff for all $a$-expansions $\mathfrak{M}'_F$ of $\mathfrak{M}_F$, $\mathfrak{M}'_F \models_F \varphi[c_a/x]$.

The definition is mostly standard. Given clauses (3)–(4), we can define the clauses for the remaining connectives in the usual way. In contrast to *objectual quantification* which interprets the quantifiers via *variable assignments*, here the quantifiers are interpreted *substitutionally*; instead of considering all the possible values for the variables, the base model $\mathfrak{M}_F$ satisfies a formula of the form '$\forall x \varphi$' if all expansions $\mathfrak{M}'_F$ satisfy '$\varphi[c_a]$' where the new constants '$c_a$' are substituted for the variable '$x$'. By Definitions 20–21, every element of the domain $D$ is considered so that the truth of '$\forall x \varphi$' does *not* depend on the particular choice of $\mathsf{Const}_{\mathcal{L}_F}$.

We can define a corresponding notion of *logical consequence* analogous to Definition 8; just substitute '$\mathcal{L}_F$' for '$\mathcal{L}_A$', '$\mathfrak{M}_F$' for '$\mathfrak{M}_\Vdash$', and '$\models_F$' for '$\Vdash$'. With that at hand, one peculiarity of the above is the following.

**Theorem 23**
Let $P \in \mathsf{Pred}^1_{\mathcal{L}_F}$. Then: $\models_F \exists x P(x)$.

*Proof.* Let $P \in \mathsf{Pred}^1_{\mathcal{L}_\mathsf{F}}$. Let $\mathfrak{M}_\mathsf{F} = \langle D, \|\cdot\|_{\mathfrak{M}_\mathsf{F}} \rangle$ be an $\mathcal{L}_\mathsf{F}$-model. By Definition 19 (2b), $\emptyset \neq \|P\|_{\mathfrak{M}_\mathsf{F}} \subseteq D$. Let $a \in \|P\|_{\mathfrak{M}_\mathsf{F}}$. Let $\mathcal{L}'_\mathsf{F}$ be an $\mathcal{L}_\mathsf{F}$-$a$-expansion of $\mathcal{L}_\mathsf{F}$, and $\mathfrak{M}'_\mathsf{F}$ be an $a$-expansion of $\mathfrak{M}_\mathsf{F}$ to $\mathcal{L}'_\mathsf{F}$. By Definition 21 (2), $\|P\|_{\mathfrak{M}_\mathsf{F}} \subseteq \|P\|_{\mathfrak{M}'_\mathsf{F}}$ so that $a \in \|P\|_{\mathfrak{M}'_\mathsf{F}}$. By Definition 21 (3), $\|c_a\|_{\mathfrak{M}'_\mathsf{F}} = a \in \|P\|_{\mathfrak{M}'_\mathsf{F}}$. Thus, by Definition 22 (1), $\mathfrak{M}'_\mathsf{F} \models_\mathsf{F} P(c_a)$. Then, there is an $a$-expansion $\mathfrak{M}'_\mathsf{F}$ of $\mathfrak{M}_\mathsf{F}$, $\mathfrak{M}'_\mathsf{F} \models_\mathsf{F} P(c_a)$. Therefore, by Definition 22 (5), $\mathfrak{M}_\mathsf{F} \models_\mathsf{F} \exists x P(x)$. $\square$

This also means that universal quantification implies the existential one.

**Corollary 24**
$\forall x P(x) \models_\mathsf{F} \exists x P(x)$.

We can also note that the quantifiers behave as expected.

**Theorem 25**
The following equivalences hold:

(1) $\models_\mathsf{F} \forall x \varphi \leftrightarrow \neg \exists x \neg \varphi$;   (2) $\models_\mathsf{F} \exists x \varphi \leftrightarrow \neg \forall x \neg \varphi$;

(3) $\models_\mathsf{F} \neg \exists x \varphi \leftrightarrow \forall x \neg \varphi$;   (4) $\models_\mathsf{F} \neg \forall x \varphi \leftrightarrow \exists x \neg \varphi$.

Furthermore, because of the non-emptiness requirement in Definition 19 (2b), analogues of conversion hold.

**Theorem 26** *(Conversion)*
The following conversions hold:

(**a**-**i**-conv$^{\models_\mathsf{F}}$) $\forall x(A(x) \to B(x)) \models_\mathsf{F} \exists x(B(x) \wedge A(x))$

(**i**-**i**-conv$^{\models_\mathsf{F}}$) $\exists x(A(x) \wedge B(x)) \models_\mathsf{F} \exists x(B(x) \wedge A(x))$

(**e**-**e**-conv$^{\models_\mathsf{F}}$) $\forall x(A(x) \to \neg B(x)) \models_\mathsf{F} \forall x(B(x) \to \neg A(x))$

*Proof.* I only show the interesting case.

(**a**-**i**-conv$^{\models_\mathsf{F}}$): Let $\mathfrak{M}_\mathsf{F}$ be an $\mathcal{L}_\mathsf{F}$-model such that $\mathfrak{M}_\mathsf{F} \models_\mathsf{F} \forall x(A(x) \to B(x))$. Then, by Definition 22 (6), for all $a$-expansions $\mathfrak{M}'_\mathsf{F}$ of $\mathfrak{M}_\mathsf{F}$, $\mathfrak{M}'_\mathsf{F} \models_\mathsf{F} A(c_a) \to B(c_a)$. By Definition 19 (2b), $\emptyset \neq \|A\|_{\mathfrak{M}_\mathsf{F}}$. Let

$a \in \|A\|_{\mathfrak{M}_\mathsf{F}}$. Then, for the $a$-expansion $\mathfrak{M}^*_\mathsf{F}$ of $\mathfrak{M}_\mathsf{F}$, $\mathfrak{M}^*_\mathsf{F} \models_\mathsf{F} A(c_a) \to B(c_a)$. By Definition 21 (2), $a \in \|A\|_{\mathfrak{M}_\mathsf{F}} \subseteq \|A\|_{\mathfrak{M}^*_\mathsf{F}}$, i.e., $a \in \|A\|_{\mathfrak{M}^*_\mathsf{F}}$. Thus, for the $a$-expansion $\mathfrak{M}^*_\mathsf{F}$ of $\mathfrak{M}_\mathsf{F}$, $\mathfrak{M}^*_\mathsf{F} \models_\mathsf{F} A(c_a)$. Also for the $a$-expansion $\mathfrak{M}^*_\mathsf{F}$ of $\mathfrak{M}_\mathsf{F}$, $\mathfrak{M}^*_\mathsf{F} \models_\mathsf{F} A(c_a) \to B(c_a)$. Therefore, for the $a$-expansion $\mathfrak{M}^*_\mathsf{F}$ of $\mathfrak{M}_\mathsf{F}$, $\mathfrak{M}^*_\mathsf{F} \models_\mathsf{F} B(c_a)$ and, so, for some $a$-expansion $\mathfrak{M}'_\mathsf{F}$ of $\mathfrak{M}_\mathsf{F}$, $\mathfrak{M}'_\mathsf{F} \models_\mathsf{F} B(c_a) \wedge A(c_a)$. By Definition 22 (5), $\mathfrak{M}_\mathsf{F} \models_\mathsf{F} \exists x(B(x) \wedge A(x))$.

□

This much suffices in terms of exposition of a Fregean language and its semantics. As we are only interested in semantics, there is no need to introduce a proof system.

## 4 Ben-Yami's **QUARC**

### 4.1 Background

In recent years, Hanoch Ben-Yami has introduced a novel logical system called the *QUantified ARgument Calculus* (**QUARC**). The underlying motivation is to find a formal system that captures more adequately the semantics of natural language.[105] Section 3.1 already suggests that Frege's main motivation is *not* to come up with a formal language to capture the semantics of natural language; however, the elegance and strength of his formal language surpassed anything else known and so was a natural candidate to be used outside its original intended range of application.

Ben-Yami introduces an early version of **QUARC** in his book *Logic & Natural Language* [13]—which is the main focus of this brief exposition. Note, though, that certain of Ben-Yami's views have developed and changed since the book was published in 2004; my concern here is not to paint an accurate picture of his current views (for some of those see [49]), though I mention some in footnotes.

---

[105] Hanoch prefers '*logic* of natural language' (personal communication). I stick to 'semantics': Where it is clear to me that natural language has a semantics (and, potentially, several), it is less clear to me that it has a *logic*. My views are not settled, but am inclined to deny that there is *the* logic of natural language.

Since the appearance of the book, Ben-Yami published an article exposing QUARC [16], and considered how it treats the Barcan formulas and necessary existence [17] as well as how QUARC compares to natural logic [18]. There have also been discussions with respect to *generalized quantifiers* (Ben-Yami [14, 15] and Westerståhl [48]).

Moreover, QUARC's logical properties have been investigated. Lanzet and Ben-Yami [28] provide an early assessment in model-theoretic terms, Raab [39, 41] considers QUARC's relationship to classical logic and so does Lanzet [27] in a three-valued setting. There are completeness results for QUARC in different settings (e.g., Lanzet and Ben-Yami [28], Raab [39], Ben-Yami and Pavlović [19]), and treatments based on many-valued truth-valuational semantics (Yin and Ben-Yami [49]).

QUARC has also been investigated proof-theoretically (Pavlović [33], Pavlović and Gratzl [34, 35, 36]) as well as axiomatically (Pascucci [32]). Moreover, Pavlović and Gratzl also consider abstract forms of quantification within QUARC [37] and investigate into decidable fragments [38]. Several further aspects of QUARC are currently investigated.

Ben-Yami [13] rejects Fregean languages when investigating the semantics of natural language. He suggests two main reasons, viz., the treatment of *reference* and *quantification*.[106] I do not go into detail with all the subtleties, but focus on some general points.

Regarding reference, Ben-Yami notes that natural language contains *plural* referring expressions. Fregean languages, on the other hand, only allow *singular* reference. In the Fregean languages, this is achieved solely via the *variables* and *individual-constants*. Thus, as detailed in Section 3.1, a sentence like 'All human beings are mortal' is captured as '$\forall x(H(x) \to M(x))$', quantifying *singularly* over everything. However, the surface structure of the sentence sees 'all human beings' as the subject of the sentence and 'human beings' refers *plurally* to *human beings* while 'all' specifies the relevant quantity of what's being referred to.

Based on the treatment of reference as singular, Ben-Yami [13, p. 2] also argues that Fregean languages misconstrue *predication* and *quantification* in natural language. Ben-Yami [13, p. 8] suggests that Fregean

---

[106] Hanoch's current views changed regarding *reference* which dropped out of the picture; indeed, he insists that the notion of reference is irrelevant for QUARC (personal communication).

languages understand singular terms to be the *sole* source of reference, and common nouns as logical predicates. Ben-Yami [13, p. 8], on the other hand, argues that common nouns are used to refer to (pluralities of) particulars too. Given this understanding, he claims to arrive, among others, at a "radically different analysis of quantification" [13, p. 12].

Ben-Yami's main point is that "quantification involves reference to a plurality" [13, p. 59]. In the example sentence above, 'human beings' refers to a plurality of *human beings*, and the quantifier 'all' specifies how much of that plurality is relevant, i.e., a "quantifier is attached to a noun that is used to refer to a plurality" [13, pp. 59f.] and these elements together "form a noun phrase" [13, p. 60]. Such noun phrases—called *quantified arguments*—can function as subjects of sentences; they can be put in the argument places of predicates. In the example sentence, 'mortal' is predicated of 'all human beings', i.e., the quantified argument 'all human beings' is put into the argument-slot of the predicate 'mortal'. Ben-Yami [13, p. 62] claims that this is in agreement with Aristotle's understanding of predication.

One topic of concern for Ben-Yami is that of the *expressive power* of systems. Ben-Yami [13, p. 78] notes that Aristotelian logic is *not* expressive enough to handle relations and nested quantification. His goal is to develop a system that is able to handle that, and suggests that "[a]ny alternative logic should have comparable power" [13, p. 78] to Fregean logic with its predicate calculus. The QUARC is meant to have that.

In order to achieve that, QUARC needs a device to establish *cross-reference*; it captures it by the incorporation of *anaphora*. Moreover, natural language contains *active* and *passive* constructions, and different ways of negating, viz., negating a whole sentence ('*it is not the case that* Socrates is mortal') and *negative predication* ('Socrates *is not* mortal'); all this is incorporated in the QUARC too. The QUARC also includes *identity*, though it treats it slightly different from the way it is in Fregean languages, as predication is understood differently [13, p. 142]. All these elements are incorporated in the formalism below.

## 4.2 A Formalism

Let me make the QUARC formalism precise. I generally follow the exposition of Section 3.2, and comment only on the QUARC-specific details of the formalism.

First, again, let's specify the underlying vocabulary.

**Definition 27 (*QUARC-Language*)**
A *QUARC-language* ($\mathcal{L}_Q$) consists of the following:

- a countably infinite set $\mathsf{Ana}_{\mathcal{L}_Q} = \{\alpha_0, \alpha_1, \alpha_2, \dots\}$ of *anaphors*,

- a countable set $\mathsf{SA}_{\mathcal{L}_Q} = \{s_0, s_1, s_2, \dots\}$ of *singular arguments*,

- for every $n > 0$, a countable set $\mathsf{Pred}^n_{\mathcal{L}_Q} = \{P_0^{1,\dots,n}, P_1^{1,\dots,n}, P_2^{1,\dots,n}, \dots\}$ of *n-ary predicate-symbols*,

- for every $n > 0$, for every $i \geq 0$, for every $P_i^{1,\dots,n} \in \mathsf{Pred}^n_{\mathcal{L}_Q}$, a set $\mathsf{Reord}^n_{\mathcal{L}_Q} = \{P_i^{\pi(1),\dots,\pi(n)} | \pi\colon \{1,\dots,n\} \to \{1,\dots,n\} \text{ a permutation}\}$ of *n-ary reorders*,

- the set of *logical symbols* including '¬', '∧', '∨', '→', '↔', '=', '∀', and '∃', and

- the set of *auxiliary symbols* including '(', ')', and ','.

For every $n \geq 1$, $\mathsf{Pred}^n_{\mathcal{L}_Q} \subseteq \mathsf{Reord}^n_{\mathcal{L}_Q}$; all other sets are assumed to be disjoint. Let $\mathsf{Pred}_{\mathcal{L}_Q} := \bigcup_{n>0} \mathsf{Pred}^n_{\mathcal{L}_Q}$ and $\mathsf{Reord}_{\mathcal{L}_Q} := \bigcup_{n>0} \mathsf{Reord}^n_{\mathcal{L}_Q}$.

Compared to Definition 17, Definition 27 is more complex. Firstly, what's analogous to the Fregean language, a QUARC-language contains *anaphors* which play a similar role to the *variables* of Fregean languages. However, Fregean languages need variables to achieve quantification, QUARC does not as witnessed by formulas of the form '$(\forall P)Q$' for $P, Q \in \mathsf{Pred}^1_{\mathcal{L}_Q}$. Moreover, the QUARC-language contains *singular arguments* which correspond to Fregean (individual-)constants. The logical and auxiliary symbols of the languages are the same. However, QUARC specifies its predicates differently. In particular, I put the members of $\mathsf{Pred}^n_{\mathcal{L}_Q}$ as '$P_i^{1,\dots,n}$', not just indicating the arity $n$, but also the order of the slots. The reason for this is that this guarantees that they are identical to *reorders*. For any $P \in \mathsf{Pred}^n_{\mathcal{L}_\mathsf{Q}}$, there are $n!$-many reorders, generated by permutations on the predicate's argument-places. However, I just write '$P^\pi$' instead of '$P^{\pi(1),\ldots,\pi(n)}$' ($P \in \mathsf{Pred}^n_{\mathcal{L}_\mathsf{Q}}$) to indicate the reorder if it is relevant.[107] Since $\mathsf{Pred}^n_{\mathcal{L}_\mathsf{Q}} \subseteq \mathsf{Reord}^n_{\mathcal{L}_\mathsf{Q}}$, we can often work with the latter in setting up the formalism; this helps reducing some complexity in specifying the QUARC-formulas.

**Definition 28** *($\mathcal{L}_\mathsf{Q}$-Formula)*
Let $\mathcal{L}_\mathsf{Q}$ be a QUARC-language. The set of $\mathcal{L}_\mathsf{Q}$-*formulas* ($\mathsf{Form}_{\mathcal{L}_\mathsf{Q}}$) is recursively defined by:

(1) given $n > 0$ $s_1, \ldots, s_n \in \mathsf{SA}_{\mathcal{L}_\mathsf{Q}}$ and $P \in \mathsf{Reord}^n_{\mathcal{L}_\mathsf{Q}}$, then $(s_1, \ldots, s_n)P \in \mathsf{Form}_{\mathcal{L}_\mathsf{Q}}$;

(2) given $s_1, s_2 \in \mathsf{SA}_{\mathcal{L}_\mathsf{Q}}$, $(s_1, s_2) = \in \mathsf{Form}_{\mathcal{L}_\mathsf{Q}}$ (usually written in the usual infix-notation '$(s_1 = s_2)$');

(3) given $n > 0$, $s_1, \ldots, s_n \in \mathsf{SA}_{\mathcal{L}_\mathsf{Q}}$, $P \in \mathsf{Reord}^n_{\mathcal{L}_\mathsf{Q}}$, and $*$ a string of negation-symbols $\neg$, $((s_1, \ldots, s_n) * P) \in \mathsf{Form}_{\mathcal{L}_\mathsf{Q}}$;

(4) if $\varphi \in \mathsf{Form}_{\mathcal{L}_\mathsf{Q}}$, then $\neg \varphi \in \mathsf{Form}_{\mathcal{L}_\mathsf{Q}}$;

(5) if $\varphi, \psi \in \mathsf{Form}_{\mathcal{L}_\mathsf{Q}}$ and $\circ \in \{\wedge, \vee, \rightarrow, \leftrightarrow\}$, then $(\varphi \circ \psi) \in \mathsf{Form}_{\mathcal{L}_\mathsf{Q}}$;

(6) if $\varphi \in \mathsf{Form}_{\mathcal{L}_\mathsf{Q}}$ contains, from left to right, $s_1, \ldots, s_n$ ($n \geq 2$) occurrences of $s \in \mathsf{SA}_{\mathcal{L}_\mathsf{Q}}$, none of which is the source of $\beta \in \mathsf{Ana}_{\mathcal{L}_\mathsf{Q}}$ that occurs in $\varphi$, and $\varphi$ does not contain $\alpha \in \mathsf{Ana}_{\mathcal{L}_\mathsf{Q}}$, then $\varphi[s_\alpha/s_1, \alpha/s_2, \ldots, \alpha/s_n] \in \mathsf{Form}_{\mathcal{L}_\mathsf{Q}}$ where $\varphi[s_\alpha/s_1, \alpha/s_2, \ldots, \alpha/s_n]$ is the result of substituting $\alpha$ for the occurrences $s_2, \ldots, s_n$ of $s$;

(7) if $\varphi[s] \in \mathsf{Form}_{\mathcal{L}_\mathsf{Q}}$, $q \in \{\forall, \exists\}$, $P \in \mathsf{Pred}^1_{\mathcal{L}_\mathsf{Q}}$, then $\varphi[qP/s] \in \mathsf{Form}_{\mathcal{L}_\mathsf{Q}}$ if $qP$ *governs* $\varphi$ (see Definition 30).

Let $\mathsf{QA}_{\mathcal{L}_\mathsf{Q}}$ be the set of *quantified arguments*, i.e., expressions of the form $qP$ for $q \in \{\forall, \exists\}$ and $P \in \mathsf{Pred}^1_{\mathcal{L}_\mathsf{Q}}$.

---

[107] Hanoch (personal communication) prefers to think of $\pi$ as an *operator* acting on predicates $P \in \mathsf{Pred}^n_{\mathcal{L}_\mathsf{Q}}$ so that the predicate stays the same, but gets reordered.

Clauses (1)–(2) correspond to Definition 18's (1)–(2); the only difference is that QUARC takes predicates from $\text{Reord}_{\mathcal{L}_Q}$ and that we write the arguments *in front of* the predicate-symbol. Moreover, clauses (4)–(5) are standard too. Let me comment on the remaining clauses.

Clause (3) is QUARC-specific. It allows arbitrarily many negation-symbols inbetween a predicate-symbol's argument-slots and predicate-sign. Thus, we allow, e.g., '$(s)\neg\neg\neg P$' as an $\mathcal{L}_Q$-formula.

Clause (6) allows for the introduction of *anaphors*. If a formula contains several occurrences of a singular argument $s$, we can replace all but the first by new anaphors. For example, we can move from '$(s,s)P$' to '$(s_\alpha, \alpha)P$'. As long as no quantified arguments are involved, these anaphors are not necessary, but they are once cross-reference is needed.

Clause (7), finally, allows the introduction of *quantified arguments*, i.e., expressions combining *quantifiers* with *unary predicates* so that quantification is understood as *plural*. These expressions can replace *singular* arguments given that they satisfy a certain condition, viz., that the quantified argument *governs* the formula—which we define below. As the quantified arguments can take the place of a singular argument that has anaphors referring to it, we also define the notion *source of anaphora*.

**Definition 29** *(Source of Anaphora)*
If an anaphor is introduced according to clause (6), then the term $s$ *is the source of* $\alpha$ (indicated as '$s_\alpha$') if it is the rightmost occurrence of $s$ that is to the left of the anaphor $\alpha$; if such a term is replaced by a $t \in \text{QA}_{\mathcal{L}_Q}$ due to an application of clause (7), then $t$ *is the source of* $\alpha$ (indicated as '$qP_\alpha$' if $t = qP$).

**Definition 30** *(Governance)*
Let $\varphi$ be a string of symbols and $t \in \text{QA}_{\mathcal{L}_Q}$. Then, $t$ *governs* $\varphi$ if it is the leftmost quantified argument and $\varphi$ does not contain any other string of symbols $\psi$ such that $\psi \in \text{Form}_{\mathcal{L}_Q}$ contains $t$ and all the anaphors of all arguments in $\psi$.

Given Definition 30, Definition 28 (7) is well-defined now. Roughly, the idea is that we can introduce *quantified arguments* if they are the *main symbol*, i.e., when breaking up the formula, one has to start with it.

As in Section 3.2, all formulas are *closed*. The mechanism to introduce anaphors and quantified expressions is via *substitution* and so we treat quantification *substitutionally*. This follows the treatment from Section 3.2. In particular, we use *models* to interpret QUARC-languages.

**Definition 31** *($\mathcal{L}_Q$-Model)*
Let $\mathcal{L}_Q$ be a QUARC-language. A *model for $\mathcal{L}_Q$* ($\mathcal{L}_Q$-model) is an ordered pair $\mathfrak{M}_Q = \langle D, \|\cdot\|_{\mathfrak{M}_Q}\rangle$ such that

(1) $D$ is a set (the *domain* of $\mathfrak{M}_Q$);

(2) $\|\cdot\|_{\mathfrak{M}_Q}$ is an *interpretation-function of* $\mathfrak{M}_Q$ such that

  (a) if $s \in \mathsf{SA}_{\mathcal{L}_Q}$, then $\|s\|_{\mathfrak{M}_Q} \in D$;
  
  (b) if $P \in \mathsf{Pred}^1_{\mathcal{L}_Q}$, then $\emptyset \neq \|P\|_{\mathfrak{M}_Q} \subseteq D$;
  
  (c) if $n > 1$ and $P \in \mathsf{Pred}^n_{\mathcal{L}_Q}$, then $\|P\|_{\mathfrak{M}_Q} \subseteq D^n$;
  
  (d) if $n \geq 1$ and $P^\pi \in \mathsf{Reord}^n_{\mathcal{L}_Q}$ for permutation $\pi\colon \{1,\ldots,n\} \to \{1,\ldots,n\}$, $\|P^\pi\|_{\mathfrak{M}_Q} =$
  $\{\langle\|s_{\pi(1)}\|_{\mathfrak{M}_Q},\ldots,\|s_{\pi(n)}\|_{\mathfrak{M}_Q}\rangle | \langle\|s_1\|_{\mathfrak{M}_Q},\ldots,\|s_n\|_{\mathfrak{M}_Q}\rangle \in \|P\|_{\mathfrak{M}_Q}\}$.

This definition corresponds to Definition 19. The only QUARC-specific part is clause (2d) which interprets *reorders* in the obvious way. As a reorder $P^\pi \in \mathsf{Reord}^n_{\mathcal{L}_Q}$ comes from reordering the argument-places of a predicate $P \in \mathsf{Pred}^n_{\mathcal{L}_Q}$, the interpretation does the same.

As before, we do not want to be held hostage to the particular choice of what individuals the language can name, i.e., to the specific $\mathsf{SA}_{\mathcal{L}_Q}$, so we expand the language (Definition 32), and specify the corresponding model expansions (Definition 33). With that, we can define the satisfaction-relation (Definition 34).

**Definition 32** *($\mathcal{L}_Q$-A-Expansion)*
Let $\mathcal{L}_Q$ be a QUARC-language and $\mathfrak{M}_Q = \langle D, \|\cdot\|_Q\rangle$ be an $\mathcal{L}_Q$-model. Let $A \subseteq D$. The *$\mathcal{L}_Q$-A-expansion of $\mathcal{L}_Q$* is the language $\mathcal{L}'_Q := \mathcal{L}_Q \cup \{s_a | a \in A\}$ where the $s_a$ are new singular arguments not contained in $\mathcal{L}_Q$.

If $A = \{a\}$, we call $\mathcal{L}'_Q$ an *$\mathcal{L}_Q$-a-expansion*.

**Definition 33 ($\mathcal{L}_Q$-Model Expansion)**
Let $\mathcal{L}_Q$ be a QUARC-language and $\mathfrak{M}_Q = \langle D, \|\cdot\|_{\mathfrak{M}_Q}\rangle$ an $\mathcal{L}_Q$-model. Let $A \subseteq D$, and $\mathcal{L}'_Q$ be an $\mathcal{L}_Q$-$A$-expansion. The $A$-*expansion of* $\mathfrak{M}_Q$ *to* $\mathcal{L}'_Q$ is the model $\mathfrak{M}'_Q = \langle D', \|\cdot\|_{\mathfrak{M}'_Q}\rangle$ such that

(1) $D' = D$;   (2) $\|\cdot\|_{\mathfrak{M}_Q} \subseteq \|\cdot\|_{\mathfrak{M}'_Q}$;

(3) $\|s_a\|_{\mathfrak{M}'_Q} = a \in A$ for every new singular argument $s_a$.

**Definition 34 (Satisfaction $\models_Q$)**
Let the QUARC *satisfaction-relation* $\mathfrak{M}_Q \models_Q \varphi$ for $\varphi \in \mathsf{Form}_{\mathcal{L}_Q}$ and $\mathcal{L}_Q$-model $\mathfrak{M}_Q = \langle D, \|\cdot\|_{\mathfrak{M}_Q}\rangle$ be recursively defined as follows:

(1) $\mathfrak{M}_Q \models_Q (s_1,\ldots,s_n)P$ iff $\langle \|s_1\|_{\mathfrak{M}_Q},\ldots,\|s_n\|_{\mathfrak{M}_Q}\rangle \in \|P\|_{\mathfrak{M}_Q}$ ($P \in \mathsf{Reord}^n_{\mathcal{L}_Q}$);

(2) $\mathfrak{M}_Q \models_Q s_1 = s_2$ iff $\|s_1\|_{\mathfrak{M}_Q} = \|s_2\|_{\mathfrak{M}_Q}$;

(3) $\mathfrak{M}_Q \models_Q \neg \varphi$ iff it is not the case that $\mathfrak{M}_Q \models_Q \varphi$ ($\mathfrak{M}_Q \not\models_Q \varphi$);

(4) $\mathfrak{M}_Q \models_Q \varphi \wedge \psi$ iff $\mathfrak{M}_Q \models_Q \varphi$ and $\mathfrak{M}_Q \models_Q \psi$;

(5) $\mathfrak{M}_Q \models_Q *((s_1,\ldots,s_n)\neg *' P)$ iff $\mathfrak{M}_Q \models_Q \neg *((s_1,\ldots,s_n) *' P)$ (where $*, *'$ are possibly empty strings of negation-symbols $\neg$);

(6) $\mathfrak{M}_Q \models_Q \varphi[s_\alpha/s_1, \alpha/s_2, \ldots, \alpha/s_n]$ iff $\mathfrak{M}_Q \models_Q \varphi$;

(7) $\mathfrak{M}_Q \models_Q \varphi[\exists P_\alpha]$ iff for some $a$-expansion $\mathfrak{M}'_Q$ of $\mathfrak{M}_Q$ such that $a \in \|P\|_{\mathfrak{M}_Q}$, $\mathfrak{M}'_Q \models_Q \varphi[(s_a)_\alpha/\exists P_\alpha]$ ($\exists P$ governs $\varphi$ and is the source of $\alpha \in \mathsf{Ana}_{\mathcal{L}_Q}$ if there is one);

(8) $\mathfrak{M}_Q \models_Q \varphi[\forall P_\alpha]$ iff for all $a$-expansions $\mathfrak{M}'_Q$ of $\mathfrak{M}_Q$ such that $a \in \|P\|_{\mathfrak{M}_Q}$, $\mathfrak{M}'_Q \models_Q \varphi[(s_a)_\alpha/\forall P_\alpha]$ ($\forall P$ governs $\varphi$ and is the source of $\alpha \in \mathsf{Ana}_{\mathcal{L}_Q}$ if there is one).

Since the QUARC-models are pretty much the same as the Fregean-models, the satisfaction-relation is quite similar too. Clauses (1)–(4) correspond to Definition 22's (1)–(4). The remaining clauses are QUARC-specific.

Clause (5) concerns *predicate-negation*. As long as no quantified arguments occur in a formula, we just move the negation-symbols from the predicate-negation into sentence-negation; the resulting formulas are in the range of clause (3).

Clause (6) concerns anaphora. As the anaphors are just referring to whatever their source refers to, we interpret them accordingly. That is, as long as no quantified arguments occur, they refer to what their source singular argument refers. That is, a model satisfies it in exactly the same circumstances as when they are replaced by their source.

Clauses (7)–(8) concern the quantified arguments. The general idea is the same as it was in the case of Fregean languages, i.e., as specified in Definition 22 (5)–(6). However, as QUARC does not allow unrestricted quantification, we have to restrict the expansions in consonance with the *quantified argument*, consisting of a quantifier and unary predicate. Thus, instead of considering *some* or *all* $a$-expansions, we *only* consider those such that $a$ is an element of the interpretation of the restricting unary predicate. If $qP \in \mathsf{QA}_{\mathcal{L}_\mathsf{Q}}$, we only consider those $a \in D$ such that $a \in \|P\|_{\mathfrak{M}_\mathsf{Q}}$, i.e., if for *some* (*all*) of these the expanded model $\mathfrak{M}'_\mathsf{Q}$ satisfies a formula $\varphi$, then the base model $\mathfrak{M}_\mathsf{Q}$ satisfies the formula involving the quantified argument $\exists P$ ($\forall P$), i.e., it satisfies that *some P* (*all P*) satisfy the formula.

Given the QUARC-language, its models, and the satisfaction-relation, we can define *logical consequence* etc. as in Definition 8, and obtain the QUARC-specific results below.

**Theorem 35**
$\models_\mathsf{Q} (\exists P)P$.

*Proof.* Let $\mathfrak{M}_\mathsf{Q} = \langle D, \|\cdot\|_{\mathfrak{M}_\mathsf{Q}} \rangle$ be an $\mathcal{L}_\mathsf{Q}$-model. By Definition 31 (2b), $\emptyset \neq \|P\|_{\mathfrak{M}_\mathsf{Q}} \subseteq D$ for $P \in \mathsf{Pred}^1_{\mathcal{L}_\mathsf{Q}}$. Let $a \in \|P\|_{\mathfrak{M}_\mathsf{Q}}$, and let $\mathcal{L}'_\mathsf{Q}$ be the $\mathcal{L}_\mathsf{Q}$-$a$-expansion of $\mathcal{L}_\mathsf{Q}$. Let $\mathfrak{M}'_\mathsf{Q}$ be the $a$-expansion of $\mathfrak{M}_\mathsf{Q}$ to $\mathcal{L}'_\mathsf{Q}$. Then, $\mathfrak{M}'_\mathsf{Q} \models_\mathsf{Q} (s_a)P$ since $\|s_a\|_{\mathfrak{M}'_\mathsf{Q}} = a \in \|P\|_{\mathfrak{M}_\mathsf{Q}} \subseteq \|P\|_{\mathfrak{M}'_\mathsf{Q}}$ by Definition 33 (2)–(3). Therefore, by Definition 34 (7), $\mathfrak{M}_\mathsf{Q} \models_\mathsf{Q} (\exists P)P$. □

**Theorem 36**
$(\forall P)Q \models_\mathsf{Q} (\exists P)Q$.

*Proof.* Let $\mathfrak{M}_Q$ be an $\mathcal{L}_Q$-model such that $\mathfrak{M}_Q \models_Q (\forall P)Q$. By Definition 34 (8), for all $a$-expansions $\mathfrak{M}'_Q$ of $\mathfrak{M}_Q$ such that $a \in \|P\|_{\mathfrak{M}_Q}$, $\mathfrak{M}'_Q \models_Q (s_a)Q$. Moreover, by Definition 31 (2b), $\|P\|_{\mathfrak{M}_Q} \neq \emptyset$. Thus, there is an $a$-expansion $\mathfrak{M}'_Q$ of $\mathfrak{M}_Q$ such that $a \in \|P\|_{\mathfrak{M}_Q}$, $\mathfrak{M}'_Q \models_Q (s_a)Q$. Then, by Definition 34 (7), $\mathfrak{M}_Q \models_Q (\exists P)Q$. □

The quantifiers still behave as one would expect them to:

**Theorem 37**
The following equivalences hold:

(1) $\models_Q (\forall P)S \leftrightarrow \neg((\exists P)\neg S)$;  (2) $\models_Q (\exists P)S \leftrightarrow \neg((\forall P)\neg S)$;

(3) $\models_Q \neg(\exists P)S \leftrightarrow (\forall P)\neg S$;  (4) $\models_Q \neg(\forall P)S \leftrightarrow (\exists P)\neg S$.

*Proof.* I only illustrate part of one case:

**(3):** Let $\mathfrak{M}_Q \models_Q \neg(\exists P)S$. Then, by Definition 34 (3), $\mathfrak{M}_Q \not\models_Q (\exists P)S$, i.e., by (7), it is not the case that for some $a$-expansion $\mathfrak{M}'_Q$ of $\mathfrak{M}_Q$ such that $a \in \|P\|_{\mathfrak{M}_Q}$, $\mathfrak{M}'_Q \models_Q (s_a)S$ iff for all $a$-expansions $\mathfrak{M}'_Q$ of $\mathfrak{M}_Q$ such that $a \in \|P\|_{\mathfrak{M}_Q}$, $\mathfrak{M}'_Q \not\models_Q (s_a)S$, i.e., by (3), for all $a$-expansions $\mathfrak{M}'_Q$ of $\mathfrak{M}_Q$ such that $a \in \|P\|_{\mathfrak{M}_Q}$, $\mathfrak{M}'_Q \models_Q \neg(s_a)S$, and so, by (5), for all $a$-expansions $\mathfrak{M}'_Q$ of $\mathfrak{M}_Q$ such that $a \in \|P\|_{\mathfrak{M}_Q}$, $\mathfrak{M}'_Q \models_Q (s_a)\neg S$. Thus, by (8), $\mathfrak{M}_Q \models_Q (\forall P)\neg S$.

□

QUARC also validates the conversions.

**Theorem 38** *(Conversion)*
The following conversions hold:

(**a-i**-conv$^{\models_Q}$)  $(\forall A)B \models_Q (\exists B)A$

(**i-i**-conv$^{\models_Q}$)  $(\exists A)B \models_Q (\exists B)A$

(**e-e**-conv$^{\models_Q}$)  $(\forall A)\neg B \models_Q (\forall B)\neg A$

*Proof.* Let $\mathfrak{M}_Q$ be an $\mathcal{L}_Q$-model.

(**a-i-conv**$^{\models_Q}$): Follows from Theorem 36 and (**i-i-conv**$^{\models_Q}$).

(**i-i-conv**$^{\models_Q}$): Let $\mathfrak{M}_Q \models_Q (\exists A)B$. By Definition 31 (2b), $\|A\|_{\mathfrak{M}_Q} \neq \emptyset \neq \|B\|_{\mathfrak{M}_Q}$. Then, by Definition 34 (7), for some $a$-expansion $\mathfrak{M}'_Q$ of $\mathfrak{M}_Q$ such that $a \in \|A\|_{\mathfrak{M}_Q}$, $\mathfrak{M}'_Q \models_Q (s_a)B$, i.e., by Definition 34 (1), $\|s_a\|_{\mathfrak{M}'_Q} \in \|B\|_{\mathfrak{M}'_Q}$. Since, by Definition 33 (3), $\|s_a\|_{\mathfrak{M}'_Q} = a$, it follows that $a \in \|B\|_{\mathfrak{M}'_Q}$. By Definition 33 (2), $\|B\|_{\mathfrak{M}_Q} = \|B\|_{\mathfrak{M}'_Q}$ and $\|A\|_{\mathfrak{M}_Q} = \|A\|_{\mathfrak{M}'_Q}$. Thus, $a \in \|A\|_{\mathfrak{M}_Q} \cap \|B\|_{\mathfrak{M}_Q} = \|A\|_{\mathfrak{M}'_Q} \cap \|B\|_{\mathfrak{M}'_Q}$, so also $\mathfrak{M}'_Q \models_Q (s_a)A$. Overall, for some $a$-expansion $\mathfrak{M}'_Q$ of $\mathfrak{M}_Q$ such that $a \in \|B\|_{\mathfrak{M}_Q}$, $\mathfrak{M}'_Q \models_Q (s_a)A$, i.e., by Definition 34 (7), $\mathfrak{M}_Q \models_Q (\exists B)A$.

(**e-e-conv**$^{\models_Q}$): Let $\mathfrak{M}_Q \models_Q (\forall A)\neg B$. Thus, by Definitions 34 (8), (5), and (3), it is not the case that for some $a$-expansion $\mathfrak{M}'_Q$ of $\mathfrak{M}_Q$ such that $a \in \|A\|_{\mathfrak{M}_Q}$, $\mathfrak{M}'_Q \models_Q (s_a)B$.

Suppose that $\mathfrak{M}_Q \models_Q (\exists B)A$. Then, by (**i-i-conv**$^{\models_Q}$), $\mathfrak{M}_Q \models (\exists A)B$, i.e., for some $a$-expansion $\mathfrak{M}'_Q$ of $\mathfrak{M}_Q$ such that $a \in \|A\|_{\mathfrak{M}_Q}$, $\mathfrak{M}'_Q \models_Q (s_a)B$, a contradiction. Therefore, $\mathfrak{M}_Q \not\models_Q (\exists B)A$, i.e., $\mathfrak{M}_Q \models_Q \neg(\exists B)A$. Thus, by Theorem 37 (3), $\mathfrak{M}_Q \models_Q (\forall B)\neg A$.

$\square$

Note also that the semantics distinguishes only between even and odd numbers of predicate-negations:

**Theorem 39**
$\models_Q ((s_1, \ldots, s_n)\neg\neg * P) \leftrightarrow ((s_1, \ldots, s_n) * P)$.

Theorem 39 generalizes to cases including quantified arguments. Applied repeatedly, we get that if '*' contains an even number of negation-symbols, then '$((s_1, \ldots, s_n) * P)$' is equivalent to '$(s_1, \ldots, s_n)P$', and if it contains an odd number, it is equivalent to '$((s_1, \ldots, s_n)\neg P)$', and so, by Definition 34 (5), to '$\neg(s_1, \ldots, s_n)P$'.

This finishes the exposition of QUARC.

## 5 Sommers's Term Logic

### 5.1 Background

Fred Sommers is also not satisfied with the common approach to the semantics of natural language. He develops his *Term Functor Logic* (**TFL**) as an alternative approach. In this brief exposition, I focus on his book *The Logic of Natural Language* [45], and only consider a few points that suggest themselves for comparison here (for a nice exposition, see Englebretsen [23]).

Sommers's conviction is that

> traditional formal logic is especially suited to the task of making perspicuous the logical form of sentences in the natural languages that are actually used in deductive reasoning and that, in virtue of this, traditional logic provides models for the study of what actually happens when we reckon the premises and arrive at conclusion. [45, p. 4]

Given that we generally reason in natural language, traditional formal logic is in a better position to make explicit how we do so; Fregean languages, with their machinery, rather distort this. In this context, Sommers emphasizes that the

> traditional logician emphasized syntactic simplicity, requiring of a canonical sentence that it have a straightforward noun-phrase verb-phrase structure (or be a compound of such 'categorical' sentences). [45, p. 9]

The simple noun-phrase verb-phrase structure can be found in Aristotle's logic, though needs to be extended to overcome the syllogistic's shortcomings. Indeed, Sommers is concerned in constructing a language that is similarly powerful as Fregean languages while maintaining the basic analysis of sentences.

The basic analysis is into noun-phrase and verb-phrase; both are considered to be *terms*. Additionally, the noun-phrase as well as all other subject expressions are assigned a *quantity* [45, p. 67]. The general form of a sentence is then 'every/some $S$ is (are)/is (are) not $P$' where 'every/some' is the quantity of the subject $S$ [45, e.g., p. 95].

In order to increase the expressive power, Sommers introduces *proterms* and allows *complex* terms. As in Aristotle's logic, terms can play the role of *both* subject and predicate in sentences [45, p. 116]. Moreover, Sommers also allows *n-ary* terms [45, p. 139], construed in a way so that the subject-predicate structure remains via *nesting* them [45, p. 148]. As terms can play several roles, Sommers [45, pp. 116f.] argues that there is no need to include *identity* in the way Fregean languages do. This also means that TFL is more parsimonious than Fregean languages are with respect to their primitive symbols.

Overall, Sommers claims that his TFL, already in a more basic form which he calls 'Primitive Term Logic' (PTL), is

> roughly equivalent to that of a standard first-order logic whose logical particles consist of the existential quantifier and the signs for conjunction, negation and identity. [45, p. 174]

He goes on to *amplify* PTL to full TFL. However, for the purposes of comparing the systems, I stick to the more basic system, though even depart from Sommers's presentation and particular claims regarding it. Moreover, I continue the model-theoretic approach which is significantly different from Sommers's algebraic treatment of term logic.

## 5.2 A Formalism

**Definition 40 *(TFL-Language)***
A *TFL*-language ($\mathcal{L}_T$) consists of the following:

- a countably infinite set $\mathsf{PTerm}_{\mathcal{L}_T} = \{\alpha_0, \alpha_1, \alpha_2, \ldots\}$ of *proterms*,

- a countable set $\mathsf{ITerm}_{\mathcal{L}_T} = \{t_0, t_1, t_2, \ldots\}$ of *individual-terms*,

- for every $n > 0$, a countable set $\mathsf{STerm}^n_{\mathcal{L}_T} = \{T^n_0, T^n_1, T^n_2, \ldots\}$ of (*simple*) *n-ary term-symbols*,

- the set of *logical symbols* including '¬', '—', '∧', '∨', '→', '↔', '∀', and '∃', and

- the set of *auxiliary symbols* including '(', ')', and ','.

All the sets are assumed to be disjoint. Let $\mathsf{STerm}_{\mathcal{L}_T} := \bigcup_{n>0} \mathsf{STerm}^n_{\mathcal{L}_T} \cup \mathsf{ITerm}_{\mathcal{L}_T}$.

The TFL-language $\mathcal{L}_T$ is different from the one Sommers actually uses, and changes certain aspects. What's left are *proterms* $\mathsf{PTerm}_{\mathcal{L}_T}$ which play a similar role to Fregean variables and QUARC-anaphora. The language does not contain anything like individual-constants or singular arguments, but only terms. One kind of term are the *individual-terms* $\mathsf{ITerm}_{\mathcal{L}_T}$—playing a similar role as individual-constants—another *n-ary terms* $\mathsf{STerm}^n_{\mathcal{L}_T}$. Similar to the language $\mathcal{L}_A$ and in contrast to $\mathcal{L}_F$ and $\mathcal{L}_Q$, $\mathcal{L}_T$ does not include an identity-symbol '=' among its logical symbols, but includes a second negation-symbol '—' which figures in the introduction of complex terms.

One important difference to Definition 1 of $\mathcal{L}_A$ is that $\mathcal{L}_T$ includes *n-ary term*. These are necessary to capture relational predications that Aristotle's syllogistic misses.

Given this basic vocabulary, we can introduce the complex terms.

**Definition 41** *(Complex $\mathcal{L}_T$-Terms)*
For each $n > 0$, the *set of complex n-ary $\mathcal{L}_T$-terms* ($\mathsf{CTerm}^n_{\mathcal{L}_T}$) is recursively defined as follows:

(1) if $t \in \mathsf{ITerm}_{\mathcal{L}_T}$, then $t \in \mathsf{CTerm}^1_{\mathcal{L}_T}$;

(2) if $A \in \mathsf{STerm}^n_{\mathcal{L}_T}$, then $A \in \mathsf{CTerm}^n_{\mathcal{L}_T}$;

(3) if $A \in \mathsf{CTerm}^n_{\mathcal{L}_T}$, then $\overline{A} \in \mathsf{CTerm}^n_{\mathcal{L}_T}$;

(4) if $A, B \in \mathsf{CTerm}^n_{\mathcal{L}_T}$, then $(A \circ B) \in \mathsf{CTerm}^n_{\mathcal{L}_T}$ ($\circ \in \{\wedge, \vee, \rightarrow, \leftrightarrow\}$);

(5) if $1 \leq i \leq n-1$, $t_1, \ldots, t_i \in \mathsf{ITerm}_{\mathcal{L}_T}$, $q_1, \ldots, q_i \in \{\forall, \exists\}$, and $A \in \mathsf{CTerm}^n_{\mathcal{L}_T}$, then $(q_1 t_1, \ldots, q_i t_i, \_{-i+1}, \ldots, \_n)A \in \mathsf{CTerm}^{n-i}_{\mathcal{L}_T}$ and so are all ways of putting the $i$ terms into the $n$ slots of $A$ (where '$\_k$' indicates the $k$th argument-slot of $A$, $1 \leq k \leq n$);

(6) if $A \in \mathsf{CTerm}^n_{\mathcal{L}_T}$ and $\pi\colon \{1, \ldots, n\} \to \{1, \ldots, n\}$ is a permutation, then $A^\pi \in \mathsf{CTerm}^n_{\mathcal{L}_T}$ where '$A^\pi$' is the result of permuting $A$'s slots according to $\pi$.

Terms generated by clause (5) are called *n-ary reduced terms* ($\mathsf{RTerm}^n_{\mathcal{L}_T}$). Let $\mathsf{RTerm}_{\mathcal{L}_T} := \bigcup_{n>1} \mathsf{RTerm}^n_{\mathcal{L}_T}$.

Clauses (2)–(4) are analogous to the clauses (1)–(3) of Definition 2 of Term$_{\mathcal{L}_A}$, just generalized from only *unary* terms to *n-ary* terms. These allow to capture relational predications in more complex settings. Moreover, clause (1) includes the individual terms among the *unary* complex terms.

Clause (5) additionally allows to form further terms, reducing an $n$-ary term $A$ to an $m$-ary term $B$ by filling up slots with elements from ITerm$_{\mathcal{L}_T}$. In the spirit of TFL, each term is assigned a *quantity*. However, as the particular quantity does not make a difference for the individual-terms, both '$\forall$' and '$\exists$' are allowed as quantities.

Note, too, that clause (5) also sticks to the QUARC convention to place the argument-places to the left of the term symbol.

Clause (6), finally, allows for *reordered* terms analogous to QUARC's reorders in Definition 27. The clause allows to reorder reorders, but it is clear that there are only $n!$-many different ones. For example, '$(\_1, \_2)A$' only leads to '$(\_2, \_1)A^\pi$' as $A^{\pi^\pi} = A$.

Given the vocabulary and the set of terms, we can define what counts as formula in a way mirroring Definition 28 of Form$_{\mathcal{L}_Q}$.

**Definition 42** *($\mathcal{L}_T$-Formula)*
Let $\mathcal{L}_T$ be a TFL-language. The *set of $\mathcal{L}_T$-formulas* (Form$_{\mathcal{L}_T}$) is recursively defined by:

(1) if $n \geq 1$, $A \in$ CTerm$_{\mathcal{L}_T}^n$, $t_1, \ldots, t_n \in$ ITerm$_{\mathcal{L}_T}$, $q_1, \ldots, q_n \in \{\forall, \exists\}$, and $*$ a possibly empty string of negation-symbols $\neg$, then $((q_1 t_1, \ldots, q_n t_n) * A) \in$ Form$_{\mathcal{L}_T}$;

(2) if $\varphi \in$ Form$_{\mathcal{L}_T}$, then $\neg\varphi \in$ Form$_{\mathcal{L}_T}$;

(3) if $\varphi, \psi \in$ Form$_{\mathcal{L}_T}$, then $(\varphi \circ \psi) \in$ Form$_{\mathcal{L}_T}$ ($\circ \in \{\wedge, \vee, \rightarrow, \leftrightarrow\}$);

(4) if $\varphi \in$ Form$_{\mathcal{L}_T}$ contains, from left to right, $t_1, \ldots, t_m$ ($m \geq 2$) occurrences of $t \in$ ITerm$_{\mathcal{L}_T}$, none of which is the source of $\beta \in$ PTerm$_{\mathcal{L}_T}$ that occurs in $\varphi$, and $\varphi$ does not contain $\alpha \in$ PTerm$_{\mathcal{L}_T}$, then $\varphi[t_\alpha/t_1, \alpha/t_2, \ldots, \alpha/t_m] \in$ Form$_{\mathcal{L}_T}$ (which is the result of substituting $\alpha$ for the occurrences $t_2, \ldots, t_m$ of $t$;

(5) if $q \in \{\forall, \exists\}$, $t \in$ ITerm$_{\mathcal{L}_T}$, $\varphi[qt] \in$ Form$_{\mathcal{L}_T}$, $A \in$ CTerm$_{\mathcal{L}_T}^1$, then $\varphi[qA/qt] \in$ Form$_{\mathcal{L}_T}$ if $qA$ *governs* $\varphi$.

Definition 42 resembles Definition 28 which defines Form$_{\mathcal{L}_Q}$. Indeed, *governance* in clause (5) is to be understood analogous to Definition 30. Moreover, an analogue of Definition 29 applies to the proterms in clause (4) and once the $t \in$ ITerm$_{\mathcal{L}_T}$ gets substituted by $A \in$ CTerm$^1_{\mathcal{L}_T}$. Also, we collapsed Definition 28 (1) and (3) into one clause (1).

What's TFL-specific in Definition 42 is the assignment of *quantities* to *all* terms. Thus, the basic formulas are $n$-ary terms applying to $n$ *individual-terms* $t_i \in$ ITerm$_{\mathcal{L}_T}$ while assigning them a quantity, i.e., one of the quantifiers. As these terms are such that the particular quantifier does not make a difference, both are allowed. Definition 42 does not introduce *wild* quantities, but just assigns *both* quantities and the rest will be taken care by the interpretation.

Moreover, we allow for the usual combination of sentences via clauses (2)–(3). Terms for which the quantity makes a difference are only introduced in the last clause (5), and they always replace individual terms for which they are substituted—and this includes individual terms used in the *reduced terms* $R \in$ RTerm$_{\mathcal{L}_T}$.

As before, all formulas are *closed*, i.e., sentences. The complexity introduced by Definition 41 is mirrored in the interpretation of terms.

**Definition 43** *($\mathcal{L}_T$-Model)*
Let $\mathcal{L}_T$ be a TFL-language. An $\mathcal{L}_T$-*model* is a tuple $\mathfrak{M}_T = \langle D, \|\cdot\|_{\mathfrak{M}_T} \rangle$ such that

(1) $D$ is a set (the *universe*)

(2) $\|\cdot\|_{\mathfrak{M}_T}$ is an *interpretation-function of* $\mathfrak{M}_T$ such that

  (a) if $t \in$ ITerm$_{\mathcal{L}_T}$, then $\|t\|_{\mathfrak{M}_T} = \{a\}$ for an $a \in D$;

  (b) if $n = 1$ and $A \in$ STerm$^n_{\mathcal{L}_T}$, then $\emptyset \neq \|A\|_{\mathfrak{M}_T} \subseteq D$;

  (c) if $n > 1$ and $A \in$ STerm$^n_{\mathcal{L}_T}$, then $\|A\|_{\mathfrak{M}_T} \subseteq D^n$;

  (d) if $n > 0$ and $A \in$ CTerm$^n_{\mathcal{L}_T}$ is of the form '$\overline{B}$' for a $B \in$ CTerm$^n_{\mathcal{L}_T}$, then $\|A\|_{\mathfrak{M}_T} = \{\langle a_1, \ldots, a_n\rangle \in D^n | \langle a_1, \ldots, a_n\rangle \notin \|B\|_{\mathfrak{M}_T}\}$;

  (e) if $n > 0$ and $A \in$ CTerm$^n_{\mathcal{L}_T}$ is of the form '$(B \circ C)$' for $B, C \in$ CTerm$^n_{\mathcal{L}_T}$, then $\|A\|_{\mathfrak{M}_T} = \{\langle a_1, \ldots, a_n\rangle \in D^n |\ \langle a_1, \ldots, a_n\rangle \in \|B\|_{\mathfrak{M}_T} \circ \langle a_1, \ldots, a_n\rangle \in \|C\|_{\mathfrak{M}_T}\}$ ($\circ \in \{\wedge, \vee, \rightarrow, \leftrightarrow\}$);

(f) if $n > 0$ and $A \in \mathsf{RTerm}_{\mathcal{L}_\mathsf{T}}^n$ stemming from $B \in \mathsf{CTerm}_{\mathcal{L}_\mathsf{T}}^m$ ($m > n$), $i = m - n$ individual terms $t_1, \ldots, t_i \in \mathsf{ITerm}_{\mathcal{L}_\mathsf{T}}$ and $q_1, \ldots, q_i \in \{\forall, \exists\}$ such that $A$ is of the form '$(q_1 t_1, \ldots, q_i t_i, \_{i+1}, \ldots, \_m)B$', then $\|A\|_{\mathfrak{M}_\mathsf{T}} = \{\langle a_1, \ldots, a_n\rangle \in D^n | \langle \bigcup \|t_1\|_{\mathfrak{M}_\mathsf{T}}, \ldots, \bigcup \|t_i\|_{\mathfrak{M}_\mathsf{T}}, a_1, \ldots, a_n\rangle \in \|B\|_{\mathfrak{M}_\mathsf{T}}\}$; similarly for all other ways of generating an $A \in \mathsf{RTerm}_{\mathcal{L}_\mathsf{T}}^n$;

(g) if $n > 0$ and $A^\pi \in \mathsf{CTerm}_{\mathcal{L}_\mathsf{T}}^n$ for permutation $\pi$, then $\|A^\pi\|_{\mathfrak{M}_\mathsf{T}} = \{\langle \bigcup \|t_{\pi(1)}\|_{\mathfrak{M}_\mathsf{T}}, \ldots, \bigcup \|t_{\pi(n)}\|_{\mathfrak{M}_\mathsf{T}}\rangle | \langle \bigcup \|t_1\|_{\mathfrak{M}_\mathsf{T}}, \ldots, \bigcup \|t_n\|_{\mathfrak{M}_\mathsf{T}}\rangle \in \|A\|_{\mathfrak{M}_\mathsf{T}}\}$.

The $\mathcal{L}_\mathsf{T}$-models $\mathfrak{M}_\mathsf{T}$ are similar to the models seen so far. However, similar to the $\mathcal{L}_\mathsf{A}$-models $\mathfrak{M}_\mathsf{A}$, they have to take care of the interpretation of the complex terms.

Clause (2a) interprets individual terms *as terms*, i.e., as a set; they are *individual* as the sets are singletons.

In line with how I introduced it before, *unary* terms are interpreted by *non-empty* sets. The reason is again to facilitate comparison with QUARC.

Complex $n$-ary terms are interpreted analogous to how $\mathcal{L}_\mathsf{A}$-models $\mathfrak{M}_\mathsf{A}$ interpreted complex *unary* terms; clause (2c) just generalizes from unary to $n$-ary terms, i.e., from subsets of the domain to $n$-ary relations on the domain.

Clause (2f) interprets the reduced terms. These are $n$-ary terms generated out of $m$-ary terms ($m > n$) by filling up slots with individual terms. These individual terms have quantities assigned, though as they are *individual*, the quantity does not make a difference. Thus, they are simply interpreted as $\bigcup \|t\|_{\mathfrak{M}_\mathsf{T}}$ ($t \in \mathsf{ITerm}_{\mathcal{L}_\mathsf{T}}$). If $\|t\|_{\mathfrak{M}_\mathsf{T}} = \{a\}$, $\bigcup \|t\|_{\mathfrak{M}_\mathsf{T}} = a$.

The last clause (2g) is analogous to Definition 31 (2d), i.e., it interprets *reorders* by considering what they reorder; simply apply the permutation $\pi$ to the $n$-tuples in the interpretation of term $A$ in order to get the interpretation of $A^\pi$.

Since we treat quantification substitutionally and $\mathsf{ITerm}_{\mathcal{L}_\mathsf{T}}$ plays the role of individual-constants, we need to make sure that the particular choice of $\mathsf{ITerm}_{\mathcal{L}_\mathsf{T}}$ does not lead to problematic results; we do that as before by expanding the language.

**Definition 44 ($\mathcal{L}_T$-$A$-Expansion)**
Let $\mathcal{L}_T$ be a TFL-language and $\mathfrak{M}_T = \langle D, \|\cdot\|_{\mathfrak{M}_T}\rangle$ be an $\mathcal{L}_T$-model. Let $A \subseteq D$. The $\mathcal{L}_T$-$A$-expansion of $\mathcal{L}_T$ is the language $\mathcal{L}'_T := \mathcal{L}_T \cup \{t_a | a \in A\}$ where the $t_a$ are new individual-terms not contained in $\mathcal{L}_T$.

If $A = \{a\}$, we call $\mathcal{L}'_T$ an $\mathcal{L}_T$-$a$-expansion.

Once the language is expanded, we need to make sure that the interpretation keeps up.

**Definition 45 ($\mathcal{L}_T$-Model Expansion)**
Let $\mathcal{L}_T$ be a TFL-language and $\mathfrak{M}_T = \langle D, \|\cdot\|_{\mathfrak{M}_T}\rangle$ be an $\mathcal{L}_T$-model. Let $A \subseteq D$ and $\mathcal{L}'_T$ be an $\mathcal{L}_T$-$A$-expansion. The $A$-expansion of $\mathfrak{M}_T$ to $\mathcal{L}'_T$ is the model $\mathfrak{M}'_T = \langle D', \|\cdot\|_{\mathfrak{M}'_T}\rangle$ such that

(1) $D' = D$;  (2) $\|\cdot\|_{\mathfrak{M}_T} \subseteq \|\cdot\|_{\mathfrak{M}'_T}$;

(3) $\|t_a\|_{\mathfrak{M}'_T} = \{a\} \subseteq A$ for every new individual term $t_a$.

As in the cases before, Definition 45 keeps the domain the same, and extends the interpretation-function to $\|\cdot\|_{\mathfrak{M}'_T}$ so that the new individual-terms $t_a$ are interpreted in alignment as they have been introduced. In accordance with Definition 43 (2a), these are not elements of the domain, but singleton-subsets.

**Definition 46 (Satisfaction $\models_T$)**
Let the *TFL satisfaction-relation* $\mathfrak{M}_T \models_T \varphi$ for $\varphi \in \mathsf{Form}_{\mathcal{L}_T}$ and $\mathcal{L}_T$-model $\mathfrak{M}_T = \langle D, \|\cdot\|_{\mathfrak{M}_T}\rangle$ be recursively defined as follows:

(1) $\mathfrak{M}_T \models_T (q_1 t_1, \ldots, q_n t_n) A$ iff $\langle \bigcup \|t_1\|_{\mathfrak{M}_T}, \ldots, \bigcup \|t_n\|_{\mathfrak{M}_T}\rangle \in \|A\|_{\mathfrak{M}_T}$;

(2) $\mathfrak{M}_T \models_T (q_1 t_1, \ldots, q_n t_n) \neg * A$ iff $\mathfrak{M}_T \models_T \neg(q_1 t_1, \ldots, q_n t_n) * A$;

(3) $\mathfrak{M}_T \models_T \neg\varphi$ iff it is not the case that $\mathfrak{M}_T \models_T \varphi$ ($\mathfrak{M}_T \not\models_T \varphi$);

(4) $\mathfrak{M}_T \models_T \varphi \wedge \psi$ iff $\mathfrak{M}_T \models_T \varphi$ and $\mathfrak{M}_T \models_T \psi$;

(5) $\mathfrak{M}_T \models_T \varphi[t_\alpha/t_1, \alpha/t_2, \ldots, \alpha/t_n]$ iff $\mathfrak{M}_T \models_T \varphi$;

(6) $\mathfrak{M}_T \models_T \varphi[\exists A]$ iff for some $a$-expansions $\mathfrak{M}'_T$ of $\mathfrak{M}_T$ such that $a \in \|A\|_{\mathfrak{M}_T}$, $\mathfrak{M}'_T \models_T \varphi[\exists t_a]$;

(7) $\mathfrak{M}_T \models_T \varphi[\forall A]$ iff for all $a$-expansion $\mathfrak{M}'_T$ of $\mathfrak{M}_T$ such that $a \in \|A\|_{\mathfrak{M}_T}$, $\mathfrak{M}'_T \models_T \varphi[\forall t_a]$.

As already done in Definition 43 (2f), individual-terms are interpreted regardless of their specific quantity as done in clause (1); individual-terms are pretty much treated as individual-constants in Definition 22 (1), as is predication.

As QUARC, TFL allows for negative predication; clause (2) is analogous to clause (5) of Definition 34. The negation-symbols ¬ are moved in front of formulas and then interpreted via clause (2) as long as only individual-terms are involved.

The remaining clauses are analogous to those of QUARC in Definition 34. In particular, we interpret quantifiers via the expansions, where, as in the QUARC-case given in Definition 34 (7)–(8), we consider appropriate expansions, i.e., expansions which expand with elements in the interpretation of the subject-term $A$ and consider as many as the quantity $q$ of $A$ specifies.

As before, we can define *logical consequence* as done in Definition 8. Given these notions, we can formulate the TFL-specific treatment of individual-terms.

**Theorem 47**
For $t \in \mathsf{ITerm}_{\mathcal{L}_T}$, $\models_T (\exists t)A \leftrightarrow (\forall t)A$.

*Proof.* Let $\mathfrak{M}_T$ be an $\mathcal{L}_T$-model and $t \in \mathsf{ITerm}_{\mathcal{L}_T}$. Let $\mathfrak{M}_T \models_T (\exists t)A$. By Definition 46 (1), $\bigcup \|t\|_{\mathfrak{M}_T} \in \|A\|_{\mathfrak{M}_T}$ and so $\mathfrak{M}_T \models_T (\forall t)A$. □

Moreover, we get a similar result regarding non-emptiness as Theorem 35, though extended to include individual-terms.

**Theorem 48**
For $A \in \mathsf{ITerm}_{\mathcal{L}_T} \cup \mathsf{STerm}^1_{\mathcal{L}_T}$, $\models_T (\exists A)A$.

*Proof.* Let $\mathfrak{M}_T$ be an $\mathcal{L}_T$-model.

- If $A \in \mathsf{ITerm}_{\mathcal{L}_T}$, by Definition 43 (2a) $\|A\|_{\mathfrak{M}_T} = \{a\}$ for an $a \in D$. Thus, $\bigcup \|A\|_{\mathfrak{M}_T} = a \in \|A\|_{\mathfrak{M}_T}$. Therefore, by Definition 46 (1),

$\mathfrak{M}_T \models_T (\exists A)A$.

- If $A \in \mathsf{STerm}^1_{\mathcal{L}_T}$, by Definition 43 (2b), $\|A\|_{\mathfrak{M}_T} \neq \emptyset$. Let $a \in \|A\|_{\mathfrak{M}_T}$. Then, for some $a$-expansion $\mathfrak{M}'_T$ of $\mathfrak{M}_T$ such that $a \in \|A\|_{\mathfrak{M}_T}$, $\mathfrak{M}'_T \models_T (\exists t_a)A$. By Definition 46 (6), $\mathfrak{M}_T \models_T (\exists A)A$.

□

However, as we allow for complex terms, this does not hold in general.

**Theorem 49**
$\not\models_T (\exists A)A$.

*Proof.* Let $\mathfrak{M}_T$ be an $\mathcal{L}_T$-model. Consider $A \in \mathsf{STerm}^1_{\mathcal{L}_T}$ such that $\|A\|_{\mathfrak{M}_T} = D$. Then, by Definition 43 (2d), $\|\overline{A}\|_{\mathfrak{M}_T} = \emptyset$. Thus, there is no $a$-expansion $\mathfrak{M}'_T$ of $\mathfrak{M}_T$ such that $a \in \|\overline{A}\|_{\mathfrak{M}_T}$, so $\mathfrak{M}_T \not\models_T (\exists \overline{A})\overline{A}$. □

For similar reasons, we get that that the universal doesn't imply the particular.

**Corollary 50**
$(\forall A)B \not\models_T (\exists A)B$.

*Proof.* Consider the model in the proof of Theorem 49. Since there is no $a$-expansion $\mathfrak{M}'_T$ of $\mathfrak{M}_T$ such that $a \in \|\overline{A}\|_{\mathfrak{M}_T}$ it follows that for all $a$-expansions $\mathfrak{M}'_T$ of $\mathfrak{M}_T$ such that $a \in \|\overline{A}\|_{\mathfrak{M}_T}$, $\mathfrak{M}'_T \models_T (\forall t_a)B$, i.e., by Definition 46 (7), $\mathfrak{M}_T \models_T (\forall \overline{A})B$. However, as there are no $a$-expansions $\mathfrak{M}'_T$ of $\mathfrak{M}_T$ such that $a \in \|\overline{A}\|_{\mathfrak{M}_T}$, $\mathfrak{M}_T \not\models_T (\exists \overline{A})B$. □

Of course, as in the case of Theorem 15, we obtain a restricted version.

**Theorem 51**
$(\exists A)A, (\forall A)B \models_T (\exists A)B$.

Overall, as was to be expected, the $\mathcal{L}_T$-models $\mathfrak{M}_T$ behave similar to the rejected non-empty $\mathcal{L}_A$-models $\mathfrak{M}_{ne}$.

Moreover, the quantifiers still behave as expected.

**Theorem 52**
The following equivalences hold:

(1) $\models_\mathsf{T} (\forall A)B \leftrightarrow \neg((\exists A)\neg B)$;   (2) $\models_\mathsf{T} (\exists A)B \leftrightarrow \neg((\forall A)\neg B)$;

(3) $\models_\mathsf{T} \neg(\exists A)B \leftrightarrow (\forall A)\neg B$;   (4) $\models_\mathsf{T} \neg(\forall A)B \leftrightarrow (\exists A)\neg B$.

Given the way term-negation '$\overline{\phantom{a}}$' is interpreted, it is equivalent to a negative predication.

**Theorem 53**
$\models_\mathsf{T} (qA)\neg B \leftrightarrow (qA)\overline{B}$ $(q \in \{\forall, \exists\})$.

*Proof.* Let $\mathfrak{M}_\mathsf{T}$ be an $\mathcal{L}_\mathsf{T}$-model such that $\mathfrak{M}_\mathsf{T} \models_\mathsf{T} (qA)\neg B$. Then, by Definition 46 (6)/(7), for some/all $a$-expansions $\mathfrak{M}'_\mathsf{T}$ of $\mathfrak{M}_\mathsf{T}$ such that $a \in \|A\|_{\mathfrak{M}_\mathsf{T}}$, $\mathfrak{M}'_\mathsf{T} \models_\mathsf{T} (q't_a)\neg B$. Thus, by Definition 46 (2), for some/all $a$-expansions $\mathfrak{M}'_\mathsf{T}$ of $\mathfrak{M}_\mathsf{T}$ such that $a \in \|A\|_{\mathfrak{M}_\mathsf{T}}$, $\mathfrak{M}'_\mathsf{T} \models_\mathsf{T} \neg((q't_a)B)$, i.e., by Definition 46 (3) and (1), $\bigcup \|t_a\|_{\mathfrak{M}'_\mathsf{T}} \notin \|B\|_{\mathfrak{M}'_\mathsf{T}}$. Then, by Definition 43 (2d), $\bigcup \|t_a\|_{\mathfrak{M}'_\mathsf{T}} \in \|\overline{B}\|_{\mathfrak{M}'_\mathsf{T}}$, i.e., for some/all $a$-expansions $\mathfrak{M}'_\mathsf{T}$ of $\mathfrak{M}_\mathsf{T}$ such that $a \in \|A\|_{\mathfrak{M}_\mathsf{T}}$, $\mathfrak{M}'_\mathsf{T} \models_\mathsf{T} (q't_a)\overline{B}$. Thus, by Definition 46 (6)/(7), $\mathfrak{M}_\mathsf{T} \models_\mathsf{T} (qA)\overline{B}$. □

Similar again to the non-empty models of Aristotelian syllogistic, only two conversions hold generally, and the third one with a restriction in place.

**Theorem 54** *(Conversion)*
The following conversions hold:

(**a**-**i**-conv$^{\models_\mathsf{T}}$ ↾ $\exists A$)   $(\exists A)A, (\forall A)B \models_\mathsf{T} (\exists B)A$

(**i**-**i**-conv$^{\models_\mathsf{T}}$)   $(\exists A)B \models_\mathsf{T} (\exists B)A$

(**e**-**e**-conv$^{\models_\mathsf{T}}$)   $(\forall A)\neg B \models_\mathsf{T} (\forall B)\neg A$

Lastly, negation works as expected as well.

**Theorem 55**
The following hold ('$*$' being a possibly empty string of negation-symbols '$\neg$'):

(1) $\models_\mathsf{T} (q_1t_1,\ldots,q_nt_n)\neg\neg * A \leftrightarrow (q_1t_1,\ldots,q_nt_n) * A$;

(2) $\models_\mathsf{T} (q_1t_1,\ldots,q_nt_n) * \overline{\overline{A}} \leftrightarrow (q_1t_1,\ldots,q_nt_n) * A$;

(3) $\models_\mathsf{T} (q_1t_1,\ldots,q_nt_n) * \neg\overline{A} \leftrightarrow (q_1t_1,\ldots,q_nt_n) * A$.

## 6 Comparison

Having sketched the different systems, let's compare them. Aristotle's syllogistic and Fregean logic function as base; we consider how QUARC and TFL compare to them and differ from each other. The comparison, however, does not account for all the subtleties and differences between QUARC and TFL, but is restricted to more general points. It also remains open to see whether QUARC can be developed along TFL-lines and vice versa. For this reason, among others, I do not argue for the superiority of either of these systems when it comes to the question of which one better captures the semantics of natural language—the underlying motivation of both QUARC and TFL. The comparison is rather meant to consider potential differences which might lead to further development of either of these approaches along the lines of the other.

### 6.1 Aristotelian Roots

As we have seen in Sections 4.1 and 5.1, both Sommers and Ben-Yami claim a strong connection to Aristotelian logic. Ben-Yami [13, p. 62] sees his understanding of predication as fundamentally in agreement with that of Aristotle, and Sommers considers several of Aristotle's points throughout the development of TFL.

In the version of TFL developed in Section 5.2, I excluded many of Sommers's more specific points that show a strong similarity to Aristotle's logical discussions. For example, I did not include *categories* and, as a consequence, excluded Sommers's discussion of *contrariety* [45, e.g., p. 80].

TFL, in contrast to QUARC, takes the subject-predicate structure of (basic) sentences to be fundamental. The formalism from Section 5.2 does not fully reflect that, though takes some steps towards it with the introduction of *reduced terms* collected in $\mathsf{RTerm}_{\mathcal{L}_\mathsf{T}}$ in Definition 41 (5). This allows to reduce $n$-ary terms to *unary* terms which can be the predicate in the subject-predicate structure. For example, a binary predicate like 'loves' can be reduced to a unary predicate 'loves $t$' ($t$ a term) serving as predicate to a subject. Similarly, we can iterate this and use reduced terms to reduce further terms. This can account for the intended nesting of terms to keep the subject-predicate structure intact [45, e.g., pp. 113ff.].

Moreover, TFL does not include *individual-constants*, but does include *individual-terms* in form of $\mathsf{ITerm}_{\mathcal{L}_\mathsf{T}}$. As all the descriptive signs are *terms*, each term can play the role of subject *and* predicate. This is reflected in the conversions (Theorem 54) which only hold in QUARC for the unary predicates.

Each term in subject position is assigned a *quantity*—indicated by a *quantifier*. In the case of individual terms, the quantity does not make a difference (Theorem 47). In the formalism of Section 5.2, a quantity is assigned, but not in the form of a *wild* quantity (as in [45, p. 18]). Given $t \in \mathsf{ITerm}_{\mathcal{L}_\mathsf{T}}$, $\mathcal{L}_\mathsf{T}$-models interpret them accordingly as *singletons* which puts them on a par with other *unary* terms. Indeed, Definition 43 (2b) allows for unary terms to be interpreted as singletons, too. The difference between an $A \in \mathsf{STerm}^1_{\mathcal{L}_\mathsf{T}}$ and a $t \in \mathsf{ITerm}_{\mathcal{L}_\mathsf{T}}$ would only show up once the system is modalized; $t$ would still be interpreted as singleton, $A$ might not.

QUARC follows the Fregean line of dividing the language into individual-constants ($\mathsf{Const}_{\mathcal{L}_\mathsf{F}}$)/singular arguments ($\mathsf{SA}_{\mathcal{L}_\mathsf{Q}}$) and $n$-ary predicates ($\mathsf{Pred}^n_{\mathcal{L}_\mathsf{F}}/\mathsf{Pred}^n_{\mathcal{L}_\mathsf{Q}}$). The Aristotelian root that Ben-Yami sees for QUARC is when it comes to predication. The sentences of the syllogistic ($\mathsf{Form}_{\mathcal{L}_\mathsf{A}}$) follow the subject-predicate pattern, where the predication can be *universally* or *particularly* and so assign the subject a *quantity*. This general structure is not kept for all the QUARC-sentences though, but only for those with *unary* predicates. In particular, only quantified sentences come with the assignment of quantities, not all sentences. TFL, in contrast, takes every sentence to come with a quantity.

Relational predications, on the other hand, are treated by QUARC as they are in Fregean languages. This contrasts with TFL-sentences which keep the subject-predicate structure also for those. However, Form$_{\mathcal{L}_T}$ also allows for sentences involving *connectives* so that complex sentences without this subject-predicate structure are included, too, but such complex sentences bottom out in sentences with subject-predicate structure in TFL; in QUARC, they do not.

## 6.2 Identity

Sommers [45, ch. 6] argues that there is no need to include an identity-symbol '=' into TFL. Rather, we can understand Aristotle's basic notion of *predicated of all/none* (Section 2.7) as providing us with a *substitution principle* so that identity becomes superfluous. This substitution principle can be taken to be a formal rendering of (B<u>a</u>rb<u>a</u>r<u>a</u>) which allows to conclude $AaC$ from $AaB$ and $BaC$. In the languages of TFL and QUARC, this can be captured as (where '$\models$' is either '$\models_T$' or '$\models_Q$')

$$(\forall C)B, (\forall B)A \models (\forall C)A.$$

However, TFL comprises more notions here as we are allowed to use individual-terms. QUARC, on the other hand, only allows unary predicates, and so the formal rending of (B<u>a</u>rb<u>a</u>r<u>a</u>) does *not* apply to individuals as such. Nevertheless, as there is nothing ruling out unary predicates which are interpreted as *singletons*—i.e., as the $t \in \text{ITerm}_{\mathcal{L}_T}$—it can be taken to apply indirectly, via establishing a connection between the singular arguments and specific predicates.

Moreover, TFL allows this substitution also in cases where the predicate is $n$-ary. For example, if $B \in \text{CTerm}^n_{\mathcal{L}_T}$, we get from

$$(q_1 A_1, \ldots, q_{i-1} A_{i-1}, \forall A_i, q_{i+1} A_{i+1}, \ldots, q_n A_n) B$$

and

$$(\forall C) A_i$$

that

$$(q_1 A_1, \ldots, q_{i-1} A_{i-1}, \forall C, q_{i+1} A_{i+1}, \ldots, q_n A_n) B.$$

Even though QUARC can validate such consequences too, $\mathcal{L}_Q$ contains an identity symbol '=' among its logical constants. Given the different understanding of *predication*, though, it behaves slightly different compared to the Fregean case. As Fregean languages quantify *unrestrictedly* over individuals, it can capture that everything is self-identical ('$\forall x(x = x)$'). QUARC, on the other hand, cannot (see Section 6.4), though identity works similar. For example, given two singular arguments $s_1, s_2 \in \mathsf{SA}_{\mathcal{L}_Q}$, '$s_1 = s_2$' is a QUARC-sentence. However, for $\alpha, \beta \in \mathsf{Ana}_{\mathcal{L}_Q}$, '$\alpha = \beta$' would not be well-formed (and neither would be '$\forall_\alpha \alpha = \beta$' or something similar). Anaphora can only be introduced by replacing singular arguments; see Definition 28 (6). Thus, '$s = s$' can lead to '$s_\alpha = \alpha$' which, in turn, can lead to '$\forall P_\alpha = \alpha$'.

As $\mathcal{L}_T$ does not contain any individual-constants or variables, identity cannot be introduced as in $\mathcal{L}_F$ or $\mathcal{L}_Q$. Nevertheless, in principle, it could be introduced as restricted to individual-terms. For example, if $t_1, t_2 \in \mathsf{ITerm}_{\mathcal{L}_T}$, '$t_1 = t_2$' could be interpreted via Definition 46 (1), i.e., an $\mathcal{L}_T$-model $\mathfrak{M}_T$ satisfies it iff $\langle \bigcup \|t_1\|_{\mathfrak{M}_T}, \bigcup \|t_2\|_{\mathfrak{M}_T} \rangle \in \| = \|_{\mathfrak{M}_T}$ (or, equivalently, $\bigcup \|t_1\|_{\mathfrak{M}_T} = \bigcup \|t_2\|_{\mathfrak{M}_T}$ or simply $\|t_1\|_{\mathfrak{M}_T} = \|t_2\|_{\mathfrak{M}_T}$). One could then also show that $(\forall t_1)t_2 \models_T t_1 = t_2$ (and so use '$(\forall t_1)t_2$' as definition of '$t_1 = t_2$'). In principle, this could also be achieved in QUARC.

### 6.3 Negation

As in the case of $\mathcal{L}_A$, several ways to negate have been introduced into the systems. In the syllogistic, *terms* can be negated ('$\overline{A}$') and sentences can be *negative* ('$(qA)\neg B$'). Fregean languages, on the other hand, only contain sentence-negations ('$\neg\varphi$').[108]

The version of QUARC presented in Section 4.2 incorporates *sentence-* and *predication*-negation. The former works as it does in Fregean languages, the latter negates predication and so compares to the negative sentences of the syllogistic ('ti apo tinos'). What is captured by predicate-negation is that a predicate such as 'friendly' can be *affirmed* or *denied*. However, as long as there is no quantification involved, these are treated as equivalent to sentence-negations as specified in Definition 34 (5).

Similarly, TFL, as presented in Section 5.2, contains both sentence-

---

[108] Or, given a different set-up, formula-negation.

and predicate-negation. Additionally, it contains negated terms as the syllogistic does. As I did not incorporate categories and contrariety into the formalism, these are also treated as equivalent as shown in Theorems 53 and 55.

There is no reason to treat predicate-negation as equivalent to sentence-negation in quantifier-free cases. Following Sommers's discussion, we can understand predicate-negation as connected to categories and category mistakes (which the formalism in Section 5.2 does *not*). For example, the number 2 is neither friendly nor not friendly; the sentences '*it is not the case that* the number 2 is friendly' (which is true) and 'the number 2 *is not* friendly' (which is false) come apart. This, too, could be incorporated into QUARC.

Given TFL's additional *negative terms*, TFL can also treat predicate-negation as introduced, and construe the negated terms as connected to categories directly. It could also understand the predicate-negation so and the term-negation as introduced. The different ways to negate open different possibilities to introduce where negation can "go wrong".

In the empty semantics for the syllogistic, on the other hand, negative predication and negated terms are not equivalent; this is shown in Theorem 16. The reason is that the $\mathcal{L}_A$-models $\mathfrak{M}_A$ allow terms with empty extensions which rule out the validation of **a**-type sentences by ($\mathbf{a}_+$). In the alternative semantics $\mathfrak{M}_{ne}$, simple terms are taken to be non-empty—as are the simple terms in TFL according to Definition 43 (2b) as well as the (Fregean/QUARC) unary predicates according to Definition 19 (2b)/Definition 31 (2b). If incorporated into TFL or QUARC, this opens different ways of interpreting the different ways to negate.

## 6.4 Quantification

As the 'QUAR' in 'QUARC' suggests, Ben-Yami considers QUARC's treatment of *quantification* as one of its major divergences from Fregean languages. Firstly, Ben-Yami [13, §9.8] argues that quantification comes with what he calls 'referential import' in his book ('instantiation' in his [16]).[109] However, in my presentation of the Fregean language, I incorpo-

---

[109] By now he prefers 'instantial import'. He also insists (personal communication) that there are two issues that are mixed together, viz., unary predicates are not empty as to keep QUARC *bivalent*, and instantial import is about quantification, viz.,

rated this already; see Definition 19 (2b), Theorem 23, and Corollary 24.

Secondly, Ben-Yami [13, §6.1] argues that Fregean languages *presuppose* a domain of quantification whereas QUARC does not. Rather, quantification in natural language is always combined with a specification as to what is quantified over, i.e., a plurality is identified and the quantifier specifies *how much* of that plurality is relevant. For example, in 'all human beings are mortal', 'human beings' refers to a plurality of human beings (i.e., reference is to be construed *plurally*) and 'all' suggests how much of that plurality is relevant. The Fregean analysis, on the other hand, quantifies over the whole domain which, therefore, has to be presupposed.[110] Since TFL considers sentences to have subject-predicate structure where the subject is assigned a quantity, the treatment aligns with that of QUARC, viz., quantification is always *restricted* by a term. Thus, insofar as QUARC's treatment differs from that of Fregean languages, TFL's does too.

However, Sommers treats 'human beings' as the *subject* of the sentence, whereas Ben-Yami takes it to be 'all human beings'. The 'all' only indicates the quantity of the subject, but does not figure as part of it in TFL. Formally, this does not make a difference, though, as can be seen by Definitions 34 (7)–(8) and 46 (6)–(7). Nevertheless, the underlying

---

a sentence of the form '$\varphi[\forall P]$' can only be true or false if there are $P$s. Hanoch points out that he has been clear about the distinction since after the publication of his [16].

[110] I have to admit that I—still; see [40, n. 29, p. 315]—don't fully grasp Ben-Yami's claim that domains are not needed. As interpreted here via Definitions 31 and 34 (7)–(8), it is true that all quantification is *restricted* by the interpretation of the quantified argument: $\mathfrak{M}_Q \models_Q \varphi[\forall/\exists P]$ iff for all/some $a$-expansions $\mathfrak{M}'_Q$ of $\mathfrak{M}_Q$ such that $a \in \|P\|_{\mathfrak{M}_Q}$, $\mathfrak{M}'_Q \models_Q \varphi[s_a]$. However, that still *presupposes* a domain in which $\|P\|_{\mathfrak{M}_Q}$ lives.

Lanzet likewise claims to develop a "domain-free semantics" [27, p. 550] and goes on to suggest that when "reference is made to the domain of an interpretation $\mathcal{M}$, what will be meant is the domain of $\mathcal{M}$ *as a function*" [27, p. 565, his emphasis]. However, unless the *function* maps *into* somewhere—its range or our domain—it is not a function, and so the model would not be well-defined.

One might suggest that the problem is the *model-theoretic* approach, but I don't see how the problem disappears by going for a *valuational semantics* (seemingly, Ben-Yami's preferred approach). Whether I presuppose for each predicate $P$ what exactly is referred to or whether I presuppose a domain and then restrict it to predicates seems to me to amount to the same (with the latter option to be in many cases more convenient and expressively richer; see Section 6.5).

understanding of *reference* is a different one, though one that I do not discuss here.

Another difference is that every sentence comes with quantities according to TFL but not to QUARC. The reason is that TFL takes all the descriptive signs to be *terms* for which one can specify quantities. QUARC, on the other hand, follows the Fregean approach. However, as treated in Definition 46 (1), the quantity does not make a difference for individual terms; we might as well reformulate Definition 42 so as to allow $t \in \mathsf{ITerm}_{\mathcal{L}_\mathsf{T}}$ to occur *without* quantifier in $\mathcal{L}_\mathsf{T}$-formulas. Similarly, we could reformulate Definition 28 (1) to *include* quantifiers which don't affect the interpretation.

## 6.5 Expressive Power

Both Sommers and Ben-Yami are concerned with the *expressive power* of their systems. Indeed, both consider expressive power as an adequacy criterion when it comes to alternatives to the Fregean approach. As Aristotelian syllogistic is clearly inferior in this respect, it fails to meet the criterion.

Both QUARC and TFL have a legitimate claim as to satisfy the criterion. QUARC achieves the expressive power by including *anaphora* and *reorders* of any arity; TFL by including *proterms* and *complex terms* of any arity. Both systems also have formal results to show their expressive power in comparison to Fregean languages. Sommers claims that "the expressive power of [PTL] is that of a standard language of modern predicate logic" [45, p. 176], i.e., of a Fregean language; see also [45, Appendix A]). Given that TFL extends PTL, it is clear that TFL does not fall behind with respect to its expressive power.

QUARC, too, has been investigated with respect to its expressive power compared to a Fregean language. Once we expand $\mathcal{L}_\mathsf{Q}$ by a unary predicate $T$ such that $\|T\|_{\mathfrak{M}_\mathsf{Q}} = D$, all $\varphi \in \mathsf{Form}_{\mathcal{L}_\mathsf{F}}$ can be translated into QUARC and vice versa. One way to introduce such a predicate is to allow complex predicates into QUARC; see [39, ch. 5] and, for a fuller treatment, [41]. For, we can then define $(\cdot)T$ as $(\cdot)(P \lor \neg P)$. This, then, allows to capture quantified sentences which don't have restricting predicates such as '$\forall x(x = x)$'; QUARC captures it as '$\forall T_\alpha = \alpha$'. A similar approach works when showing that TFL can capture all $\varphi \in \mathcal{L}_\mathsf{F}$.

What has not been investigated is how exactly TFL and QUARC compare. Once translations between the systems and a Fregean language have been introduced, they can be used to establish the relation between them. However, this has not been done yet. Nevertheless, if the formal systems that have been introduced here are adequate representations of the intended systems, translations between them suggest themselves. Since the presentation of TFL has been quite diminished compared to Sommers's developments, I would think that TFL is the most expressive systems among those considered here. However, there does not seem to be a principled reason to suggest that QUARC couldn't similarly developed further to match this expressive richness.

# 7 Conclusion

I have developed four formalisms here, one for each of Aristotelian syllogistic, Fregean languages, QUARC, and TFL. Both QUARC and TFL are meant to favourably compare to Aristotle's logic. QUARC's understanding of predication and quantification and TFL's understanding of terms and the subject-predicate structure of basic sentences is claimed to be close to Aristotle's understanding of these. Moreover, both systems have been developed as a better way to the semantics of natural language compared to what Fregean languages are capable. Again, it's the Aristotelian root that does much of the heavy lifting.

The expressive power of Fregean languages remains one of the main arguments to adopt the Fregean approach. However, QUARC and TFL have a claim to match this power, and so undermine at least the argument from expressive power. On the other hand, the availability of translations of both TFL and QUARC into Fregean languages also shows that the expressive power alone cannot decide here. One major way in which the case is made for either QUARC or TFL is by the *syntactic* similarity of their formal grammars compared to that of natural language. Given that those formal grammars differ from one another while claiming to fit that of natural language well, it needs to be seen in which ways these formalisms can be extended to capture more and more of natural language. But even once that is done, if we can establish the precise relationship between these systems, it might well be that both can be developed to incorporate parts of the other so that nothing might decide between the two. As it stands, it's focus on the terms and the subject-predicate structure of basic sentences means that TFL is a more radical alternative to Fregean languages; whether it is a better one than QUARC, I leave the readers to decide for themselves.

## Acknowledgements

I gratefully acknowledge the Irish Research Council's funding (project-number GOIPD/2022/635). I would also like to thank Vasilis Politis, Norbert Gratzl, and Hanoch Ben-Yami.

## References

[1] Aristotelis (1949). *Categoriae et Liber de Interpretatione*. Recognovervnt brevique adnotatione critica instrvxervnt L. Minio-Paluello. Oxford: Clarendon Press.

[2] Aristotelis (1957). *Metaphysica*. Recognovit brevique adnotatione critica instrvxit W. Jaeger. Oxford: Clarendon Press.

[3] Aristotelis (1958). *Topica et Sophistici Elenchi*. Recensvit brevique adnotatione critica instrvxit W. D. Ross. Oxford: Clarendon Press.

[4] Aristotelis (1964). *Analytica Priora et Posteriora*. Recensvit brevique adnotatione critica instrvxit W. D. Ross, praefatione et appendice avxit L. Minio-Paluello. Oxford: Clarendon Press.

[5] Aristotle (1963). Categories *and* De Interpretatione. Translated with Notes by J. L. Ackrill. Oxford: Clarendon Press.

[6] Aristotle (1993a). *Posterior Analytics*. Translated with a Commentary by Jonathan Barnes. Second Edition. Oxford: Clarendon Press.

[7] Aristotle (1993b). *Metaphysics*, Books Γ, Δ, and E. Translated with Notes by Christopher Kirwan. Second Edition. Oxford: Clarendon Press.

[8] Aristotle (1994). *Metaphysics*, Books Z and H. Translated with a Commentary by David Bostock. Oxford: Clarendon Press.

[9] Aristotle (1997). *Topics*. Books I and VIII with excerpts from related texts. Translated with a Commentary by Robin Smith. Oxford: Clarendon Press.

[10] Aristotle (2009). *Prior Analytics*. Book I. Translated with an Introduction and Commentary by Gisela Striker. Oxford: Clarendon Press.

[11] Ayer, A. J. (ed.) (1959). *Logical Positivism*. Glencoe: The Free Press.

[12] Barnes, Jonathan (ed.) (1995). *The Complete Works of Aristotle, The Revised Oxford Translation*. 2 Volumes. Sixth Printing, with Corrections. Princeton: Princeton University Press.

[13] Ben-Yami, Hanoch (2004). *Logic & Natural Language: On Plural Reference and Its Semantic and Logical Significance*. First Edition. London: Routledge.

[14] Ben-Yami, Hanoch (2009). Generalized Quantifiers, and Beyond. *Logique et Analyse* 52(208), 309–326.

[15] Ben-Yami, Hanoch (2012). Response to Westerståhl. *Logique et Analyse* 55(217), 47–55.

[16] Ben-Yami, Hanoch (2014). The Quantified Argument Calculus. *The Review of Symbolic Logic* 7(1), 120–146.

[17] Ben-Yami, Hanoch (2020a). The Barcan Formulas and Necessary Existence: The View from Quarc. *Synthese* 198(11), 11029–11064.

[18] Ben-Yami, Hanoch (2020b). The Quantified Argument Calculus and Natural Logic. *Dialectica* 74(2), 35–70.

[19] Ben-Yami, Hanoch and Edi Pavlović (2022). Completeness of the Quantified Argument Calculus on the Truth-Valuational Approach. In: [20, pp. 53–77].

[20] Berčić, Boran, Aleksandra Golubović, and Majda Trobok (eds.) (2022). *Human Rationality: Festschrift for Nenad Smokrović*. Rijeka: Faculty of Humanities and Social Sciences, University of Rijeka (https://repository.ffri.uniri.hr/islandora/object/ffri%3A3517/datastream/FILE0/view).

[21] Carnap, Rudolf (1930/31/59). The Old and the New Logic. Translated by Isaac Levi. In: [11, pp. 133–145].

[22] Carnap, Rudolf (1963). Replies and Systematic Exposition. In: [44, pp. 859–1013].

[23] Englebretsen, George (2016). Fred Sommers' Contributions to Formal Logic. *History and Philosophy of Logic* 37(3), 269–291.

[24] Frege, Gottlob (1879). *Begriffsschrift*, a Formula Language, Modeled upon that of Arithmetic, for Pure Thought. In: [47, pp. 1–82].

[25] Hendricks, V., F. Neuhaus, S. A. Pedersen, U. Scheffler, and Heinrich Wansing (eds.) (2004). *First-Order Logic Revisited*. Berlin: Logos Verlag.

[26] Kneale, William and Martha Kneale (1962). *The Development of Logic*. Oxford: Clarendon Press.

[27] Lanzet, Ran (2017). A three-valued Quantified Argument Calculus: Domain-free Model-theory, Completeness, and Embedding of FOL. *The Review of Symbolic Logic* 10(3), 549–582.

[28] Lanzet, Ran and Hanoch Ben-Yami (2004). Logical Inquiries into a New Formal System with Plural Reference. In: [25, pp. 173–223].

[29] Leibniz, Gottfried Wilhelm (1966). *Logical Papers. A Selection*. Translated and edited with an Introduction by G. H. R. Parkinson. Oxford: Clarendon Press.

[30] Link, Godehard (2009). *Collegium Logicum. Logische Grundlagen der Philosophie und der Wissenschaften*. Band 1. Paderborn: Mentis Verlag.

[31] Mras, Gabriele M., Paul Weingartner, and Bernhard Ritter (eds.) (2019). *Philosophy of Logic and Mathematics: Proceedings of the 41st International Ludwig Wittgenstein Symposium*. Berlin: de Gruyter.

[32] Pascucci, Matteo (2023). An Axiomatic Approach to the Quantified Argument Calculus. *Erkenntnis* 88(8), 3605–3630.

[33] Pavlović, Edi (2017). *The Quantified Argument Calculus: An Inquiry into its Logical Properties and Applications*. PhD Thesis, Central European University, Budapest.

[34] Pavlović, Edi and Norbert Gratzl (2019a). Proof-Theoretic Analysis of the Quantified Argument Calculus. *The Review of Symbolic Logic* 12(4), 607–636.

[35] Pavlović, Edi and Norbert Gratzl (2019b). Free Logic and the Quantified Argument Calculus. In: [31, pp. 105–116].

[36] Pavlović, Edi and Norbert Gratzl (2021a). A More Unified Approach to Free Logics. *Journal of Philosophical Logic* 50(1), 117–148.

[37] Pavlović, Edi and Norbert Gratzl (2023a). Abstract Forms of Quantification in the Quantified Argument Calculus. *The Review of Symbolic Logic* 16(2), 449–479.

[38] Pavlović, Edi and Norbert Gratzl (2023b). Decidable Fragments of the Quantified Argument Calculus. *The Review of Symbolic Logic*, forthcoming, doi: 10.1017/S175502032300031X.

[39] Raab, Jonas (2016). *The Relationship of QUARC and Classical Logic*. Master's thesis, Ludwig-Maximilians-Universität München.

[40] Raab, Jonas (2018). Aristotle, Logic, and QUARC. *History and Philosophy of Logic* 39(4), 305–340.

[41] Raab, Jonas (ms). QUARC and Classical Logic. Manuscript (based on Raab [39]).

[42] Read, Stephen (ms). Aristotle's Theory of the Assertoric Syllogism. Manuscript (available online at: https://philpapers.org/archive/REAATO-5.pdf; last checked on May 7, 2024).

[43] Russell, Bertrand (1946/2004). *History of Western Philosophy*. London: Routledge.

[44] Schilpp, Paul A. (ed.) (1963). *The Philosophy of Rudolf Carnap*. La Salle: Open Court.
[45] Sommers, Fred (1982). *The Logic of Natural Language*. Oxford: Clarendon Press.
[46] Tugendhat, Ernst (1958/2003). *TI KATA TINOΣ. Eine Untersuchung zu Struktur und Ursprung aristotelischer Grundbegriffe*. 5. Auflage. Studienausgabe mit einem neuen Nachwort. Freiburg/München: Alber Symposion.
[47] van Heijenoort, Jean (ed.) (1967). *From Frege to Gödel. A Source Book in Mathematical Logic, 1879–1931*. Cambridge: Harvard University Press.
[48] Westerståhl, Dag (2012). Explaining Quantifier Restriction: Reply to Ben-Yami. *Logique et Analyse* 55(217), 109–120.
[49] Yin, Hongkai and Hanoch Ben-Yami (2023). The Quantified Argument Calculus with Two- and Three-Valued Truth-Valuational Semantics. *Studia Logica* 111(2), 281–320.

# Index Locorum

**Analytica Posteriora**
A
    01 71a23–24, 15
    02 72a13–14, 4
    04 73b05–10, 16
    22 83b24–31, 13
B
    13 97b25, 16
**Analytica Priora**
A
    01 24a11–15, 16
    01 24a16–17, 16
    01 24a17, 5, 17
    01 24a18–22, 17
    01 24b16, 3
    01 24b16–18, 17
    01 24b18–20, 17
    01 24b20–22, 18
    01 24b26–30, 18

    02 25a05–06, 19
    02 25a05–13, 19
    04 25b35–36, 20
    04 25b36–37, 20
    05 26b34–37, 21
    06 28a10–13, 21
    27 43a25–27, 12
    27 43a32–36, 12
    27 43a33, 13
    27 43a36–37, 13
    27 43a40–43, 13
    27 43a41–42, 14
    27 43a42, 13
    33 47b15–18, 15
    33 47b18–29, 14
    33 47b29–39, 14
B
    27 70a03–04, 15
    27 70a10, 15

27 70a16–20, 14
**Categoriae**
02
   1a16–17, 10
   1a21–22, 11
   1a25–27, 11
   1b01–03, 11
   1b04–05, 11
05
   2a11–14, 11
   2a14–16, 11
**De Interpretatione**
01
   16a01, 3
   16a13–16, 3
03
   16b19–20, 3
04
   16b26–27, 4
   16b28–30, 4
   17a03–04, 3
05
   17a09–10, 4
   17a20–21, 4
   17a20–22, 5
   17a21–22, 4
06
   17a25–26, 4
07
   17a38–17b01, 5, 16
   17b01–03, 6
   17b03–05, 6
   17b05–06, 6
   17b08–10, 6
   17b11–12, 7
   17b12–15, 7
   17b15–16, 7
   17b16–20, 23
   17b20–23, 23
   17b23–29, 23
   17b26–29, 7
   17b38–18a01, 7
   18a02–03, 8
08
   18a19–23, 9
09
   18a39–18b03, 8
10
   19b12, 4
   19b14–18, 8
   19b28, 9
   19b37, 9
   20a09–10, 7
   20a20–23, 10
   20a39–40, 10
**De Sophisticis Elenchis**
01
   164b27–165a02, 18
06
   168b31–32, 34
**Metaphysica**
B
   03 998b22, 13
Γ
   07 1011b25–27, 8
Z
   03 1028b33–37, 11
   04 1029b25–28, 9
   17 1041a20–23, 4
**Topica**
A
   18 108b02–04, 34

B
  01 109a03–06, 23
  03 110a32–37, 24
E
  06 136a33–34, 9

H
  01 152a31–32, 34
  01 152b25–29, 34
  01 152b34–35, 34

# SINGULAR POWER: ON THE LOGIC OF SINGULAR TERMS IN TERM LOGIC

GEORGE ENGLEBRETSEN

[T]he fixation on first-order logic as the proper vehicle for analyzing the "logical forms" or the "meaning" of sentences in natural language is mistaken.    P. Suppes

The widespread but baseless belief that the logic of terms is essentially weaker in inference power than modern predicate logic ... has been largely responsible for the unseemly abandonment of a terminist tradition that is one of the intellectual glories of medieval philosophy.    F. Sommers

[Leibniz's] numerous scattered thoughts can be collected into a coherent presentation of term logic. Moreover, a neo-Leibnizian system has recently been elaborated [by Sommers] in technical detail and with considerable sophistication. [Today's] sentence logicians may refer to term logic as a dead Titan, but a requiem would be premature.    J. Barnes

## 1  Terms vs Predicates

The formal language favoured by today's standard first-order predicate calculus is awash in singular terms. In fact, the only kinds of expressions that can play the role of arguments for function expressions, predicates, are singular personal pronouns and proper names. A natural language sentence like 'Every cat is grey' would be first paraphrased (regimented) as 'Every thing (in the universe of discourse) is such that if it is a cat then it is grey'. It is then translated into the formal language as '$\forall x(Cx \supset Gx)$'. Note that the antecedent and consequent of the imbedded conditional are "atomic" and their arguments are singular pronouns, now rendered as individual variables. Moreover, those variables are "bound" by the universal quantifier, since an expression using one or more free variables (an "open" formula) cannot be used to formulate a sentence so as to have a truth-value (be true or false). A sentence such as 'Felix is grey' is formulated as '$Gf$', but can be given a truth-value since 'Felix' (in the guise of '$f$' here), is not a variable. It's a name of an individual. Modern Predicate Logic (MPL) takes the distinction between singular expressions and general expressions very seriously. They are quite different species of logical animals. Singulars are restricted in many ways open only to generals. Nevertheless, as I've said, MPL is saturated with singular terms – pronouns and names. Obviously, natural language is far less profligate, as well as restrictive, when it comes to the employment of singular terms.

There are other ways to account for the logical formulations of natural language expressions. As far as we know, the first, most venerable way was Aristotle's. Before he became fully engaged in the task of building a system of formal logic, he followed his teacher in thinking that the grammar of his Greek sentences would reveal logical form. A simple sentence (rendered in English) such as 'Socrates walks' is, from a grammatical point of view, a noun (in this case a name) somehow attached to a verb. But just how do they attach? How do they form a unified expression rather than just a list of terms? Plato had claimed that nouns and verbs adhere by virtue of their corresponding Forms merging. Naturally, things get even more complicated when more complex sentences are considered (e.g. 'Some philosopher walks' and 'No philosopher walks'). By the time Aristotle was ready to offer a better way of logically analyzing sentences used in the construction of deductions, *syllogisms* (in *Prior Analytics*) and then demonstrations, deductions from premises known to be true (in *Posterior Analytics*), he had hit on a number of ideas that made his syllogistic system viable. He abandoned the analysis of statements into nouns and verbs (along with the notion that they could form a unified expression via Form merging). He simply took all meaningful expressions as *terms*. This meant that the lexicon of his formal system would be homogeneous, consisting of just terms (nouns, verbs, adjectives, singulars, generals, mass, count, simple, complex, relational, clausal, propositional, etc.). He embraced the notion that terms can be negative ("indefinite"). He analyzed sentences (when used as premises or conclusions of syllogisms) as pairs of terms bound together by a unifying *logical copula*, formative expressions with no meaning on their own (*syncategoremata*, as opposed to expressions that have meaning on their own, *categoremata*). There were four such copulae, English versions of which were 'belongs to every', 'belongs to no', 'belongs to some', and 'does not belong to some'. Any sentence formulated by a copula standing between two terms was a *categorical*. As well, the copula functioned on both terms simultaneously; it indicated at once both the *quality* (whether the categorical was an affirmative or a denial) and its *quantity* (whether it was universal or particular). He made use of a semi-symbolic formal language by formalizing terms as uppercase Greek letters. Aristotle's syllogistic logic was a formal logic of logically copulated pairs of terms – it was a *term logic*. Aristotle was the first term logician, the first symbolic logician, the first formal logician, and perhaps the first logician.

In some ways, Aristotle's syllogistic term logic can do a better job of modeling those aspects of our natural language that are involved in our everyday logical reckonings than MPL can. For one thing, it was built to do that. In contrast, modern predicate logic had been built to serve as the foundation of mathematics (though it has been shown, via Gödel's incompleteness theorems, to fail to do so). Syllogistic also seems to fare better in accounting for propositional unity. Aristotle abandoned the lessons from Greek grammar (and Plato) that did this in terms of nouns and verbs somehow just being fit for each other to result in sentences. His account was the logical copulation of pairs of terms. MPL accounts for propositional unity in a way oddly similar to Plato's. In this case, however, the clues come not from grammar but from mathematics. Frege, the father of MPL, was a mathematician well versed in mathematical analysis, the "grammar" of which makes use of the distinction between *arguments* and *functions* on them. And, just as Plato looked to his ontology of Forms, for inspiration, Frege looked to his own ontology of *objects* and *concepts* for his.

Yet, even though syllogistic logic (now generally known as *traditional logic*) had more than two millennia in the limelight, there were many things it could not do as a system of formal logic. It could offer no clear, convincing account of the logic of singular terms. This was due in part to Aristotle's view that the kinds of demonstrations required by science would only involve the universal features of nature and could safely sideline individual objects. However, it must be noted that singular terms were nonetheless found in several places in *Prior Analytics*, (28a24-26, 43a34-35, 43a41-42, 70a26-28), especially in his account of *echthesis*. (For more, see [246, 247, 70].) It also had great difficulty accounting for the logic of relational expression (e.g. 'loves', 'greater than', 'identical to', 'gives ... to ...', etc.). Finally, and in spite of the logic developed by the Stoics in the generation or two after Aristotle, syllogistic offered no satisfactory account of the logic of compound, unanalyzed propositions (e.g. 'If every student attends the class, then Kim will pass' and 'Maria will attend or the party will be boring'). The hegemony of MPL today is due to its ability to provide convincing accounts of each of these. The pronouns and names of natural language are reflected in the bound variables and individual constants of the formal language. Where the old logic considered only monadic (non-relational) predicates, MPL recognizes, and fully exploits, predicates (function expressions) that apply to any number of arguments. And, perhaps, most importantly, the new system takes the logic of compound, unanalyzed propositions (the propositional/sentential calculus) as the foundation of the entire enterprise.

Aristotle's version of logic, modified in many ways over the following centuries, was the first term logic, and had many advantages, but it was certainly not fated to be the last. From the time of Leibniz in the 17$^{\text{th}}$ century to Boole and the algebraic logicians in the 19$^{\text{th}}$ century, mathematics gained more and more inroads into logic. Yet, even after Frege's revolution, there are logicians eager to show that the old logic is far from the last word to be heard from term logic. Fred Sommers began developing his version of term logic, *term-functor logic* (TFL) in the late 1960s. It is a system of formal logic that goes well beyond traditional term logic. In fact, it was a system purposely initiated to rival MPL. The challenge faced by TFL was to offer a formal logic of natural language (something generally rejected or ignored by MPL), while also matching the latter in both expressive and deductive powers. The result was a logic that was *cognitively veridical*, potentially revealing the principles governing the ways we actually reason. TFL makes use of a formal language that accounts for sentential unity in the way Aristotle did, viz. by treating all complex expressions as logically copulated pairs of terms. Relational expressions are treated in the same way. Sentences (including sentential clauses) are treated as terms as well. In fact, the sentential/propositional logic is shown to simply be a part of the general logic of terms. After all, any meaningful expression is a term in TFL. Furthermore, TFL needs no special treatment of identity, unlike MPL, which requires special notation ('='), special interpretation, and special rules of deduction to deal with identity (that's why the standard system is officially 'the first-order predicate calculus, with identity'). Thus far, Sommers' system has answered the challenge to traditional logic to account for relational terms and sentential terms. It also accounts for singular terms, which will be the topic of most of what follows in this essay.

Before offering our analysis of the logical powers enjoyed by singular terms in TFL, it should be added here that TFL is equipped with a simple, yet elegant formal language, a notational system making use of only term symbols (alphabetic letters), the familiar plus (+) and minus (-) signs for functors, and various parenthetical marks for punctuation. In addition, TFL has a semantic theory of terms, one that reveals the nature of term sense, the distinction between denotation and reference, and the notion of propositional truth. And there is one further point that should be kept in mind in what follows. Aristotle's term logic logically parsed statements as pairs of terms unified by a logical copula. *The copula determines at once both the quantity and the quality of the statement as a whole.* Later medieval logicians found it convenient to split the copula into two fragments, a quantifier and a qualifier (often misleadingly called a copula), with the first now attached to the second term and the other attached to the first term. The two results were then reordered (to make the results closer to natural language), yielding what from then on was seen as the Subject and Predicate. Term-functor logic takes all complex terms as copulated pairs of terms, and sometimes the copula is split – other times it is not.

> Of things there are some universal and some individual or singular, according, I mean, as their nature is such that they can or they cannot be predicates of numerous subjects, as 'man,' for example, and 'Callias.'
>
> Aristotle

## 2 Singular Terms

Singular terms are familiar denizens of the language(s) we use every day. We often make use of them when talking about persons, places, things. We say, 'Bob is foolish', 'This city is dangerous', 'Mars is red'. We use singular personal pronouns when talking about some person or thing that we or our audience has already mentioned. We say, 'He's invited' in answer to our friend's question, 'Will Jan be at the party?' We often use singular expressions other than proper names and pronouns. We use definite descriptions ('the first dog in space', 'the Queen of Mobius', 'the only child in Neverland'); and we use demonstratives such as 'this' and 'that' ('this place', 'that man'). Yet, while our language is well-supplied with various kinds of singular expressions, it is not saturated with them in the way MPL would legislate. As we noted above, today's standard logic puts a great deal of weight on the distinction between singular and general expressions, a distinction reflected in MPL's distinction between individual variables (pronouns) and names on the one hand and predicates (function expressions) on the other. The distinction is further reflected in the notation: lowercase letter for the individual variables $(x, y, \ldots)$ or names $(a, b, \ldots)$ and uppercase letter for predicates $(P, Q, \ldots)$. So we have: '$Fb$' for 'Bob is foolish', '$Ix$' for 'He's invited', and '$\forall x(Sx \supset Ix)$' for 'Every student is invited'. Even definite descriptions require pronouns. 'The one in charge is brave' is paraphrased as 'Someone is in charge, no one else is in charge, and she is brave', and then symbolized as '$\exists x(Cx \& \forall y(Cy \supset y = x) \& Bx))$'. One must remember that, in this logical system at least, singular terms are restricted in a number of ways. In particular, they can only play the logical role of arguments – never predicates. They cannot be negated; they cannot be conjoined or otherwise combined with one another to form syntactically more complex expressions. In the case of names, they cannot be bound by any quantifier. Such things as being negated, conjoined, etc. can apply only to entire propositions (sentences or sentential clauses). So how do singular terms make a place for themselves in TFL?

There is, of course, a difference between singular and general terms. MPL makes that difference a matter of syntax and accordingly assigns two different notations. These two, now formulated as argument expressions and function expressions are only fit for their assigned logical roles; neither can play the role of the other. They are formally asymmetric (the A*symmetry Thesis*, more of which later). For TFL, the difference is simply a matter of semantics. Singular terms and general terms only differ in their denotations: singulars denote just one thing, general terms might denote any number of things. We will emphasize the distinction between denotation and *reference*.

Here is a very brief account of the basic semantic theory. First, any *statement* is a sentence used to make a truth claim (which is normally implicit). To assert 'Some horse has wings' is to expect the audience to understand that in doing so a claim is being made as well as that it is true. Moreover, statements are made relative to an understood (by speaker – and hopefully audience) *universe/domain of discourse*. A domain is a totality of things. It might be the actual world, a salient part of the world, a possible world, a fictitious world, the contents of my house, the students in my class, the ingredients in the soup I ate for lunch, etc. Presumably, when one says, 'Some horse has wings' the domain is not the actual world (or even just the equine part of it) but the mythological world of ancient Greece. This notion of a universe of discourse is not the one familiar to most modern logicians, for whom it refers to a set of individuals that can be the "values" of bound variables. The semantic theory embedded in TFL claims that any term used (not merely mentioned) in a statement (relative to a specifiable domain) is said to *express* a *concept*, *signify* a *property*, and *denote* some *object*(s). Consider a statement made by the appropriate use of the sentence 'No horse has wings'. A common assumption would be that it is made relative to the actual world (or an appropriate part thereof). The term 'horse' here expresses *the concept of horse* (abbreviated as [horse]), signifies *the property of being a horse* (abbreviated as <horse>), and denotes all the horses in the domain (i.e., actual horses). There is much more to this semantic theory. For example, the ideas of concept, property, denotation, etc. will need to be fully spelled out, but this will do for now.

## 3  Negative Singulars

> [N]egative terms are of great advantage to traditional logic. On the one hand they increase the power and scope of the system. On the other hand they result in a genuine simplification of the system. ... Moreover, the rules of traditional logic with negative terms are at once more powerful and simpler than the old rules: more powerful since they test a wider range of forms of argument, and simpler since they require a smaller apparatus of concepts.  Miller [194, p. 93]

TFL treats singular terms as just *terms*. They can do what any term can do, and they are restricted only in the ways any term is restricted. But, can they be negated (see [279, pp. 36, 92])? Let's begin with a brief sketch of the formal language of terms and term-functors. Terms, when taken to be simple, are symbolized by letters. All terms ultimately come in *charged* pairs, one positive ($+X$) and the other negative ($-X$). Examples of such pairs in English are 'hopeful/hopeless', 'massive/massless', 'fed/unfed', 'in the Agora/not in the Agora', 'finished the race/failed to finish the race'. Notice that in English some positively charged terms have no explicit evidence of charge. But it must be kept in mind that each term in the formal language *is* charged (explicitly or implicitly). So, singular terms come in charged pairs. And right here is a case where singular terms have one of their powers. Consider a name like 'Bob'. Though it has no explicit indication of charge, it is positive (like 'fed' and 'in the Agora'). But what of its negative mate? To accept that it even has a negative partner is to admit that names might be negative – negated. 'Bob' names (thus denotes) just Bob. But few things bother most logicians more than the idea of a negating a name. What could a negated name possibly name?

The Asymmetry Thesis entails that general terms enjoy logical powers which singular terms are denied. Not only are general terms *predicable*, they can be negative; neither is so for singular terms. However, it should immediately be made clear that even though a general term can be negative ('Bess is uninvited'), and can be negated in natural language ('Bess is not invited'), in the latter case (according to the standard predicate logic) the negation actually properly applies to the entire sentence ('It is not the case that Bess is invited'). The "regimented" form of the sentence is '$\sim (Ib)$' and it is never '$(\sim I)b$'. After all, so the story continues, what could it mean to use a negative singular term? Who (or what) could 'nonBess' be? Keep in mind that TFL takes it as given that any statement is made relative to a specifiable domain of discourse. Now 'NonBess was invited' would likely be used relative to the small domain constituted by the people the host considered inviting. In such a case, it would seem reasonable to interpret the sentence as saying that every person who is not Bess is invited. Everyone other than Bess is invited, and it would be reasonably understood that Bess is not invited. The lesson here is that, in this logic of terms, *negating a singular term yields not another singular term (viz. its oppositely charged partner) but a general term*, a term that can denote any number of other things (in the case at hand, the invited guests). In our everyday language we negate singulars to give us general terms in a variety of ways (for example, think of: 'Every one *other than* Bess was invited', 'All the solar planets *but* Mars', 'Every politician, *except* ,the president, will be investigated', 'Every book, *unless* the one written by Kim, will be sold'). Needless to say, such linguistic devices are not only applied to singular expressions.

Why would modern mathematical logicians reject the idea that singular terms can be negated? Well, one reason is that they believe that a negated singular term must in some way denote, name, refer to, etc., a negative object. Yet, according to this idea, a negative object would have to have all the properties that fail to hold of the object denoted by the corresponding un-negated term. And that would be logically impossible, since the set of properties that would be predicable of such a negative object would be inconsistent, some contrary to others. Of course, they are correct. The very idea of a negative object is nonsense. Yet, their argument is logically invalid. Consider the fact that Socrates was Greek. The argument claims that nonSocrates would then have to have such properties as French *and* Russian *and* Roman *and* Egytian *and* .... But surely that is not so. All that would follow would be that nonSocrates was French *or* Russian *or* Roman *or* Egyptian *or* ...; and that's just what nonSocrates (anyone in the appropriate domain other than Socrates) would be. TFL rejects the idea that the negation of a singular term is itself a singular term denoting one object – and certainly not any putative negative object. So, there are good reasons to recognize a power attributable to singular terms that has been denied them by MPL:

1. *A singular term, like any term, can be negated - the result is a general term.*

# 4 Identity

MPL, as noted already, is burdened with a special appendage – *identity theory*. Because it takes versions of the infinitive verb 'to be' to play no essential logical role in atomic propositions, which consist of nothing other than an incomplete predicate that is completed by an appropriate number of names (pronouns, individual variables), 'is', 'are', 'was', etc. can be dispensed with. However, the Asymmetry Thesis demands that names can never be predicates, which raises an issue for sentences like 'Twain is Clemens'. Where is the predicate? All we find is 'Twain', 'is', and 'Clemens'. Sentential unity depends on a predicate. In such cases, the poor rejected 'is' is rescued to do just one logical job – be a predicate. It is now a special binary relational term, a term of *identity*. In effect, it goes proxy for 'is identical with'. It is given its own special notation, along with special rules of inference as well as conditions on its semantics. Needless to say, TFL takes a different approach. As Sommers asked, [267], "Do We Need Identity?"

Traditional term logicians tended to see sentences with singular terms as subjects as having a hidden quantifier (making them categorical in form). Most often, that quantifier was taken to be universal, which allowed the singular term in such a sentence to be distributed. Needless to say, that won't do for the modern mathematical logician. Even a logician well-versed in Aristotle's logical works, wrote this:

> The history of logic even to this day is replete with embarrassingly desperate attempts to force logical experience into inflexible paradigms. Many of these attempts were based on

> partial understanding or misunderstanding of the relevant paradigm. [adding a footnote]: One of the more ridiculous was to insist that a singular such as 'Socrates is a Greek' was an ellipsis for a universal 'Every Socrates is a Greek'. This absurdity was designed to perpetuate the illusion that Aristotle's paradigm required that every proposition is categorical. This illusion was based on mistaking Aristotle's particular illustration of his general theory of deduction to be that general theory. Corcoran [59, p. 14]

In a very brief study written after 1690, Leibniz [162, pp. 115–121] raised a question about the proper way to think about the logical quantity of singular subjects. Three centuries later, his answer prompted Sommers' further question Traditional logicians had tended to treat sentences with such subjects as implicitly universals. He wrote:

> Some logical difficulties worth solution have occurred to me. How is it that opposition is valid in the case of singular propositions – e.g. 'The Apostle Peter is a soldier' and 'The Apostle Peter is not a soldier' – since elsewhere a universal affirmative and a particular negative are opposed? Should we say that a singular proposition is equivalent to a particular, since the conclusion in the third figure must be particular, and can nevertheless be singular; e.g. 'Every writer is a man, some writer is the Apostle Peter, therefore the Apostle Peter is a man'. I reply that here also the conclusion is really particular, and it is as if we had drawn the conclusion 'Some Apostle Peter is a man'. For 'some Apostle Peter' and 'every Apostle Peter' coincide, since the term is singular.

What Leibniz was suggesting as a solution to his difficulties was that a singular subject can be construed logically as particularly quantified *and* as universally quantified. The two differently quantified subjects "coincide." In other words, the subjects of singular propositions have either quantity depending on the requirements at hand (i.e. the validation of inferences in which they are used). They have what Sommers called *wild quantity* (see, for example [266, 269, 279, 63], also [11, pp.1 54–165] and [13, pp. 118-119, p. 167, n. 70]); They are arbitrarily quantified as required. The freedom to treat singular subjects as wild in quantity turns out to give singular terms another power denied them by MPL. It is an important power, since it allows TFL to analyze identity statements as categoricals, without the need to annex the special logical devices required by modern first-order predicate logic (with identity). To see why this is so, one must show how a categorical treatment of such sentences preserves the formal features required for any equivalence relation (such as identity): reflexivity, symmetry, and transitivity. That process will first involve recognition of another power of singular terms. Again, singular terms are terms, thus having among their logical features those that any term has. They can be used as subject terms, which means that they can be quantified (wildly, as we've seen). But like any term, they can be *qualified*, affirmed or denied of a subject. They are predicable; they can be predicate terms. Frege had insisted [121, p. 43] that a singular term, viz., a proper name "is quite incapable of being used as a grammatical predicate." (But see Ashworth [7], then Lockwood [176] and Mendelsohn [192].) According to Jonathan Barnes [11, p. 114], "In Peripatetic logic, an item is a predicable if and only if it is a subjectible." Later he added (p. 158), "Aristotle was in fact committed to the thesis that singular terms may function as predicates" (see for example *Prior Analytics* 43a34-35; for an analysis, see [70]). Barnes was correct in reporting on Aristotle's acceptance of singular predicate terms, but Barnes went on to hold that Aristotle was wrong to do so, saying (p.162), "Now there can be no doubt that it is a far far better thing to be a Fregean than to be an Aristotelian." Geach, good Fregean that he was, would have agreed with that [127, p. 48]:

> It is logically impossible for a term to shift about between subject and predicate positions without undergoing a change in sense as well as a change of role. Only a name can be a logical subject; and a name cannot retain the role of a name if it becomes a logical predicate.

However, Leibniz's '[S]ome writer is the Apostle Peter' is an example of a singular term in the role of a predicate term while not undergoing a "change in sense" in whatever way Geach had in mind. We can even think of this as another logical power of singular terms.

2. *Singular terms can be qualified (affirmed or denied). They can be predicated.*

That power can be coupled with the power of singular terms to be (wildly) quantified:

3. *Singular subject terms can be quantified either universally or particularly; they can have wild quantity.*

These two powers allow us to show how TFL can analyze identity statements as categorical while preserving the features of reflexivity, symmetry, and transitivity.

Tautologies, in TFL, are universally quantified affirmative statements that logically copulate two tokens of a given term (e.g. 'Every cat is a cat'). They have the form: $-C + C$. Let '$f$' be the name of my cat (lowercase initials of names have no special role here except to help remind us that the term is a name). Since '$f$' is singular, it can be given either quantifier. Making it universal results in '$-f + f$', a tautology. Given that this is the case for any singular term, reflexivity is exhibited. Now consider the case in which I often call Felix 'Missy'. Felix is Missy, so Missy is Felix. Formally: '$+f + m$' entails '$+m + f$', which holds no matter what terms replace '$f$' and '$m$'. Generally, then, '$+X + Y$' entails '$+Y + X$', guaranteeing symmetry. Finally, take Felix to be golden. Since Felix is golden and Missy is Felix, then Missy is golden. Formally: $+m + f, -f + G \therefore +m + G$. The argument is formally valid according to TFL (an argument is formally valid if and only if the algebraic sum of its premises equals its conclusion and the number of particular premises equals the number of particular conclusions, i.e. 1 or 0). Thus, transitivity is exhibited, along with symmetry and reflexivity for singular statements analyzed as categorical. The result is this additional power of singular terms:

4. *The powers of singular terms to be predicated and to be (wildly) quantified gives them the resulting power to allow identity statements to be categorical in form, while preserving all the formal features of an equivalence relation.*

And there is more. Sommers pointed out [276, p. 597] that once the facts that singular terms can be predicated and can be quantified (wildly), there is no need for what the standard predicate calculus takes to be essential rules governing quantified formulas. The two rules, Existential Generalization and Universal Instantiation, are meant to account for inferences of the forms: '$Gm \therefore \exists x Gx$' and '$\forall x Gx \therefore Gm$'. Each of these can be treated as simple syllogisms, where there is the suppressed benign premise '$m$ is a thing' and each token of '$m$' is given wild quantity.

## 5  Team Terms

The Asymmetry Thesis is widely accepted today and was strongly defended by prominent philosophers of logic like Strawson [305] and Geach [127]. It holds that the inviolable distinction between singular and general terms entails that only the latter can be logically negated (and that negation is then shifted to apply to the entire sentence in which the term is used). This is because, for MPL, singular terms (viz. pronouns and names) carry the full "burden of reference", but have no logical power; they can play no other logical roles. All logical power is assigned to general terms, predicates. It also holds that only general terms can be combined to form compound terms. In particular, 'wealthy' and 'powerful' can be combined to yield 'wealthy and powerful'. By contrast, per the thesis, singular terms cannot be so combined. We have 'Russell' and we have 'Whitehead', but 'Russell and Whitehead' does not constitute the name of anyone. 'Russell and Whitehead were British' would be logically paraphrased as 'Russell was British and Whitehead was British'. So far, so good. But now what of 'Russell and Whitehead wrote *Principia Mathematica*'? Perhaps one could formulate it as 'Russell wrote *Principia Mathematica* and Whitehead wrote *Principia Mathematica*, but that's not quite correct. To be more precise, they wrote their book *together*, neither was alone the author, they were co-authors. Thus, an accurate paraphrase might be 'Russell and Whitehead co-authored *Principia Mathematica*'. Obviously no single tennis player ever won a doubles match. Venus and Serena won a Wimbledon Doubles Championship, not individually, but as a doubles *team*. Modern logicians are wed to the Asymmetry Thesis because the failure to restrict the logical roles of singular terms would mean a commitment not only to the nonsense of negative objects via negative singular terms, but also a commitment to the nonsense of compound objects via compound singular terms. Just what do they think a compound object could be? Generally, they argue that a combination of singular terms such as 'Russell and Whitehead' must itself be a singular term, and the object denoted by that term must have all the properties that hold of Russell as well as all the properties that hold of Whitehead. The obvious problem is that there could be no such object, since it would have to be both a member of the House of Lords and not a member of the (House of Lords. Such a thing, just like poor nonSocrates, would be logically impossible nonsense.

There are many expressions in our language that can only be sensibly predicated of quantified terms with collective denotation (e.g., 'together', 'meet', 'surround'). There is a growing literature devoted to the debate surrounding the logic of so-called "plural predication" (see, for example, [16, 19, 139, 170, 171, 172, 173, 174, 187, 191, 217, 218, 219, 244, 245, 256, 331, 332]). In addition, terms with a collective denotation are also familiar in our native language. Some teams (clubs, co-ops, organizations, even mobs, etc.) are given special names ('Bourbaki', 'the Williams Sisters', 'The Marx Brothers', 'NATO', 'the New York Yankees', 'Shakespeare's comedies', 'Amazon.com', 'The Beatles', 'New England', 'the January 6 Capitol insurrectionists'). Others often get by simply listing their members ('Peter, Paul, and Mary', , '2, 3, 5, 7, and all the other primes less than 1,000,000', 'the sun, the moon, the wandering stars, and the fixed stars', 'Hardy-Littlewood'). In the case of the last of these, recall Harald Bohr's 1947 *bon mot*, "Nowadays there are only three great English Mathematicians: Hardy, Littlewood, and Hardy-Littlewood."

Sometimes a conjoined pair of names is just that, two terms applying to two different individuals. Other times, in appropriate contexts, they operate as a single expression (so called *phrasal conjunctions*) applying to a collection of the individuals to which their constituent names apply. That is just the case with our examples above. Let's say that in those kinds of cases the resulting expression is a *team term*. Such terms "are neither odd nor frightening" [13, p. 122, n. 192; also p. 147, n. 2]. 'Bourbaki' is the name of a team of mathematicians, for example. Suppose the term 'philosopher' is used in a statement (made relative to a specifiable domain). It denotes the constituents of the domain that are philosophers (have the property <philosopher>). Now suppose the statement is universally quantified, with 'philosopher' as the subject term ('Every philosopher is comedian'). In other sentential contexts it could just as well be an object term of a relational expression ('Heloise loved every philosopher'). In such cases, the *subject term* denotes all the philosophers in the domain, and the *subject*, the quantifier plus term, *refers* to everything the term denotes. Next, consider the quantifier to be particular. In that case, the term 'philosopher' still denotes all the philosophers in the domain, but the subject ('some philosopher') refers to only an undetermined part (perhaps the whole) of the constituents of the domain which are philosophers. Finally, suppose that we could name every philosopher in the domain (to make things simple, take the domain to have only famous ancient Greek philosophers among its constituents. In that case, we can replace the universally quantified subject with a *conjunction* of names and the particular quantified subject with a *disjunction* of names. Most, though not all, of this has been familiar to both traditional and modern logicians. We can say that the subject terms, rendered as conjunctions or disjunctions of appropriate singular terms (names), have *explicit denotation* and are taken as subjects (quantified terms) by virtue of their forms as either conjunctive or disjunctive.

Let '[Russell, Whitehead]' be a term with explicit denotation. There are true statements of the form 'Every [Russell, Whitehead] is $X$' and of the form 'Some [Russell, Whitehead] is $Y$'. It is true that some [Russell, Whitehead] was a member of the House of Lords. As well, it is true that every [Russell, Whitehead] was British, since Russell and Whitehead were British. Russell was, and so was Whitehead. But it is not true that Russell wrote *Principia Mathematica* and Whitehead wrote *Principia Mathematica*. What is true is that they co-authored *Principia Mathematica*. In this case, the term of explicit denotation is operating as a team term. We can formalize it as '{Russell, Whitehead}'. Now what is the quantity of such a term when used as a subject term (or object term)? The best way is to think of such terms as singular, with wild quantity when used in referential (subject or object) expressions. We could say 'The team (group, collective, etc.) constituted by Russell and Whitehead wrote *Principia Mathematics*.' We could even imagine giving that team a proper name, just as the mathematicians who teamed together to form Bourbaki did. Unlike 'Bourbaki', a team term whose denotation is rarely made explicit, the team term '{Russell, Whitehead}' does have explicit denotation. Nevertheless, both these terms, 'Bourbaki' and '{Russell, Whitehead}' are singular terms. It's important to emphasize that *denotation* applies to terms and reference applies to quantified terms (logical subject or object expressions), which is dependent on both the denotation of the term involved and the quantifier.

A term of the form '$\{a_1, a_2, \ldots, a_n\}$' differs in important ways from one of the form '$[a_1, a_2, \ldots, a_n]$'. In the latter case, such terms of explicit denotation are singular just when $n = 0$; in the former case, such phrasal conjunctions are always singular, team terms. Also, in the latter case, it is analytic that every $[a_1, a_2, \ldots, a_n]$ is $[a_1, a_2, \ldots, a_n, a_{n+1}]$; in the former, it is not the case that every $\{a_1, a_2, \ldots, a_n\}$ is $\{a_1, a_2, \ldots, a_n, a_{n+1}\}$. Finally, it is analytic that every $[a_1, a_2, \ldots, a_n]$ is $F$ if and only if $a_1$ is $F$ and $a_2$ is $F$, and $\ldots a_n$ is $F$. However, while it is analytic that, for any singular term '$a$', $\{a\}$ is $F$ if and only if $a$ is $F$, it is not analytic that if $\{a_1, a_2, \ldots, a_n\}$ is $F$ then $a_1$ is $F$ and $a_2$ is $F$ and $\ldots a_n$ is $F$ (consequently the corresponding bi-conditional also fails to hold). To illustrate these principles, consider the fact that (relative to the appropriate domain) every player who won a match at the tennis tournament was a qualified entrant in the tournament, but it is not the case the Williams sisters doubles team was a member of the U.S. tennis team (even though both Venus and Serena were members of the U.S. team). Finally, while everyone who won a match got a prize means that each player who won a match got a prize, even if the Williams doubles team won a match, it could have happened that both of them lost all their singles matches.

Strawson was among those who was most distressed by not only negative names but phrasal conjunctions of names as well. In each case he argued (see especially [305, 306]) that such expression could never be logical subjects. (He was under the spell of the *Fregean Dogma* (see [266]), the conviction that all subjects are singular), since the referent of such an expression would (like nonSocrates), be impossible nonsense. For example, the putative referent of a phrasal conjunction, a team name like ('the Williams sisters doubles tennis team' or '{Venus, Serena}', would, according to Strawson, have to have all the properties of each member, which is impossible (Venus would have to be both older than and younger than Serena). Such ideas, as we noted above, are due in large measure to the unreflective adherence to the Asymmetry Thesis coupled with the Fregean Dogma. Abandoning both leaves one free to explore a term logic that differs in important ways from the standard logic now in place. One can then recognize yet another power of singular terms:

> 5. *The ability to form a new singular term, with collective denotation from several singular terms, yields team terms having all the other powers of any singular term, and allowing a simple analysis of phrasal conjunctions.*

# 6 Propositional Terms

According to the principles underlying MPL, the logic of propositions, the logic of atomic sentences and their truth-functional combinations, is *primary*. So, the first-order predicate calculus (with identity) is secondary, depending on the primary logic. This is obvious from reflecting on their formal language. More often than not, quantifiers range over open formulas that are conditionals or conjunctions. And, of course, all logical negation is construed as sentential. Nonetheless, such a formal system is able to offer an account of the logic of propositions as well as the logic of predicates, showing their relation to one another, thereby achieving a single formal logic, a primary system underlying the rest. Traditional term logic saw the challenge of somehow joining their system with a logic of proposition. Yet they had little success. In the 17$^{\text{th}}$ century, Leibniz, a prophetic logician in so many ways, was eager to find a way to incorporate an account of the logic of propositions into the traditional logic of terms. His central idea was that this project would require treating entire propositions as themselves terms. He wrote [162, p. 55]

> If, as I hope, I can conceive all propositions as terms, and if I can treat all propositions universally, this promises a

wonderful ease in my symbolism and analysis of concepts and will be a discovery of greatest importance.

Lebniz's intent was to incorporate the logic of propositions into the logic of terms. The first step would be conceiving of entire propositions as terms; the second step would be to insure that the result of doing so would be an account of compound propositions (conditional, conjunctive, etc.) as categorical in form – a pair of terms bound together by a logical copula. It is certainly easy to see how a conditional can be construed logically as a universal proposition. For example (letting lowercase letters from the middle of the alphabet symbolize unanalyzed propositions), 'If $p$ then $q$' shares certain formal features with 'Every $A$ is $B$'. The formatives 'if ... then' and 'every ... is' are reflexive and transitive (but not symmetric). Thus, 'If $p$ then $q$, if $q$ then $r$, therefore if $p$ then $r$' and 'Every $A$ is $B$, every $B$ is $C$, therefore every $A$ is $C$' are both valid. Also, 'Both $p$ and $q$' and 'Some $A$ is $B$' share the formal feature of symmetry (but not transitivity and reflexivity). All this has been well-known for a very long time because it is so easy to see. A logic of terms, something like what Leibniz had in mind, would start then with taking any meaningful expression (no matter how simple or complex) as a term. So, any propositional expression (a proposition or a propositional clause) is a term. Like any complex term it can be analyzed into its components, but it need not be. It goes without saying, that, like any term, propositional terms are charged, they come in oppositely charged pairs (just like 'happy/unhappy').

However, it looks as though the categorialization of compound sentences, as obvious as it may appear, is not so straightforward. It can be agreed that 'Both $p$ and $q$' and 'Some $A$ is $B$' can be said to share a common syntax, there is a disanology between them. The pair, 'Some $A$ is $B$' and 'Some $A$ is not $B$', are logically compatible; the formally corresponding pair, 'Both $p$ and $q$' and 'Both $p$ and not $q$', are not. An additional disanology is that while 'Both $p$ and $q$' formally entails 'If $p$ then $q$', 'Some $A$ is $B$' does not entail 'Every $A$ is $B$'. However, TFL provides a way to show that each disanalogy really is only a matter of appearances.

One of the many advances in perfecting a logic of terms is seen in the way TFL carried out Leibniz's program for dealing with the logic of propositions (in the terminology of TFL, statements, or sentential terms). Sommers built this in a number of places (especially in Sommers [286]). The key to seeing his results is to note that statements (sentential terms) are sentences used (relative to a domain) to make a truth claim. They are complex terms, and, as such, they are logically copulated term pairs. So, they share the same kinds of semantic features common to all terms. As we have already said, any term (used relative to a specifiable domain of discourse) expresses a concept, signifies a property, and denotes some object(s). When it comes to these notions of *concept*, *property*, and *object*, TFL gives each a special name in order to highlight for later on what makes such sentential terms special. A statement, $p$ (used relative to a specifiable domain, $D$) expresses a *proposition*, signifies a *constitutive property*, and denotes $D$. A proposition, now, is the sense of the statement, what is understood by the speaker and, ideally, the audience. A constitutive property is a property that a totality of objects has by virtue of what is, or is not, a constituent of that totality. Domains are totalities. Not all properties of a totality are constituent. The dramatis personae in *Hamlet* are English speaking and in Denmark, which are *non-constitutive* properties of the *Hamlet* characters. Constitutive properties are either *positive*, determined by some object(s) *being in* (presence) the totality, or *negative*, determined by some object(s) *not being in* (absence) the totality. The presence of grave-diggers and Polonius and the absence of elephants and Sherlock Holmes are constitutive properties of the *Hamlet* cast of characters. The salad I had with my dinner this evening was nutritious and delicious, but it also had such constitutive properties as the presence of Romaine lettuce and red onions and the absence of olives and tomatoes.

This bit of semantic theory is important for showing how sentential logic, the logic of sentential terms, can be seen as a "special branch" of term logic. Consider again a statement, $p$, made relative to $D$. What it denotes is $D$ itself. Totalities, like $D$, are constituted by individual objects, but totalities themselves are individual objects. We might go so far as to say they are special team terms. Since they have singular denotation, they are *singular terms* (by definition). This means that when a sentential term is used as a subject term (or an object term) it has wild quantity. A happy consequence of this is that the two disanalogies noted above are easily resolved once it is understood that sentential terms, when quantified, used as implicitly quantified subjects (or objects), have wild quantity. As it happens, although he came close, Leibniz never got to the point where he would say that "if a proposition is true ... it will be made true by ... just one fact." As Castañeda added, this was unfortunate because "Leibniz apparently never put together the special principles required by singular propositions [propositions seen as singular terms] with the general principles of his logical system" [thus allowing them to be given wild quantity like any singular term, (see [26, p. 487]).

6. *Because sentential terms are singular terms, with all the other powers, they have the power to make possible the logic of propositions as nothing more than a special part of term logic.*

Of course, there is much more to say about what treating the logic of statements as a special branch contributes to the overall expressive and inferential powers of TFL, how it resolves the issued traditional logicians like Leibniz faced in unifying the treatment of terms and sentences in a single formal logic. In addition, it provides an elegant way to give a formal treatment of truth, falsity, existence, nonexistence, etc. But we have seen, thus far, the many costumes that singular terms wear, and the often unexpected logical powers they have. Still, there is another costume they can wear.

# 7 Pronouns

As we said early on, modern first-order predicate logic (with identity), MPL, makes use of a formal language that is pronoun-saturated. Every atomic sentence has as its referential expressions either names or pronouns, and the latter are in the guise of individual free variables that are fit for being bound by quantifiers. We have gone some way in rehearsing how singular terms, particularly names, are treated in TFL in a manner that better reflects their uses in our natural language. They are used when called for, but they are certainly not ubiquitous. Suffice it to say, pronominal terms are terms, fit for any role that any kind of term can play. See: [279, pp. 67–119]; [273, 300, 78, 105].

Quine was an ardent promoter of MPL during the last century. He accepted fairly uncritically the idea that the standard system has the advantage, not only over traditional logic, but over our usual understanding of much of our own everyday language, in that it can give a clear account of pronouns. He presumed both the Asymmetry Thesis and the Fregean Dogma, believing that any simple, atomic sentence requires a part that does the job of referring and another that does the job of predicating (he called it the "basic combination") – and the two jobs require two different kinds of expressions. MPL requires reference to be something only a singular term can do (in fact, the *only* thing it can do). It is what bound variables and names do. Moreover, Quine saw no difficulty in eliminating the latter, leaving (singular) pronouns to do the heavy lifting of reference (heavy, because what can be referred to at all are *bare particulars* (mere objects before being characterized by use of a predicate). He wrote [236, p. 165]

> The pronoun is the tenable linguistic counterpart of the untenable old metaphysical notion of a bare particular. The variable is the legitimate latter-day embodiment of the incoherent old idea of a bare particular.

Most of us would unreflectively think that we make reference by using expressions such as 'all politicians', 'some cats', 'Bob', and definite descriptions like 'the only even prime number', but also pronouns such as 'it' and 'she'. Note that all of these kinds of expressions really *are* referential when they are explicitly or implicitly quantified. Their constituent terms denote, but the quantifier-term combination is logically a subject or object expression. Traditional logicians, as well has contemporary term logicians, generally agree. Now if pronouns can be used to make reference, just what are their referents? Particulars, of course, but are these clothed or bare? If clothed, then they must have at least some property, something predicated of them. Yet, Quine has precluded that choice by separating the referential from the predicational roles, blocking pronouns from being clothed. Their referents must be bare particulars – an "untenable old metaphysical notion ... an incoherent old idea" that has now been made "legitimate" and fitted out as an individual, bound variable. Modesty has been preserved. For an excellent take on this affair, see D.S. Oderberg's "Predicate Logic and Bare Particulars" [216, pp. 183–210].

Grammatically, a pronoun is an expression that is meant to somehow replace a referential expression (a name or a quantified term) that has or will be used in a given linguistic context. Since our topic is the logic of singular terms, plural pronouns ('they', 'them', 'those', etc.) will not be considered here. Most of the time a (singular) pronoun is an anaphor, picking up reference found earlier, most often (but not always) in the same sentence. Examples are: 'Anna took a long walk, so she got home late', 'A person in the garden said she was lost', 'Some man broke the window, and then he ran away', 'Every soprano in the opera said she was ready'. Of course, anaphora can be multiple even in a single sentence ('Tom owns a dog, but he can't house-train it', 'A man gave a gift to a woman who rejected both the gift and him'). Consider the first example. 'Anna' is the subject (in this case a wildly quantified singular term). When not used as a logical subject or object, 'Anna' simply denotes an individual in the domain relative to which the sentence is being used. When quantified, as in this case, 'Anna' makes reference to some/every Anna (which is just Anna). Unquantified 'Anna' denotes; quantified 'Anna' refers. Now what of 'she'?

A look at the second example will help. Here the subject is overtly quantified, 'a person'. The term 'person' now denotes people (in the domain of discourse at hand); 'a person' makes indefinite (undetermined) reference to at least one of them. We are tempted to say that the anaphoric pronoun, 'she', either denotes what 'person' denotes or refers to what 'a person' refers to. In addressing this issue, Sommers wrote [274, pp. 605–605]

> According to the traditional doctrine, every logical subject is of form 'every $x$' or 'some $x$'. We apply this doctrine to pronouns and distinguish between the logical sign of the pronoun and the term (or proterm) that follows the sign. Thus [an anaphoric pronoun] can be taken as a whole subject and its logical form will include a sign of quantity or it can be taken as the term that follows the (implicit) sign of quantity. The ambiguity of [the pronoun] is similar to the ambiguity of 'Socrates' in subject position. For 'Socrates' can be taken

as a whole subject with implicit quantity or it can be taken simply as a singular term. Because every logical subject consists of a sign of quantity followed by a term, the traditional theory distinguishes between the modes of signification of the whole subject and the term it contains. Let *reference* be the mode of signification of the subject and let *denotation* be the mode of signification of a term. In 'some man is at the door' the subject 'some man' *refers* to some man in a non-identifying way; the term 'man' *denotes* all men. ... Taken as a whole subject 'he' refers to whatever the antecedent refers, i.e., it refers to "the man in question". Taken as the term of this subject (the proterm), 'he' also *denotes* what is referred to by the antecedent subject refers to. This is generally true of the proterm: it is defined to denote what the antecedent subject *refers* to. ... In effect, the pronoun for a definite subject 'some $x$' will always be an expression with "wild" quantity. Pronouns that cross-refer to 'every $x$' are another matter. They of course will have universal quantity.

Sommers [279, p. 67] calls his account of pronomonialization "the proterm theory" in contrast with the "bound variable theory" of MPL. The denotation/reference distinction is fundamental in TFL. An anaphoric pronominal expression refers; its proterm denotes. A pronoun, then, logically consists of a quantifier and a term (viz., a proterm). What such a proterm denotes is just what its antecedent subject has referred to. The essential nature of a proterm [279, p. 93]: "is a term introduced for the special purpose of denoting what was antecedently referred to. As with any other term, one can form its contrary [i.e., negate it] and it can serve as the subject term of an antecedent referring expression for further pronominalization." Perhaps the most surprising element of the theory of anaphora found in TFL is this claim [279, p. 230]: "*proper names are pronouns.*" That is because [279, p. 231] "Pronominalization always precedes nominalization." He adds [279, p. 234], "A proper name may be explicitly introduced by an anaphoric description." And [279, p. 246] "... proper names are anaphoric pronominal expressions and any explanation of their special characteristics is to be found in the theory of anaphoric reference."

My focus has been on singular terms. A relatively simple summary of the logic of singular anaphora would then be this. Pronominal expressions are introduced into discourse in order to make reference to something that has already been referred to (in that very sentence or a previous sentence). In the normal case, that antecedent reference was made by an indefinite logical subject (or object) of the form 'some $A$'. The pronoun is an implicitly quantified proterm ('it', 'she', etc.) *denoting* the referent of the antecedent ('some $A$'), and the pronoun refers to every thing the antecedent referred to. The quantity of the pronoun is wild because the pronoun inherits the particular quantity of the antecedent and also has its own universal quantity by virtue of its reference to every referent of the antecedent. We introduce proper names into our natural language discourse (by fiat) for convenience only. "Names are special-duty pronouns – they are *pro-pronouns*. It's interesting to note that this is the term Quine used when he went about eliminating proper names (e.g., 'Socrates') by replacing them with predicates (e.g. 'socratises') and pronouns, bound variables for Quine ([228] and [98, p. 179]). Why are they pronouns at all?

Consider the fact that, on a terminist analysis, pronouns are links in anaphoric chains. Often such chains are quite short ('Some witness perjured herself', 'A man is at my door. He is armed'). But they can be longer, making use of various expressions that can pick up the preceding references ('A man is at my door. The man at my door is knocking and he is armed', 'A leopard escaped. It's dangerous, so beware. If spotted, call authorities'). Note that the referents at each link are accumulated by subsequent links, "each one being more comprehensive than its predecessor" [279, p. 256]. That is because "the proterm of a pronoun is designed to denote what was antecedently referred to. In this sense the pronoun is bound to the antecedent sentence" (Sommers 1982, p. 67). In general, anaphoric chains are inter-sentential. The pronoun in 'A man is at my door. He is armed' is grammatically bound to 'A man' but semantically bound to 'A man ... at my door' Though such anaphoric chains are initiated by an indefinite description (a particularly quantified term), subsequent anaphora are definite (definite descriptions like 'the man at my door', or demonstratives like 'that' in 'that is threatening', or pronouns like 'he'). On certain occasions (such as baptisms, ship launches, christenings, etc.), at some point in the chain, a speaker might have reason to assign a special name to the referent at hand (say, 'Mr. Gunman'). In this last case, the name 'Mr. Gunman' picks up the reference of the preceding 'he', showing that it is itself a pronoun (viz., a pro-pronoun). For a clear analysis of such anaphoric chains, see Chastain [52, especially pp. 206-214]. We might now add one more power enjoyed by singular terms.

> 7. *Singular terms in the form of anaphoric pronouns or pro-pronouns (names) have all the powers of any singular term, and they allow for a clear analysis of anaphora.*

# 8 Some Terminal Remarks

How did modern logic go astray here? Why accept the Asymmetry Thesis? The foundation of the thesis, as alluded to earlier, is the presumption that the logical roles that any expression can be allowed to play are fully determined by the syntactical natures of those expressions. Frege had insisted that in any simple atomic sentence one term must be unsaturated, incomplete (having one or more holes); they are function expressions. Any other term must be complete, an argument of the function. A statement completed by the appropriate number of complete expressions was itself complete (viz. completed). An incomplete expression is a predicate; complete expressions are names, singular expressions. These have distinct natures that determine their logical roles here. Names *refer* to objects; predicates somehow express or "refer" to concepts. One result was what Sommers [266, 275, 279] called the "Fregean Dogma" (the assumption that subjects must always be logically singular). The object/concept distinction leads to the singular/general distinction, which in turn leads to the Asymmetry Thesis. There was some pushback. A century ago, Frank Ramsey [243, p. 404], challenged the subject/predicate asymmetry found in both traditional and modern logic, writing:

> Both the disputed theories make an important assumption which to my mind, has only to be questioned to be doubted. They assume a fundamental antithesis between subject and

predicate, that if a proposition consists of two terms copulated, these two terms must be functioning in different ways, one as subject, the other as predicate. ...Hence there is no essential distinction between the subject of a proposition and its predicate, and no fundamental classification of objects can be based on such a distinction.

The fact is that being a predicate is not what an expression *is* but what it *does* (i.e., what it can be used to do). From the terminist point of view, any term, regardless of it denotation (singular or nonsingular), is a *predicable* term. Any term can occur as the predicate term of a sentence or sentential term. Likewise, any term can occur as the subject term of a sentence or sentential term. In other words, any term is a *subjectable* term. That's what Ramsey saw.

Now for one final remark about singular terms, namely, names. In the 19$^{th}$ century, J.S. Mill argued that names denote/designate individual objects, but they do not connote, they have no descriptive content [193, p. 41]. Frege, Russell, and the majority of logicians who followed them, rejected Mill's view and held that names are descriptive. Eventually, S. Kripke argued for a Mill-like thesis: a name is not descriptive but designates – rigidly; it designates the very same object on each occasion of its use, including in modal contexts [158, pp. 48 ff.]. Something of Mill's thesis survives in TFL. A name does designate/denote an individual. However, in contrast with both Mill and Kripke, TFL holds that a name is nothing more than a special-duty pronoun (a quantified pro-pronoun) that accumulates descriptive content as it is passed along an anaphoric chain. And that is part of the revelation of the powers of singular terms.

# References and Some Related Works

[1] Ademollo, F., 2015, "Names, Verbs, and Sentences in Ancient Greek Philosophy," Linguistic *Content. New Essays on the History and Philosophy of Language*, M. Cameron and R.J. Stainton (eds.), Oxford: Oxford University Press, pp. 33-54.

[2] Alvarez, E. and Correia, M., 2012, "Syllogistic with Indefinite Terms," *History and Philosophy of Logic*, 4: 297-306.

[3] Aristotle, 1962, *Aristotle: The Categories, On Interpretation, Prior Analytics*, H.P. Cooke and H. Trendennick (tranlators.), Cambridge, MA: Harvard University Press.

[4] Aristotle, 1963, *Aristotle's* Categories *and* De Interpretatione, J.L. Ackrill (transl.), Oxford: Clarendon Press.

[5] Aristotle, 1989, *Aristotle: Prior Analytics*, R. Smith (transl.), Indianapolis and Cambridge: Hackett.

[6] Ashworth, E., 2019, "Medieval Theories of Singular Terms," *The Stanford Encyclopedia of Philosophy*, winter edition, E.N. Zalta (ed.).

[7] Ashworth, E., 2004, "Singular Terms and Predication in some Late Fifteenth and Sixteenth Century Thomistic Logicians," *Medieval Theories on Assertive and Non-Assertive Language. Acts of the 14$^{th}$ European Symposium on Medieval Logic and Semantics. Rome, June 11-15, 2002*, A. Maierù and L. Valente (eds.), Florence: Olschki.

[8] Bacon, J., 1985, "The Completeness of Predicate Functor Logic," *Journal of Philosophical Logic*, 50: 323-339.

[9] Bacon, J., 1987, "Sommers and Modern Logic," *The New Syllogistic*, G. Englebretsen (ed.), pp. 121-160.

[10] Bacon, J., 2000, *Syllogistica Carolina Rediviva*, Uppsala: Uppsala University.

[11] Barnes, J., 2007, *Truth, etc.: Six Lectures on Ancient Logic*, Oxford: Clarendon Press.

[12] Barnes, J., 2009, "Notes on the Copula," *Dianoia*, 14: 27-62.

[13] Barnes, J., 2012, *Logical Matters: Essays in Ancient Philosophy II*, Oxford: Clarendon Pres.

[14] Barwise, J. and Cooper, R., 1981, "Generalized Quantifiers and Natural Language," *Linguistics and Philosophy*, 4: 159-219.

[15] Ben-Yami, H., 2004, *Logic & Natural Language: On Plural Reference and Its Semantic and Logical Significance*, Aldershot: Ashgate.

[16] Ben-Yami, H., 2009, "Plural Quantification: A Critical Appraisal," *Review of Symbolic Logic*, 2: 208-232.

[17] Ben-Yami, H., 2020, "The Quantified Argument Calculus and Natural Logic," *Dialectica*, 74: 35-70.

[18] Boër, s.E., 1975, "Proper Names as Predicates," *Philosophical Studies*, 27: 389-400.

[19] Boolos, G., 1989, "To Be is to Be the Value of a Variable (or to Be Some Value of Some Variables)," *Logic, Logic, and Logic*, Cambridge, MA: Harvard University Press, p. 54-72.

[20] Böttner, M. and Thümmel, W., "Introduction," *Variable-free Sematics*, Böttner and W. Thümmel (eds.), Osnabrück: Secolo Verlag, pp. 8-19.

[21] Bradley, M.C., 1986, "Geach and Strawson on Negating Names," *Philosophical Quarterly*, 36: 16-28.

[22] Burge, T., 1973, "Reference and Proper Names," *Journal of Philosophy*, 70: 425-439.

[23] Burge, T., 1974, "Truth and Singular Terms," *Noûs*, 8: 309-325.

[24] Burgess, J.P., 2004, "*E Pluribus Unum*: Plural Logic and Set Theory," *Philosophia Mathematica*, 12: 193-221.

[25] Carroll, Lewis (C. L. Dodgson), 1895, "What the Tortoise Said to Achilles," *Mind*, 14: 278-280.

[26] Castañeda, H.-N., 1976, "Leibniz's Syllogistico-Propositional Calculus," *Notre Dame Journal of Formal Logic*, 4: 481-500.

[27] Castro-Manzano, J.-M, 2018, "A Tableaux Method for Term Logic," *LAN-MAR: Languages, Algorithms and New Methods of Reasoning*, 1-14.

[28] Castro-Manzano, J.-M 2018, "Leibniz, Sommers y Englebretsen," *Leibniz en Mexico*, R.C. Garcia and P.O. Reyes-Cádenas (eds.), Mexico, MX: Editorial Torres Asociados, pp. 59-74.

[29] Castro-Manzano, J.-M, 2018, "Term Functor Logic Tableaux," *South American Journal of Logic*, 4: 29-50.

[30] Castro-Manzano, J.-M, 2019, "Algunas Lógicas de Términos Contemporáneas," *Argumentatción: Lógica, Argumentación y Pensamíento Crítica*, G.H. Decidero, R. Casales-Garcia, and J.-M. Castro-Manzano (eds.), Puebla, MX: UPAEP, pp. 119-136.

[31] Castro-Manzano, J.-M, 2019, "An Intermediate Term Functor Logic," *Argumentos Revists de Filosofia*, 11: 17-31.

[32] Castro-Manzano, J.-M 2020, "Murphree's Numerical Term Logic Tableaux," *Elecronic Notes in Theoretical Computer Science*, 354: 17-28.

[33] Castro-Manzano, J.-M, 2020, "Intermediate Syllogistic, Terms, and Trees," *Tópicos*, 56: 209-237.

[34] Castro-Manzano, J.-M, 2020, "Distribution Tableaux, Distribution Models," *Axioms*, 9: 1-7.

[35] Castro-Manzano, J.-M, 2020, "¿Es la Lógica de Términos una Lógica Libre?" *Stoa*, 11: 98-109.

[36] Castro-Manzano, J.-M, 2020, "A Tableaux Method for Modal Term Logic," *Open Insight*, 11: 165-180.

[37] Castro-Manzano, J.-M, 2021, "Traditional Logic and Computational Thinking," *Philosophies*, 6: 12.

[38] Castro-Manzano, J.-M, 2021, "Silogística Estadística Usando Términos," *Universitas Philosophica*, 38: 171-187.

[39] Castro-Manzano, J.-M, 2022, "Mixing Colors, Mixing Logics," *Diagrammatic Respresentation and Inference. Diagrams 2022*, (Lecture Notes in Artificial Intelligence, vol. 13462), V. Giardino *et al.* Cham: Springer Nature, pp. 70-77.

[40] Castro-Manzano, J.-M, 2022, "On Mixing Term Logics," *Logics for New-Generation AI*, B. Liao, R. Markovich, and Y.N. Yang (eds.), London: College Publications, pp. 6-23.

[41] Castro-Manzano, J.-M, 2022, "Towards a Relevance Term Logic," *Computación y Sistemas*, 26: 761-768.

[42] Castro-Manzano, J.-M, 2022, "Lógica Tradicional para Razonamiento no Tradicional," *Research in Computer Science*, 5: 115-127.

[43] Castro-Manzano, J.-M, 2023, "Una Jerarquía de Lógicas Terministas (A Hierarchy of Term Logics)," *Stoa*, 14: 61-76.

[44] Castro-Manzano, J.-M, 2023, "Interpolation in Term Functor Logic," *Crítica: Revista Hispanoamericana de Filosofía*, 55: 53-69.

[45] Castro-Manzano, J.-M, 2023, "Una Revisión de los Diagramas de Murphree," *Open Insight*, 14: 123-136.

[46] Castro-Manzano, J.-M, 2024, "On Line Diagrams Plus Modality," *Logics*, 2: 1-10.

[47] Castro-Manzano, J.-M, 2024, "On Sommersian Concept Analysis: Sobre el análisis sommersian de conceptos," *Metafísica y Persona*, 16: 25-36.

[48] Castro-Manzano, J.-M, forthcoming, "Una Traduccíon Terminista de la Lógica Natural de Moss," *Daimon. Revista Internacional de Filosofia*. htt://dx.doi.org/10.6018/daimon.466741

[49] Castro-Manzano, J.-M, and Flores-Martínez, E., 2021, "Deliberative Semantics for Term Functor Logic," *Philosophies*, forthcoming.

[50] Castro-Manzano, J.-M,, Lozanos-Cabos, L.I., and Reyes-Cárdena, P.-O., 2018, "Programming with Term Logic," *BRAIN: Broad Research in Artificial Intelligence and Neuroscience*, 9: 22-36.

[51] Castro-Manzano, J.-M, and Reyes-Cardenas, P.-O., 2018, "Term Functor Logic Tableaux," *South American Journal of Logic*, 4: 29-50.

[52] Chastain, C., 1975, "Reference and Context," *Language, Mind, and Knowledge*, K. Gunderson, (ed.), Minneapolis: University of Minneapolis Press, pp. 194-269.

[53] Clark, D.S., 1983, "Negating the Subject," *Philosophical Studies*, 43: 349-353.

[54] Cocchiarella, N., 2005, "Denoting Concepts, Reference, and the Logic of

Names, Classes as Many, and Plurals," *Linguistics and Philosophy*, 28: 135-179.

[55] Cooper, W.S., 2001, *The Evolution of Reason: Logic as a Branch of Biology*, Cambridge: Cambridge University Press.

[56] Corcoran, J., 1972, "Aristotle's Natural Deduction System," *Journal of Symbolic Logic*, 37: 437.

[57] Corcoran, J., 1974, *Ancient Logic and Its Modern Interpretations*, Dordrecht: Reidel.

[58] Corcoran, J., 1974, "Aristotelian Syllogisms: Valid Arguments or Generalized Conditionals?" *Mind*, 83: 278-281.

[59] Corcoran, J., 2009, "Aristotle's Demonstrative Logic," *History and Philosophy of Logic*, 30: 1-20.

[60] Correia, M., 2003, *La Lógica de Aristóteles: Lecciones sobe el Origen del Pensamient,Lógico en la Antigüedad*, Santiago: Ediciones Universidad Católica de Chile.

[61] Correia, M, 2006, "¿Es lo mismos Ser No-Justo que Ser Injusto? Aristóteles y sus Comentaristas," *Méthexis*, 19: 41-56.

[62] Correia, M, 2017, "La Lógica Aristotélica y sus Perspectivas," *Pensamiento*, 73: 5-19.

[63] Czeżowski, T., 1955, "On Certain Peculiarities of Singular Propositions," *Mind*, 64: 392-395.

[64] Desclés, J.-P. and Guentchéva, Z., 2000, "Quantification Without Bound Variables," *Variable-Free Semantics*, M. Böttner and W. Thümmel (eds.), Osnabrück: secolo Verlag, pp. 210-233.

[65] Devitt, M., 1974, "Singular Terms," *Journal of Philosophy*, 71: 183-205.

[66] Devitt, M., 2012, "Should Proper Names Still Seem So Problematic?" *New Essays on Reference*, A. Bianchi (ed.), Oxford: Oxford University Press.

[67] Díez, G.F., 2010, "A Note on Plural Logic," *Organon*, 2: 150-162.

[68] DiLeo, J.R., 1997, " Charles Peirce's Theory of Proper Names," *Studies in the Logic of Charles Sanders Peirce*, N. Houser, D.D. Roberts, and J. van Evra (eds.), Bloomington: Indiana University Press, pp. 574-594.

[69] Englebretsen, G., 1974, "A Note on Contrariety," *Notre Dame Journal of Formal Logic*," 4: 613-614.

[70] Englebretsen, G.,1980a, "Singular Terms and the Syllogistic," *The New Scholasticism*, 54: 68-74.

[71] Englebretsen, G., 1980b, "Denotation and Reference," *Philosophical Studies* (Ire.), 27: 229-236.

[72] Englebretsen, G., 1981a, "Prediates, Predicables and Names," *Crítica*, 8:

105-108.
[73] Englebretsen, G., 1981b, "A Note on Identity, Reference and Logical Form," *Crítica*, 75-81.
[74] Englebretsen, G., 1981c, *Three Logicians: Aristotle, Leibniz and Sommers, and the Syllogistic*, Assen: Van Gorcum.
[75] Englebretsen, G., 1981d, *Logical Negation*, Assen: Van Gorcum.
[76] Englebretsen, G., 1982a, "Do We Need Relative Identity?" *Notre Dame Journal of Formal Logic*, 23: 91-93.
[77] Englebretsen, G., 1982b, "Leibniz on Logical Syntax," *Studia Leibniziana*,: 14: 119-126.
[78] Englebretsen, G., 1983, "Reference, Anaphora, and Singular Quantity," *Dialogos*: 41: 67-72.
[79] Englebretsen, G., 1984a, "Opposition," *Notre Dame Journal of Formal Logic*, 25: 79-85.
[80] Englebretsen, G., 1984b, "Despre Logica Conjuncţiiler Intrapropoziţionale," *Revistei de Filosofie*, 31: 141-147.
[81] Englebretsen, G., 1985a, "On the Proper Treatment of Negative Names," *Journal of Critical Analysis*, 8: 109-115.
[82] Englebretsen, G., 1985b, "Negative Names," *Philosophia*, 15: 133-136.
[83] Englebretsen, G., 1985c, "Geach on Logical Syntax," *The New Scholasticism*, 59: 168-174.
[84] Englebretsen, G., 1985d, "Defending Distribution," *Dialogos*, 45: 157-159.
[85] Englebretsen, G., 1985e, "Semantic Considerations for Sommers' Logic," *Philosophy Research Archives*, 11: 281-318.
[86] Englebretsen, G., 1986a, "Singular/General," *Notre Dame Journal of Formal Logic*, 27: 104-107.
[87] Englebretsen, G., 1986b, "Czeżowski on Wild Quantity," *Notre Dame Journal of Formal Logic*, 27: 62-65.
[88] Englebretsen, G., 1987a, (ed.), *The New Syllogistic*, NY and Bern: Peter Lang.
[89] Englebretsen, G., 1987b, "Natural Syntax and Sommers' Theory of Logical Form," in Englebretsen 1987a, pp. 245-272.
[90] Englebretsen, G., 1987c, "Subjects," *Studia Leibnitiana*, 19: 85-90.
[91] Englebretsen, G., 1987d, "Logical Polarity," *The New Syllogistic*, G. Englebretsen (ed.), pp. 305-311.
[92] Englebretsen, G., 1988a, "A Note on Leibniz's Wild Quantity Thesis," *Studia Leibnitiana*, 20: 87-89.
[93] Englebretsen, G., 1988b, "Preliminary Notes on a New Modal Syllogistic,"

*Notre Dame Journal of Formal Logic*, 29: 381-395.

[94] Englebretsen, G., 1989, "Formatives," *Notre Dame Journal of Formal Logic*, 30: 382-389.

[95] Englebretsen, G., 1990, "A Note on Copulae and Qualifiers," *Linguistic Analysis*, 20: 82-86.

[96] Englebretsen, G., 1992, "Linear Diagrams for Syllogisms (with Relationals), *Notre Dame Journal of Formal Logic*, 33: 37-69.

[97] Englebretsen, G., 1992, "Parry and Hacker's *Aristotelian Logic*," *Informal Logic*, 14: 75-82.

[98] Englebretsen, G., 1996, *Something to Reckon With: The Logic of Terms*, Ottawa: University of Ottawa Press.

[99] Englebretsen, G., 1997 [1999], "The Unifying Copula," *Logique et Analyse*, 159: 255-259.

[100] Englebretsen, G., 2000, "Preliminaries for a Term-Functor Logic," *Variable-Free Semantics*, M. Böttner and W. Thümmel (eds.), Osnabrück: secolo Verlag, pp. 90-99.

[101] Englebretsen, G., 2004, "Predicate Logic, Predicates, and Terms," *First-Order Logic Revisited*, V. Hendricks, F. Neuhaus, S. A. Pedersen, U. Scheffler, H. Wansing (eds.), Berlin: Logos Verlag, pp. 75-88.

[102] Englebretsen, G., 2005, "Trees, Terms, and Truth: The Philosophy of Fred Sommers," *The Old New Logic: Essays on the Philosophy of Fred Sommers*, D. Oderberg (ed.), Cambridge, MA: MIT Press, pp. 25-48.

[103] Englebretsen, G., 2006, *Bare Facts and Naked Truths: A New Correspondence Theory of Truth*, Aldershot: Ashgate.

[104] Englebretsen, G., 2012, *Robust Reality: An Essay in Formal Ontology*, Frankfurt: Ontos.

[105] Englebretsen, G., 2015, *Exploring Topics in the History and Philosophy of Logic*, Boston and Berlin: De Gruyter.

[106] Englebretsen, G., 2016a, "Fred Sommers' Contributions to Formal Logic," *History and Philosophy of Logic*, 37: 269-291.

[107] Englebretsen, G., 2016b, "La Quadrature du Carré," *Soyons Logiques*, F. Schang, A. Moktefi, et A. Moretti (eds.), London: College Publications.

[108] Englebretsen, G., 2019, "Is Natural Logic Part of Naturalized Logic?" *Natural Arguments: A Tribute to John Woods*, D. Gabbay, L. Magnani, W. Park, and A.-V. Pietarinen (eds.), London: College Publications, pp. 593-620.

[109] Englebretsen, G., 2021, *Carrollian Notes*, London: College Publications.

[110] Englebretsen, G., Castro-Manzano, J.-M., and Pacheco-Montes, J.R.,

2020, *Figuring It Out: Logic Diagrams*, Boston and Berlin: De Gruyter.

[111] Englebretsen, G. and Sayward, C., 2011, *Philosophical Logic: An Introduction to Advanced Topics*, London: Continuum

[112] Englebretsen, G. and Sommers, F., 2000, *An Invitation to Formal Reasoning*, Aldershot: Ashgate.

[113] Evans, G., 1977, "Pronouns, Quantifiers, and Relative Clauses, (I)" *Canadian Journal of Philosophy*, 7: 467-536.

[114] Evans, G., 1980, "Pronouns," *Linguistic Inquiry*, 11: 337-362.

[115] Evans, G., 1982, *Varieties of Reference*, Oxford: Oxford University Press.

[116] Fara, D.G., 2015, "Names Are Predicates," *Philosophical Review*, 124: 59-117.

[117] Frederick, D., 2013, "Singular Terms, Predicates and the Spurious 'is' of Identity," *Dialectica*, 67:325-343.

[118] Frege, G., 1879, *Begriffsschrift, eine der arithmetischen nachbildete Formelsprache des reinen Denkens*, Halle: L. Nebert.

[119] Frege, G., 1964, *The Basic Laws of Arithmetic*, M. Furth (transl. and ed.), Berkeley: University of California Press.

[120] Frege, G., 1979, *Frege: Posthumous Writings*, H. Hermes, F. Kambartel, and F. Kaulbach (eds.), Oxford: Basil Blackwell.

[121] Frege, G., 1980, *Translations from the Philosophical Writings of Gottlob Frege*. P. Geach and M. Black (eds.), third edition, Oxford: B.H. Blackwell.

[122] Frege, G., 1997, *The Frege Reader*, M. Beaney (ed.), Oxford: Blackwell.

[123] Geach, P., 1950, "Subject and Predicate," *Mind*, 59: 461-482.

[124] Geach, P., 1962, *Reference and Generality*, Ithaca, NY: Cornell University Press.

[125] Geach, P., 1963, "Mr Strawson on Symbolic and Traditional Logic," *Mind*, 72: 125-128 (reprinteded in Geach 1972, pp. 66-70).

[126] Geach, P., 1966, "Comment on Mr Grimm," *Analysis*, 26: 146.

[127] Geach, P., 1968, "History of the Corruptions of Logic" (reprinted in Geach 1972, pp. 44-61).

[128] Geach, P., 1972, *Logic Matters,* Oxford: Basil Blackwell.

[129] Geach, P., 1975, "Names and Identity," *Mind and Language*, S. Guttenplan (ed.), Oxford: Oxford University Press, pp. 139-158.

[130] Geach, P., 1980, "Some Problems about the Sense and Reference of Proper Names," *Canadian Journal of Philosophy*, Supplementary Volume, 6: 83-96.

[131] Geach, P., 1980, "Strawson on Subject and Predicate," *Philosophical Subjects*, Z. van Straaten (ed.), Oxford: Oxford University Press, pp. 174-

188.

[132] Geach, P., 1987, "Relative Identity," *Proceedings of the 30$^{th}$ Conference on the History of Logic*, Cracow: Jagiellonian University.

[133] Friedman, W.H., 1978, "Uncertainties over Distribution Dispelled," *Notre Dame Journal of Formal Logic*, 19: 653-662.

[134] Friedman, W.H., 1980, "Calculemus," *Notre Dame Journal of Formal Logic*, 21: 166-174.

[135] Friedman, W.H., 1987, "Algebraic Rules for Syllogisms and Antilogisms," *The New Syllogistic*, G. Englebretsen (ed.), pp. 213-221.

[136] Friedman, W.H., 1995, "Bradley's Regress, the Copula, and the Unity of the Proposition," *Philosophical Quarterly*, 45: 161-180.

[137] Friedman, W.H., 2008, *The Unity of the Proposition*, Oxford: Oxford University Press.

[138] Geach, P., 1950, "Subject and Predicate," *Mind* 59: 461-482.

[139] Gleitman, J., 1965, "Coordinating Conjunction in English," *Language*, 51: 177-184.

[140] Grandy, R., 1977, "Predication and Singular Terms," *Noûs*, 11: 163-167.

[141] Grandy, R., 1985, "On the Logic of Singular Terms," *Grazer Philosophische Studien*, 25: 285-296.

[142] Grimau, B., 2021, "Structured Plurality Reconsidered," *Journal of Semantics*, 38: 145-193.

[143] Grimm, R.H., 1966, "Names and Predicables," *Analysis*, 26: 138-146.

[144] Haack, S., 1998, "Quantifiers," *Contemporary Readings in Metaphysics*, Oxford: Blackwell, pp. 55-68.

[145] Hale, B., 1979, "Strawson, Geach and Dummett on Singular Terms," *Synthese*, 42: 275-295.

[146] Heitz, J. 1973, *Subjects and Predicables: A Study in Subject-Predicate Asymmetry*, The Hague and Paris: De Gruyter Mouton.

[147] Hodges, W., 2009, "Traditional Logic, Modern Logic and Natural Language" *Journal of Philosophical Logic*, 38: 589-606.

[148] Hofstadter, D., 1979, *Gödel, Escher, Bach: An Eternal Golden Braid*, NY: Basic Books.

[149] Horden, J. and López de Sa, D., 2012, "Groups as Pluralities," *Synthese*, 198: 10237-10271.

[150] Horn, L., 1989, *A Natural History of Negation*, Chicago: University of Chicago Press. Reissued 2001, Stanford, CA, CSLI Publications.

[151] Horn, L. and Wansing, H., 2022, "Negation," *The Stanford Encyclopedia of Philosophy*, winter edition, E. Zalta (ed.).

[152] Hornsby, J., 1976, "Proper Names: A Defense of Burge," *Philosophical Studies*, 30: 227-234.
[153] Jeshion, R., 2015, "Referentialism and Predicativism about Proper Names," *Erkenntnis*, 80: 363-404.
[154] Justice, J., 2007, "Unified Semantics of Singular Terms," *Philosophical Quarterly*, 57: 363-373.
[155] Katz, B.D. and Martinich, A.P., 1976, "The Distribution of Terms," *Notre Dame Journal of Formal Logic*, 17: 279-283.
[156] Kelley, D., 1994, *The Art of Reasoning*, $2^{nd}$ expanded edition, NY: W.W. Norton.
[157] King, J.C., 2006, "Singular Terms, Reference, and Methodology in Semantics," *Philosophical Issues*,16: 141-161.
[158] Kripke, S., 1980, *Naming and Necessity*, Cambridge, MA: Harvard University Press.
[159] La Palme Reyes, M., Macnamara, J., and Reyes, G.E., 1995, "A Category-Theoretic Approach to Aristotle's Term Logic, with Special Reference to Syllogisms," *Québec Studies in the Philosophy of Science I*, M. Marion and R.S. Cohen (eds.), Dordrecht: Kluwer, pp. 57-68.
[160] La Palme Reyes, M., Macnamara, J., Reyes, G.E., and Zolfaghari, H., 1994, "A Category-Theoretic Approach to Aristotle's Term Logic, with Special Reference to Negation," *Actas del Primer Congreso Internacional de Ontología*, V. Gómez Pin (coordinador), Bellaterra: Universitat Autònoma de Barcelona, pp. 241-249.
[161] Leckie, G., 2013, "The Double Life of Names," *Philosophical Studies*, 165: 1139-1160.
[162] Leibniz, G., 1966, "A Paper on 'Some Logical Difficulties'," *Leibniz: Logical Papers*, G.H.R. Parkinson (ed.), Oxford: Oxford University Press, pp. 115-121.
[163] Lenzen, W., 1986, " 'Non est' non est 'est non'," *Studia Leibnitiana*, 18: 1-37.
[164] Lenzen, W., 1989, "Concepts vs. Predicates – Leibniz's Challenge to Modern Logic," *The Leibniz Renaissance*, Florence: Centro Fiorentino di Storia e Filosofia della Scienzia, pp. 153-172.
[165] Lenzen, W., 1990, *Das System der Leibnizchen Logik*, Berlin: De Gruyter.
[166] Lenzen, W., 1991, "Leibniz on Privative and Primitive Terms," *Theoria*, 14-15: 83-96.
[167] Lenzen, W., 1992, "Leibniz on Properties and Individuals," *Language, Truth and Ontology*, K. Mulligan (ed.), Boston: Kluwer, pp. 193-204.
[168] Lenzen, W., 1995, "Frege und Leibniz," *Logik und Mathematik*, I. Mak

(ed.), Berlin: De Gruyter, pp. 82-92.

[169] Lenzen, W., 2004, *Calculus Universalis. Studien zur Logik von G.W. Leibniz*, Panborn: mentis Verlag.

[170] Linnebo, Ø., 2003, "Plural Quantification Exposed," *Noûs*, 37: 71-92.

[171] Linnebo, Ø., 2009, "Plural Quantification," *The Stanford Encyclopedia of Philosophy*, E. Zalta (ed.), Spring 2009 Edition.

[172] Linnebo, Ø., 2010, "Pluralities and Sets," *Journal of Philosophy*, 107: 144-164.

[173] Linnebo, Ø. and Florio, S., 2020, "Critical Plural Logic," *Philosophica Mathematica*, 28: 172-203.

[174] Linnebo, Ø. and Florio, S., 2021, *The Many and the One: A Philosophical Study of Plural Logic*, Oxford: Oxford University Press.

[175] Loar, B., 1976, "The Semantics of Singular Terms," *Philosophical Studies*, 30: 353-377.

[176] Lockwood, M., 1975, "On Predicating Proper Names," *Philosophical Review* 84: 471-498.

[177] Lockwood, M., 1982, "The Logical Syntax of TFL," in Sommers 1982, Appendix G, pp. 426-456.

[178] Lockwood, M., 1987, "Proofs and Pronouns: Extending the System," *The New Syllogistic*, G. Englebretsen (ed.), pp. 161-211.

[179] Makinson, D., 1969, "Remarks on the Concept of Distribution in Traditional Logic," *Noûs*, 3: 103-108.

[180] Marques, E., 2013, "A quantificação das expressōs singulars em Leibniz: reflexes acerca de uma tese interpretative de Georg Englebretsen," *Analytica*, 17: 265-277.

[181] Martin, J. N., 2001, "Proclus and Neoplatonic Syllogistic," *Journal of Philosophical Logic*, 30: 187-240.

[182] Martin, J. N., *Themes in Neoplatonic and Aristotelian Logic*, Aldershot: Ashgate.

[183] Martin, J. N., 2004, "Ecthesis and Existence in the Syllogistic," *Themes in Neoplatonic and Aristotelian Logic: Order, Negation and Abstraction*, London: Ashgate, pp. 19-24.

[184] Martin, J. N., 2012, "Existential Commitment and the Cartesian Semantics of the Port Royal Logic," *New Perspectives on the Square of Opposition*, J.-Y. Beziau (ed.), Bern: Peter Lang.

[185] Martin, J. N., 2013, "Distributive Terms, Truth, and the Port Royal Logic," *History and Philosophy of Logic*, 34: 133-154.

[186] Martin, J. N., 2016, "Privative Negation and *The Port Royal Logic*,"

*Review of Symbolic Logic*, 9: 664-685.

[187] Massey, G., 1976, "Tom, Dick and Harry and All the King's Men," *American Philosophical Quarterly*, 13: 89-107.

[188] Massie, D., 2015, "Computer Implementation of Term Functor Logic (TFL) Based on Directed Graph Representation of TFL," U.S. Ptent Application No. 13/987,835, filed March 12, 2015.

[189] McCulloch, G., 1984, "Frege, Sommers, Singular Reference," *Philosophical Quarterly*, 34: 295-310.

[190] McIntosh, C., 1982, "Appendix F," in Sommers 1982, pp. 387-425.

[191] McKay, T.J., 2006, *Plural Predication*, Oxford: Clarendon Press.

[192] Mendelsohhn, R.L., 1987, "Frege's Two Senses of 'Is'," *Notre Dame Journal of Formal Logic*, 28: 139-160.

[193] Mill, J.S., 1843, *A System of Logic*, London: J.W. Parker.

[194] Miller, J.W., 1938, *The Structure of Aristotelian Logic*, London: Kegan Paul, Trench, Trubner & Co. Ltd. (reprinted: 2016, London and NY: Routledge).

[195] Montague, R., 1974, *Formal Philosophy: Selected Papers of Richard Montague*, R. Thomason (ed.), New Haven: Yale University Press.

[196] Morrissey, C.S., 2015, "A Logic without Nominalism," *The American Journal of Semiotics*, 31: 183-202.

[197] Morrissey, C.S., 2018, *The Way of Logic*, Nanjing: Nanjing Normal University Press.

[198] Moss, L., 2008, "Completeness Theorems for Syllogistic Fragments," *Logic for Linguistic Structures*, 29: 143-173.

[199] Moss, L., 2010, "Syllogistic Logic with Verbs," *Journal of Logic and Computation*, 14: 761-793.

[200] Moss, L., 2010, "Natural Logic and Semantics," *Logic, Language and Meaning: $17^{th}$ Amsterdam Collogquium, 2009, Revised Selected Papers*, Berlin and Heidelburg: Springer, pp. 84-93.

[201] Moss, L., 2011, "Syllogistic Logic with Complements," *Games, Norms and Reasons: Logic at the Crossroads*, Synthese Library, J. van Benthem, A. Gupta, and E. Pacuit (eds.), Berlin, Springer, pp. 179-197.

[202] Moss, L., 2011, "Syllogistic Logic with Comparative Adjectives," *Journal of Logic, Language and Information*, 20: 397-417.

[203] Moss, L., 2015, "Natural Logic," *The Handbook of Contemporary Semantic Theory*, S. Lappin and C. Fox (eds.), London: Wiley, pp.559-592.

[204] Moss, L. and Kruckman, A., 2021, "Exploring the Landscape of Syllogistic Logics," *Review of Symbolic Logic*, 14: 728-765.

[205] Moss, L. and Pratt-Hartmann, I., 2009, "Logics for Relational Syllogistics," *Review of Symbolic Logic*, 2: 647-683.

[206] Murphree, W., 1991, *Numerically Exceptive Logic: A Reduction of the Classical Syllogistic*, NY and Bern: Peter Lang.

[207] Murphree, W., 1993, "Expanding the Traditional Syllogism," *Logique et Analyse*, 141-142: 105-120.

[208] Murphree, W., 1994, "The Irrelevance of Distribution for the Syllogism," *Notre Dame Journal of Formal Logic*, 35: 433-439.

[209] Murphree, W., 1997, "The Numerical Syllogism and Existential Presupposition," *Notre Dame Journal of Formal Logic*, 38: 49-64.

[210] Murphree, W., 1998, "Numerical Term Logic," *Notre Dame Journal of Formal Logic*, 39: 346-362.

[211] Nemirow, R.L., 1979, "No Argument Against Ramsey," *Analysis* 39: 201-209.

[212] Noah, A., 1973, *Singular Terms and Predication*, PhD thesis, Brandeis Universtiy.

[213] Noah, A., 1980, "Predicate Functors and the Limits of Decidability," *Notre Dame Journal of Formal Logic*, 21: 701-707.

[214] Noah, A., 1982, "Quine's Version of Term Logic and its Relation to TFL," *The Logic of Natural Language*, F. Sommers, 1982, Appendix E, pp. 372-385.

[215] Noah, A., 1987, "The Two Term Theory of Predication," *The New Syllogistic*, G. Englebretsen (ed.), pp. 223-243.

[216] Oderberg, D.,(ed.), 2005, *The Old New Logic:Essays on the Philosophy of Fred Sommers*, Cambridge, MA: M.I.T. Press.

[217] Oliver, A. and Smiley, T., 2006, "A Modest Logic of Plurals," *Journal of Philosophical Logic*, 35: 317-348.

[218] Oliver, A. and Smiley, T., 2008, "Is Plural Denotation Collective?" *Analysis*, 68: 22-33.

[219] Oliver, A. and Smiley, T., 2013, *Plural Logic*, Oxford: Oxford University Press.

[220] Paasch, JT, 2023, "Ockham on Metalanguage," researchgate.net/publication/374553532_Ockham_on_Metalanguage, accessed 16/10/2023.

[221] Parsons, T., 2006, "The Doctrine of Distribution," *History and Philosophy of Logic*, 27: 59-74.

[222] Parsons, T., 2014, *Articulating Medieval Logic*, Oxford: Oxford University Press.

[223] Peterson, P.L., 2000, *Intermediate Quantifiers: Logic, Linguistics, and*

*Aristotelian Semantics*, Aldershot: Ashgate.

[224] Priest, G., 2008, "The Closing of the Mind: How the Particular Quantifier Became Existentially Loaded Behind Our Backs," *Review of Symbolic Logic*, 1: 42-55.

[225] Purdy, W.C., 1991, "A Logic for Natural Language," *Notre Dame Journal of Formal Logic*, 32: 409-425.

[226] Purdy, W.C., 1992, "On the Question 'Do We Need Identity?'" *Notre Dame Journal of Formal Logic*, 33:593-603.

[227] Purdy, W.C., 2000, "Surrogate Variables in Natural Language," *Variable-Free Semantics*, M. Böttner and W. Thümmel (eds.), Osnabrück: secolo Verlag, pp. 22-45.

[228] Quine, W.v.O., 1948, "On What There Is," *The Review of Metaphysics*, 2: 21-38 (reprinted in Quine 1953, pp. 1-19).

[229] Quine, W.v.O., 1953, *From a Logical Point of View*, New York and Evanston: Harper & Row.

[230] Quine, W.v.O., 1959, "Eliminating Variables Without Applying Functions to Functions," *Journal of Symbolic Logic*, 24: 324-325.

[231] Quine, W.v.O., 1960, "Variables Explained Away," *Proceedings of the American Philosophical Association*, 104:343-347.

[232] Quine, W.v.O., 1960, *Word and Object*, Cambridge: MIT Press.

[233] Quine, W.v.O., 1971, "Predicate Functor Logic," *Proceedings of the Second Scandanavian Logic Symposium*, J. Fenstand (ed.), Amsterdam: North-Holland, pp. 309-315.

[234] Quine, W.v.O., 1972, "Algebraic Logic and Predicate Functors," *Logic and Art: Essays in Honor of Nelson Goodman*, R. Rudner and I. Sheffler (eds.), Indianapolis: Bobbs-Merrill, pp. 214-238 (reprinted in *The Ways of Paradox and Other Essays*, revised and enlarged edition, Cambridge, MA: Harvard University Press, pp. 283-307).

[235] Quine, W.v.O., 1976, "The Variable," *The Ways of Paradox and Other Essays*, revised and enlarged edition, Cambridge, MA: Harvard University Press, pp. 272-282.

[236] Quine, W.v.O., 1980, "The Variable and its Place in Reference," *Philosophical Subjects: Essays Presented to P.F. Strawson*, Z. van Straaten (ed.), Oxford: Clarendon Press.

[237] Quine, W.v.O., 1981, "Predicate Functors Revisited," *Journal of Symbolic Logic*, 46: 649-652.

[238] Purtill, R., 1987, "Some Practical and Theoretical Features of Sommers' Cancellation Method," *The New Syllogistic*, G. Englebretsen (ed.), pp. 273-281.

[239] Raab, J., 2018, "Aristotle, Logic, and QUARC," *History and Philosophy of Logic*, 39: 305-340.
[240] Rami, D., 2014, "On the Unifiction Argument for the Predicative View on Proper Names," *Synthese*, 191: 841-862.
[241] Rami, D, 2015, "The Multiple Uses of Proper Nouns,"*Erkenntnis*, 80 (sup.): 405-432.
[242] Rami, D, 2023, *Names and Context*, London and Berlin: Bloomsbury.
[243] Ramsey, F., 1925, "Universals," *Mind*, 34: 401-417.
[244] Rayo, A., 2002, "Words and Objects," *Noûs*, 36: 436-454.
[245] Rayo, A., 2007, "Plurals," *Compass*, 2: 411-427.
[246] Read, S., 2015, "Aristotle and Łukasiewicz on Existential Import," *Journal of the American Philosophical Association*, 1: 535-544.
[247] Read, S, 2017, "Aristotle's Theory of the Assertoric Syllogism," Ms (19/6/2017). On line: https://philpapers.org/archive/REAATO-4,pdf
[248] Rearden, M., 1984, "The Distribution of Terms," *Modern Schoolman*, 61: 187-195.
[249] Sainsbury, R.M., 2002, *Departing from Frege: Essays in the Philosophy of Language*, London: Routledge.
[250] Sanford, D., 1966, "Negative Terms," *Analysis*, 27: 201-205.
[251] Sautter, F.T., 2012, "Termos Negativos e Diagrama," 1er Congreso de la Sociedad Filosófica del Uruguay Simposio Diagramas.
[252] Sayward, C., 1987, "Some Problems with TFL," *The New Syllogistic*, G. Englebretsen (ed.), pp. 283-297.
[253] Searle, J.R., 1967, "Proper Names," *Philosophical Logic*, P.F. Strawson (ed.), London: Oxford University Press.
[254] Sedlár, I. and Šebela, K., 2019, "Term Negation in First-order Logic," *Logique et Analyse*, 247: 265-284.
[255] Segal, G., 2001, "Two Theories of Proper Names," *Mind and Language*, 15: 547-563.
[256] Simons, P., 1982, "Plural Reference and Set Theory," *Parts and Moments: Studies in Logic and Formal Philosophy*, B. Smith (ed.), Munich: Philosophia Verlag, pp. 199-260.
[257] Simons, P., 1989, "Combinators and Categorial Grammar," *Notre Dame Journal of Formal Logic*, 30: 241-261.
[258] Simons, P., 1992, "Leśniewskian Term Logic," *Lingua e Stile*, 27: 23-46.
[259] Simons, P., 2010, "Relations and Truthmaking," *Aristotelian Society Supplementary*, 84: 199-213.
[260] Simons, P., 2020, "Term Logic," *Axioms*, 9: 18,

https://doi.org/10.3390/axioms9010018
[261] Slater, H., 1979, "Singular Subjects," *Dialogue*, 18: 362-372.
[262] Slater, H., 1979, "Internal and External Negation," *Mind*, 88: 588-591.
[263] Slater, H., 1987, "Back to Leibniz or on from Frege?" *The New Syllogistic*, G. Englebretsen (ed.), pp. 87-98.
[264] Smiley, T., 1973, "What is a Syllogism?" *Journal of Philosophical Logic*, 2: 136-154.
[265] Smith, B., 2005, "Against Fantology," *Experience and Analysis*, M. Reicher and J. Marek (eds.), Vienna: ÖBV&HPT, pp. 153-170.
[266] Sommers, F., 1967, "On a Fregean Dogma," *Problems in the Philosophy of Mathematics*, I. Lakatos (ed.), Amsterdam: North-Holland.
[267] Sommers, F., 1969a, "On Concepts of Truth in Natural Languages," *Review of Metaphysics*, 23: 259-286.
[268] Sommers, F., 1969b, "Do We Need Identity?" *Journal of Philosophy*, 66: 499-504.
[269] Sommers, F., 1970, "The Calculus of Terms," *Mind*, 79: 1-39 (reprinted in *The New Syllogistic*, G. Englebretsen, (ed.), 1987, pp. 11- 56.
[270] Sommers, F., 1973, "Existence and Predication" *Logic and Ontology*, M. Munitz (ed.), NY: New York University Press, pp. 159-174.
[271] Sommers, F., 1974, "The Logical and the Extra-Logical," *Boston Studies in the Philosophy of Science*, 14: 235-252, pp. 235-252.
[272] Sommers, F., 1975, "Distribution Matters," *Mind*, 84: 27-46.
[273] Sommers, F., 1976a, "Frege or Leibniz?" *Studies on Frege, III*, M. Schirn (ed.), Stuttgart, Formmann-Holzboog, pp. 11-34.
[274] Sommers, F., 1976b,"Logical Syntax and Natural Language," *Issues in the Philosophy of Language*, A. MacKay and D. Merrill (eds.), New Haven, Yale University Press, pp. 11-41.
[275] Sommers, F., 1976c, "On Predication and Logical Syntax," *Language in Focus*, A. Kasher (ed.), Dordrecht: D. Reidel, pp. 41-53.
[276] Sommers, F., 1976d, "Leibniz's Program for the Development of Logic" *Essays in Memory of Imre Lakatos*, R.S. Cohen, P.K. Feyerabend, and M.W. Wartofsky (eds.), Dordrecht: D. Reidel, pp. 589-615.
[277] Sommers, F., 1978, "The Grammar of Thought," *Journal of Social and Biological Structures*, 1: 39-51.
[278] Sommers, F., 1981, "Are There Atomic Propositions?" *Midwest Studies in Philosophy*, 6: 59-68.
[279] Sommers, F., 1982, *The Logic of Natural Language*, Oxford: Clarendon Press.

[280] Sommers, F., 1983a, "Linguistic Grammar and Logical Grammar," *How Many Questions?Essays in Honor of Sidney Morgenbesser*, L. Cauman, et al (eds.), Indianapolis, Hackett, pp. 180-194.

[281] Sommers, F., 1983b, 14 January, "The Logic of Natural Language: A Reply to Geach," *Times Literary Supplement*.

[282] Sommers, F., 1983c, 18 February, "The Logic of Natural Language: A Reply to Geach," *Times Literary Supplement*.

[283] Sommers, F., 1983d, "The Grammar of Thought: A Reply to Dauer," *Journal of Social and Biological Structures*, 6: 37-44.

[284] Sommers, F., 1987, "Truth and Existence," *The New Syllogistic*, G. Englebretsen (ed.), pp. 299-304.

[285] Sommers, F., 1990, "Predication in the Logic of Terms," *Notre Dame Journal of Formal Logic*, 31: 106-126.

[286] Sommers, F., 1993, "The World, the Facts, and Primary Logic," *Notre Dame Journal of Formal Logic*, 34: 169-182.

[287] Sommers, F., 1994, "Naturalism and Realism," *Midwes Studies in Philosophy*, 19: 22-38.

[288] Sommers, F., 1996, "Existence and Correspondence to Facts," *Formal Ontology*, R. Poli and P. Simons (eds.), Dordrecht: Kluwer, pp. 131-158.

[289] Sommers, F., 1997, "Putnam's Born-Again Realism," *Journal of Philosophy*, 94: 453-471.

[290] Sommers, F., 2000, "Term Functor Grammars," *Variable-Free Semantics*, M. Böttner and W. Thümmel (eds.), Osnabrück: Secolo Verlag, pp. 68-89.

[291] Sommers, F., 2002, "On the Future of Logic Instruction," *American Philosophical Association Newsletter on Teaching Philosophy*, 01: 176-180.

[292] Sommers, F., 2005a, "Belief de Mundo," *American PhilosophicalQuarterly*, 42: 117-124.

[293] Sommers, F., 2005b, "Bar-Hillel's Complaint," *Philosophia*, 33: 55-68.

[294] Sommers, F., 2005c, "Intellectual Autobiography," in Oderberg 2005, pp. 1-23.

[295] Sommers, F., 2005d, "Comments and Replies," in Oderberg 2005, pp. 211-231.

[296] Sommers, F., 2008a, "Reasoning: How We're Doing It," *The Reasoner*, 2: 5-7.

[297] Sommers, F., 2008b, "Ratiocination: An Empirical Account," *Ratio*, 21: 115-133. Reprinted in Englebretsen 2015, pp. 147-164.

[298] Sommers, F., 2008c, "Dissonant Belief," *Analysis*, 69: 267-274.

[299] Sommers, F., 2015, "Ryle's Way With the Liar," in Englebretsen 2015,

pp. 15-31.
[300] Sommers, F. and Englebretsen, G., 2000, *An Invitation to Formal Reasoning*, Aldershot: Ashgate.
[301] Sosa, E., 1973, "What is a Logical Constant?" *Boston Studies in the Philosophy of Science*, XIV, pp. 253-256.
[302] Stalnaker, R., "Complex Predicates," *Monist*, 60: 327-339.
[303] Steedman, M., 2000, "Does Grammar Make Use of Bound Variables?" *Variable-Free Semantics*, M. Böttner and W. Thümmel (eds.), Osnabrück: secolo Verlag, pp. 200-209.
[304] Strawson, P.F., 1961, "Singular Terms and Predication," *Journal of Philosophy*, 58: 393-412 (reprinted in *Logico-Linguistic Papers*, 1971, London: Methuen, pp. 53-74).
[305] Strawson, P.F., 1970, "The Asymmetry of Subjects and Predicates," *Language, Belief, and Metaphysics*, H.E. Kiefer and M. Munitz (eds.), Albany: State University of New York Press, pp. 69-86.
[306] Strawson, P.F., 1974, *Subjects and Predicates in Logic and Grammar*, London: Methuen.
[307] Strawson, P.F., 1982, "Review: *The Logic of Natural Language*," *Journal of Philosophy*, 79: 786-790 (reprinted in *The New Syllogistic*, 1987, G. Englebretsen (ed.), pp. 99-104).
[308] Szabolcsi, L., 1987, *Numerical Term Logic* (with a Foreword by Fred Sommers), G. Englebretsen (ed.), NY: Mellen.
[309] Taylor, K., 2015, "Names as Devices of Explicit Co-reference," *Erkenntnis*, (sup.) 80: 235-262.
[310] Toms, E., 1965, "Mr. Geach on Distribution," *Mind*, 74: 428-431.
[311] Valencia, V.M.S., 1991, *Studies on Natural Logic and Categorial Grammar*, PhD thesis, University of Amsterdam.
[312] van Benthem, J., 1986, "The Ubiquity of Logic in Natural Language," *The Task of Contemporary Philosophy*, W. Leinfellner and F. Wuketits (eds.), Vienna: Höder-Pichler-Temsky, pp. 177-186.
[313] van Benthem, J., 2006, "Where is Logic Going, and Should it? *Topoi*, 25: 117-122.
[314] van Benthem, J., 2007, "Logic in Philosophy, *Handbook of the Philosophy of Logic*, D. Jacquette (ed.), Amsterdam: Elsevier, pp. 65-99.
[315] van Benthem, J., 2008, "A Brief History of Natural Logic," *Logic, Navya-Nyaya and Applications: Homage to Bimal Matilal*, M. Chakraborty et al (eds.), London: College Publications.
[316] van Benthem, J., 2008, "Logic and Reasoning: Do the Facts Matter?"

*Studia Logica*, 88: 67-84.

[317] van Bennekom, R., 1986, "Aristotle and the Copula," *Journal of the History of Philosophy*, 24: 1-18.

[318] van Eijck, J., 2007, "Natural Logic for Natural Language," *Logic, Language, and Computation,TbiLLC 2005. Lecture Notes in Computer Science(s)*, B.D. ten Cate and H.W. Zeevat (eds.), Berlin: Springer, pp. 216-230.

[319] van Rooij, R., 2012, "The Propositional and Relational Syllogistic," *Logique et Analyse*, 55: 86-101.

[320] van Rooij, R, 2014, "Leibnizian Intensional Semantics for Syllogistic Reasoning," *Recent Trends in Philosophical Logic*, H. Wansing, C. Ciuni, and C. Willkommen (eds.), Cham: Springer, pp. 179-194.

[321] Varzi, A.C., 2022, "On Perceiving Abs nces," *Gestalt Theory*, 44: 231-241.

[322] Wald, J., 1979, "Geach on Atomicity and Singular Propositions," *Notre Dame Journal of Formal Logic*, 20: 285-294.

[323] Wang, P., 1995, *Non-Axiomatic Reasoning System: Exploring the Essence if Intelligence*, doctoral dissertation, Indiana University.

[324] Wang, P., 2004,"Cognitive Logic versus Mathematical Logic," *Proceedings of the Third International Seminar on Logic and Cognition*, China: Guangzhou, pp. 1-10.

[325] Wang, P., 2006, *Rigid Flexibility: The Logic of Intelligence*, Dordrect: Springer.

[326] Wang, P., 2013, *Nonaxiomatic Logic: A Model of Intelligence*, Singapore: World Scientific.

[327] Wang, P., 2019, "Toward a Logic of Everyday Reasoning," *Blended Cognition: The Robotic Challenge*, J. Vellverdú and V.C. Müller (eds.), Cham: Springer, pp. 275-302.

[328] Wansing, H., 2017, "Negation," *The Blackwell Guide to Philosophical Logic*, L. Goble (ed.), London: Blackwell Publishers, pp. 415-436.

[329] Williamson, C., 1971, "Traditional Logic as a Logic of Distribution Values," *Logique et Analyse*, 14: 729-746.

[330] Wilson, F., 1987, "The Distribution of Terms: A Defense of the Traditional Doctrine," *Notre Dame Journal of Formal Logic*, 28: 349-454.

[331] Yi, B., 2005, "The Logic and Meaning of Plurals. Part I," *Journal of Philosophical Logic*, 34: 459-506.

[332] Yi, B., 2006, "The Logic and Meaning of Plurals. Part II," *Journal of Philosophical Logic*, 35: 239-288.

[333] Zemach, E., 1981, "Names and Particulars," *Philosophia*, 10: 271-223.

[334] Zemach, E., 1985, "On Negative Names," *Philosophia*, 15: 137-138.

www.ingramcontent.com/pod-product-compliance
Lightning Source LLC
Chambersburg PA
CBHW071934220426
43662CB00009B/904